Beam-Wave Interaction in Periodic and Quasi-Periodic Structures

Particle Acceleration and Detection

springer.com

The series *Particle Acceleration and Detection* is devoted to monograph texts dealing with all aspects of particle acceleration and detection research and advanced teaching. The scope also includes topics such as beam physics and instrumentation as well as applications. Presentations should strongly emphasize the underlying physical and engineering sciences. Of particular interest are

- contributions which relate fundamental research to new applications beyond the immediate realm of the original field of research
- contributions which connect fundamental research in the aforementioned fields to fundamental research in related physical or engineering sciences
- concise accounts of newly emerging important topics that are embedded in a broader framework in order to provide quick but readable access of very new material to a larger audience

The books forming this collection will be of importance for graduate students and active researchers alike.

For further volumes:
http://www.springer.com/series/5267

Levi Schächter

Beam-Wave Interaction in Periodic and Quasi-Periodic Structures

Second Edition

Springer

Professor Levi Schächter
Israel Institute of Technology
Department of Electrical Engineering
Haifa 32000, Israel
levi@ee.technion.ac.il

ISSN 1611-1052
ISBN 978-3-642-26899-1 ISBN 978-3-642-19848-9 (eBook)
DOI 10.1007/978-3-642-19848-9
Springer Heidelberg Dordrecht London New York

Softcover reprint of the hardcover 2nd edition 2011

Cover design: SPi Publisher Services

Printed on acid-free paper

Springer is part of Springer Science+Business Media (www.springer.com)

To my parents, wife and children

Preface

In the second edition of the monograph published originally in 1997, I preserved the spirit of the first edition whereby the focus was on *analytic* and detailed analysis the fundamental concepts associated with beam–wave interaction in various radiation sources and accelerators. While the general framework of the material remained the same, each chapter has been improved and expanded. In the following paragraphs, I present the updated structure of the various chapters, skipping the introductory chapter – Chap. 1.

Chapter 2 is dedicated to the basic electromagnetic theory. After discussing *Maxwell equations* in general, I present simple homogeneous solutions corresponding to the TEM, TM, TE, and hybrid modes. When the current density is present, it is useful to employ Green's function method for solution of the electromagnetic field. Its formal description is accompanied by two examples which illustrate the Cerenkov radiation in free space and in a waveguide. The coherent process is also examined. Several finite length effects are considered, as well as edge effects. Scattered waves phenomena are also discussed.

Chapter 3: All topics considered throughout the text rely on *classical mechanics* and the basics are briefly discussed. Some of the methods of electrons generation are considered followed by some principles of beam transport – including limiting laws. Basic measures of beam quality are introduced. Space–charge waves are investigated, as well as a few instabilities that may develop when these waves are excited. Various radiation phenomena associated with accelerated charges conclude this chapter.

Chapter 4: In this chapter, I consider the fundamentals of beam–wave interaction in a *distributed* slow-wave structure. A dielectric loaded waveguide was chosen as the basic model in the first sections because it enables to illustrate the essence of the interaction without the complications associated with complex boundary conditions. Pierce's theory for the traveling wave amplifier extended to the relativistic regime is formulated and I present the operation of an oscillator in the context of finite length effects. While the dynamics of the beam in all these topics was considered in the framework of the hydrodynamic approximation, I gradually elevate the level of description of particles dynamics by using the macroparticle approach, namely, by representing the ensemble by a large number of clusters of electrons. This formalism enables to examine the interaction in phase-space either in the linear regime of operation or close to saturation. It also facilitates investigation of tapered structures and analysis of the interaction of prebunched beams in tapered structures. Further extension of the macroparticle approach to include the effect of reflections enables to describe the operation of an amplifier and an oscillator and obviously the transition from the former to the latter. A discussion on the interaction with hybrid modes concludes this chapter.

Chapter 5: In this chapter I present various characteristics of *periodic structures* with emphasis on those aspects relevant to interaction with electrons. Both closed and open periodic structures are considered. Smith–Purcell effect is analyzed as a particular case of a Green's function calculation for an open structure and a simple scattering problem is also considered. Planar and cylindrical Bragg waveguides are analyzed, paving the way to optical acceleration structures. Two dimensional periodic structures and some applications are considered. This chapter concludes with three examples of a transient solution in periodic structures.

Chapter 6: *Quasi-periodic structures* are the focus of this chapter. They are essential whenever it is required to maintain an interacting bunch in resonance with the wave if high efficiency is a must. Nonadiabatic change of geometry dictates a wide spatial spectrum, in which case the formulation of the interaction in terms of a single wave with a varying amplitude and phase is inadequate and I present an alternative approach. This is applicable to extraction sections in TWTs, klystron, or gyrotrons, or for evaluating roughness effect on beam quality in the case of advanced light sources. The chapter concludes with a discussion on a photoinjector.

Chapter 7: This chapter deals with the principles of *free electron laser*. Starting with the spontaneous emission as an electron traverses an ideal wiggler, I investigate coherent interaction in the low-gain and high-gain Compton regime. As in the TWT case, the macroparticle approach is introduced and I conclude the chapter with a brief overview of the various alternative schemes of free electron lasers and a special section dedicated to X-ray sources.

Chapter 8: One of the important systems where beam–wave interaction in periodic structures plays a crucial role is the *particle accelerator*. In the first part of this chapter I discuss some basics of linear acceleration. In the second part, I discuss various advanced acceleration concepts including, for example, acceleration in plasma by either laser or electrons wake, as well as acceleration at optical wavelengths.

Whenever possible I have referred the reader to experimental results although this was definitely not my main goal because of two reasons: first, as stated initially, my focus was to describe fundamental concepts and the models that can be represented by analytic or quasi-analytic expressions to readers. Second, the progress in the advanced acceleration physics is so rapid that by the time the monograph would be actually printed, some of the experimental results would become history. Nevertheless, I have definitely addressed the reader to the relevant reference whenever conceptually new ideas are supported by experiments.

Many new exercises were added in all chapters and a small fraction of the new material was formulated as exercises relying on the formulation in the specific chapter. Most of the exercises are not trivial and they fit to an advanced graduate course.

Now, to a few words of gratitude. Two mentors have influenced my research activity as reflected in this monograph: Professor John A. Nation of Cornell University during more than a decade of collaboration and Professor David Schieber of the Technion during almost three decades of discussions and deliberations. They both deserve my deepest gratitude. Three other individuals have impacted my decisions at some key junctions: the late Professor Norman M. Kroll of UCSD to whom I owe the encouragement to pursue the PASER concept at its early stages. To Professor Ilan Ben Zvi I am in debt for diffusing my hesitations as a theoretician to actually making an experiment and eventually performing the PASER experiment at Brookhaven National Laboratory. Last but definitely not least, Professor Maury Tigner of Cornell University. During a few discussions, he brought me to think seriously about energy recovery in optical accelerators which I believe will be an essential concept in the future.

Based on the first edition of the book, several mini-courses have been delivered out of which I wish to express my gratitude for the opportunity to Professor Koji Takata who organized the mini-course at KEK (Japan). To Mr. Guobin Fan the Director of the Institute of Applied Electronics, China Academy of Engineering Physics, Mianyang, Sichuan (China), and Dr. Zhuo Xu his deputy. They gathered a bunch of researchers from all over the country to this 1 week mini-course.

A couple of years ago after the first edition of this monograph was already out-of-print and even the copies available at Amazon were sold out, people from all over the world started to contact me directly for available exemplars. At this point I forwarded Dr. Chris Caron, the Topical Editor of Springer-Verlag, part of these letters with the suggestion to print out a few hundreds of exemplars. After a while, he had written to me with the suggestion to write a new edition. I was quite embarrassed because, on the one hand, when I finished the first edition I promised myself never to repeat the mistake of writing another book. On the other hand, it is difficult to ignore the readers. Naturally, after almost 15 years I had only the LaTeX version of the manuscript while the drawings were gone. To make a long story short, the commitment of Mrs Lesley Price and Mrs Hanna Bismut to assist had convinced me to commit to the laborious project. The first has agreed to convert the LaTeX to MS Word while the second agreed to redo all the drawings, as well as to make the new ones – about two hundreds overall. They both deserve my sincere appreciation, without their help, I would have never completed the project. Obviously, Dr. Caron deserves the credit for his initiative, without him, this edition would have never materialized.

Throughout the years, many students took courses that rely on the original text and their valuable questions lead to this revised version. In fact, part of these students, those whom I directly supervised during their graduate studies, had an important impact on several topics added in this edition. Four of them I wish to mention in the chronological order of their contributions: Dr. Samer Banna,

Dr. Assaf Lahav, Dr. Amit Mizrahi, and Mr. Vadim Karagodsky. To all I thank for the dedication and diligence manifested during our collaboration that is only partially reflected in this edition.

Once the draft was completed, I have asked two colleagues to review the book and spell out their criticism. Dr. Eric Colby (SLAC) and Dr. Wayne D. Kimura (STI) did a wonderful job and their numerous suggestions have improved significantly the monograph. Their willingness to help, as well as the precious time they dedicated to the laborious task, is highly appreciated. Last but not least there are the members of the editorial board of the series "Particle Acceleration and Detection", in which this book is published: their input and suggestions were greatly appreciated.

Throughout the past 20 years and more, several agencies have directly supported my research activity as reflected in this manuscript. I am listing them in chronologic order: Rothschild Foundation, United States Department of Energy, United States Air Force, Bi-National United States – Israel Foundation, Israel Science Foundation, and the Kidron Foundation. Among these, four individuals from the US DoE deserve personal appreciation for their support throughout the years: Dr. David Sutter, Dr. Gerald Peters, Dr. Bruce Strauss, and Dr. L.K. Lin.

Finally, I wish to express my deepest gratitude for the support I was fortunate to get from my parents Genia and Izu, my wife Tal, and my three children Michal, Roy, and Yuval.

Haifa/Israel, June 2011 Levi Schächter

Contents

Chapter 1
Introduction

This monograph aims to provide the reader with the foundation of rigorous analysis of microwave and other radiation sources based on free electrons as well as some basic electrons acceleration concepts. While for communication solid-state sources provide all needs of mobile communication, in case of satellite, radar and a few other applications vacuum tubes are virtually the only option.

Two major scientific programs rely on microwave or millimeter waves as part of their operation. Millimeter waves heat up electrons that in turn raise the temperature of the hydrogen-based plasma in order to facilitate fusion for future power plants. Modern particle accelerators rely on the acceleration experienced by a relativistic particle as it moves in the presence of a wave which propagates at the speed of light. In fact, the analysis presented in this book relies on the experience gained from a research program whose goal is to develop high power microwave radiation using distributed interaction in quasi-periodic structures for particles accelerators.

Future plans may present great new challenges for the designer of radiation sources. High-power radiation sources may contribute to repairing the ozone layer which is so vital to life on earth. Another possible future application is to launch, in low orbit, loads which would then be used to construct the international space station. The latter approach may have a substantial advantage over chemical rockets in which the weight of the load is a small fraction of the total rocket since in the case of electromagnetic propulsion the load is the majority of the weight.

The heart of all the applications mentioned above and many others, is the radiation source which can be of many kinds and a few of which are briefly overviewed in Sect 1.2. In all cases, the radiation is generated by converting kinetic energy from electrons. These electrons form a beam which propagates in vacuum where it interacts with electromagnetic waves in the presence of an auxiliary structure. The development of these sources started at the beginning of the twentieth century with the magnetron, followed by the klystron in the thirties, the traveling wave tube in the late 1940s, the gyrotron in the early 1960s and the free electron laser in the mid 1970s. These are a small fraction of the devices which have been developed during the years and have played a crucial role in defense, communication and research. Over the years, with the better understanding of their

L. Schächter, *Beam-Wave Interaction in Periodic and Quasi-Periodic Structures*,
Particle Acceleration and Detection, DOI 10.1007/978-3-642-19848-9_1,

operation principles, their performance improved and with it, the demand of the systems' designers. Consequently, the regime these tubes operate has broadened. For example, the first traveling wave tubes operated in a continuous mode with currents of the order of mA's and beam voltages of kV's whereas today, in addition to these kind of tubes, one can find high-power devices which are driven by kA's beams (and sometime tens of kA's) with voltages on the order of 1MV – in a pulse mode. The six order of magnitude increase in current and three in the voltage, correspond to an increase of nine orders of magnitude in the power level. Consequently, entire new varieties of conceptual and technological problems evolved. Obviously, power is not the only design parameter. Frequency, bandwidth, tunability, stability and repetition rate are only a few of the considerations, which should be taken into account while designing a radio frequency generator or amplifier.

In this text we present a detailed description and analysis of the concepts involved in the interaction of electromagnetic waves and electrons. Since we intend to present a rigorous analysis, within the limits of a reasonably sized volume, we chose a small fraction of the existing devices. Special attention is paid to various aspects of the interaction in periodic or quasi-periodic structures. We start with some basic concepts of electron-wave interaction.

1.1 Single-Particle Interaction

On its own, an electron cannot transfer energy via a linear process to a monochromatic electromagnetic wave in vacuum if the interaction extends over a very long region. In this introductory chapter we limit the discussion to *single*-particle schemes. *Collective* effects, where the current is sufficiently high to affect the electromagnetic field, are discussed in Chaps. 4, 6 and 7.

1.1.1 Infinite Length of Interaction

Far away from its source, in vacuum, an electromagnetic wave forms a plane wave which is characterized by a wave-number whose magnitude equals the angular frequency, ω, of the source divided by $c = 299,792,458$m/s, the phase velocity of the plane wave in vacuum, and its direction of propagation is perpendicular to both the electric and magnetic field. For the sake of simplicity let us assume that such a wave propagates in the z direction and the component of the electric field is parallel to the x axis i.e.,

$$E_x(z,t) = E_0 \cos\left[\omega\left(t - \frac{z}{c}\right)\right]. \tag{1.1.1}$$

If a charged particle moves at a uniform velocity v parallel to z axis, then the electric field this charge experiences (neglecting the effect of the charge on the wave) is given by

$$E_x[z(t),t] = E_0 \cos\left\{\omega\left[t - \frac{z(t)}{c}\right]\right\}. \tag{1.1.2}$$

A crude estimate for the particle's trajectory may be assumed to be $z(t) \simeq \mathrm{v}t$, therefore, if the charge moves in the presence of this wave from $t \to -\infty$ to $t \to \infty$ then the average electric field it experiences is *zero*,

$$\int_{-\infty}^{\infty} dt \cos\left[\omega t\left(1 - \frac{\mathrm{v}}{c}\right)\right] = 0, \tag{1.1.3}$$

even if the particle is highly relativistic (Pantell 1981). The lack of interaction can be illustrated in a clearer way by superimposing the dispersion relation of the wave and the particle on the same diagram – see Fig. 1.1. Explicitly, the relation between energy and momentum for an electron, $E = c\sqrt{p^2 + (mc)^2}$, where $m = 9.1094 \times 10^{-31}$ Kg is the rest mass of the electron and the corresponding relation for a photon in free space $E = cp$ are plotted on the same diagram. For the interaction to take place the electron has to change its initial state, subscript i, denoted by $(E_\mathrm{i}, p_\mathrm{i})$ along the dispersion relation to the final, subscript f, denoted by $(E_\mathrm{f}, p_\mathrm{f})$ in such a way that the resulting photon in case of emission or absorbed photon for absorption, has exactly the same difference of energy and momentum i.e.,

$$E_\mathrm{i} = E_\mathrm{f} + E_\mathrm{ph}, \tag{1.1.4}$$

and

$$p_\mathrm{i} = p_\mathrm{f} + p_\mathrm{ph}. \tag{1.1.5}$$

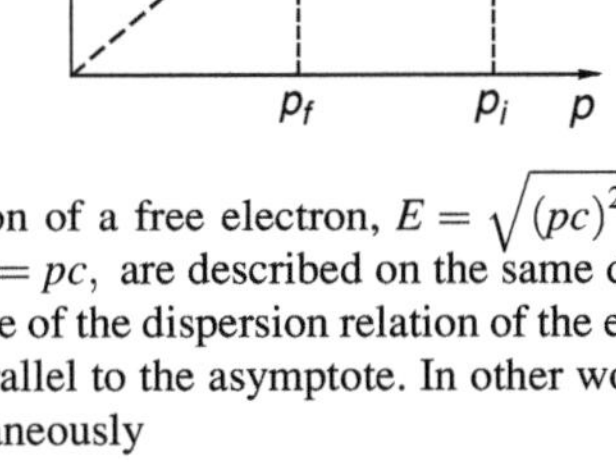

Fig. 1.1 The dispersion relation of a free electron, $E = \sqrt{(pc)^2 + (mc^2)^2}$, and an electromagnetic plane wave in vacuum, $E = pc$, are described on the same diagram. The dispersion relation of the wave is also the asymptote of the dispersion relation of the electron. Consequently, it cannot change its state along a line parallel to the asymptote. In other words, the energy and momentum laws cannot be satisfied simultaneously

In the case of vacuum this is impossible. Figure 1.1 reveals this fact graphically. The expression, $E = cp$, which describes the photon's dispersion relation, is parallel to the *asymptote* of the electron's dispersion relation. Thus, if we start from one point on the latter, a line parallel to $E = cp$ will never intersect the particle's line again. In other words, energy and momentum can not be conserved simultaneously in vacuum.

1.1.2 Finite Length of Interaction

If we go back to (1.1.3) we observe that if the electron spends only a finite time in the interaction region then it can experience a net electric field. Let us denote by $-T$ the time the electron enters the interaction region and by T the exit time. The average electric field experienced by the electron (subject to the same assumptions indicated above) is

$$\langle E\rangle = E_0\frac{1}{2T}\int_{-T}^{T} \mathrm{d}t \cos\left[\omega t\left(1-\frac{\mathrm{v}}{c}\right)\right] = E_0 \,\mathrm{sinc}\left[\omega T\left(1-\frac{\mathrm{v}}{c}\right)\right] \tag{1.1.6}$$

here $\mathrm{sinc}(x) = \sin(x)/x$. This is to say that if the time the electron spends in the interaction region, as measured in its frame of reference, is small on the scale of the radiation period $T_0 = 2\pi/\omega$ then the net electric field it experiences is not zero. From the perspective of the conservation laws, the interaction is possible since although the energy conservation remains unchanged i.e.,

$$E_\mathrm{i} = E_\mathrm{f} + \hbar\omega, \tag{1.1.7}$$

the constraint on momentum conservation was released somewhat and it reads

$$\left|p_\mathrm{i} - p_\mathrm{f} - \hbar\frac{\omega}{c}\right| < \frac{\hbar}{cT}, \tag{1.1.8}$$

which clearly is less stringent than in (1.1.5) as also illustrated in Fig. 1.2; $\hbar = 1.05457 \times 10^{-34}\mathrm{J}\cdot\mathrm{sec}$ is the Planck constant. The operation of the klystron relies

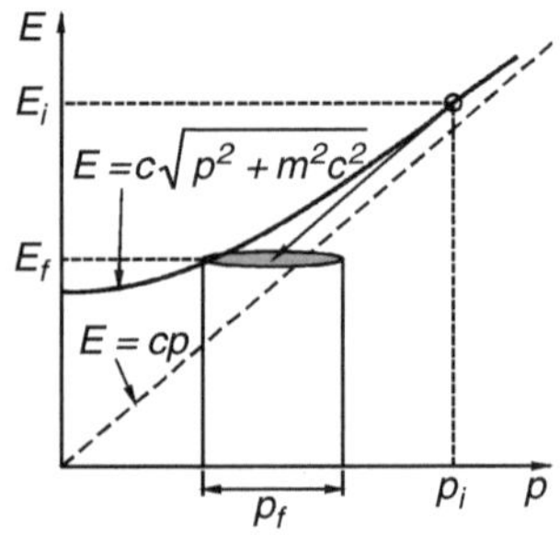

Fig. 1.2 The dispersion relation of a free electron, $E = \sqrt{(pc)^2 + (mc^2)^2}$, and an electromagnetic plane wave in vacuum, $E = pc$, are described on the same diagram. The constraint on the momentum conservation is less stringent because the interaction occurs in a finite length

on the interaction of an electron with a wave in a region which is shorter than the radiation wavelength.

1.1.3 Cerenkov Interaction

It was previously indicated that since the dispersion curve of the photon is parallel to the asymptote of the electron's dispersion relation, the interaction is not possible in an infinite domain. However, it is possible to change the *slope* of the photon, namely to change its phase velocity – see Fig. 1.3. The easiest way to do so is by "loading" the medium where the wave propagates with a material whose dielectric coefficient is larger than one. Denoting the refraction coefficient by n, the dispersion relation of the photon is given by

$$E_{\mathrm{ph}} = \frac{c}{n} p_{\mathrm{ph}}, \tag{1.1.9}$$

while the dispersion relation of the electron remains unchanged. Substituting in the expressions for the energy and the momentum conservation laws we find that the condition for the interaction to occur is

$$\frac{c}{n} = \mathrm{v}, \tag{1.1.10}$$

where it was assumed that the electron's recoil is relatively small i.e., $\hbar\omega/mc^2 \ll 1$. The result in (1.1.10) indicates that for the interaction to occur, the *phase-velocity in the medium has to equal the velocity of the particle*. This is the so-called Cerenkov condition in the 1D case. Although dielectric loading is conceptually simple, it is not always practical because of electric charges that accumulate on the surface and of a relatively low breakdown threshold, which is critical in high-power devices. For these reasons, the phase velocity is typically slowed down using metallic structures with periodic boundaries. The operation of traveling wave tubes

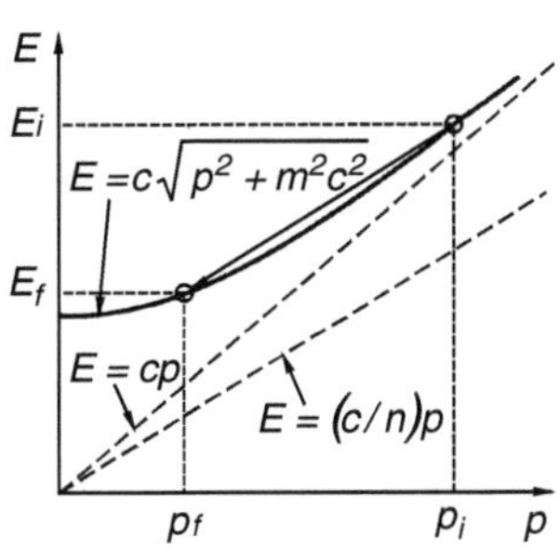

Fig. 1.3 The interaction of an electron with an electromagnetic wave whose phase velocity is smaller than c is possible

(or backward wave oscillators) relies on this concept and it will be discussed extensively in Chaps. 4, 5 and 6.

1.1.4 Compton Scattering: Static Fields

Not only a structure with periodic boundaries facilitates the interaction between electrons and electromagnetic waves, but also periodic fields. For example, if a *magneto-static field* of periodicity L is applied on the electron in the interaction region, then this field serves as a momentum "reservoir" which can supply momentum quanta of $n\hbar(2\pi/L)$ where $n = 0, \pm1, \pm2, \ldots$; see Fig. 1.4. The energy conservation law remains unchanged i.e.,

$$E_{\rm i} = E_{\rm f} + E_{\rm ph}, \tag{1.1.11}$$

but the momentum is balanced by the applied static field

$$p_{\rm i} = p_{\rm f} + p_{\rm ph} + \hbar\frac{2\pi}{L}n. \tag{1.1.12}$$

For a relativistic particle ($\beta \simeq 1$) and when the electron's recoil is assumed to be small, these two expressions determine the so-called resonance condition which reads

$$\omega \simeq 2\gamma^2\left(\frac{2\pi c}{L}n\right), \tag{1.1.13}$$

where $\gamma \equiv [1-(v/c)^2]^{-1/2}$. Note that the frequency of the emitted photon depends on the velocity of the electron, which means that by varying the velocity we can change the operating frequency. A radiation source that possesses this feature is a tunable source. Identical result is achievable if we assume a periodic electrostatic field and both field configurations are employed in free electron lasers discussed in Chap. 7.

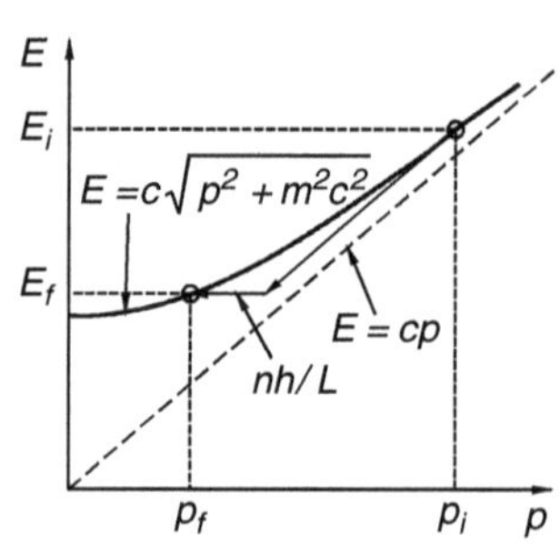

Fig. 1.4 The interaction of an electron with an electromagnetic wave in a periodic static field whose periodicity is L

1.1.5 Compton Scattering: Dynamic Fields

Static electric or magnetic field can be conceived as limiting cases of a dynamic field of zero or vanishingly small frequency and we indicated above that they facilitate the interaction between an electron and a wave. Consequently, we may expect that the interaction of an electron with a wave will occur in the presence of another wave. Indeed, if we have an initial wave of frequency ω_1 and the emitted wave is at a frequency ω_2 the conservation laws read

$$E_{\rm i} + \hbar\omega_1 = E_{\rm f} + \hbar\omega_2, \tag{1.1.14}$$

and

$$p_{\rm i} = p_{\rm f} + \hbar\frac{\omega_1}{c} + \hbar\frac{\omega_2}{c}. \tag{1.1.15}$$

Following the same procedure as above, the ratio between the frequencies of the two waves is

$$\frac{\omega_2}{\omega_1} \simeq 4\gamma^2, \tag{1.1.16}$$

which is by a factor of 2 larger than in the static case. Figure 1.5 illustrates this process and it will be elaborated in more detail in Chap. 8.

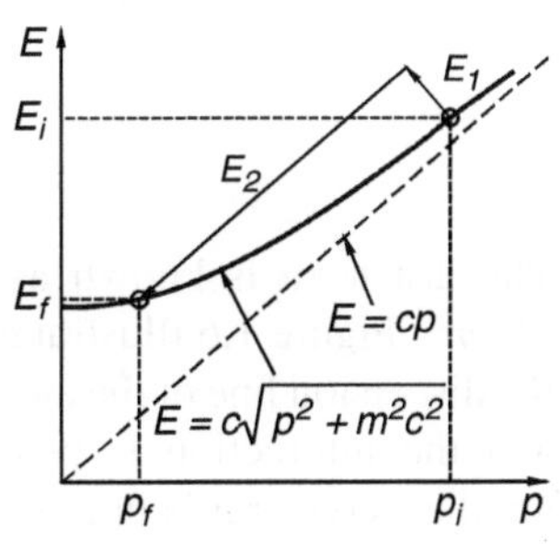

Fig. 1.5 The interaction of an electron with an electromagnetic wave in the presence of another electromagnetic wave

1.1.6 Uniform Magnetic Field

A periodic magnetic field can provide quanta of momentum necessary to satisfy the conservation law. It does not affect the average energy of the particle. The opposite happens when the electron moves in a uniform magnetic field (B): there

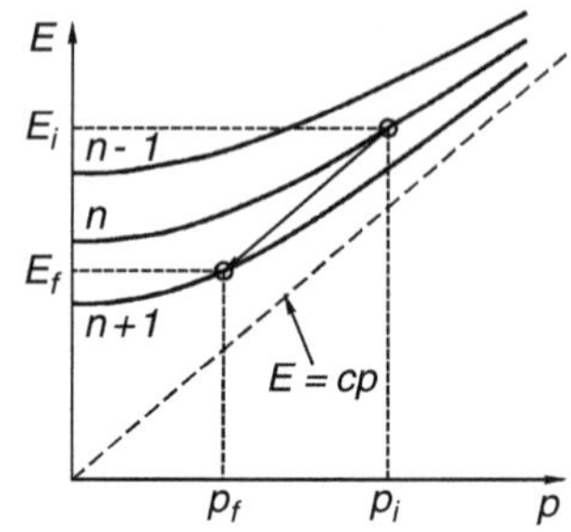

Fig. 1.6 The interaction of an electron with an electromagnetic wave in the presence of a uniform magnetic field $E = c\sqrt{p^2 + m^2c^2 - 2neB\hbar}$

is no change in the momentum of the particle whereas its energy may vary according to

$$E_n = c\sqrt{p^2 + (mc)^2 - 2n\hbar eB}, \qquad (1.1.17)$$

where $e = 1.6022 \times 10^{-19}$Cb is the charge of the electron and $n = 0, \pm 1, \pm 2....$

For most practical purposes the energy associated with the magnetic field is much smaller than the energy of the electron therefore we can approximate

$$E_{\rm i} - n_1\hbar\frac{ec^2B}{E_{\rm i}} = E_{\rm f} - n_2\hbar\frac{ec^2B}{E_{\rm f}} + E_{\rm ph}, \qquad (1.1.18)$$

and the momentum conservation remains unchanged i.e.,

$$p_{\rm i} = p_{\rm f} + p_{\rm ph}. \qquad (1.1.19)$$

From these two equations we find that the frequency of the emitted photon is

$$\omega = 2\gamma\frac{eB}{m} = 2\gamma^2\left(\frac{eB}{m\gamma}\right). \qquad (1.1.20)$$

The last term is known as the relativistic cyclotron angular frequency, $\omega_{\rm c,rel} \equiv eB/m\gamma$. Figure 1.6 illustrates schematically this type of interaction. It indicates that the dispersion line of the electron is split by the magnetic field in many lines (index n) and the interaction is possible since the electron can move from one line to another. Gyrotron's operation relies on this mechanism and it is discussed briefly in the next section.

1.2 Radiation Sources: Brief Overview

There are numerous types of radiation sources driven by electron beams. Our purpose in this section is to continue the *qualitative* discussion from the previous section and briefly describe the operation principles of one "member" of each class of what we consider the main classes of radiation sources. A few comments on experimental work will be made but for further details, the reader is referred to recent review studies. The discussion continues with the classification of the major radiation sources according to several criteria which we found to be instructive.

1.2.1 The Klystron

The klystron was one of the first radiation sources to be developed (Varian and Varian 1939). It is a device in which the interaction between the particle and the wave is localized to the close vicinity of a gap of a cavity, as illustrated in Fig. 1.7. Electrons move along a drift tube whose geometry is chosen such that at the frequency of interest it does not allow the electromagnetic wave to propagate. The latter is confined to cavities attached to the drift tube. The wave which feeds the first cavity modulates the velocity of the otherwise uniform beam. This means that after the cavity, half of the electrons have a velocity slightly larger than the average beam velocity whereas the second half has a smaller velocity. According to the change in the velocity of the electrons the beam becomes bunched down the stream since *accelerated* electrons from one period of the electromagnetic wave catch up with the *decelerated* electrons from the previous period. When this bunch enters the gap of another cavity, it may generate radiation very efficiently.

The operation of a klystron driven by a *relativistic* electron beam is different from that described above for a *non-relativistic* beam. If we were to use the same implementation in the case of a relativistic beam, then the distance the beam has to propagate in order to become density modulated is prohibitively long since the change in velocity is relatively small. What comes to our aid in the relativistic case is the fact that the current is much higher comparing to the non-relativistic case and when bunching the beam, we generate the, so-called, *space-charge waves* (Nation 1970). Fortunately, the velocity modulation from the input cavity translates in a density modulation in a quarter period of the plasma wave number (defined in Chap. 3) which is inversely-proportional to the square root of the current. Consequently, if the current is sufficiently high, then the distance between two cavities again becomes reasonable.

For efficient modulation of the beam, the quality factor of the cavities has to be high and therefore in general the klystron is not a tunable device. In high power

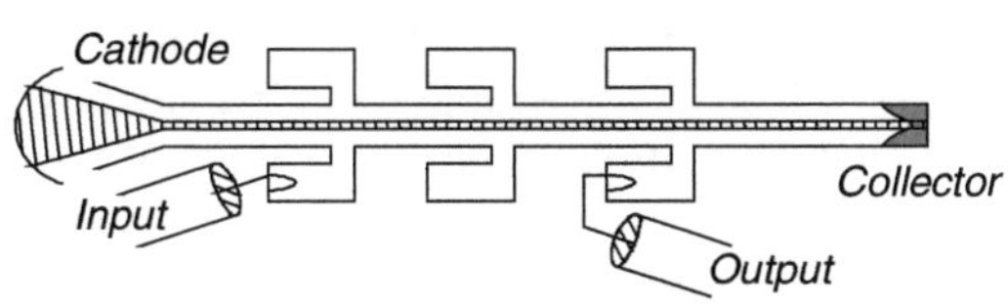

Fig. 1.7 The basic configuration of a klystron: the first cavity bunches the beam, the second amplifies the modulation and the third extracts power from the beam and converts it into radiation power

devices the choice of geometry is a trade-off between a small cavity gap required for good modulation and a large gap required to sustain the large electric field in the gap associated with high power levels. The most recent generation of high power klystron operates at 11.4 GHz it is driven by 440 kV, 500 A beam (Caryotakis 1994) and the goal is to generate power levels of the order of 100 MW in a 1.5 μs pulse for the Next Linear Collider (NLC) developed at SLAC National Accelerator Laboratory. Another class of high power klystrons was developed during the 1980s and 1990s by Friedman and Serlin (1985). Operating at relatively low frequencies (~1 GHz) the transverse geometry was chosen to be sufficiently large such that large amounts of current can be injected before reaching the limiting current (to be discussed in Chap. 3). In this case (Lau 1989), annular beams were shown to have significant advantages in generating multi-gigawatt pulses (Serlin and Friedman 1994).

1.2.2 The Traveling Wave Tube

The traveling wave tube (TWT) is a Cerenkov device, namely the phase velocity of the interacting wave is smaller than c and contrary to the klystron where the interaction occurs in the close vicinity of the cavity's gap, the interaction is *distributed* along many wavelengths. Generally speaking, as the beam and the wave advance, the beam gets modulated by the electric field of the wave and in turn, the modulated beam increases the amplitude of the electric field. In this process, both the beam modulation and the radiation field grow exponentially in space. The coupling between the wave and the beam is determined by the interaction impedance, which is a measure of the electric field acting on the electrons (E) for a given total electromagnetic power (P) flowing in the system

$$Z_{\text{int}} = \frac{E^2}{2k^2P}, \tag{1.2.1}$$

where k is the wave-number. This definition introduced first by Pierce (1950) is the basis of his theory of the TWT, which is in very good agreement with experiments in uniform and low power devices. Nation (1970) first introduced the concept of using space-charge waves in order to generate high power microwave radiation with traveling wave structures.

The TWT can be designed to be a broad-band device and it can occur in various configurations: helix, disk-loaded waveguide (coupled cavities), dielectric loaded waveguide, gratings, dielectrically coated metal and others. Several of these configurations are illustrated in Fig. 1.8a–e. Whenever the electromagnetic wave can propagate parallel to the beam, it means that a wave can also propagate in the opposite direction. Therefore the input is not isolated from the output, and in amplifiers, this problem can be detrimental. At the same time this is the basis for the design of an oscillator.

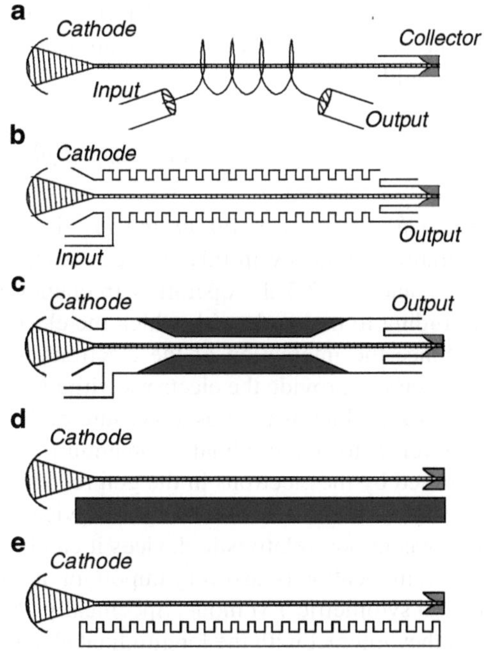

Fig. 1.8 (**a**–**e**) The basic configurtion of a traveling wave amplifier: (**a**) helix (**b**) coupled cavity structure (**c**) based on a dielectric loaded waveguide (**d**) based on an open dielectric structure and (**e**) based on an open periodic structure

In the interaction process the electron oscillates primarily along the major axis (z-direction) and the interaction is with the parallel component of the electric field. Correspondingly, the interaction occurs here with the transverse magnetic (TM) mode. This device will be extensively treated due to its relative simplicity and relevance to a wide range of other devices (gyrotrons, free electron lasers).

1.2.3 The Gyrotron

The gyrotron relies on the interaction between an *annular* beam, *gyrating* around the axis of symmetry due to an applied magnetic field, and a transverse electric (TE) mode. The concept of generating coherent radiation from electrons gyrating in a magnetic field was proposed independently by three different researchers in the late 1950s, Twiss (1958), Schneider (1959) and Gaponov (1959), and it has attracted substantial attention due to its potential to generate millimeter and sub-millimeter radiation.

Electrons move azimuthally and they get bunched by the corresponding azimuthal electric field. As in the case of the TWT the bunches act back on the field and amplify it. In contrast to traveling wave tubes or klystrons in which the beam typically interacts with the lowest mode, in the gyrotron the interaction is with high modes therefore various suppression techniques are employed in order to obtain coherent operation with a single mode.

The operation frequency is determined by the applied magnetic field, the energy of the electrons and, in cases of high mode operation, also by the radius of the waveguide:

$$\omega = \omega_c \gamma + \gamma\beta\sqrt{\omega_c^2 + \omega_{co}^2}, \tag{1.2.2}$$

where $\beta = v/c$, $\omega_c = eB/m$ and ω_{co} is the cutoff frequency of the mode. The operating frequency in this case can reach very high values: for a magnetic field of 1T and $\gamma \simeq 2.5$ the operation frequency is of the order of 150 GHz or higher according to the mode with which the electrons interact.

Since the interaction of the electrons is with an azimuthal electric field, it is necessary to provide the electrons with maximum momentum in this direction. The parameter which is used as a measure of the injected momentum is the ratio of the transverse to longitudinal momentum $\alpha \equiv v_\perp/v_z$. This transverse motion is acquired by the electrons in the gun region as can be deduced from the schematics illustrated in Fig. 1.9. In relativistic devices this ratio is typically smaller than unity whereas in non-relativistic devices it can be somewhat larger than one.

Beam location is also very important. In the TWT case the interaction is with the lowest symmetric TM mode. Specifically, the electrons usually form a pencil beam and they interact with the longitudinal electric field, which has a maximum on axis. We indicated that gyrotrons operate with high TE modes and the higher the mode, the higher the number of nulls the azimuthal electric field has along the radial direction. Between each two nulls there is a peak value of this field. It is crucial to have the annular beam on one of these peaks for an efficient interaction to take place.

Reviews of gyrotrons have been given by Flaygin et al. (1977) and Hirshfield and Granatstein (1977). An instructive overview of gyrotron theory was published by Baird (1987) and the experimental results were reviewed by Granatstein (1987). Two updated textbooks on gyrotrons were s published in the past few years: one (Kartikeyan et al. 2004) that has a more experimental flavor and the second, Nusinovich (2004), which is more theory oriented. Two important experimental results were reported recently. Sakamoto reported (Sakamoto et al. 2007) generation of 1 MW, cw operation at 170 GHz, for gyrotron plasma heating and current drive for ITER (International Thermonuclear Experimental Reactor).

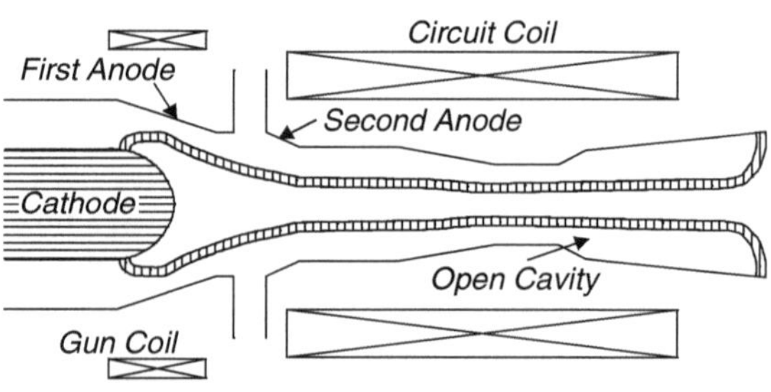

Fig. 1.9 The basic configuration of a gyratron. The magnetic insulated gun (MIG) generates electrons which spin azimuthally therefore they are suitable for interaction with a transverse electric (TE) mode

More recently 1.5 kW of 1 THz radiation were generated with a pulsed magnetic field magnetron were generated at the Institute of Applied Physics, Nizhny Novgorod (Glyavin et al. 2008).

1.2.4 The Free Electron Laser

The free electron laser (FEL) will be discussed in detail in Chap. 7. As the gyrotron, it is a fast-wave device in the sense that the interacting electromagnetic wave has a phase velocity larger or equal to c but instead of a uniform magnetic field it has a periodic magnetic field. The "conventional" free electron laser has a magnetic field perpendicular to the main component of the beam velocity. As a result, the electrons undergo a *transverse* oscillatory motion, which is suitable for interaction with either a TE or a TEM mode. The oscillation of electrons is in the transverse direction but the bunching is longitudinal and in this last regard the process is similar to the one in the traveling wave tube. However, its major advantage is the fact that it does not require a metallic structure for the interaction to take place. Consequently, it has the potential to either generate very high power at which the contact of radiation with metallic walls would create very serious problems, or produce radiation at UV, XUV or X-ray where there are no other coherent radiation sources. Figure 1.10 illustrates the basic configuration.

1.2.5 The Magnetron

The magnetron was invented at the beginning of the twentieth century (Hull 1921a, b) and it played a pivotal role in the radar development in WWII due to its relative high efficiency. Because of its complexity there is no analytical model which can describe its operation adequately as a whole. In recent years, great progress has been made in the understanding of the various processes with the aid of particle in cell (PIC) codes. Its operation combines potential and kinetic energy conversion. Figure 1.11 illustrates the basic configuration. Electrons are generated on the cathode (inner surface) and since a perpendicular magnetic field is applied, they form a flow which rotates azimuthally. The magnetic field and the voltage applied on the anode are chosen in such a way that, in equilibrium, the average velocity of the electrons equals the phase velocity of the wave supported by the periodic structure at the frequency of interest.

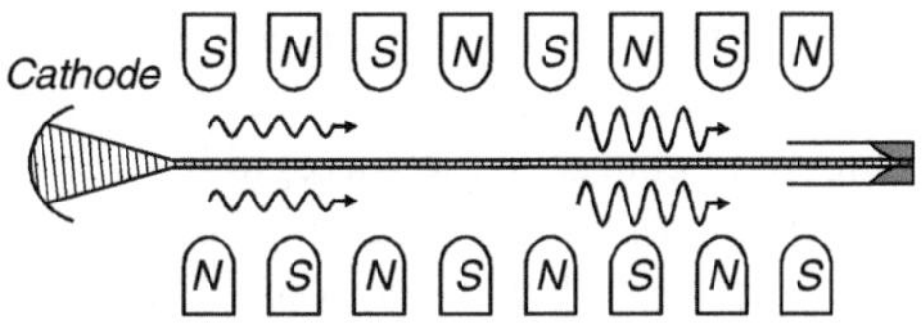

Fig. 1.10 The schematic of a free electron laser

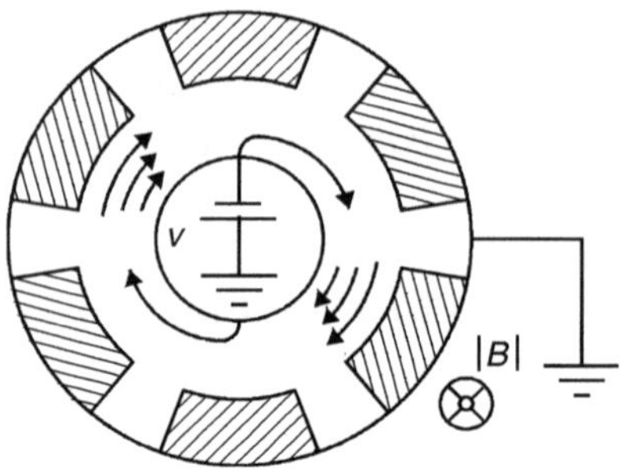

Fig. 1.11 The schematic of a magnetron

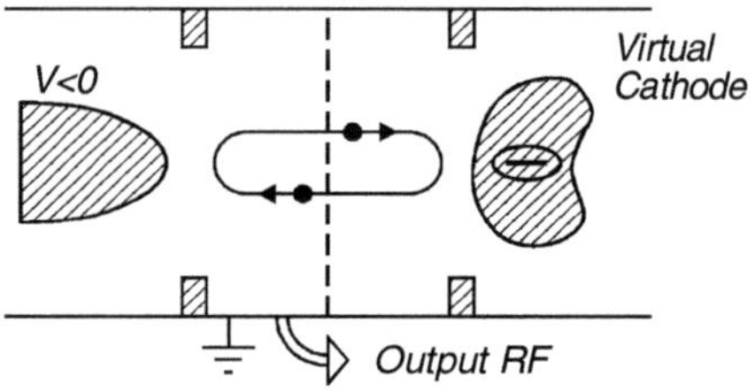

Fig. 1.12 The schematic of a vircator

A simplistic picture of the interaction can be conceived in the following way: electrons which lose energy to the wave via the Cerenkov type interaction, move in upward trajectories – closer to the anode. Consequently, two processes occur. Firstly, the closer the electron is to the periodic surface the stronger the radiation field and therefore the deceleration is larger, causing a further motion upwards. Secondly, as it moves upwards its (dc) potential energy varies. Again, this is converted into electromagnetic energy.

Two major differences between the magnetron and other radiation sources mentioned above, are evident: (1) in the magnetron, the beam generation, acceleration and collection occur all in the same region where the interaction takes place. (2) The potential energy associated with the presence of the charge in the gap plays an important role in the interaction; the other device where this is important is the vircator briefly discussed in the next sub-section. High power magnetrons are primarily used as drivers for medical accelerators and recently deposition of thin layers of various materials (sputtering) – see (Sarakinos et al. 2010).

1.2.6 The Vircator

The vircator takes advantage of the fact that the amount of current generated by a given voltage that can be injected into a grounded metallic waveguide is limited. Any current injected above this limit is reflected, but on average there is a finite amount of charge in the waveguide – see Fig. 1.12. This charge forms what is called a *virtual cathode* (i.e. negative potential) which can be conceived as the reason for the reflection of the electrons. These oscillate between the real and the virtual cathode at a frequency which is directly related to the electrons' density (plasma frequency). A review of the vircator's theory has been given by Sullivan et al. (1987) and later Alyokhin et al. (1994) presented a review of the various studies.

1.2.7 Classification Criteria

The variety of operation principles and consequently of devices does not allow to define a single criterion for their classification. We start our discussion with the trivial observation that any radiation source consists of at least two components: *electron beam* and *electromagnetic wave*. From the *wave* perspective the first question we should ask is whether it is guided or confined by metallic walls as is the case in most sources, or if it can propagate in free space as is the case in a few of the free electron laser schemes. If it is guided, then the next question is whether its phase velocity is smaller or greater than *c*. The first category is that of *slow*-wave devices and its main members are the traveling wave tube and the backward *wave* oscillator (BWO). The second category that of *fast-wave* devices, consists of the gyrotron, cyclotron auto-resonance maser (CARM) and the free electron laser. Among the slow-wave structures, there is room for an additional subdivision since there are *closed* or *open* slow-wave structures. Although the great majority of today's systems rely on closed structures, the continuous demand for high frequency sources will continuously enhance the number of devices that have open structures as their main component; primarily because of the relatively limited number of modes that may develop.

Still in the context of the electromagnetic wave, the various sources can be classified according to the interacting *mode*. In TWTs the interaction is always with the transverse magnetic (TM) mode whereas in gyrotrons the interaction is always with the transverse electric (TE) mode. FELs, on the other hand, may interact with either TE or TEM mode. Combinations of TE and TM modes (hybrid) are, in general destructive – as happens in acceleration sections where the hybrid mode (HE_{11}) causes beam-breakup. However, this effect can be utilized for constructive purposes in particular when a highly relativistic beam has to be dumped to the wall. Beam break-up is discussed in Chap. 8.

Even in two devices in which the interaction is with the same mode, say TM mode such as in the case of TWT and klystron, there is room for additional subdivision regarding the character of the interaction. In the TWT (as in FEL, gyrotron and magnetron) the interaction is *distributed* and it occurs over many wavelengths. On the other hand, in a klystron, the interaction is *localized* and it is limited to the close vicinity of the cavity's gap – which is typically a fraction of the wavelength.

The electromagnetic structure determines whether there are *reflections* in the system and, as we shall see in Chaps. 4 and 6, these determine if the system operates as an *amplifier* or an *oscillator*. In the case of metallic periodic structures, the feedback can be designed to be part of the electromagnetic characteristics of the structure as happens in the case of the backward wave oscillator. Furthermore, the transverse dimension of the structure determines the number of electromagnetic modes in the structure. In most cases, the geometry is chosen such that a single mode is supported at a given frequency but there are cases where frequency or power impose large geometry therefore the system becomes a multi-mode device. This is the case for gyrotrons and also a few Cerenkov devices, e.g., Bugaev et al. (1990).

If we examine the sources from the point of view of the *electron beam* there are also many possible classifications. High-power devices utilize typically relativistic beams and devices like the free electron laser have meaningful operations primarily in this regime. Others like the TWT, gyrotron and klystron can operate either with relativistic or non-relativistic electrons. Relativistic beams in many cases are associated not only with high voltages (>200 kV) but also high currents (>250 A) which implies high power levels (>50 MW). These can be sustained for relatively short periods of time; typically of the order of 1 μs or shorter. In many of the cases of interest, several such pulses are fired per second and this is referred to as the *repetition rate*. For example, in case of the SLAC X-band klystron mentioned above, the tube is driven by 1.5 μs long electron pulses at a repetition rate of 180 Hz. At the other extreme, there are continuous wave (CW) sources such as magnetrons, gyrotrons and TWTs, which operate at high average power (>1 kW). Repetitive pulse sources, such as the one driven by the 50 MW beam mentioned above, may provide a maximum average power (assuming 50% rf efficiency conversion) of 2.5 W if the pulse duration is 100 ns and the repetition rate is 1 Hz. At 1 μs and 10 Hz the average power goes up to 250 W.

Without exception the beam has to be guided, otherwise the electrons blow apart and they are of no use for energy exchange. In most cases, the beam is guided by a uniform magnetic field and in a small fraction by a permanent periodic magnetic field. In the gyrotron or the free electron laser this field plays a crucial role in the interaction process itself. Furthermore, in cross-field devices the uniform magnetic field is accompanied by a perpendicular electric field that also contributes to the interaction.

Beam quality, which is associated with fluctuations in the energy around the average value, is another classification criterion. This topic is addressed in Sect. 3.4.4 and it is of particular interest in accelerators and in free electron lasers. In the former because the electrons have to travel very long distances and ultimately have to be focused with great precision, therefore both the transverse and longitudinal momentum are important. In free electron lasers, this parameter is important as we go up in frequency and in fact beam quality is one of the major limitations of today's free electron lasers – at least with regard to optical or shorter wavelengths.

Energy conversion efficiency brings us to another possible way of classification, based on whether the initial beam is uniform or pre-bunched. In the latter case, the efficiency of energy extraction can be very high. There are basically two ways to pre-bunch a beam: either in a two (or more) stage system as in a klystron or to form the bunches at the same place where the electrons are generated, namely to produce bunches in which all electrons have the same velocity. In this regard the way the electrons are generated is critical and may have a dramatic impact on the performance of the device.

Finally, the amount of current injected into the system can also be used for classification of sources. SLAC klystrons, for example, operate well below the limiting current whereas the relativistic klystron amplifier (RKA) developed at NRL by Friedman and Serlin (1985) operates close to the limiting current. At the extreme, the vircator operates well above the limiting current.

1.3 Accelerators

It is virtually impossible to cover in depth the variety of topics involved in the operation of modern accelerators in a single monograph therefore, the compromise we adopt is to focus on a few basic phenomena which are directly related to beam-wave interaction in periodic or quasi-periodic structures. Motivated by the essentials of beam-wave interaction, as in the previous section, we focus on several aspects of particle acceleration. Specifically, in this sub-section, we discuss superficially several topics: rf photo-injectors and linear accelerators, circular machines and damping ring and some essentials of advanced acceleration concepts. If the reader is interested in topics that are not covered by this text, there is a large variety of books that cover different aspects of accelerator physics. Starting from the handbook by Chao and Tigner (1998), to milestone articles edited by Pellegrini and Sessler (1995) and didactic volumes of Wiedemann (1999a, b) or Lee (2004) and others that I omitted and I owe them my apology.

1.3.1 RF Photo Injector

A linear collider or an FEL based light source require very good quality electron beam in particular for generating short wavelengths moreover, a high peak current is necessary for a reasonable gain. After the demonstration of infra-red (IR) radiation amplification by electrons in a wiggler by Elias et al. (1976), these requirements were already met by the best conventional injectors (dc gun + buncher). Westenskow and Madey (1984) proposed to put a thermionic cathode in an rf cavity. This new gun named "rf gun" or "microwave gun" was used as a bright electron source for the Mark III FEL. Meanwhile, it was found that very high current densities could be obtained from semiconductor photo-cathodes and in the late 1980s the first demonstration of a FEL driven by electrons from a laser irradiated photocathode was reported at Stanford by Curtin et al. (1990). Since then, rf photo-injectors became the standard in any linear accelerator design.

Conceptually, an rf-photo-injector consists of a photocathode placed in an rf cavity illuminated by a laser which, in turn, delivers short pulses and since their duration is much shorter than the rf period, as they leave the cathode, the emerging electrons are already bunched. Comparing to the normal thermionic cathodes that for extended life-time are limited to about 10 A cm^{-2}, photo-cathodes may deliver

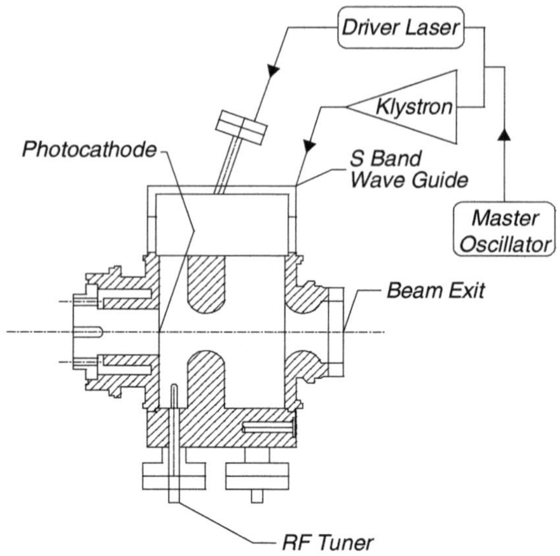

Fig. 1.13 The schematic of an rf photo-injector

very high current densities (hundreds of A cm^{-2}) but the actual current density is determined by the required life-time and the beam quality necessary for the specific application. A high-power (SLAC's 2,856 MHz) klystron feeds a relatively short (1.5 cell) acceleration structure and the rf field accelerates the electrons – McDonald et al. (1988). Typically 5–10 ps pulses of 1 nC are available and the outgoing electrons have an energy of less than 10 MeV. Schematic of the Brookhaven National Laboratory (BNL) design is illustrated in Fig. 1.13 and a review of the state of the art was compiled by Russell (2003).

1.3.2 Damping Ring

One or more acceleration modules according to the specifications of the required system follow a photo-injector. Such a module consists of either a series of coupled cavity structure or disk-loaded waveguide. A wave that has a longitudinal electric field and moves at the speed of light in vacuum is supported by this kind of structure. As a result, a relativistic electron experiences a constant electric field. A detail of the processes associated with this acceleration process is discussed in more detail in Chap. 8. In this section we adopt a system-oriented approach and consider the next central component of a linear collider, this is the damping ring. It was pointed out above that an electron moving at a constant velocity in vacuum does not radiate however, if it accelerates then it may radiate. In fact, if the acceleration is perpendicular to the trajectory, the power emitted (synchrotron radiation) is proportional to γ^4 thus the electron is decelerated. This effect plays a crucial role in damping rings which, in turn, is vital for reducing the "temperature" of relativistic electrons. Low "temperature" is essential when nanometer size bunches need to collide after being accelerated along many kilometers in a linear collider. In order to assess the difficulty in the latter case we need to remember that charged particles tend to repel each other therefore, if no action is taken, the bunch will spread out leading to fewer collisions (events) than a narrowly focused one. To avoid this spread-out, the electron and positron bunches are injected into damping rings (DR). Here the particles' trajectory is bended by vertical magnetic fields and they are decelerated according to their energy. Along the circumference of the ring, acceleration modules recuperate the deficit in the momentum but this is provided only in the *longitudinal* direction. In other words, as the bunches circulate in the damping ring, they lose energy by synchrotron radiation and are re-accelerated each time they pass through a cavity. The synchrotron radiation reduces the motion in *all* direction, while the cavity re-accelerates only those in the *desired* direction. Thus, the bunch of electrons or positrons becomes more and more parallel in motion as the radiation "damps out" motion in the undesired directions. As this monograph is being written significant efforts are dedicated by the accelerator physics community to the design of the damping rings of the International Linear Collider (ILC). Although this is remote from the focus of this monograph, it should be also mentioned that circular machines are employed for the acceleration of ions since the synchrotron radiation by heavy particle is negligible comparing to that emitted by electrons. As a result, ions can be accelerated when following a circular trajectory without a significant energy loss.

Beyond the injector, the damping ring and the main accelerator body, a linear collider consists of a set of magnet(s) that focus the electrons and positrons to the interaction point. The dynamics of the particles in this region is not in our scope in the framework of this monograph.

1.3.3 Advanced Acceleration Concepts

Charged particles are accelerated in either a cavity or slow-wave structure by an electromagnetic wave. The latter's frequency varies between 500 MHz to 30 GHz and the power levels scale, as a rough estimate for a given gradient, as $P \propto f^{-2}$ therefore it is just natural to aim for higher frequencies since the necessary power is reduced accordingly. Moreover, the threshold for breakdown at higher frequencies is elevated facilitating higher gradients thus shorter structures. Beyond 30 GHz the availability of high-power radiation sources is limited until we get into the optical range 10–0.5 μm. Reducing the operating wavelength by four orders of magnitude entails a reduction of about 2 orders of magnitude in the longitudinal dimension – assuming that all the other components (optics, cryogenics and vacuum systems) can be scaled like the acceleration structure. As a result, the 30 km long ILC may be accommodated within the 2-miles long tunnel at SLAC if the operating wavelength is 1 μm. Obviously nature is not as generous with us and many obstacles ought to be removed before this wishful thinking will become a feasible alternative and in this text we shall discuss a few.

Laser acceleration is very appealing since the technology is already being developed by the communication industry and the drastic change in the operating frequency leads to a few revolutionary alternatives. For example, in this frequency range metals have much higher loss comparing to dielectrics therefore, the acceleration structure should be of dielectric material which for ages was a taboo in the accelerators community. An even more revolutionary concept is the use of plasma. From the first days of vacuum tubes, it was clear that in order to have a reasonable control of the charged particles they must move in *vacuum* in order to avoid decoherence due to scattering with surrounding atoms/molecules (hydrogen). The powerful laser available today facilitate to generate intense plasma waves (gradients of 200 GV/m!!) which may be employed to accelerate electrons. In this text we briefly consider acceleration in plasma but we will consider laser acceleration in dielectric structures. Readers interested in laser driven acceleration in plasma can consider a recent review article by Esarey et al. (2009).

1.4 Choice and Organization of the Material

With such a variety of concepts and paradigms we owe the reader an explanation of how the material was selected, and why we chose to present one topic, whereas another, which might be as important, was left out. The principle that directed me was to have a coherent and thorough presentation of the beam-wave interaction in a few modern devices with most mathematical details that enable a simple description of their operation.

From the very beginning, it was clear that it is virtually impossible to meet the requirement of detailed presentation and encompass the whole variety of sources and interaction schemes, discussed above, in one reasonably-sized volume. During the nineties, I was actively involved in the development of high-power, high-efficiency traveling wave amplifiers whereas during the last decade my focus has shifted to advanced acceleration concepts. These facts have biased the choice of presentation towards the interaction in *periodic* and *quasi-periodic structures*.

The book can be divided into three parts. The first includes Chaps. 2 and 3 which present some of the elementary concepts in the electromagnetic theory and electrons' dynamics which are relevant to beam-wave interaction. The second part includes Chaps. 4, 5 and 6. It addresses the interaction in periodic (and quasi-periodic) metallic structures. The third part (Chaps. 7 and 8) focuses on free electron laser and an introduction the linear acceleration. Let us now briefly review the content of the various chapters.

Chapter 2. After we discuss *Maxwell equations* in general we present simple homogeneous solutions corresponding to the TEM, TM, TE and hybrid modes. When the current-density is present it is useful to use Green's function method for solution of the electromagnetic field. Its formal description is accompanied by two examples which illustrate the Cerenkov radiation in free space and in a waveguide. Several finite length effects are considered as well as edge effects. Scattered waves phenomena are also discussed.

Chapter 3. All topics considered throughout the text rely on *classical mechanics* and the basics are briefly discussed. Some of the methods of electrons generation are considered followed by some principle of beam transport – including limiting laws (e.g. Child-Langmuir). In this context, some of the basic measures (emittance and brightness) are introduced. Space-charge waves are introduced and some of the fundamental instabilities that may develop when these waves are excited. Various radiation phenomena conclude this chapter.

Chapter 4. In this chapter we investigate the fundamentals of beam-wave interaction in a *distributed* slow-wave structure. A dielectric loaded waveguide was chosen as the basic model in the first sections because it enables us to illustrate the essence of the interaction without the complications associated with complex boundary conditions. In the first section, we present part of Pierce's theory for the traveling wave amplifier applied to dielectric loaded structure and extended to the relativistic regime. Finite length effects are considered and the operation of an oscillator is described. While the dynamics of the beam in all these topics was considered in the framework of the hydrodynamic approximation, we elevate the level of description of particles dynamics by employing the macro-particle approach namely, by representing the ensemble by a large number of clusters of electrons. This formalism enables us to examine the interaction in phase-space either in the linear regime of operation or close to saturation. It also facilitates investigation of tapered structures and analysis of the interaction of pre-bunched beams in tapered structures. Further extension of the macro-particle approach to include the effect of reflections enables us to describe the operation of an amplifier and an oscillator and obviously the transition from the former to the latter. A discussion on the interaction with hybrid modes concludes this chapter.

Chapter 5. This chapter presents various characteristics of *periodic structures* with emphasis on those aspects relevant to interaction with electrons. First we present the basic theorem of periodic structures, namely the Floquet theorem. Following this theorem we bring an investigation of closed and open periodic structures. Smith-Purcell effect is considered as a particular case of a Green's function calculation for an open structure and a simple scattering problem is also considered. Planar and cylindrical Bragg waveguides are being analyzed, paving the way to optical acceleration structures. Two dimensional periodic structures are being considered and some applications are considered. The chapter concludes with three examples of a transient solution in periodic structures.

Chapter 6. This chapter deals with metallic *quasi-periodic structures* which, among others, are required in order to maintain an interacting bunch in resonance with the wave when high efficiency is required. Non-adiabatic change of geometry dictates a wide spatial spectrum, in which case the formulation of the interaction in terms of a single wave with a varying amplitude and phase is inadequate. In fact, the electromagnetic field cannot be expressed in a simple (analytic) form if substantial geometric variations occur from one cell to another. To be more specific: in uniform or weakly tapered structures the beam-wave interaction is analyzed assuming that the general functional form of the electromagnetic wave is known i.e., $A \times (z) \cos[\omega t - kz - \phi(z)]$ and the beam affects the amplitude $A(z)$ and the phase, $\phi(z)$. Furthermore, it is assumed that the variation due to the interaction is *small* on the scale of one wavelength of the radiation. Both assumptions are not acceptable in the case of a structure designed for high efficiency interaction. In order to overcome this difficulty and others, we present an analytic technique which has been developed in order to design and analyze quasi-periodic metallic structures of the type discussed in Chap. 5. The method relies on a model which consists of a cylindrical waveguide to which a number of pill-box cavities and radial arms are attached. In principle, the number of cavities and arms is arbitrary. Surface roughness effect may be described using this model and the Chapter concludes with a discussion on a photo-injector.

In the third part of this book we consider the beam-wave interaction in periodic and quasi-periodic structures different from that in the second part, namely free electron lasers and particle accelerators. These two topics have been extensively discussed in literature and they are the subject of many articles, books and conferences. Therefore, our approach in this part combines our approach of detailed analysis used in the previous chapters with a general discussion of alternative concepts and configurations.

Chapter 7. This chapter deals with the principles of *free electron laser*. In the first section we consider the spontaneous emission as an electron traverses an ideal wiggler. It is followed by the investigation of coherent interaction in the low-gain Compton regime and subsequently we present the high-gain Compton regime which includes cold and warm beam operations. The macro-particle approach is introduced and we conclude the chapter with a brief overview of the various alternative schemes of free electron lasers and a special section dedicated to X-ray sources.

Chapter 8. One of the important systems where beam-wave interaction in periodic structures plays a crucial role is the *particle accelerator*. In the first part of this chapter we discuss some basics of linear acceleration. In the second part we discuss various advanced acceleration concepts including two-beam acceleration, acceleration in plasma by either laser or electrons wake as well as acceleration at optical wavelengths.

Chapter 2
Elementary Electromagnetic Phenomena

All the effects discussed in this text rely on the presence of electric, magnetic or electro-magnetic fields in the system. It is therefore natural to discuss first the governing equations and some basic electromagnetic phenomena. With this regard, "elementary" in the title of this chapter refers to subjects related to beam-wave interaction and not necessarily to undergraduate-level topics, though we discuss a few elementary concepts in the first two subsections.

2.1 Maxwell's Equations

At the foundations for the analysis of all electro-magnetic phenomena are Maxwell's equations that relate the electric (**E**) and magnetic (**H**) field, the electric (**D**) and magnetic (**B**) inductions with the current (**J**) and charge (ρ) densities:

$$\nabla \times \mathbf{E}(\mathbf{r},t) + \frac{\partial}{\partial t}\mathbf{B}(\mathbf{r},t) = 0, \tag{2.1.1}$$

$$\nabla \times \mathbf{H}(\mathbf{r},t) - \frac{\partial}{\partial t}\mathbf{D}(\mathbf{r},t) = \mathbf{J}(\mathbf{r},t), \tag{2.1.2}$$

$$\nabla \cdot \mathbf{D}(\mathbf{r},t) = \rho(\mathbf{r},t), \tag{2.1.3}$$

$$\nabla \cdot \mathbf{B}(\mathbf{r},t) = 0. \tag{2.1.4}$$

This set of equations determines the electromagnetic field at any point in space and in time provided that the source terms (ρ and **J**), are known. In addition, the initial and boundary conditions have to be determined together with the constitutive relations of the medium, i.e., the relation between the inductions (**B** and **D**) and the field components (**H** and **E**).

L. Schächter, *Beam-Wave Interaction in Periodic and Quasi-Periodic Structures*, Particle Acceleration and Detection, DOI 10.1007/978-3-642-19848-9_2,

2.1.1 Constitutive Relations

Matter reacts to the presence of an electromagnetic field and the constitutive relations characterize this reaction. In general, these relations are non-linear and they couple all the components of the electromagnetic field. In many of the cases of interest, the constitutive relations are linear and scalar

$$\mathbf{B}(\mathbf{r},t) = \mu_0\mu_r\mathbf{H}(\mathbf{r},t), \tag{2.1.5}$$

$$\mathbf{D}(\mathbf{r},t) = \varepsilon_0\varepsilon_r\mathbf{E}(\mathbf{r},t), \tag{2.1.6}$$

and in case of a metal Ohm law's reads

$$\mathbf{J}(\mathbf{r},t) = \sigma\mathbf{E}(\mathbf{r},t); \tag{2.1.7}$$

here $\varepsilon_0 = 8.85 \times 10^{-12}$ farad/m and $\mu_0 = 4\pi \times 10^{-7}$ henry/m are the vacuum permittivity and permeability respectively. The *relative* dielectric coefficient ε_r and its permeability counterpart μ_r characterize the material. In vacuum, $\varepsilon_r \equiv 1$, $\mu_r \equiv 1$ and $\sigma = 0$, i.e.,

$$\nabla \times \mathbf{E}(\mathbf{r},t) + \frac{\partial}{\partial t}\mu_0\mathbf{H}(\mathbf{r},t) = 0, \tag{2.1.8}$$

$$\nabla \times \mathbf{H}(\mathbf{r},t) - \frac{\partial}{\partial t}\varepsilon_0\mathbf{E}(\mathbf{r},t) = \mathbf{J}(\mathbf{r},t), \tag{2.1.9}$$

$$\nabla \cdot \varepsilon_0\mathbf{E}(\mathbf{r},t) = \rho(\mathbf{r},t), \tag{2.1.10}$$

$$\nabla \cdot \mu_0\mathbf{H}(\mathbf{r},t) = 0. \tag{2.1.11}$$

Assuming that we know the source terms (ρ and $\mathbf{J}$) it is sufficient to use the first two equations (2.1.7)–(2.1.8) in conjunction with the charge conservation,

$$\nabla \cdot \mathbf{J}(\mathbf{r},t) + \frac{\partial}{\partial t}\rho(\mathbf{r},t) = 0, \tag{2.1.12}$$

in order to solve the electromagnetic field. This statement can be examined by applying $\nabla\cdot$ on both (2.1.7) and (2.1.8). Since any vector function $\mathbf{V}$ satisfies $\nabla \cdot (\nabla \times \mathbf{V}) \equiv 0$, one obtains (2.1.10) from (2.1.7) and (2.1.9) from (2.1.8).

2.1.2 Boundary Conditions

At sharp discontinuities the differential operators are not defined therefore an integral approach has to be adopted. Alternatively, Maxwell's equations can be solved in each region separately, away of the discontinuity, and the question that needs to be addressed is the relation between the various field components from both sides of a discontinuity. Consider two regions (subscripts 1 and 2) separated by a surface which is locally characterized by its local normal **n**. The boundary condition associated with (2.1.1) is deduced from its integral form as

$$\mathbf{n} \times (\mathbf{E}_1 - \mathbf{E}_2) = 0. \tag{2.1.13}$$

Similarly, from the integral form of (2.1.1) we conclude that

$$\mathbf{n} \times (\mathbf{H}_1 - \mathbf{H}_2) = \mathbf{J}_s, \tag{2.1.14}$$

from (2.1.2)

$$\mathbf{n} \cdot (\mathbf{D}_1 - \mathbf{D}_2) = \rho_s, \tag{2.1.15}$$

and finally, from the integral form of (2.1.3) we can deduce that

$$\mathbf{n} \cdot (\mathbf{B}_1 - \mathbf{B}_2) = 0. \tag{2.1.16}$$

Here $\mathbf{J}_s$ is the surface current density and ρ_s is the surface charge density.

Equation (2.1.12) indicates that the tangential component of the electric field, at any time, has to be continuous at the transition between two discontinuities. In a similar way, the tangential component of the magnetic field can be discontinuous only if there is a *surface current density* ($\mathbf{J}_s$) – see (2.1.13). The other two expressions indicate that any discontinuity in the normal component of the electric induction is due to surface charge density (ρ_s) and the normal component of the magnetic induction is always continuous.

Comment 2.1. As in the case of Maxwell's equations, it is sufficient to use the first two sets of boundary conditions since the latter two are then automatically satisfied.

Comment 2.2. One outcome of the boundary conditions as formulated above is that at the surface of an ideal metal ($\sigma \to \infty$) the tangential electric field vanishes. This is because the electric field is zero in the metal and the tangential electric field has to be continuous.

2.1.3 Poynting's Theorem

The energy conservation associated with the electromagnetic field can be deduced from Maxwell's equations by multiplying (scalarly) (2.1.1) by **H**, (2.1.1) by **E** and subtracting the latter from the former. In a linear medium, the result reads

$$\nabla \cdot \mathbf{S} + \frac{\partial}{\partial t}\left[\frac{1}{2}\varepsilon_0\varepsilon_r \mathbf{E}\cdot\mathbf{E} + \frac{1}{2}\mu_0\mu_r \mathbf{H}\cdot\mathbf{H}\right] = -\mathbf{J}\cdot\mathbf{E}, \tag{2.1.17}$$

where

$$\mathbf{S}(\mathbf{r},t) \equiv \mathbf{E}(\mathbf{r},t) \times \mathbf{H}(\mathbf{r},t) \tag{2.1.18}$$

is the instantaneous *Poynting vector* which represents the energy flux (power per unit surface) in the vector direction. The second term,

$$\mathrm{w}(\mathbf{r},t) \equiv \frac{1}{2}\varepsilon_0\varepsilon_r \mathbf{E}(\mathbf{r},t)\cdot\mathbf{E}(\mathbf{r},t) + \frac{1}{2}\mu_0\mu_r \mathbf{H}(\mathbf{r},t)\cdot\mathbf{H}(\mathbf{r},t), \tag{2.1.19}$$

represents the instantaneous *energy density* stored in the electric and magnetic field respectively. And the right-hand side term in (2.1.16) represents the coupling between the electromagnetic field and the sources (or sinks) in the system.

Gauss's theorem can be used to formulate Poynting's theorem in its *integral form*. We integrate over a volume V whose boundary is denoted by **a**; the result is

$$\frac{\mathrm{d}}{\mathrm{d}t}W(t) = -\oiint \mathbf{da}\cdot S - \int_v \mathrm{d}V \mathbf{J}\cdot\mathbf{E}, \tag{2.1.20}$$

where for a linear medium

$$W(t) \equiv \int_V \mathrm{d}V\left[\frac{1}{2}\varepsilon_0\varepsilon_r \mathbf{E}\cdot\mathbf{E} + \frac{1}{2}\mu_0\mu_r \mathbf{H}\cdot\mathbf{H}\right], \tag{2.1.21}$$

is the total energy stored in the volume V. Explicitly (2.1.19) reveals that the change in the energy stored in the volume is either due to energy flux flowing *through* the surrounding envelope or due to sources in the volume (or both).

One important aspect to emphasize at this stage is that the electromagnetic power is carried by the field and not by the metallic boundaries; the latter only guide the energy flow. This is an important observation since subsequently, we discuss the propagation of electromagnetic waves of hundreds of megawatts and all this power propagates in vacuum. To illustrate the process let us consider an elementary electric circuit consisting of a battery, two parallel lossless wires, and a resistor at the end as illustrated in Fig. 2.1.

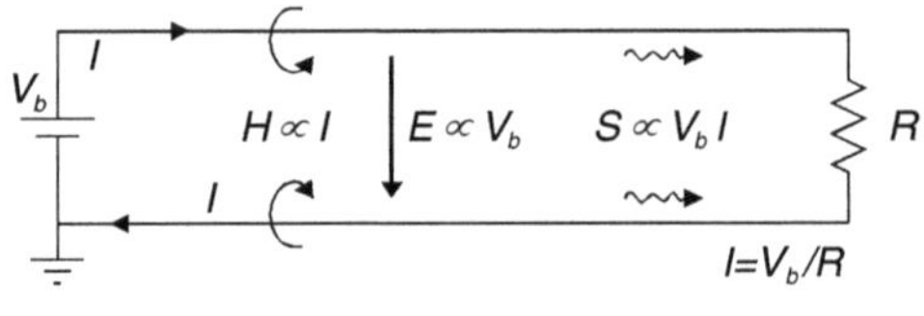

Fig. 2.1 Energy flow in a simple circuit. The power flows in the air and is guided by the wires

Firstly, we examine the Poynting vector term of (2.1.19): the voltage V_b is determined by the battery whereas the current is determined by the resistor (R) namely, $I = V_b/R$. Since the distance between the two wires is d, the typical electric field between the two wires is V_b/d moreover, the azimuthal magnetic field generated by one wire at the location of the other is proportional to the current I. Consequently, the Poynting vector is parallel to the wires and it is proportional to the product of the two field components $S \propto IV_b$. The power which propagates from the battery towards the resistor is proportional to Poynting vector thus as expected, the power is proportional to IV_b or V_b^2/R. Since there are no time variations the energy term in Poynting theorem vanishes whereas the second term in the right-hand side of (2.1.19) can be readily calculated to show that the power dissipated in the resistor is V_b^2/R. For further discussion see Chap. 11 in the text book of Haus and Melcher (1989).

2.1.4 Steady-State Regime

In many cases of interest all the components of the electromagnetic field oscillates at a single angular frequency (ω) thus all components have the following functional form

$$F(\mathbf{r},t) = f(\mathbf{r})\cos[\omega t + \psi(\mathbf{r})]. \tag{2.1.22}$$

It is convenient to omit the time dependence and represent the function $F(\mathbf{r},t)$ using a complex notation, namely we introduce the imaginary number $j \equiv \sqrt{-1}$ and utilize the fact that $\exp(j\xi) \equiv \cos(\xi) + j\sin(\xi)$ the function F

$$F(\mathbf{r},t) = \frac{1}{2}\{f(\mathbf{r})\exp[j\psi(\mathbf{r})]\exp(j\omega t) + f(\mathbf{r})\exp[-j\psi(\mathbf{r})]\exp(-j\omega t)\}. \tag{2.1.23}$$

With this notation, it is convenient to define

$$\bar{F}(\mathbf{r},\omega) \equiv f(\mathbf{r})e^{j\psi(\mathbf{r})}, \tag{2.1.24}$$

which permits us to use this function instead of $F(\mathbf{r},t)$ and consequently,

$$F(\mathbf{r},t) = \mathrm{Re}\left[\bar{F}(\mathbf{r},\omega)e^{j\omega t}\right]; \tag{2.1.25}$$

$\bar{F}(\mathbf{r},\omega)$ is called the *phasor* associated with the function $F(\mathbf{r},t)$. To illustrate the use of this notation, Maxwell's equations read

$$\nabla \times \overline{\mathbf{E}} + j\omega\overline{\mathbf{B}} = 0, \tag{2.1.26}$$

$$\nabla \times \overline{\mathbf{H}} - j\omega\overline{\mathbf{D}} = \overline{\mathbf{J}}, \tag{2.1.27}$$

$$\nabla \cdot \overline{\mathbf{D}} = \overline{\rho}, \tag{2.1.28}$$

$$\nabla \cdot \overline{\mathbf{B}} = 0. \tag{2.1.29}$$

The main advantage of this notation is now evident since the differential operator $\partial/\partial t$ was replaced by a simple algebraic operator $j\omega$.

2.1.5 Complex Poynting's Theorem

The phasor notation, as introduced above, cannot be directly applied to Poynting's theorem since all quantities are quadratic in the electromagnetic field. In principle, we have two options: (1) transform the field components to the time domain and then substitute in Poynting's theorem as defined in (2.1.16) – abandoning in the process the phasor notation. (2) Limit the information to the *average energy* and *average power* – but preserving the phasor notation. Since in the former case there is no real advantage to the new notation, we next pursue the latter option.

When we consider the product of two oscillating quantities, we have

$$\begin{aligned} &A_1 \cos(\omega t + \psi_1) A_2 \cos(\omega t + \psi_2) \\ &= \frac{1}{4}\left[\bar{A}_1 \exp(j\omega t) + \bar{A}_1^* \exp(-j\omega t)\right]\left[\bar{A}_2 \exp(j\omega t) + \bar{A}_2^* \exp(-j\omega t)\right] \end{aligned} \tag{2.1.30}$$

the *average* of the product of these two oscillating functions corresponds to the non-oscillating term in the expression above i.e.,

$$\frac{1}{4}\left[\bar{A}_1\bar{A}_2^* + \bar{A}_1^*\bar{A}_2\right] = \frac{1}{2}A_1A_2\cos(\psi_1 - \psi_2). \tag{2.1.31}$$

We use this fact in order to formulate the *complex Poynting's theorem*. First (2.1.25) is multiplied scalarly by the complex conjugate of the magnetic field phasor ($\overline{\mathbf{H}}^*$). From the product we subtract the complex conjugate of (2.1.26) multiplied by the electric field; the result reads

$$\nabla \cdot \overline{\mathbf{S}} + 2j\omega[\overline{\mathrm{w}}_\mathrm{M} - \overline{\mathrm{w}}_\mathrm{E}] = -\frac{1}{2}\overline{\mathbf{E}} \cdot \overline{\mathbf{J}}^*, \tag{2.1.32}$$

wherein $\overline{\mathbf{S}} = \overline{\mathbf{E}} \times \overline{\mathbf{H}}^*/2$ is the *complex Poynting vector*, $\overline{\mathrm{w}}_\mathrm{M} = \mu_0\mu_\mathrm{r}\overline{\mathbf{H}} \cdot \overline{\mathbf{H}}^*/4$ is the average (in time) magnetic energy density and $\overline{\mathrm{w}}_\mathrm{E} = \varepsilon_0\varepsilon_\mathrm{r}\overline{\mathbf{E}} \cdot \overline{\mathbf{E}}^*/4$ is the electric counterpart.

Energy conversion is associated with the *real* part of the Poynting vector whereas the *imaginary* component is associated with electro-magnetic energy stored in the system. Throughout the text we omit the bar from the phasor quantities, except if ambiguities may occur.

2.1.6 Potentials

It is convenient, instead of solving a couple of first order differential equations, to solve a single second-order differential equation. For this purpose we benefit from the fact that the divergence of the magnetic induction is zero ($\nabla \cdot \mathbf{B} = 0$) and introduce the magnetic vector potential $\mathbf{A}$ which determines the magnetic induction through

$$\mathbf{B} = \nabla \times \mathbf{A} \tag{2.1.33}$$

By virtue of this definition, the equation $\nabla \cdot \mathbf{B} = 0$ becomes an identity. Substituting this definition in Faraday's law (2.1.25) we obtain

$$\nabla \times (\mathbf{E} + j\omega\mathbf{A}) = 0. \tag{2.1.34}$$

Further using the fact that $\nabla \times (\nabla\Phi) \equiv 0$ we conclude that

$$\mathbf{E} = -j\omega\mathbf{A} - \nabla\Phi, \tag{2.1.35}$$

wherein Φ is the *scalar electric potential*. Both potentials satisfy, in a Cartesian coordinate system and in a linear medium ($\mu_r = 1$ and $\varepsilon_r > 1$), the non-homogeneous wave equation:

$$\left[\nabla^2 + \varepsilon_r \frac{\omega^2}{c^2}\right]\mathbf{A} = -\mu_0 \mathbf{J}, \tag{2.1.36}$$

and

$$\left[\nabla^2 + \varepsilon_r \frac{\omega^2}{c^2}\right]\Phi = -\frac{1}{\varepsilon_0 \varepsilon_r}\rho, \tag{2.1.37}$$

provided that the divergence of the vector function $\mathbf{A}$ is chosen to be

$$\nabla \cdot \mathbf{A} + j\omega \frac{\varepsilon_r}{c^2}\Phi = 0. \tag{2.1.38}$$

This is the so-called *Lorentz gauge*; $c \equiv 1/\sqrt{\mu_0 \varepsilon_0}$ is the phase velocity of a plane electromagnetic wave in vacuum.

2.1.7 Edge Effect

In addition to the boundary conditions discussed above in the context of sharp discontinuity, we need to consider the field and the energy near an edge. It is demonstrated in what follows that while near an edge, the electric field diverges, the energy stored is finite.

With this purpose in mind, consider a simple configuration where the radius of curvature of a realistic edge is much smaller than the characteristic wavelength of the electromagnetic field in its vicinity ($\lambda \gg R$). Based on this assumption, the electric field in the vicinity of an ideal edge ($R = 0$) as the one schematically illustrated in Fig. 2.2, is a solution of the Laplace's equation and further assuming that the system is infinite in the z-direction, then

$$\left[\frac{1}{r}\frac{\partial}{\partial r}r\frac{\partial}{\partial r}+\frac{1}{r^2}\frac{\partial^2}{\partial \phi^2}\right]\Phi = 0 \qquad (2.1.39)$$

is the equation to be solved subject to the zero potential condition on the metallic walls

$$\Phi\left(r, \phi = \frac{\alpha}{2}\right) = 0 \quad \text{and} \quad \Phi\left(r, \phi = 2\pi - \frac{\alpha}{2}\right) = 0. \qquad (2.1.40)$$

Its solution has the form $\Phi \sim Ae^{j\nu\phi}r^{\nu} + Be^{-j\nu\phi}r^{\nu}$, thus imposing the boundary conditions namely, $\Phi(r, \phi = \alpha/2) = 0$ and $\Phi(r, \phi = 2\pi - \alpha/2) = 0$ we conclude that a non-trivial solution is possible if $\sin[\nu(2\pi - \alpha)] = 0$, implying that the radius of curvature of the field (ν) is given by

$$\nu = \frac{\pi}{2\pi - \alpha}n \qquad n = 1, 2, 3\ldots \qquad (2.1.41)$$

and consequently,

$$\Phi\left(r, \frac{\alpha}{2} \le \phi \le 2\pi - \frac{\alpha}{2}\right) = \sum_{n=1}^{\infty} A_n \sin\left[\frac{\pi n}{2\pi - \alpha}\left(\phi - \frac{\alpha}{2}\right)\right] r^{\frac{\pi n}{2\pi - \alpha}}. \qquad (2.1.42)$$

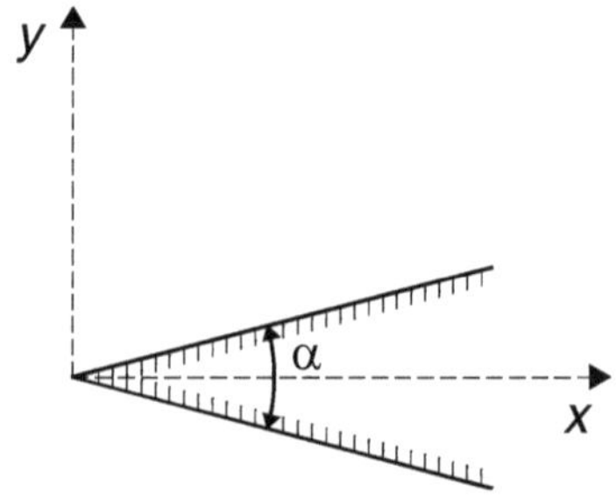

Fig. 2.2 In the vicinity of an ideal edge the curvature of the electric field is determined by its angle α

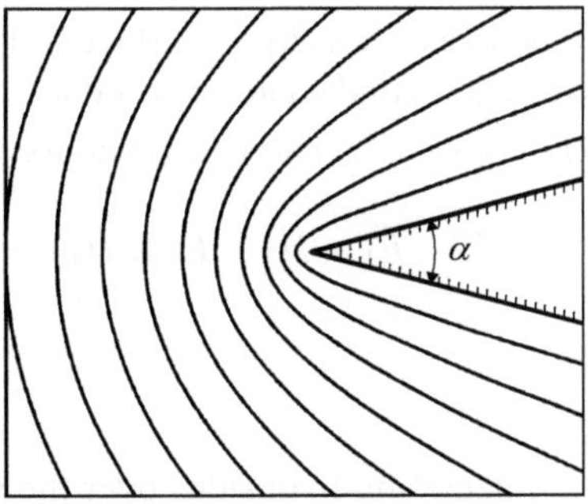

Fig. 2.3 Contours of constant potential ($n = 1$) in the vicinity of an ideal edge

In order to demonstrate the previous statement, let us consider the first harmonic ($n = 1$) illustrated in Fig. 2.3 for $\alpha = \pi/6$. The corresponding field components are

$$E_r = -\frac{\partial \Phi_1}{\partial r} = -\frac{\pi}{2\pi - \alpha} A_1 \sin\left[\frac{\pi}{2\pi - \alpha}\left(\phi - \frac{\alpha}{2}\right)\right] r^{\frac{\pi}{2\pi-\alpha}-1}$$
$$E_\phi = -\frac{1}{r}\frac{\partial \Phi_1}{\partial \phi} = -\frac{\pi}{2\pi - \alpha} A_1 \cos\left[\frac{\pi}{2\pi - \alpha}\left(\phi - \frac{\alpha}{2}\right)\right] r^{\frac{\pi}{2\pi-\alpha}-1} \quad (2.1.43)$$

revealing that at the limit $r \to 0$, if $\alpha < \pi$, then the electric field *diverges*. Nevertheless, the energy, stored in a volume of radius R and length Δ_z, is *finite* as can be deduced from the explicit expression for the stored energy

$$W_E = \Delta_z \int_{\alpha/2}^{2\pi-\alpha/2} d\phi \int_0^R drr \left[\frac{1}{2}\varepsilon_0 E_r^2 + \frac{1}{2}\varepsilon_0 E_\phi^2\right] \propto A_1^2 R \frac{2\pi}{2\pi - \alpha}. \quad (2.1.44)$$

Comment 2.3. A similar approach may be followed to investigate the field distribution in the vicinity of a *dielectric* edge. In this case the curvature of the field (ν) is determined by both the angle of the edge (α) as well as the dielectric coefficient (ε_r) and it is a solution of

$$\varepsilon_r \tan\left(\nu \frac{\alpha}{2}\right) + \tan\left[\nu\left(\pi - \frac{\alpha}{2}\right)\right] = 0. \quad (2.1.45)$$

2.1.8 Reciprocity Theorem

The Lorentz reciprocity theorem is a useful theorem for solution of electromagnetic problems, since it may be used to deduce a number of fundamental properties of practical devices. It provides the basis for demonstrating the reciprocal properties of electronic microwave circuits and for showing that the receiving and transmitting characteristics of antennas are the same. To derive the theorem, consider a volume

V bounded by a closed surface A. Let a current source $\vec{J}_1$ in V produce a field $\vec{E}_1, \vec{H}_1$ while a second source $\vec{J}_2$ produces a field $\vec{E}_2, \vec{H}_2$. Expanding the relation $\nabla \cdot (\vec{E}_1 \times \vec{H}_2 - \vec{E}_2 \times \vec{H}_1)$ and using Maxwell's equation it can be shown that

$$\begin{aligned}\nabla \cdot (\vec{E}_1 \times \vec{H}_2 - \vec{E}_2 \times \vec{H}_1) &= (\nabla \times \vec{E}_1) \cdot \vec{H}_2 - (\nabla \times \vec{H}_2) \cdot \vec{E}_1 \\ &\quad - (\nabla \times \vec{E}_2) \cdot \vec{H}_1 + (\nabla \times \vec{H}_1) \cdot \vec{E}_2 \\ &= -\vec{J}_2 \cdot \vec{E}_1 + \vec{J}_1 \cdot \vec{E}_2.\end{aligned} \tag{2.1.46}$$

Integrating both sides over the volume V and using Gauss' theorem

$$\begin{aligned}\int_V \nabla \cdot (\vec{E}_1 \times \vec{H}_2 - \vec{E}_2 \times \vec{H}_1) dV &= \oiint_A (\vec{E}_1 \times \vec{H}_2 - \vec{E}_2 \times \vec{H}_1) \cdot \vec{n} dA \\ &= \int_V (\vec{E}_2 \cdot \vec{J}_1 - \vec{E}_1 \cdot \vec{J}_2) dV,\end{aligned} \tag{2.1.47}$$

where $\vec{n}$ is the unit outward normal to A.

There are at least two important cases where the surface integral vanishes: in the first case of radiating fields (to be discussed subsequently) and in the case of quasi-state fields when $E \propto r^{-2}$ and $H \propto r^{-2}$. Since the surface of integration is proportional to r^2 at the limit $r \to \infty$ the surface integral clearly vanishes, therefore (2.1.46) reduces to

$$\int_V \vec{E}_1 \cdot \vec{J}_2 dV = \int_V \vec{E}_2 \cdot \vec{J}_1 dV. \tag{2.1.48}$$

If $\vec{J}_1$ and $\vec{J}_2$ are *infinitesimal* current elements this is to say that the variations of the electric field of the other source are negligible in the region of the source, then

$$\vec{E}_1(\mathbf{r}_2) \cdot \vec{J}_2(\mathbf{r}_2) = \vec{E}_2(\mathbf{r}_1) \cdot \vec{J}_1(\mathbf{r}_1), \tag{2.1.49}$$

which states that the field $\vec{E}_1$ produced by $\vec{J}_1$ has a component along $\vec{J}_2$ that is equal to the component along $\vec{J}_1$ of the field generated by $\vec{J}_2$ when $\vec{J}_1$ and $\vec{J}_2$ have unit magnitude. The form (2.1.48) is essentially the reciprocity principle used in circuit analysis except that $\vec{E}$ and $\vec{J}$ are replaced by the voltage V and current I.

2.2 Simple Wave Phenomena

In this section, we present solutions of the wave equation for several simple cases. A few of the examples presented here will be used subsequently to develop models which in turn enable the investigation of complex structures.

2.2.1 Simple Propagating Waves

With the source terms, constitutive relations and boundary conditions determined, one could proceed towards solution of a few simple wave phenomena. For simplicity we consider a scalar function $\psi(\mathbf{r})$ which oscillates at an angular frequency ω (i.e., we assume a steady-state regime of the form $\exp j\omega t$) and which is a solution of

$$\left[\nabla^2 + \frac{\omega^2}{c^2}\right]\psi(\mathbf{r}) = 0. \tag{2.2.1}$$

As a first stage, we examine waves propagating in *one dimension*. In a Cartesian system (x, y, z) we consider a system in which all variations are only in the z direction ($\partial/\partial x \sim 0$ and $\partial/\partial y \sim 0$), and the homogeneous wave equation reads

$$\left[\frac{\mathrm{d}^2}{\mathrm{d}z^2} + \frac{\omega^2}{c^2}\right]\psi(z) = 0. \tag{2.2.2}$$

A second order differential equation, has two solutions:

$$\psi(z) = A_+ \exp\left(-j\frac{\omega}{c}z\right) + A_- \exp\left(j\frac{\omega}{c}z\right); \tag{2.2.3}$$

these represent plane waves since the phase is constant, in the plane defined by $z = \text{const}$. The first term describes a wave propagating in the z-direction whereas the second represents a wave propagating in the opposite direction.

In a cylindrical coordinate system (r, ϕ, z), ignoring azimuthal and longitudinal variations ($\partial^2/\partial\phi^2 \sim 0$ and $\partial^2/\partial z^2 \sim 0$), the wave equation reads

$$\left[\frac{1}{r}\frac{\mathrm{d}}{\mathrm{d}r}r\frac{\mathrm{d}}{\mathrm{d}r} + \frac{\omega^2}{c^2}\right]\psi(r) = 0. \tag{2.2.4}$$

Its solution is

$$\psi(r) = A_+ \mathrm{H}_0^{(2)}\left(\frac{\omega}{c}r\right) + A_- \mathrm{H}_0^{(1)}\left(\frac{\omega}{c}r\right), \tag{2.2.5}$$

where $\mathrm{H}_0^{(1)}(\xi)$ and $\mathrm{H}_0^{(2)}(\xi)$ are the zero order Hankel function of the first and second kind; they are related to Bessel functions of the first and second kind by $\mathrm{H}_0^{(1)}(x) \equiv \mathrm{J}_0(x) + j\mathrm{Y}_0(x)$ and $\mathrm{H}_0^{(2)}(x) \equiv \mathrm{J}_0(x) - j\mathrm{Y}_0(x)$. As in the previous case, the first term represents a wave propagating from the axis outwards and the second term describes a wave propagating inwards. For completeness, we also present the solution in a spherical coordinate system (r, ϕ, θ). Ignoring all angular variations the wave equation is given by

$$\left[\frac{1}{r}\frac{\mathrm{d}^2}{\mathrm{d}r^2}r+\frac{\omega^2}{c^2}\right]\psi(r)=0, \tag{2.2.6}$$

and its solution is

$$\psi(r)=A_+\left(\frac{\omega}{c}r\right)^{-1}\exp\left(-j\frac{\omega}{c}r\right)+A_-\left(\frac{\omega}{c}r\right)^{-1}\exp\left(j\frac{\omega}{c}r\right), \tag{2.2.7}$$

where the first term represents a spherical wave propagating outwards (from the center out) whereas the second represents an inward flow.

2.2.2 The Radiation Condition

From the pure mathematical point of view, the two waves in each one of the solutions of above are a direct result of the fact that the wave equation is a second order differential equation. However, in absence of obstacles, our daily experience dictates a wave which propagates from the source outwards; this implies that in all three cases there are no "advanced" waves i.e., $A_-\equiv 0$. This is one possible interpretation of the so-called the *radiation condition* and it can be considered an additional boundary condition which is a byproduct of the causality constraint imposed on the solutions of the wave equation.

This formulation relies on the simple solutions presented above; however, the general trend is valid for solutions that are more complex. In the case of *cylindrical* azimuthally non-symmetric waves, the radiation condition implies for a solution $\psi(r,\phi,z)$, that the limit

$$\left[\psi(r,\phi,z)\exp(j\omega r/c)r^{1/2}\right]_{r\to\infty}, \tag{2.2.8}$$

is finite and it is r independent. In a similar way, for *spherical* waves described by a function $\psi(r,\phi,\theta)$, the limit

$$\left[\psi(r,\phi,\theta)\mathrm{e}^{j(\omega/c)r}r\right]_{r\to\infty}, \tag{2.2.9}$$

is finite and r independent. While this condition looks straightforward in the analytic examples presented above, it is not as trivial to impose it in numerical solvers in particular in a broad frequency range and/or when the mode configuration cannot be explicitly specified.

Wheeler and Feynman (1945) have used advanced solutions of the wave equation in order to explain the source of the so-called radiation reaction force. It is well known that electromagnetic power is emitted by a particle when it is accelerated. This power is emitted from the particle outwards and comes at the expense of its kinetic energy. Since this change in the kinetic energy of the particle can be conceived as an effective force this is also referred to as the *radiation reaction force*.

2.2.3 Evanescent Waves

So far we have presented only waves which vary and propagate in one dimension (1D), namely solutions of the wave equation either in a Cartesian, cylindrical or spherical system of coordinate. At this point, the level of complexity is slightly elevated to include waves that vary in two dimensions. First, consider a Cartesian coordinate system in which we ignore variations in the y direction. The wave equation in this case reads

$$\left[\frac{\partial^2}{\partial x^2}+\frac{\partial^2}{\partial z^2}+\frac{\omega^2}{c^2}\right]\psi(x,z,\omega)=0, \tag{2.2.10}$$

and its formal solution, assuming a propagating behavior in the z-direction, is given by

$$\psi(x,z,\omega)=\exp(-jkz)\left[A_+\exp\left(-\sqrt{k^2-\frac{\omega^2}{c^2}}x\right)+A_-\exp\left(\sqrt{k^2-\frac{\omega^2}{c^2}}x\right)\right]. \tag{2.2.11}$$

However in the half-plane defined by $x>0$ the solution is

$$\psi(x,z,\omega)=A_+\exp(-jkz)\exp\left(-\sqrt{k^2-\frac{\omega^2}{c^2}}x\right), \tag{2.2.12}$$

since otherwise the solution diverges at $x\to\infty$. For $|k|c>\omega$ the wave decays exponentially in the x direction. This is an *evanescent wave*: it propagates in one direction and decays exponentially in another. In the opposite case, for $|k|c<\omega$, the wave propagates at an angle $\theta=\cos^{-1}(kc/\omega)$ relative to the z axis.

It is instructive to examine (2.2.12) in the time domain. Assuming zero phase for A_+ then

$$\psi(x>0,z,t)=A_+\cos(\omega t-kz)\exp\left[-\sqrt{k^2-\left(\frac{\omega}{c}\right)^2}x\right]. \tag{2.2.13}$$

Based on this expression it is convenient to introduce the concept of *phase velocity*: this is the velocity at which an imaginary observer has to move, in order to measure a constant phase ($\omega t-kz=\text{const}$); explicitly, this reads

$$\mathrm{v}_{\mathrm{ph}} \equiv \frac{\omega}{k}. \tag{2.2.14}$$

With this definition, we observe that in a two dimensional case, an evanescent wave is characterized by a phase velocity smaller than c.

2.2.4 *Waves of a Moving Charge*

Evanescent waves play an important role in the interaction process of particles and waves. The simplest manifestation of their role is the representation of the spectrum of a moving charge in the laboratory frame of reference. For this purpose, we examine now the waves associated with a point charge (e) moving in the z direction at a constant velocity v_0 in vacuum; no boundaries are involved and the system is azimuthally symmetric ($\partial/\partial\phi = 0$). The current distribution in this case is given by

$$\mathbf{J}(\mathbf{r}, t) = -e\mathrm{v}_0 \frac{1}{2\pi r}\delta(r)\delta(z - \mathrm{v}_0 t)\mathbf{1}_z, \tag{2.2.15}$$

where $\mathbf{1}_z$ is a unit vector in the z direction. This current distribution excites the z component of the magnetic vector potential that in turn satisfies

$$\left[\frac{1}{r}\frac{\partial}{\partial r}r\frac{\partial}{\partial r} + \frac{\partial^2}{\partial z^2} - \frac{1}{c^2}\frac{\partial^2}{\partial t^2}\right]A_z(r, z, t) = -\mu_0 J_z(r, z, t); \tag{2.2.16}$$

its solution is assumed to have the form

$$A_z(r, z, t) = \int_{-\infty}^{\infty} \mathrm{d}\omega \exp(j\omega t) \int_{-\infty}^{\infty} \mathrm{d}k \exp(-jkz) a_z(r, k, \omega), \tag{2.2.17}$$

where $a_z(r, k, \omega)$ satisfies

$$\left[\frac{1}{r}\frac{\mathrm{d}}{\mathrm{d}r}r\frac{\mathrm{d}}{\mathrm{d}r} - \Gamma^2\right]a_z(r, k, \omega) = \frac{e\mathrm{v}_0\mu_0}{(2\pi)^2 r}\delta(r)\delta(\omega - k\mathrm{v}_0), \tag{2.2.18}$$

and $\Gamma^2 = k^2 - \omega^2/c^2$. Off-axis the solution of this equation is

$$a_z(r, k, \omega) = A_+(k, \omega)\mathrm{K}_0(\Gamma r), \tag{2.2.19}$$

where $\mathrm{K}_0(\xi)$ is the zero order modified Bessel function of the second kind. In order to determine the amplitude A_+ there are two ways to proceed: (1) calculate the azimuthal magnetic field and then impose the boundary conditions at $r = 0$. An alternative way is to (2) integrate (2.2.18) from $r = 0$ to $r = \delta \to 0$. At this point we prefer the latter primarily because this approach will be utilized extensively subsequently. For small arguments the modified Bessel function behaves as $K_0(\xi) \simeq -\ln(\xi)$ (see Abramowitz and Stegun 1968, p. 375) and consequently,

$$A_+(k,\omega) = -\frac{e\mathrm{v}_0\mu_0}{(2\pi)^2}\,\delta(\omega - k\mathrm{v}_0). \tag{2.2.20}$$

Substituting this result in (2.2.17), (2.2.19) we obtain

$$A_z(r,z,t) = -\frac{e\mu_0}{(2\pi)^2}\int_{-\infty}^{\infty} \mathrm{d}\omega \exp\left[j\omega\left(t - \frac{z}{\mathrm{v}_0}\right)\right] K_0\left(\frac{\omega}{c} r \frac{1}{\gamma\beta}\right), \tag{2.2.21}$$

where $\beta = \mathrm{v}_0/c$ and $\gamma = [1-\beta^2]^{-1/2}$. Using the Lorentz gauge one can determine the scalar electric potential

$$\Phi(r,z,t) = -\frac{e}{4\pi\varepsilon_0}\frac{1}{\mathrm{v}_0}\frac{1}{\pi}\int_{-\infty}^{\infty} \mathrm{d}\omega \exp\left[j\omega\left(t - \frac{z}{\mathrm{v}_0}\right)\right] K_0\left(\frac{\omega}{c} r \frac{1}{\gamma\beta}\right). \tag{2.2.22}$$

This expression indicates that the field associated with a moving charge is a superposition of cylindrical evanescent waves (for large arguments the modified Bessel function decays exponentially following $K_0(\xi) \simeq \exp(-\xi)\sqrt{\pi/2\xi}$ Abramowitz and Stegun 1968, p. 378). There is no electromagnetic *average* power emitted by this particle in the radial direction however, this average power is non-zero in the direction parallel to the particle's motion – see Exercise 2.2. When scattered by periodic structures, the evanescent waves can be "converted" into propagating waves as we shall see when the Smith-Purcell effect will be discussed in Chap. 5.

2.3 Guided Waves

In all the solutions presented above, no boundaries were involved, while in many of the topics to be considered, the electromagnetic wave is guided by either a metallic or dielectric structure. In addition to the injection of electromagnetic power into the system, metallic/dielectric structures facilitate the storage, the interaction process itself and ultimately, they allow extraction of the power out of the system.

2.3.1 *Transverse Electromagnetic Mode*

The simplest mode, which may develop when two metallic surfaces are present, is the *transverse electro-magnetic* (TEM) mode. In the first part of this subsection we consider the way this mode is excited. In conjunction with the electromagnetic field generated by a moving charge let us consider a *radial transmission line* consisting of two parallel lossless plates; the distance between the plates is denoted by d and it is much smaller than the (vacuum) wavelength i.e., $\lambda(\equiv 2\pi c/\omega) \gg d$. Subject to this condition, we ignore the longitudinal variations ($\partial^2/\partial z^2 \simeq 0$) therefore, for an azimuthally symmetric system the wave equation reads

$$\left[\frac{1}{r}\frac{\mathrm{d}}{\mathrm{d}r}r\frac{\mathrm{d}}{\mathrm{d}r}+\frac{\omega^2}{c^2}\right]A_z(r,\omega) = -\mu_0 J_z(r,\omega). \tag{2.3.1}$$

An infinitely thin "wire" located on axis carries an oscillatory (ω) current, excites the magnetic vector potential; the corresponding current density is

$$J_z(r,\omega) = I\frac{1}{2\pi r}\delta(r). \tag{2.3.2}$$

Figure 2.4a illustrates schematically the system under consideration. A solution of the homogeneous wave equation, which satisfies the radiation condition, is given by

$$A_z(r,\omega) = A_+ \mathrm{H}_0^{(2)}\left(\frac{\omega}{c}r\right), \tag{2.3.3}$$

and A_+ is determined by the discontinuity at $r = 0$. Integrating (2.3.1) in the close vicinity of $r = 0$,

$$\left[r\frac{\mathrm{d}}{\mathrm{d}r}A_z(r,\omega)\right]_{r=0^+} = -\frac{\mu_0}{2\pi}I, \tag{2.3.4}$$

and using the expression for Hankel function for small arguments i.e., $\mathrm{H}_0^{(2)}(x) \simeq -j\ln(x)2/\pi$ (Abramowitz and Stegun 1968, p. 360), we obtain $A_+ = -jI\mu_0/4$.

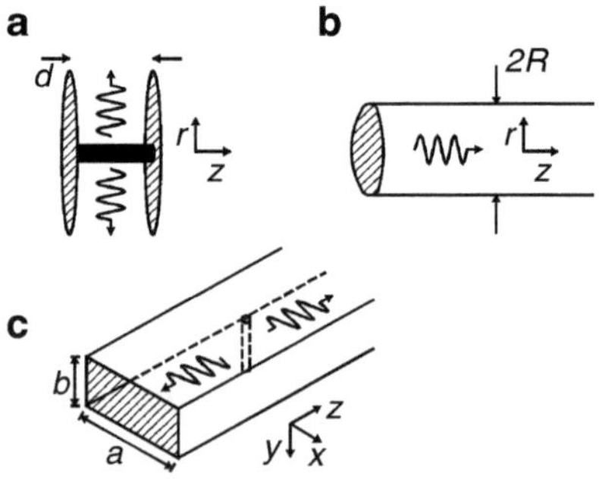

Fig. 2.4 (**a**) Propagation of transverse electro-magnetic (TEM) mode in a radial transmission line$\lambda \gg d$. (**b**) Propagation of a transverse magnetic (TM) mode in a circular waveguide – see Sect. 2.3.2. (**c**) Propagation of transverse electric (TE) mode in rectangular waveguide – Sect. 2.3.4; the curled arrows represent the direction of propagation of the waves

The corresponding longitudinal component of the electric field and the azimuthal counterpart of the magnetic field are

$$E_z(r,\omega) = -j\omega A_z(r,\omega) = -j\omega A_+ \mathrm{H}_0^{(2)}\left(\frac{\omega}{c}r\right),$$
$$H_\phi(r,\omega) = -\frac{1}{\mu_0}\frac{\mathrm{d}}{\mathrm{d}r}A_z(r,\omega) = \frac{1}{\mu_0}\frac{\omega}{c}A_+\mathrm{H}_1^{(2)}\left(\frac{\omega}{c}r\right). \tag{2.3.5}$$

With these two components, the radial component of the Poynting vector is

$$S_r(r) = -\frac{1}{2}E_z(r)H_\phi^*(r), \tag{2.3.6}$$

and consequently, the total power radiated is

$$P = \mathrm{Re}[2\pi r d S_r(r)] = \frac{1}{8}\left(\frac{\omega}{c}d\right)\eta_0 I^2. \tag{2.3.7}$$

In the last expression, we used the asymptotic approximation for large arguments of Hankel function i.e., $\mathrm{H}_0^{(2)}(x) \simeq \exp(-jx)\sqrt{2/\pi x}$ (see Abramowitz and Stegun 1968, p. 364). Bearing mind that in steady state the average power dissipated on a resistor carrying a current I is $P = RI^2/2$, the impedance associated with the radiation process is

$$R_{\mathrm{rad,TEM}} \equiv \frac{P}{I^2/2} = \frac{1}{4}\eta_0\left(\frac{\omega}{c}d\right); \tag{2.3.8}$$

in this expression $\eta_0 \equiv \sqrt{\mu_0/\varepsilon_0}$ is the *vacuum impedance* of a plane wave. At 9 GHz and for $d = 5\,\mathrm{mm}$ the impedance is $90[\Omega]$ which is 5 times larger (for the same parameters) than the radiation impedance in free-space defined as $R_{\mathrm{rad}} = \eta_0(\omega d/c)^2/6\pi \sim 18[\Omega]$. The radiation impedance is a measure, extensively used in antenna theory, which represents the effect of the surroundings on the radiation emitted by a source.

2.3.2 *Transverse Magnetic Mode*

Transverse magnetic (TM) modes can develop in the radial system discussed previously and their characteristics will be further investigated in Chap. 4, in the context of periodic structures. Here we review the characteristics of these modes for a circular cylindrical waveguide of radius R filled with a dielectric material of relative permittivity ε_r; the relative permeability is taken to unity ($\mu_r = 1$). We assume that the walls of the waveguide are made of an ideal conducting material ($\sigma \to \infty$) therefore, the tangential electric field at the walls vanishes. To this configuration, a cylindrical system of coordinates (r, ϕ, z) is attached – see Fig. 2.4b and the waves are assumed to be excited by an *azimuthally symmetric* source therefore we may take $\partial/\partial\phi = 0$.

The electromagnetic field in the waveguide has two contributions. One is from the z component of the magnetic vector potential

$$A_z(r, z, \omega) = \sum_{s=1}^{\infty} A_s \mathrm{J}_0\left(p_s \frac{r}{R}\right) \mathrm{e}^{-\Gamma_s z}, \tag{2.3.9}$$

where

$$\Gamma_s^2 = \frac{p_s^2}{R^2} - \varepsilon_\mathrm{r} \frac{\omega^2}{c^2}, \tag{2.3.10}$$

$\mathrm{J}_0(\xi)$ is the zero order Bessel function of the first kind and p_s are the zeros of this function ($p_1 = 2.4048, p_2 = 5.52\ldots$). The second, is from the scalar electric potential Φ

$$\Phi(r, z, \omega) = \sum_{s=1}^{\infty} \Phi_s \mathrm{J}_0\left(p_s \frac{r}{R}\right) \mathrm{e}^{-\Gamma_s z}. \tag{2.3.11}$$

Lorentz gauge (2.1.38) correlates the two amplitudes, namely

$$\Phi_s = \frac{c^2 \Gamma_s}{j\omega\varepsilon_\mathrm{r}} A_s. \tag{2.3.12}$$

In this solution, the waves are assumed to propagate from the source without obstacles thus no reflected waves were included.

The three non-trivial components of the electromagnetic field are: the azimuthal magnetic field

$$H_\phi = -\frac{1}{\mu_0} \frac{\partial A_z}{\partial r} = \frac{1}{\mu_0} \sum_{s=1}^{\infty} A_s \frac{p_s}{R} \mathrm{J}_1\left(p_s \frac{r}{R}\right) \mathrm{e}^{-\Gamma_s z} \tag{2.3.13}$$

the radial electric field

$$E_r = -\frac{\partial \Phi}{\partial r} = \sum_{s=1}^{\infty} A_s \frac{c^2 \Gamma_s}{j\omega\varepsilon_\mathrm{r}} \frac{p_s}{R} \mathrm{J}_1\left(p_s \frac{r}{R}\right) \mathrm{e}^{-\Gamma_s z}, \tag{2.3.14}$$

and the longitudinal electric field

$$E_z = -\frac{\partial \Phi}{\partial z} - j\omega A_z = \sum_{s=1}^{\infty} A_s \frac{c^2}{j\omega\varepsilon_\mathrm{r}} \left(\frac{p_s}{R}\right)^2 \mathrm{J}_0\left(p_s \frac{r}{R}\right) \mathrm{e}^{-\Gamma_s z} \tag{2.3.15}$$

With the electromagnetic field determined, the average magnetic and electric energy per unit length can be calculated. These are given by

$$\begin{aligned} W_{\mathrm{M}} &= \frac{1}{4}\mu_0 2\pi \int_0^R \mathrm{d}rr|H_\phi|^2 = \frac{\pi}{2\mu_0}\sum_{s=1}^{\infty}|A_s|^2 \frac{p_s^2}{R^2}\left[\frac{R^2}{2}\mathrm{J}_1^2(p_s)\right]\mathrm{e}^{-(\Gamma_s+\Gamma_s^*)z}, \\ W_{\mathrm{E}} &= \frac{1}{4}\varepsilon_0\varepsilon_{\mathrm{r}}2\pi\int_0^R \mathrm{d}rr\left[|E_r|^2+|E_z|^2\right] \\ &= \frac{\pi}{2}\varepsilon_0\varepsilon_{\mathrm{r}}\sum_{s=1}^{\infty}|A_s|^2\left[\frac{c^4}{\omega^2\varepsilon_{\mathrm{r}}^2}\frac{p_s^2}{R^2}\right]\left[\Gamma_s\Gamma_s^*+\frac{p_s^2}{R^2}\right]\left[\frac{R^2}{2}\mathrm{J}_1^2(p_s)\right]\mathrm{e}^{-(\Gamma_s+\Gamma_s^*)z} \end{aligned} \tag{2.3.16}$$

In these expressions the orthogonality of the Bessel functions was used i.e.,

$$\int_0^R \mathrm{d}rr\mathrm{J}_0\left(p_s\frac{r}{R}\right)\mathrm{J}_0\left(p_{s'}\frac{r}{R}\right) = \frac{1}{2}R^2\mathrm{J}_1^2(p_s)\delta_{s,s'}. \tag{2.3.17}$$

In a similar way, we can determine the total average power that flows in the waveguide:

$$\begin{aligned} P &= \mathrm{Re}\left[2\pi\int_0^R \mathrm{d}rr\frac{1}{2}E_rH_\phi^*\right] \\ &= \frac{\pi}{\mu_0}\sum_{s=1}^{\infty}|A_s|^2\frac{p_s^2}{R^2}\left[\frac{R^2}{2}\mathrm{J}_1^2(p_s)\right]\mathrm{Re}\left[\mathrm{e}^{-(\Gamma_s+\Gamma_s^*)z}\frac{c^2\Gamma_s}{j\omega\varepsilon_{\mathrm{r}}}\right]. \end{aligned} \tag{2.3.18}$$

According to this expression, we observe that power is carried along the waveguide only by the propagating modes namely those which satisfy

$$\Gamma_s^2 = \frac{p_s^2}{R^2} - \varepsilon_{\mathrm{r}}\frac{\omega^2}{c^2} < 0. \tag{2.3.19}$$

The remainders are below cut-off and they do not carry any (real) power. The situation is different when reflections are present.

2.3.3 Velocities and Impedances

Energy Velocity. In the context of power-flow presented above it is convenient to define several parameters that help to characterize the interaction of waves and electron beams in various configurations. For a relatively narrow band signal, the energy velocity is a measure of the power flow in the system relative to the total energy stored per unit length namely,

$$\mathrm{v}_{\mathrm{en}} = \frac{P}{W_{\mathrm{M}} + W_{\mathrm{E}}}. \quad (2.3.20)$$

In a circular cylindrical waveguide with a single propagating mode ($s = 1$), the energy velocity reads

$$\mathrm{v}_{\mathrm{en}} = c\frac{1}{\varepsilon_{\mathrm{r}}}\sqrt{\varepsilon_{\mathrm{r}} - \left(\frac{p_1 c}{\omega R}\right)^2}. \quad (2.3.21)$$

From the definition of the energy velocity (2.3.20) it is evident that whenever more than one mode propagates in the waveguide the energy velocity is dependent on the *relative* amplitudes of the various modes. Another point which should be emphasized since it will be encountered again later in this text is the fact that even if only one mode propagates and there is a substantial amount of energy stored in the higher non-propagating modes, the energy velocity will be much slower than indicated by the expression in (2.3.21).

Phase Velocity. A general definition of this quantity was introduced in Sect. 2.2.3 (2.2.14). In a cylindrical waveguide with no dielectric, the phase velocity is always larger than c. However if $\varepsilon_{\mathrm{r}} > 1 + (p_1 c/\omega R)^2$ the phase velocity is smaller than c. In fact, for high frequencies ($\omega R/c \gg p_1$) the phase velocity is determined entirely by the medium: $\mathrm{v}_{\mathrm{ph}} \sim c/\sqrt{\varepsilon_{\mathrm{r}}}$.

Group Velocity. This is a kinematical quantity indicative of the propagation of a relatively smooth spectrum of waves. To envision the meaning of the group velocity, imagine that a system is fed by two waves oscillating at adjacent frequencies $\omega_1 = \omega + \Delta\omega$, $\omega_2 = \omega - \Delta\omega$ having the form

$$f(z,t) = \cos(\omega_1 t - K_1 z) + \cos(\omega_2 t - K_2 z), \quad (2.3.22)$$

where the wave-numbers $K_1 = k + \Delta k$, $K_2 = k - \Delta k$ are the corresponding wave-numbers with $k = \sqrt{(\omega/c)^2 - (p_1/R)^2}$. Explicitly we can now write the expression in (2.3.22) as

$$f(z,t) = 2\cos(\Delta\omega t - \Delta k z)\cos(\omega t - kz). \quad (2.3.23)$$

Assuming that $|\Delta\omega| \ll \omega$, we can consider the first trigonometric function as a slow varying amplitude. As such, we can ask what has to be the velocity of an observer in order to experience a constant *amplitude* i.e., $\Delta\omega\delta t - \Delta k\delta z = 0$; in this case, the answer will be $\mathrm{v}_{\mathrm{gr}} \equiv \frac{\Delta\omega}{\Delta k}$ or at the limit of $\Delta\omega \to 0$,

$$\mathrm{v}_{\mathrm{gr}} \equiv \frac{\partial\omega}{\partial k}. \quad (2.3.24)$$

If the dielectric coefficient is not frequency dependent, the group velocity of a propagating TM mode is $\mathrm{v}_{\mathrm{gr}} = c^2 k/\omega\varepsilon_{\mathrm{r}}$ and it satisfies

$$\mathrm{v_{gr}}\mathrm{v_{ph}} = \frac{c^2}{\varepsilon_r}. \tag{2.3.25}$$

Although this relation is valid only for uniformly filled waveguide it provides information about the general trend in the variation of the group velocity as the (effective) dielectric coefficient changes in partially loaded systems.

Characteristic Impedance. There are several kinds of impedances that can be defined. Two of which will be defined here and a third one, will be defined in Chap. 8. The first is basically oriented towards the propagation of the electromagnetic mode in the structure and this is the characteristic impedance which is the ratio between the two transverse components of the field, E_r and H_ϕ, it reads

$$Z_{\mathrm{ch}} \equiv \frac{E_r}{H_\phi} = \eta_0 \frac{c\Gamma_s}{j\omega\varepsilon_{\mathrm{r}}}. \tag{2.3.26}$$

Interaction Impedance. The second impedance is indicative of the electric field which a *thin* pencil or annular beam experiences as it traverses the waveguide. For this purpose, we define the effective longitudinal electric field in the region where the electron beam will be injected. For a pencil beam ($0 \le r \le R_{\mathrm{b}}$) this is given by

$$|E(z)|^2 \equiv \frac{2}{R_b^2} \int_0^{R_{\mathrm{b}}} \mathrm{d}rr|E_z(r,z,\omega)|^2, \tag{2.3.27}$$

whereas for an annular beam ($R_{\mathrm{b}} - \Delta/2 \leqslant r \leqslant R_{\mathrm{b}} + \Delta/2$) it reads

$$|E(z)|^2 \equiv \frac{1}{\Delta R_{\mathrm{b}}} \int_{R_{\mathrm{b}}-\Delta/2}^{R_{\mathrm{b}}+\Delta/2} \mathrm{d}rr|E_z(r,z,\omega)|^2. \tag{2.3.28}$$

For either one of the cases we define the interaction impedance as

$$Z_{\mathrm{int}} \equiv \frac{1}{2}|E(z)|^2 \pi R^2 \frac{1}{P(z)}. \tag{2.3.29}$$

Note that although we are motivated by the presence of a beam of electrons, all the quantities in the definition of the interaction impedance are "cold" quantities namely, they do not account for the presence of the beam. It should be pointed out that the definition introduced here differs from Pierce's [Pierce (1947)] definition, $Z_{\mathrm{int}} = |E|^2/2k^2P$ by the factor k^2 which was replaced by the inverse of the area where the wave propagates, $1/\pi R^2$. This definition is in particular useful in tapered structures where the internal radius of the system is kept constant but the other geometric parameters may vary in space such that the phase velocity varies.

For our particular system the interaction impedance reads

$$Z_{\mathrm{int}} = \eta_0 \left[\frac{p_1}{\varepsilon_r}\frac{c}{\omega R}\right]^2 \frac{\mathrm{J}_0^2(p_1R_{\mathrm{b}}/R) + \mathrm{J}_1^2(p_1R_{\mathrm{b}}/R)}{\mathrm{J}_1^2(p_1)} \frac{1}{\beta_{\mathrm{en}}}; \tag{2.3.30}$$

here $\beta_{en} = v_{en}/c$ is the normalized energy velocity which in many cases is equal or close to the group velocity (in this particular case it is equal). One may expect to achieve maximum efficiency when the longitudinal electric field $[E(z)]$ experienced by the electron is maximum. Therefore, according to the definition in (2.3.29), from the point of view of the beam-wave interaction, the purpose should be to design a structure with the highest interaction impedance. According to (2.3.30) there are three possibilities: (1) operate at low frequency, which in many cases is not desirable, (2) have a structure with small radius which might be acceptable or (3) design a structure with low energy (group) velocity. It should be pointed out that these three possibilities are interdependent since for example, the energy velocity depends on both frequency and radius. One possibility to design a low group velocity structure is to have a small radius.

Interaction Dielectric Coefficient. This quantity is indicative of the total average electromagnetic energy stored per unit length in terms of the longitudinal component of the electric field experienced by a thin annular/pencil beam:

$$\varepsilon_{\rm int} \equiv W(z)\left[\frac{1}{2}\varepsilon_0|E(z)|^2\pi R^2\right]^{-1}. \tag{2.3.31}$$

In our particular case it reads

$$\varepsilon_{\rm int} = \left[\frac{\varepsilon_{\rm r}}{p_1}\frac{\omega}{c}R\right]^2 \frac{{\rm J}_1^2(p_1)}{{\rm J}_0^2(p_1R_{\rm b}/R) + {\rm J}_1^2(p_1R_{\rm b}/R)}. \tag{2.3.32}$$

Note that according to the definitions of the interaction impedance (2.3.29) and the effective dielectric coefficient (2.3.31) their product is inversely proportional to the energy velocity:

$$Z_{\rm int}\varepsilon_{\rm int} = \eta_0\frac{1}{\beta_{\rm en}}. \tag{2.3.33}$$

Since the definitions above (2.3.29) and (2.3.31) are general, as long as there is only one dominant mode in the system, the result in the last expression is also general.

2.3.4 *Transverse Electric Mode*

In many cases, electromagnetic power is transferred along a waveguide in the transverse electric (TE) mode due to its low loss (Ramo et al. 1965, p. 424). In many devices, power is extracted using rectangular waveguides, therefore we consider next the characteristics of such a waveguide. In Sect. 2.3.1 we examined the radiation emitted from a dipole oscillating in azimuthally symmetric radial transmission line. In this geometry, the main mode generated was the transverse electro-magnetic (TEM) mode. In this section, we consider the same problem in a rectangular waveguide whose wide dimension is a and the narrow one is b – see Fig. 2.4c. Variations along the narrow dimension are neglected ($\partial/\partial y \sim 0$). An infinitesimally thin "wire" (dipole) is located in the center of the waveguide and it prescribes a current density given by

$$J_y(x, z, \omega) = I\delta\left(x - \frac{a}{2}\right)\delta(z). \tag{2.3.34}$$

It excites the transverse electric field $E_y(x, z, \omega)$ that satisfies

$$\left[\frac{\partial^2}{\partial x^2} + \frac{\partial^2}{\partial z^2} + \frac{\omega^2}{c^2}\right]E_y(x, z, \omega) = j\omega\mu_0 J_y(x, z, \omega), \tag{2.3.35}$$

subject to the boundary conditions: $E_y(x = 0, z, \omega) = 0$ and $E_y(x = a, z, \omega) = 0$. The solution can be represented as a superposition of trigonometric functions i.e.,

$$E_y(x, z, \omega) = \sum_{n=1}^{\infty} E_n(z, \omega)\sin\left(\frac{\pi n}{a}x\right), \tag{2.3.36}$$

where $E_n(z, \omega)$ satisfies

$$\left[\frac{\mathrm{d}^2}{\mathrm{d}z^2} - \left(\frac{\pi n}{a}\right)^2 + \frac{\omega^2}{c^2}\right]E_n(z, \omega) = j\omega\mu_0 I\sin\left(\frac{\pi}{2}n\right)\frac{2}{a}\delta(z) \equiv I_n\delta(z). \tag{2.3.37}$$

For $z > 0$ the solution of this equation is

$$E_n(z > 0) = A_+\mathrm{e}^{-\Gamma_n z}, \tag{2.3.38}$$

and for $z < 0$

$$E_n(z < 0) = A_-\mathrm{e}^{\Gamma_n z}, \tag{2.3.39}$$

where $\Gamma_n^2 = (\pi n/a)^2 - (\omega/c)^2$. The transverse electric field has to be continuous at $z = 0$ thus

$$A_+ = A_-, \tag{2.3.40}$$

whereas its derivative is discontinuous. The discontinuity is determined by the Dirac delta function in (2.3.37) therefore by integrating the latter we obtain

$$\left[\frac{\mathrm{d}}{\mathrm{d}z}E_n(z)\right]_{z=0^+} - \left[\frac{\mathrm{d}}{\mathrm{d}z}E_n(z)\right]_{z=0^-} = I_n, \tag{2.3.41}$$

hence

$$-\Gamma_n A_+ - \Gamma_n A_- = I_n. \tag{2.3.42}$$

From (2.3.40), (2.3.42) we conclude that the transverse electric field reads

$$E_y(x,z,\omega) = -\sum_{n=1}^{\infty} \frac{I_n}{2\Gamma_n} e^{-\Gamma_n|z|} \sin\left(\frac{\pi n}{a}x\right). \tag{2.3.43}$$

As in Sect. 2.3.1 we next calculate the power generated by the current distribution in (2.3.34). For this purpose the transverse magnetic field is calculated since it is the only component which contributes to the longitudinal component of the Poynting vector; H_x for $z>0$ reads

$$H_x(x,z>0,\omega) = -\sum_{n=1}^{\infty} \frac{\Gamma_n}{j\omega\mu_0} \frac{I_n}{2\Gamma_n} e^{-\Gamma_n z} \sin\left(\frac{\pi n}{a}x\right). \tag{2.3.44}$$

Before proceeding note that similar to the transverse magnetic mode, the phase velocity (for $\omega > \pi nc/a$ and $\varepsilon_r = 1$) is always larger than c. Nevertheless, the characteristic impedance (in vacuum) of the nth propagating mode,

$$Z_{\text{ch,TE}} \equiv \frac{E_y}{H_x} = \frac{j\omega\mu_0}{\Gamma_n} \tag{2.3.45}$$

is always larger than the vacuum impedance (η_0), in contrast to the TM mode, where the characteristic impedance is always smaller than η_0.

Now we can focus our attention to the power flow: the average power which flows in the positive z direction, assuming a single mode above cut-off, is given by

$$P_+ = \frac{1}{2} \frac{I_1}{2\sqrt{(\omega/c)^2 - (\pi/a)^2}} \frac{I_1^*}{2\omega\mu_0} \frac{1}{2}ab. \tag{2.3.46}$$

The radiation impedance is determined by the power emitted in both directions divided by $\frac{1}{2}|I|^2$ and it reads

$$R_{\text{rad,TE}} \equiv \frac{P_+ + P_-}{\frac{1}{2}|I|^2} = \eta_0 \frac{\omega b/c}{\sqrt{(\omega a/c)^2 - \pi^2}} = \frac{b}{a} Z_{\text{ch,TE}}. \tag{2.3.47}$$

At 9 GHz, and for $a = 2.5$ cm, $b = 0.5$ cm, this impedance is 100 Ω which is close to that calculated in the case of the radial transmission line as calculated in Sect. 2.3.1.

2.3.5 TE, TM and Hybrid Modes in a Dielectric Waveguide

Pure TM or TE modes are possible only in a limited set of geometries. In most cases these modes are coupled and in this section we present a well-known configuration that supports either TE, TM or hybrid modes – this is the dielectric waveguide. In its simplest configuration it consists of a dielectric (ε_r) fiber of radius R. For small-diameter rods, the field extends for a considerable distance beyond the surface, and the axial propagation constant k_z is only slightly larger than ω/c. At the limit of an infinite radius $k_z = \omega\sqrt{\varepsilon_r}/c$. The field components, omitting the term $\exp(-jn\phi - j\beta z)$, are determined in Table 2.1; $\Lambda^2 = \varepsilon_r\omega^2/c^2 - k_z^2$ and $\Gamma^2 = k_z^2 - \omega^2/c^2$; the prime indicates differentiation with respect to the arguments of the corresponding Bessel functions.

Imposing of the boundary conditions at $r = R$ leads to the dispersion relation

$$\left[\frac{\varepsilon J'_n(a)}{aJ_n(a)} - \frac{K'_n(b)}{bK_n(b)}\right]\left[\frac{J'_n(a)}{aJ_n(a)} + \frac{K'_n(b)}{bK_n(b)}\right] = \left[n\frac{ck_z}{\omega}\frac{(b^2+a^2)}{a^2b^2}\right]^2 \tag{2.3.48}$$

where $a = \Lambda R,\ \ b = \Gamma R$. When $n = 0$, the right-hand side vanishes, and each factor on the left-hand side must equal zero. These two terms determine the dispersion of the axially symmetric TM and TE modes:

$$\begin{aligned}\text{TM modes}: \quad & \frac{\varepsilon_r J'_0(a)}{aJ_0(a)} = \frac{K'_0(b)}{bK_0(b)} \\ \text{TE modes}: \quad & \frac{J'_0(a)}{aJ_0(a)} = -\frac{K'_0(b)}{bK_0(b)}.\end{aligned} \tag{2.3.49}$$

Based on their definitions, a and b are related by $a^2 + b^2 = (\varepsilon_r - 1)(\omega R/c)^2$. Clearly, pure TM or TE modes are possible only if the field is independent of the azimuthal coordinate ϕ namely, $n = 0$. As the radius of the rod increases, the number of TM and TE modes also increases. All modes with angular dependence are a combination of a TM and a TE mode, and are classified as hybrid modes.

Table 2.1 Field components in a cylindrical dielectric waveguide

$r<R$	$r>R$
$E_z = A_nJ_n(\Lambda r)$	$E_z = C_nK_n(\Gamma r)$
$E_r = -\frac{jk_z}{\Lambda}A_nJ'_n - \frac{n\omega\mu_0}{\Lambda^2 r}B_nJ_n$	$E_r = \frac{jk_z}{\Gamma}C_nK'_n + \frac{n\omega\mu_0}{\Gamma^2 r}D_nK_n$
$E_\phi = -\frac{nk_z}{\Lambda^2 r}A_nJ_n + \frac{j\omega\mu_0}{\Lambda}B_nJ'_n$	$E_\phi = -\frac{nk_z}{\Gamma^2 r}C_nK_n - \frac{j\omega\mu_0}{\Gamma}D_nK'_n$
$H_z = B_nJ_n(\Lambda r)$	$H_z = D_nK_n(\Gamma r)$
$H_r = \frac{n\omega\varepsilon}{\Lambda^2 r}A_nJ_n - \frac{jk_z}{\Lambda}B_nJ'_n$	$H_r = -\frac{n\omega\varepsilon_0}{\Gamma^2 r}C_nK_n - \frac{jk_z}{\Gamma}D_nK'_n$
$H_\phi = -\frac{j\omega\varepsilon}{\Lambda}A_nJ'_n - \frac{nk_z}{\Lambda^2 r}B_nJ_n$	$H_\phi = \frac{j\omega\varepsilon_0}{\Gamma}C_nK'_n + \frac{nk_z}{\Gamma^2 r}D_nK_n$

Comment 2.4. Contrary to metallic waveguide the HE_{11} mode, for example, has no low-frequency cutoff.

2.4 Green's Scalar Theorem

Green's function is a useful tool for calculation of electromagnetic field generated by a distributed source (particles) subject to the boundary conditions imposed by the structure. The logic behind the method presented below is the following: instead of solving for an arbitrary source we solve for a point source and by virtue of the linearity of Maxwell's equations, the field at a given location is a *superposition* of all the point sources that constitute the real source.

Let us assume that we have to solve the non-homogeneous wave equation:

$$\left[\nabla^2 + \frac{\omega^2}{c^2}\right]\psi(\mathbf{r}) = -s(\mathbf{r}), \tag{2.4.1}$$

where $s(\mathbf{r})$ is an arbitrary source which is assumed to be known. Instead of solving this equation let us assume for the moment that we know how to solve a simpler problem namely,

$$\left[\nabla^2 + \frac{\omega^2}{c^2}\right]G(\mathbf{r}|\mathbf{r}') = -\delta(\mathbf{r} - \mathbf{r}'), \tag{2.4.2}$$

where the coefficient of the Dirac delta function on the right-hand side was chosen such that the result of the integration over the entire space is unity. We can then multiply (2.4.1) by $G(\mathbf{r}|\mathbf{r}')$ and (2.4.2) by $\psi(\mathbf{r}')$, subtract the two results to obtain

$$G(\mathbf{r}|\mathbf{r}')\nabla^2\psi(\mathbf{r}') - \psi(\mathbf{r}')\nabla^2 G(\mathbf{r}|\mathbf{r}') = -G(\mathbf{r}|\mathbf{r}')s(\mathbf{r}') + \psi(\mathbf{r}')\delta(\mathbf{r} - \mathbf{r}'). \tag{2.4.3}$$

Integrating over the entire and using Gauss' theorem, we get

$$\psi(\mathbf{r}) = \int_V \mathrm{d}V' G(\mathbf{r}|\mathbf{r}')s(\mathbf{r}') + \oint\!\!\oint \mathrm{d}\mathbf{a}' \cdot [G(\mathbf{r}|\mathbf{r}')\nabla'\psi(\mathbf{r}') - \psi(\mathbf{r}')\nabla' G(\mathbf{r}|\mathbf{r}')]; \tag{2.4.4}$$

$\oint\!\!\oint$ is a surface integral which encloses the volume V. This is the scalar Green's theorem. In free space Green's theorem reads

$$\psi(\mathbf{r}) = \int_V \mathrm{d}V' G(\mathbf{r}|\mathbf{r}')s(\mathbf{r}'). \tag{2.4.5}$$

Next, we employ Green's theorem for the calculation of the Cerenkov effect in two cases: firstly, in a boundless system and secondly in a waveguide.

2.4.1 Cerenkov Radiation in the Boundless Case

Let us examine the electromagnetic field generated by a charge (e) as it moves in gas a medium which is characterized by a dielectric coefficient larger than unity, $\varepsilon_r > 1$; its velocity is v. For simplicity sake, it will be assumed that the dielectric coefficient is frequency-independent.

A current density described by the same expression as in (2.2.15) drives the system and for an azimuthally symmetric medium the wave equation is

$$\left[\frac{1}{r}\frac{\partial}{\partial r}r\frac{\partial}{\partial r}+\frac{\partial^2}{\partial z^2}-\varepsilon_r\frac{1}{c^2}\frac{\partial^2}{\partial t^2}\right]A_z(r,z,t)=-\mu_0 J_z(r,z,t); \tag{2.4.6}$$

the other two components of the magnetic vector potential are zero and the electric scalar potential can be determined using Lorentz gauge. The time Fourier transform of the magnetic vector potential is defined by

$$A_z(r,z,t)=\int_{-\infty}^{\infty}\mathrm{d}\omega \mathrm{e}^{j\omega t}A_z(r,z,\omega), \tag{2.4.7}$$

where $A_z(r,z,\omega)$ satisfies

$$\left[\frac{1}{r}\frac{\partial}{\partial r}r\frac{\partial}{\partial r}+\frac{\partial^2}{\partial z^2}+\varepsilon_r\frac{\omega^2}{c^2}\right]A_z(r,z,\omega)=-\mu_0 J_z(r,z,\omega), \tag{2.4.8}$$

and the time Fourier transform of the current density in (2.2.15) is

$$J_z(r,z,\omega)=-\frac{e}{(2\pi)^2 r}\delta(r)\exp\left(-j\frac{\omega}{\mathrm{v}}z\right). \tag{2.4.9}$$

Green's function associated with this problem is a solution of

$$\left[\frac{1}{r}\frac{\partial}{\partial r}r\frac{\partial}{\partial r}+\frac{\partial^2}{\partial z^2}+\varepsilon_r\frac{\omega^2}{c^2}\right]G(r,z|r',z')=\frac{-1}{2\pi r}\delta(r-r')\delta(z-z') \tag{2.4.10}$$

which can be represented by

$$G(r,z|r',z')=\int_{-\infty}^{\infty}\mathrm{d}k g_k(r|r')\exp[-jk(z-z')], \tag{2.4.11}$$

and $g_k(r|r')$ satisfies

$$\left[\frac{1}{r}\frac{\mathrm{d}}{\mathrm{d}r}r\frac{\mathrm{d}}{\mathrm{d}r}-\Gamma^2\right]g_k(r|r')=-\frac{1}{(2\pi)^2 r}\delta(r-r'), \tag{2.4.12}$$

where

$$\Gamma^2 = k^2 - \varepsilon_r \frac{\omega^2}{c^2}. \tag{2.4.13}$$

The solution of this equation for $r > r' > 0$ is

$$g_k(r|r' < r) = F_1(r')\mathrm{K}_0(\Gamma r), \tag{2.4.14}$$

and for $r' > r > 0$ it reads

$$g_k(r < r'|r') = F_2(r')\mathrm{I}_0(\Gamma r). \tag{2.4.15}$$

The function $g_k(r|r')$ has to be continuous at $r = r'$ i.e.,

$$F_1(r')\mathrm{K}_0(\Gamma r') = F_2(r')\mathrm{I}_0(\Gamma r'), \tag{2.4.16}$$

whereas its derivative is discontinuous at the same location. To determine the discontinuity we integrate (2.4.12)

$$\left[r\frac{\mathrm{d}}{\mathrm{d}r}g(r|r')\right]_{r=r'+0} - \left[r\frac{\mathrm{d}}{\mathrm{d}r}g(r|r')\right]_{r=r'-0} = -\frac{1}{(2\pi)^2}, \tag{2.4.17}$$

hence

$$-r'F_1(r')\Gamma\mathrm{K}_1(\Gamma r') - r'F_2(r')\Gamma\mathrm{I}_1(\Gamma r') = -\frac{1}{(2\pi)^2}. \tag{2.4.18}$$

From (2.4.16), (2.4.18) and using the fact that $\mathrm{K}_0(\xi)\mathrm{I}_1(\xi) + \mathrm{K}_1(\xi)\mathrm{I}_0(\xi) = 1/\xi$ (see Abramowitz and Stegun 1968, p. 375) we finally obtain

$$g_k(r|r') = \frac{1}{(2\pi)^2}\begin{cases} \mathrm{I}_0(\Gamma r)\mathrm{K}_0(\Gamma r') & \text{for} \quad 0 \le r \le r' < \infty, \\ \mathrm{K}_0(\Gamma r)\mathrm{I}_0(\Gamma r') & \text{for} \quad 0 \le r' \le r < \infty. \end{cases} \tag{2.4.19}$$

This expression together with (2.4.11) determine Green's function in a boundless space.

With this function, Green's theorem (2.4.5) and the current density as given in (2.4.9), we can determine the magnetic vector potential. It reads

$$A_z(r, z, \omega) = -\frac{e\mu_0}{(2\pi)^2}\mathrm{K}_0\left(\frac{\omega}{c}r\sqrt{\beta^{-2} - n^2}\right)\exp\left(-j\frac{\omega}{\mathrm{v}}z\right), \tag{2.4.20}$$

where $n \equiv \sqrt{\varepsilon_r}$ is the refractive index of the medium. If we examine this solution far away from the source and use the asymptotic value for large arguments

$\left[(\omega/c)r|\sqrt{\beta^{-2}-n^2}|\gg 1\right]$ of the modified Bessel function, the magnetic vector potential reads

$$A_z(r,z,\omega) \propto \exp\left(-\frac{\omega}{c}r\sqrt{\beta^{-2}-n^2}\right)\exp\left(-j\frac{\omega}{\mathrm{v}}z\right). \tag{2.4.21}$$

If n is smaller than $1/\beta$ the field decays exponentially in the radial direction since, as in vacuum, this is an *evanescent* wave.

When the velocity of the particle, $\mathrm{v} = \beta c$, is larger than the phase velocity of a plane wave in the medium (c/n) i.e., $\beta > 1/n$, the expression above represents a *propagating* wave – this is called Cerenkov radiation. The emitted wave is not parallel to the electron's trajectory but it propagates at an angle θ relative to this direction (z axis) given by

$$k_z = \frac{\omega}{c}n\cos\theta = \frac{\omega}{c}\frac{1}{\beta} \tag{2.4.22}$$

This determines what it is known as the Cerenkov radiation angle, θ_c

$$\theta_c = \cos^{-1}\left(\frac{1}{n\beta}\right). \tag{2.4.23}$$

Since the phase velocity of the wave is smaller than that of the particle, clearly, the radiation lags behind the particle. This fact will become evident in the next subsection. However, before proceeding, it is important to make a comment regarding Cerenkov radiation emitted by a single particle and an *ensemble* of N electrons: by virtue of the linearity of Maxwell's equation the total field is a superposition of the contributions of all electrons. For wavelengths significantly *longer* than the bunch-length, the various contributions add up coherently and since the power is proportional to the square of the field, the emitted power is proportional to the square of the number of electrons ($P \propto N^2$) – this is also referred to as *coherent radiation.* For wavelengths *shorter* than the bunch, the average field vanishes therefore, the total power is a product of the power emitted by a single electron and the number of electrons. The proof is left to the reader and the details are phrased as an Exercise 2.7 at the end of this chapter.

2.4.2 Cerenkov Radiation in a Cylindrical Waveguide

In this subsection we consider the electromagnetic field associated with the symmetric transverse magnetic (TM) mode in a *dielectric* filled waveguide. As in the previous subsection, the source of this field is a particle moving at a velocity v, however, the main difference is that the solution has a constraint since on the waveguide's wall ($r = R$) the tangential electric field vanishes. Therefore, we calculate Green's function subject to the condition $G(r = R, z|r', z') = 0$. We assume a solution of the form

$$G(r, z|r', z') = \sum_{s=1}^{\infty} G_s(z|r', z') \mathrm{J}_0\left(p_s \frac{r}{R}\right), \tag{2.4.24}$$

substitute in (2.4.10) and use the orthogonality of the Bessel functions we find that

$$G_s(z|r', z') = \mathrm{J}_0\left(p_s \frac{r'}{R}\right) \frac{1}{\frac{1}{2} R^2 \mathrm{J}_1^2(p_s)} g_s(z|z'), \tag{2.4.25}$$

where $g_s(z|z')$ satisfies

$$\left[\frac{\mathrm{d}^2}{\mathrm{d}z^2} - \Gamma_s^2\right] g_s(z|z') = -\frac{1}{2\pi} \delta(z - z'), \tag{2.4.26}$$

and $\Gamma_s^2 = p_s^2/R^2 - \varepsilon_r \omega^2/c^2$. For $z > z'$ the solution of (2.4.26) is

$$g_s(z|z') = A_+ \mathrm{e}^{-\Gamma_s (z - z')}, \tag{2.4.27}$$

and for $z < z'$ the solution is

$$g_s(z|z') = A_- \mathrm{e}^{\Gamma_s (z - z')}. \tag{2.4.28}$$

Green's function is continuous at $z = z'$ i.e.,

$$A_+ = A_-, \tag{2.4.29}$$

and its first derivative is discontinuous. The discontinuity is determined by integrating (2.4.26) from $z = z' - 0$ to $z = z' + 0$ i.e.,

$$\left[\frac{\mathrm{d}}{\mathrm{d}z} g_s(z|z')\right]_{z=z'+0} - \left[\frac{\mathrm{d}}{\mathrm{d}z} g_s(z|z')\right]_{z=z'-0} = -\frac{1}{2\pi}. \tag{2.4.30}$$

Substituting the two solutions introduced above, and using (2.4.29) we obtain

$$g_s(z|z') = \frac{1}{4\pi\Gamma_s} \exp(-\Gamma_s |z - z'|). \tag{2.4.31}$$

Finally, the explicit expression for the Green's function corresponding to azimuthally symmetric TM modes in a circular waveguide is given by

$$G(r,z|r',z') = \sum_{s=1}^{\infty} \frac{\mathrm{J}_0(p_s r/R)\mathrm{J}_0(p_s r'/R)}{\frac{1}{2}R^2\mathrm{J}_1^2(p_s)} \frac{1}{4\pi\Gamma_s} \exp(-\Gamma_s|z-z'|). \tag{2.4.32}$$

In this expression, we tacitly assumed that $\omega > 0$ and Γ_s is non-zero.

With Green's function established, we can calculate the magnetic vector potential generated by the current distribution described in (2.4.9); the result is

$$\begin{aligned} A_z(r,z,\omega) &= 2\pi\mu_0 \int_0^R \mathrm{d}r'r' \int_{-\infty}^{\infty} \mathrm{d}z' G(r,z|r',z') J_z(r',z') \\ &= -\frac{e\mu_0}{8\pi^2} \sum_{s=1}^{\infty} \frac{\mathrm{J}_0(p_s r/R)}{\frac{1}{2}R^2\mathrm{J}_1^2(p_s)} \frac{2}{\Gamma_s^2 + \omega^2/\mathrm{v}^2} \mathrm{e}^{-j(\omega/\mathrm{v})z}. \end{aligned} \tag{2.4.33}$$

It will be instructive to examine this expression in the *time domain*; the Fourier transform is

$$A_z(r,z,t) = -\frac{e}{2\pi^2\varepsilon_0 R^2} \frac{\beta^2}{1-n^2\beta^2} \sum_{s=1}^{\infty} \frac{\mathrm{J}_0(p_s r/R)}{\mathrm{J}_1^2(p_s)} \int_{-\infty}^{\infty} \mathrm{d}\omega \frac{\mathrm{e}^{j\omega(t-z/\mathrm{v})}}{\omega^2 + \Omega_s^2}, \tag{2.4.34}$$

where

$$\Omega_s^2 = \left(\frac{p_s c}{R}\right)^2 \frac{\beta^2}{1-n^2\beta^2}. \tag{2.4.35}$$

Equivalently, this result may be interpreted as the interception of the dispersion relation $-k_z^2 - (p_s/R)^2 + \varepsilon_r(\omega/c)^2 = 0$ and the "beam-line" $k_z = \omega/\mathrm{v}$. With this definition, the problem has been now simplified to the evaluation of the integral

$$F_s(\tau = t - z/\mathrm{v}) \equiv \int_{-\infty}^{\infty} \mathrm{d}\omega \frac{\mathrm{e}^{j\omega\tau}}{\omega^2 + \Omega_s^2}, \tag{2.4.36}$$

which in turn is equivalent to the solution of the following differential equation

$$\left[\frac{\mathrm{d}^2}{\mathrm{d}\tau^2} - \Omega_s^2\right] F_s(\tau) = -2\pi\delta(\tau). \tag{2.4.37}$$

Case I: If the particle's velocity is *slower* than the phase velocity of a plane wave in the medium ($n\beta < 1$) then $\Omega_s^2 > 0$ and the solution for $\tau > 0$ is

$$F_s(\tau > 0) = A_+ \mathrm{e}^{-\Omega_s \tau}, \tag{2.4.38}$$

or

$$F_s(\tau < 0) = A_- e^{\Omega_s \tau}. \tag{2.4.39}$$

As previously, in the case of Green's function, $F_s(\tau)$ has to be continuous at $\tau = 0$ and its derivative is discontinuous:

$$\left(\frac{d}{d\tau} F_s(\tau)\right)_{\tau=0^+} - \left(\frac{d}{d\tau} F_s(\tau)\right)_{\tau=0^-} = -2\pi. \tag{2.4.40}$$

When the velocity of the particle is smaller than c/n (i.e., $n\beta < 1$) the characteristic frequency Ω_s is real, therefore

$$F_s(\tau) = \frac{\pi}{\Omega_s} e^{-\Omega_s |\tau|}, \tag{2.4.41}$$

and consequently,

$$A_z(r, z, t) = -\frac{e}{2\pi\varepsilon_0 R^2} \frac{\beta^2}{1 - \beta^2 n^2} \sum_{s=1}^{\infty} \frac{J_0(p_s r/R)}{J_1^2(p_s)\Omega_s} e^{-\Omega_s |t - z/v|}. \tag{2.4.42}$$

This expression represents a *discrete* superposition of evanescent modes attached to the particle.

Case II: If the particle's velocity is *faster* than the phase velocity of a plane wave in the medium ($n\beta > 1$) then $\Omega_s^2 < 0$. In this case the waves are slower than the particle and there is no electromagnetic field in front of the particle i.e.,

$$F_s(\tau < 0) = 0. \tag{2.4.43}$$

By virtue of the continuity at $\tau = 0$ we have for $\tau > 0$

$$F_s(\tau > 0) = A_+ \sin(|\Omega_s| \tau). \tag{2.4.44}$$

Substituting these two expressions in (2.4.40) we obtain

$$F_s(\tau) = -\frac{2\pi}{|\Omega_s|} \sin(|\Omega_s| \tau) h(\tau), \tag{2.4.45}$$

and the magnetic vector potential reads

$$A_z(r, z, t) = -\frac{e}{\pi\varepsilon_0 R^2} \frac{\beta^2}{n^2\beta^2 - 1} \sum_{s=1} \frac{J_0(p_s r/R)}{J_1^2(p_s)|\Omega_s|} \sin\left[|\Omega_s| \left(t - \frac{z}{v}\right)\right] h\left(t - \frac{z}{v}\right), \tag{2.4.46}$$

where $h(\xi)$ is the Heaviside step function. This expression indicates that when the velocity of the particle is larger than c/n, there is a discrete superposition of *propagating* waves traveling behind the particle. Furthermore, all the waves have the same phase velocity which is identical with the velocity of the particle, v. It is important to bear in mind that this result was obtained after tacitly assuming that ε_r is frequency independent which generally is not the case, therefore the summation is limited to a finite number of modes. The modes which contribute are determined by the Cerenkov condition $n(\omega = \Omega_s)\beta > 1$.

After we established the magnetic vector potential, let us now calculate the *average power* which trails behind the particle. Firstly, the azimuthal magnetic field is given by

$$\begin{aligned} H_\phi(r,z,t) &= -\frac{1}{\mu_0}\frac{\partial}{\partial r}A_z(r,z,t) \\ &= \frac{1}{\mu_0}\sum_{s=1} A_s \frac{p_s}{R} \mathrm{J}_1\left(p_s\frac{r}{R}\right) \sin\left[|\Omega_s|\left(t-\frac{z}{\mathrm{v}}\right)\right] h\left(t-\frac{z}{\mathrm{v}}\right), \end{aligned} \tag{2.4.47}$$

where

$$A_s = -\frac{e}{\pi\varepsilon_0 R^2}\frac{\beta^2}{n^2\beta^2-1}\frac{1}{\mathrm{J}_1^2(p_s)|\Omega_s|}. \tag{2.4.48}$$

Secondly, the radial electric field is determined by the electric scalar potential, which in turn is calculated using the Lorentz gauge, and it reads

$$\begin{aligned} E_r(r,z,t) &= -\frac{\partial}{\partial r}\Phi(r,z,t) \\ &= \frac{c^2}{\varepsilon_r \mathrm{v}}\sum_{s=1} A_s \frac{p_s}{R} \mathrm{J}_1\left(p_s\frac{r}{R}\right) \sin\left[|\Omega_s|\left(t-\frac{z}{\mathrm{v}}\right)\right] h\left(t-\frac{z}{\mathrm{v}}\right). \end{aligned} \tag{2.4.49}$$

With these expressions, we can calculate the average electromagnetic power trailing the particle. It is given by

$$P = \frac{e^2\beta c}{2\pi\varepsilon_0\varepsilon_r R^2}\frac{1}{\varepsilon_r\beta^2-1}\sum_s \frac{1}{\mathrm{J}_1^2(p_s)}. \tag{2.4.50}$$

Note that for ultra relativistic particle ($\beta \to 1$) the power is independent of the particle's energy. In order to have a measure of the radiation emitted consider a very *small bunch* of $N \sim 10^{11}$ electrons injected in a waveguide whose radius is 9.2 mm. The waveguide is filled with a material whose dielectric coefficient is $\varepsilon_r = 2.6$ and all electrons have the same energy 450 keV. If we were able to keep their velocity constant, then 23 MW of power at 11.4 GHz (first mode, $s = 1$) will trail the bunch.

Further examining this expression we note that the average power is quadratic with the frequency i.e.,

$$P \equiv \sum_{s=1} P_s = \frac{(Ne)^2}{2\pi\varepsilon_0\varepsilon_r\beta c}\sum_{s=1}\frac{|\Omega_s|^2}{[p_s J_1(p_s)]^2}. \tag{2.4.51}$$

In addition, based on the definition of the Fourier transform of the current density in (2.4.9), we conclude that the current which this macro-particle excites in the s'th mode is $I_s = eN\Omega_s/2\pi$. With this expression, the radiation impedance of the first mode ($s = 1$) is

$$R_{C,1} = \frac{P_1}{\frac{1}{2}|I_1|^2} = \eta_0\frac{4\pi}{\varepsilon_r\beta[p_1 J_1(p_1)]^2}. \tag{2.4.52}$$

For a relativistic particle, $\beta \simeq 1$, a dielectric medium $\varepsilon_r = 2.6$ the radiation impedance corresponding to the first mode is $\simeq 1{,}200\,\Omega$ which is one order of magnitude larger than that of a dipole in free space or between two plates. Note that this impedance is independent of the geometry of the waveguide and for an ultra-relativistic particle it is independent of the particle's energy.

2.4.3 Coherent Cerenkov Radiation

Once we established the radiation from a single bunch, it is possible to proceed and investigate a distribution of electrons rather than a point-charge. For an ensemble of electrons the field components are

$$\begin{aligned}
H_\phi &= \frac{1}{\mu_0}\sum_{s=1} A_s\frac{p_s}{R}J_1\left(p_s\frac{r}{R}\right)\left\langle J_0\left(p_s\frac{r_i}{R}\right)\sin\left[|\Omega_s|\left(t-\frac{z-z_i}{\mathrm{v}}\right)\right]h\left(t-\frac{z-z_i}{\mathrm{v}}\right)\right\rangle_i \\
E_r &= \frac{c^2}{\varepsilon_r \mathrm{v}}\sum_{s=1} A_s\frac{p_s}{R}J_1\left(p_s\frac{r}{R}\right)\left\langle J_0\left(p_s\frac{r_i}{R}\right)\sin\left[|\Omega_s|\left(t-\frac{z-z_i}{\mathrm{v}}\right)\right]h\left(t-\frac{z-z_i}{\mathrm{v}}\right)\right\rangle_i \\
A_s &= -\frac{eN}{\pi\varepsilon_0 R^2}\frac{\beta^2}{n^2\beta^2-1}\frac{1}{J_1^2(p_s)|\Omega_s|}.
\end{aligned} \tag{2.4.53}$$

wherein $\langle....\rangle$ represents the ensemble average; $N$is the total number of electrons in the bunch. For simplicity sake, we assume that the electrons are uniformly distributed in the radial direction $(0 < r < R_b)$ and the transverse and longitudinal distributions are independent thus

$$H_\phi = \frac{1}{\mu_0}\sum_{s=1} A_s \frac{p_s}{R} \mathrm{J}_1\left(p_s \frac{r}{R}\right)\left\langle \sin\left[|\Omega_s|\left(t - \frac{z-z_i}{\mathrm{v}}\right)\right] h\left(t - \frac{z-z_i}{\mathrm{v}}\right)\right\rangle_i$$

$$E_r = \frac{c^2}{\varepsilon_\mathrm{r} \mathrm{v}}\sum_{s=1} A_s \frac{p_s}{R} \mathrm{J}_1\left(p_s \frac{r}{R}\right)\left\langle \sin\left[|\Omega_s|\left(t - \frac{z-z_i}{\mathrm{v}}\right)\right] h\left(t - \frac{z-z_i}{\mathrm{v}}\right)\right\rangle_i$$

$$A_s = -\frac{eN}{\pi\varepsilon_0 R^2}\frac{\beta^2}{n^2\beta^2 - 1}\frac{1}{\mathrm{J}_1^2(p_s)|\Omega_s|}\frac{2\mathrm{J}_1(p_s R_b/R)}{(p_s R_b/R)} . \quad (2.4.54)$$

Defining $\zeta = \mathrm{v}t - z$, the power emitted is given by

$$P(\zeta) = \frac{(eN)^2}{\pi\varepsilon_0\varepsilon_\mathrm{r} R^2}\frac{\mathrm{v}}{n^2\beta^2 - 1}\sum_s \left[\frac{2\mathrm{J}_1(p_s R_b/R)}{(p_s R_b/R)\mathrm{J}_1(p_s)}\right]^2 \left\langle \sin\left[\frac{|\Omega_s|}{\mathrm{v}}(\zeta + z_i)\right] h(\zeta + z_i)\right\rangle_i^2 \quad (2.4.55)$$

In case of a *single bunch* of length Δ, the trailing power is

$$\frac{P_\mathrm{av}}{\frac{(eN)^2 \mathrm{v}}{2\pi\varepsilon_0 R^2}} = \frac{1/\varepsilon_\mathrm{r}}{\varepsilon_\mathrm{r}\beta^2 - 1}\sum_s \left[\frac{2\mathrm{J}_1(p_s R_b/R)}{(p_s R_b/R)\mathrm{J}_1(p_s)}\operatorname{sinc}\left(\frac{p_s}{2}\frac{\Delta/R}{\sqrt{\varepsilon_\mathrm{r}\beta^2 - 1}}\right)\right]^2 \quad (2.4.56)$$

Figure 2.5 illustrates the normalized spectrum as expressed above as a function of s; note that it decreases rapidly thus the convergence is expected to be quick.

Analysis of the maximum average power trailing behind the bunch reveals that

$$\frac{P_\mathrm{av}}{\frac{(eN)^2 \mathrm{v}}{2\pi\varepsilon_0 R^2}} \simeq \frac{\frac{2/\varepsilon_\mathrm{r}}{\varepsilon_\mathrm{r}\beta^2 - 1}}{c_1 + c_2\left(\frac{R_b}{R}\right)^2 + \left(c_3 + c_4\sqrt{\frac{R_b}{R}}\right)\left(\frac{\Delta}{R}\right)^2 \frac{1}{\varepsilon_\mathrm{r}\beta^2 - 1}} \quad (2.4.57)$$

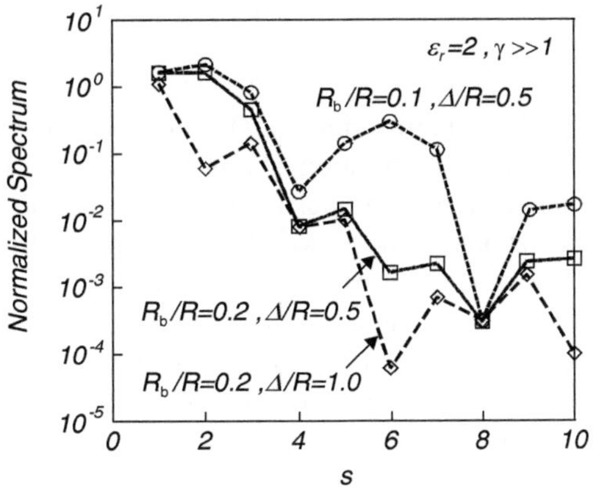

Fig. 2.5 Normalized spectrum of Cerenkov radiation emitted by a finite size azimuthally symmetric bunch in a dielectric filled waveguide of radius R. Only the first 10 modes are shown

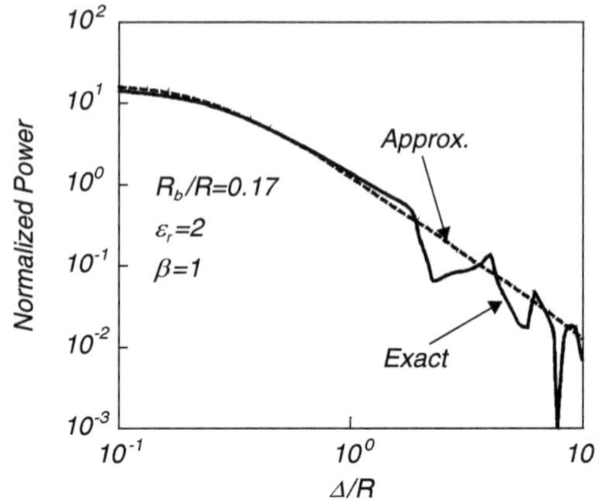

Fig. 2.6 Normalized average power of Cerenkov radiation emitted by a finite size azimuthally symmetric bunch in a dielectric filled waveguide of radius R as a function of the length of the bunch. For the exact expression the first 100 modes have been used. The expression developed reveals an excellent upper value approximation

is an excellent approximation. In this expression $c_1 \simeq 0.0048$, $c_2 \simeq 1.747$, $c_3 \simeq 0.259$ and $c_4 \simeq 1.271$; $0.005 < R_b/R < 0.2$ and $0.1 < \Delta/R < 10$. For a quantitative comparison, Fig. 2.6 shows the exact and the approximate average power generated by a finite size bunch. Some other interesting features are formulated as an Exercise 2.8 at the end of this chapter. In particular, one may investigate ways to *suppress* the coherent radiation.

Taking the same number of electrons (N) but splitting them into a *train of bunches* (M) the discrete spectrum excited in the waveguide undergoes an additional selection associated with the bunches spacing. As in the single bunch case, the bunches are identical in size (radius R_b and length Δ) their spacing is L. A similar approach as above, results into the following expression for the average power trailing behind the train

$$\frac{P_{\text{av}}}{\dfrac{(eN)^2 \text{v}}{2\pi\varepsilon_0 R^2}} = \frac{1/\varepsilon_{\text{r}}}{\varepsilon_{\text{r}}\beta^2 - 1} \sum_s \left[\frac{2\text{J}_1\left(p_s \dfrac{R_b}{R}\right)}{p_s \dfrac{R_b}{R} \text{J}_1(p_s)} \text{sinc}\left(\frac{p_s \Delta/2R}{\sqrt{\varepsilon_{\text{r}}\beta^2 - 1}}\right) \frac{\text{sinc}\left(\dfrac{p_s L/2R}{\sqrt{\varepsilon_{\text{r}}\beta^2 - 1}} M\right)}{\text{sinc}\left(\dfrac{p_s L/2R}{\sqrt{\varepsilon_{\text{r}}\beta^2 - 1}}\right)} \right]^2 . \tag{2.4.58}$$

The last term, $\text{sinc}^2(uM)/\text{sinc}^2(u)$, is responsible to the selection associated with the train configuration. If u is not an integer number of π (off resonance condition), then the term is proportional to M^{-2} implying that the total power is reduced by this factor and there is no advantage in splitting the bunch into a train of bunches. However, if we can ensure resonance namely, for a given s the bunch spacing is chosen such that $u \equiv \Omega_s L/2\text{v} = \pi n$, it is possible to generate a total average power that is of the same order of magnitude as if all the electrons were forming a single bunch. In order to reveal this selection associated with the train's configuration it is convenient to normalize the average power to the case of a single bunch $(M = 1)$ – this is illustrated in Fig. 2.7.

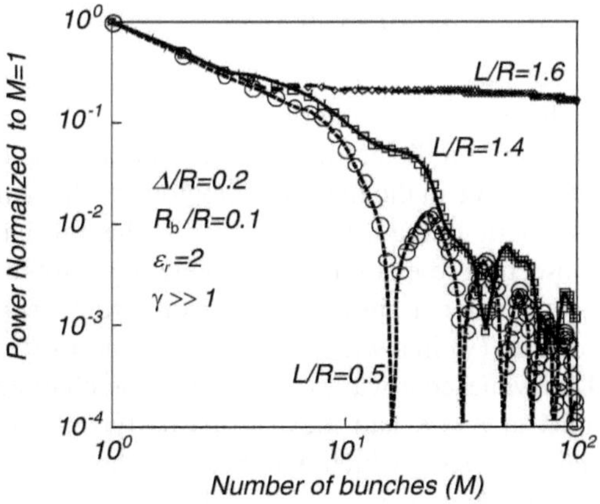

Fig. 2.7 The power normalized to the $M = 1$ case as a function of the number of bunches in the train. Off resonance ($L/R = 0.5, 1.4$), the average power is roughly proportional to M^{-2}. At resonance ($L/R = 1.6$) the power becomes virtually independent of the number of bunches and for large M, this normalized power is of order of unity

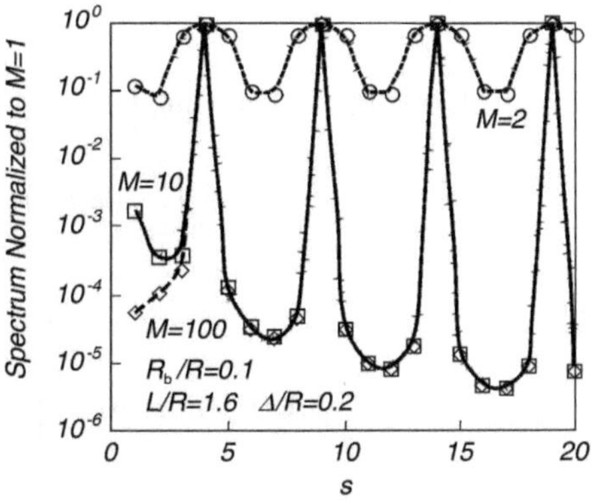

Fig. 2.8 The spectrum normalized to the $M = 1$ case as a funtion of the mode index at resonance ($L/R = 1.6$). At resonant frequencies (corresponding to the frequencies $s = 4, 9, 14.....$) the emitted power is identical to that of the single bunch

When scanning the normalized power there are many possible values of Lthat facilitate power levels of the order of that generated when all the electrons form a single bunch. In the example illustrated in Fig. 2.7, the spacing choice has taken into consideration the fact that for $s \gg 1$, $p_{s+1} - p_s \sim \pi$. This fact facilitates: $u_{s=4} = 3\pi$, $u_{s=9} = 7\pi$, $u_{s=14} = 11\pi$ .... which is reflected in the following plot (Fig. 2.8) of the spectrum of the first 20 modes.

It is evident that at resonance the spectrum is identical to that of a single bunch and in parallel, the spectrum of the off resonance frequencies is significantly suppressed. Note that there is no significant difference between the case $M = 10$ and $M = 100$. For a different choice of bunch-spacing, at resonance, it is possible to have one or at least a few resonant peaks and still to get a substantial fraction of the power generated by a single bunch. The reader is referred to Exercise 2.9 in order to examine additional options associated with the choice of parameters.

2.4.4 Cerenkov Force

In the previous subsection we examined the radiation trailing one or more bunches moving in a dielectric medium with a velocity larger than the phase velocity of a plane wave in the material. Obviously, this emitted energy comes at the expense of its kinetic energy. In other words, the particle is decelerated. It is the goal of this subsection to examine this decelerating force in detail. With this purpose in mind we consider a simple model consisting of a charge $(-e)$ moving at a constant velocity (v) in a vacuum channel of radius R surrounded by a dielectric medium ε_r. The evanescent waves attached to the charged particle impinge upon the discontinuity at $r = R$ and they are partially reflected and partially transmitted. It is the *reflected wave* which acts back on the electron decelerating it; the corresponding current density is described by (2.2.15) whereas its time Fourier transform by (2.4.9). Correspondingly, this current density generates a magnetic vector potential determined by

$$A_z(r<R,z,\omega) = 2\pi\mu_0 \int_{-\infty}^{\infty} dz' \int_0^R dr'r'G(r,z|r',z')J_z(r',z',\omega) + \int_{-\infty}^{\infty} dk\rho(k)e^{-jkz}I_0(\Gamma r), \tag{2.4.59}$$

and

$$A_z(r>R,z,\omega) = \int_{-\infty}^{\infty} dk\tau(k)e^{-jkz}K_0(\Lambda r), \tag{2.4.60}$$

where $\Gamma^2 = k^2 - (\omega/c)^2$, $\Lambda^2 = k^2 - \varepsilon_r(\omega/c)^2$, $G(r',z'|r,z)$ is the boundless Green's function as defined in (2.4.11), (2.4.19) but for vacuum i.e.,

$$G(r',z'|r,z) = \frac{1}{(2\pi)^2} \int_{-\infty}^{\infty} dke^{-jk(z-z')} \begin{cases} I_0(\Gamma r)K_0(\Gamma r') & r<r', \\ K_0(\Gamma r)I_0(\Gamma r') & r'<r. \end{cases} \tag{2.4.61}$$

The amplitudes ρ and τ represent the reflected and transmitted waves correspondingly. In order to determine these amplitudes we have to impose the boundary conditions at $r = R$. For this purpose, it is convenient to write the solution of the magnetic vector potential off-axis as

$$A_z(0<r<R,z,\omega) = \int_{-\infty}^{\infty} dke^{-jkz}[\rho(k)I_0(\Gamma r) + \alpha(k)K_0(\Gamma r)], \tag{2.4.62}$$

where

$$\alpha(k) = -\frac{e\mu_0}{(2\pi)^2}\delta\left(k - \frac{\omega}{v}\right). \tag{2.4.63}$$

From the continuity of the longitudinal electric field (E_z) we conclude that

$$\frac{c^2}{j\omega}\left[\frac{\omega^2}{c^2}-k^2\right][\rho(k)\mathrm{I}_0(\Gamma R)+\alpha(k)\mathrm{K}_0(\Gamma R)]=\frac{c^2}{j\omega\varepsilon_r}\left[\varepsilon_r\frac{\omega^2}{c^2}-k^2\right]\tau(k)K_0(\Lambda R). \tag{2.4.64}$$

In a similar way the continuity of the azimuthal magnetic field implies

$$\Gamma[\rho(k)\mathrm{I}_1(\Gamma R)-\alpha(k)\mathrm{K}_1(\Gamma R)]=-\Lambda\tau(k)\mathrm{K}_1(\Gamma R). \tag{2.4.65}$$

At this stage, we introduce the (normalized) impedances ratio

$$\zeta\equiv\frac{1}{\varepsilon_r}\frac{\Lambda}{\Gamma}\frac{\mathrm{K}_0(\Lambda R)}{\mathrm{K}_1(\Lambda R)}, \tag{2.4.66}$$

by whose means the amplitudes of the reflected waves are given by

$$\rho=\alpha\frac{\zeta\mathrm{K}_1(\Gamma R)-\mathrm{K}_0(\Gamma R)}{\zeta\mathrm{I}_1(\Gamma R)+\mathrm{I}_0(\Gamma R)}. \tag{2.4.67}$$

On axis, the only non-zero field is the longitudinal electric field and only the waves "reflected" from the radial discontinuity contribute to the force that acts on the particle, therefore

$$E_z(r=0,z=v_0t,t)=\int_{-\infty}^{\infty}\mathrm{d}\omega dk\frac{c^2}{j\omega}\left[\frac{\omega^2}{c^2}-k^2\right]\rho(\omega,k)\mathrm{e}^{j(\omega-kv)t}. \tag{2.4.68}$$

Substituting the explicit expression for ρ and using the integral over the Dirac delta function [see (2.4.63)] and defining $x=\omega R/c\beta\gamma$, we obtain

$$E_z(r=0,z=v_0t,t)=\frac{-je}{(2\pi)^2\varepsilon_0R^2}\int_{-\infty}^{\infty}\mathrm{d}xx\frac{\zeta(x)\mathrm{K}_1(|x|)-\mathrm{K}_0(|x|)}{\zeta(x)\mathrm{I}_1(|x|)+\mathrm{I}_0(|x|)}. \tag{2.4.69}$$

At this point, it is convenient to define the normalized field that acts on the particle as

$$\begin{aligned}\mathcal{E}&\equiv E_z(r=0,z=\mathrm{v}t,t)\left(\frac{e}{4\pi\varepsilon_0R^2}\right)^{-1}\\&=\frac{2}{\pi}\int_0^{\infty}\mathrm{d}xx\mathrm{Re}\left[\frac{1}{j}\frac{\zeta(x)\mathrm{K}_1(|x|)-\mathrm{K}_0(|x|)}{\zeta(x)\mathrm{I}_1(|x|)+\mathrm{I}_0(|x|)}\right].\end{aligned} \tag{2.4.70}$$

Clearly, from this representation we observe that, for a non-zero force to act on the particle, the impedance ratio ζ has to be complex since the argument of the modified Bessel functions is real.

We can make one step further and simplify this expression by defining

$$\zeta(x) \equiv |\zeta(x)|e^{j\psi(x)}, \tag{2.4.71}$$

and using $K_0(x)I_1(x) + K_1(x)I_0(x) = 1/x$, we obtain

$$\mathcal{E} = \frac{2}{\pi}\int_0^\infty dx \frac{|\zeta(x)|\sin\psi(x)}{I_0^2(x) + |\zeta(x)|^2 I_1^2(x) + 2|\zeta(x)|I_0(x)I_1(x)\cos\psi(x)}. \tag{2.4.72}$$

In order to evaluate this integral for a dielectric medium and a particle whose velocity βc is larger than $c/\sqrt{\varepsilon_r}$, we go back to (2.4.66) which now reads

$$\zeta(x) = j\frac{\gamma}{\varepsilon_r}\sqrt{\varepsilon_r\beta^2 - 1}\,\frac{K_0\left(jx\gamma\sqrt{\varepsilon_r\beta^2 - 1}\right)}{K_1\left(jx\gamma\sqrt{\varepsilon_r\beta^2 - 1}\right)}, \tag{2.4.73}$$

and it can be further simplified if we assume that the main contribution occurs for large arguments of the Bessel function (i.e., $\gamma \gg 1$) thus

$$\zeta(x) \simeq j\frac{\gamma}{\varepsilon_r}\sqrt{\varepsilon_r\beta^2 - 1}. \tag{2.4.74}$$

Since subject to this approximation $\psi = \pi/2$ and $|\zeta|$ is constant we can evaluate $\mathcal{E}$,

$$\mathcal{E} = \frac{2}{\pi}\int_0^\infty dx \frac{|\zeta|}{I_0^2(x) + |\zeta|^2 I_1^2(x)}, \tag{2.4.75}$$

for two regimes: firstly when $|\zeta| \gg 1$ i.e., $\gamma \gg 1$, the contribution to the integral is primarily from small values of x thus

$$\mathcal{E} \simeq \frac{2}{\pi}\int_0^\infty dx \frac{|\zeta|}{1 + |\zeta|^2 x^2/4} \simeq \frac{4}{\pi}\int_0^\infty du \frac{1}{1+u^2} \simeq 2 \tag{2.4.76}$$

At the other extreme ($|\zeta| \ll 1$) the normalized impedance has to be re-calculated and the result is

$$\mathcal{E} \simeq \frac{\gamma^2(\varepsilon_r\beta^2 - 1)}{\varepsilon_r}\int_0^\infty dx \frac{x}{I_0^2(x)} \simeq 1.263\frac{\gamma^2(\varepsilon_r\beta^2 - 1)}{\varepsilon_r}, \tag{2.4.77}$$

and we can summarize

$$\mathcal{E} \simeq \begin{cases} 0 & \text{for} \quad \beta < 1/\sqrt{\varepsilon_r}, \\ 1.263\gamma^2\sqrt{\varepsilon_r\beta^2 - 1}/\varepsilon_r & \text{for} \quad \gamma \ll \varepsilon_r/\sqrt{\varepsilon_r\beta^2 - 1}, \\ 2 & \text{for} \quad \gamma \gg \varepsilon_r/\sqrt{\varepsilon_r\beta^2 - 1}. \end{cases} \tag{2.4.78}$$

It is interesting to note that for ultra-relativistic electrons the decelerating Cerenkov force reaches an asymptotic value which is independent of γ and the dielectric coefficient; it is given by $E = -e/2\pi\varepsilon_0 R^2$. In addition, we observe that the normalized impedance (ζ) determines the force.

2.4.5 *Ohm Force*

If in the Cerenkov case the charged particle has to exceed a certain velocity in order to generate radiation and therefore to experience a decelerating force, in the case of a lossy medium, the moving electron experiences a decelerating force starting from a vanishingly low speed. This is because it excites currents in the surrounding walls and as a result, power is dissipated – which is equivalent to the emitted power in the Cerenkov case. The source of this power is the $\mathbf{J}\cdot\mathbf{E}$ [see (2.1.16)] term which infers the existence of a decelerating force acting on the electron. In order to evaluate this force we use the same formulation as in the previous subsection only that in this case, the dielectric coefficient is complex and it is given by

$$\varepsilon_r = 1 - j\frac{\sigma}{\varepsilon_0 \omega}, \tag{2.4.79}$$

where σ is the (finite) conductivity of the surrounding medium. It is convenient to use the same notation as above, therefore the normalized impedance ζ from (2.4.74) is replaced by

$$\zeta \simeq \frac{1}{1 - j\bar{\sigma}/x}\sqrt{1 + j(\gamma\beta)^2\frac{\bar{\sigma}}{x}}. \tag{2.4.80}$$

In this expression $\bar{\sigma} \equiv \sigma\eta_0 R/\gamma\beta$ which for typical metals and $R \sim 1$ cm is of the order of $10^8/\gamma\beta$ thus for any practical purpose $\bar{\sigma} \gg 1$ hence

$$\zeta \simeq \gamma\beta\sqrt{\frac{x}{\bar{\sigma}}}\exp\left(j\frac{3\pi}{4}\right). \tag{2.4.81}$$

Note that the phase of the normalized impedance is $\psi = 3\pi/4$. Substituting this expression in (2.4.72) and defining the characteristic angular frequency $\omega_0 \equiv 2c/R(\gamma\beta)^3$ as well as the skin-depth $\delta = \sqrt{2/\sigma\mu_0\omega_0}$, we obtain

$$\mathcal{E} = \frac{2}{\pi}\frac{\sqrt{2}}{2}\frac{\delta}{R}\int_0^\infty \mathrm{d}x \frac{\sqrt{x}}{\mathrm{I}_0^2(x) + x\mathrm{I}_1^2(x)\dfrac{\delta^2}{R^2} - \sqrt{2x}\mathrm{I}_0(x)\mathrm{I}_1(x)\dfrac{\delta}{R}} \tag{2.4.82}$$

which can be evaluated analytically for two extreme regimes: in the first case the (normalized) momentum of the particle is much smaller than the normalized conductivity term i.e., the skin-depth is much smaller than the radius of the tunnel $\delta^2 \ll R^2$ in which case

$$\mathcal{E} \simeq \frac{2}{\pi} \frac{\sqrt{2}}{2} \frac{\delta}{R} \int_0^\infty \mathrm{d}x \frac{\sqrt{x}}{\mathrm{I}_0^2(x)} \simeq 0.54 \frac{\delta}{R}. \tag{2.4.83}$$

The second case corresponds to a highly relativistic particle i.e., $(\gamma\beta)^3 \gg \sigma\eta_0 R$ or $\delta^2 \gg R^2$ implying that the main contribution to the integral is from the small values of x which justifies the expansion of the modified Bessel functions in Taylor series. Redefining $y^2 \equiv (x\gamma\beta)^3/4\sigma\eta_0 R$ we have

$$\mathcal{E} \simeq \frac{4\sqrt{2}}{3\pi} \int_0^\infty \mathrm{d}y \frac{1}{1 + y^2 - y\sqrt{2}} \simeq 2. \tag{2.4.84}$$

In fact a best fit to the exact expression in (2.4.82) reveals that

$$\mathcal{E} \simeq \left(0.54 \frac{\delta}{R} + 2\frac{\delta^2}{R^2}\right) \left[1 + \left(\frac{\delta^2}{R^2}\right)^{\frac{6}{10}}\right]^{-\frac{10}{6}} \tag{2.4.85}$$

is an excellent approximation – the integrated $(0 < \delta/R < 20)$ relative error is less than 0.02% . Clearly, as in the Cerenkov case, for ultra-relativistic particles ($\gamma^3 \gg \sigma\eta_0 R$ or $\delta \gg R$) the decelerating force is independent of γ and of the material's characteristics. However, the critical γ for operating in this regime is much higher comparing to the Cerenkov case.

The characteristic angular frequency (ω_0) is low for relativistic electrons and consequently, the skin-depth is much larger than the radius and all the bulk material "participates" in the deceleration process. On the other hand, if the frequency is high, then the skin-depth is small (comparing to the radius) and only a thin layer dissipates power, therefore the loss is proportional to δ.

Finally, imagine an interesting situation whereby the conductivity of the material is negative, this is to say that the medium is active, then the phase in (2.4.81) is $\psi = 5\pi/4$ and the force is *accelerating* which means that energy can be transferred from the medium to the electron. We will further elaborate this topic in Chap. 8 in the context of advanced acceleration concepts.

2.5 Finite Length Effects

In all the effects discussed so far, we assumed an infinite system with no reflected waves. In this section, we consider several systems and phenomena associated with reflected waves. When both forward and backward propagating waves coexist, there is a frequency selection associated with the interference of the two. Another byproduct of reflections is tunneling of the field in a region where the wave is below cutoff. We also examine the radiation generated by a particle as it traverses a geometric discontinuity in a waveguide. We conclude with the evaluation of a wake field generated by a particle in a cavity.

2.5.1 *Impedance Discontinuities*

In most cases of interest, the waveguide is not uniform and as a result, more than one wave occurs. In order to illustrate the effect of discontinuities we consider next the following problem: a cylindrical waveguide of radius R but, instead of being uniformly filled with one dielectric material, there are three different dielectrics in three different regions

$$\varepsilon_r(z) = \begin{cases} \varepsilon_1 & \text{for} \quad -\infty < z < 0, \\ \varepsilon_2 & \text{for} \quad 0 \le z \le d, \\ \varepsilon_3 & \text{for} \quad d < z < \infty, \end{cases} \tag{2.5.1}$$

as illustrated in Fig. 2.9.

A wave is launched from $z \to -\infty$ towards the discontinuity at $z = 0$. For simplicity we assume that this wave is composed of a single mode (TM_{01} i.e., $s = 1$). The z component of the magnetic vector in the first region $(-\infty < z < 0)$ is given by

$$A_z(r, -\infty < z < 0, \omega) = \left[A_{\text{in}} e^{-\Gamma_1^{(1)} z} + A_\rho e^{\Gamma_1^{(1)} z}\right] J_0\left(p_1 \frac{r}{R}\right), \tag{2.5.2}$$

where A_{in} is the amplitude of the incoming wave and A_ρ is the amplitude of the reflected wave; $\Gamma_1^{(1)} = \sqrt{(p_1/R)^2 - \varepsilon_1(\omega/c)^2}$. Between the two discontinuities at $0 < z < d$ the solution has a similar form

$$A_z(r, 0 \le z \le d, \omega) = \left[A_+ e^{-\Gamma_1^{(2)} z} + A_- e^{\Gamma_1^{(2)} z}\right] J_0\left(p_1 \frac{r}{R}\right), \tag{2.5.3}$$

where $\Gamma_1^{(2)} = \sqrt{(p_1/R)^2 - \varepsilon_2(\omega/c)^2}$ In the third region, there is no reflected wave therefore

$$A_z(r, d < z < \infty, \omega) = A_\tau e^{-\Gamma_1^{(3)}(z-d)} J_0\left(p_1 \frac{r}{R}\right), \tag{2.5.4}$$

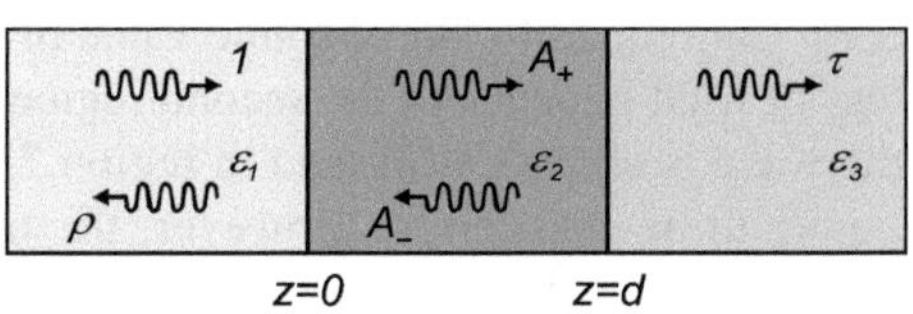

Fig. 2.9 Schematics of the system used to examine the reflected waves resulting from characteristic impedance discontinuities

and as above $\Gamma_1^{(3)} = \sqrt{(p_1/R)^2 - \varepsilon_3(\omega/c)^2}$; A_τ is the amplitude of the transmitted wave. The four as yet unknown amplitudes A_ρ, A_τ, A_+ and A_- are determined by imposing the boundary conditions at $z = 0, d$. Continuity of E_r at $z = 0$ implies

$$Z_1\left(A_{\text{in}} - A_\rho\right) = Z_2(A_+ - A_-); \tag{2.5.5}$$

Z_1 and Z_2 are the characteristic impedances (2.3.26) in the first and second regions respectively. In a similar way the continuity of H_ϕ implies

$$A_{\text{in}} + A_\rho = A_+ + A_-. \tag{2.5.6}$$

An additional set of equations is found imposing the continuity of the same components at $z = d$:

$$Z_2[A_+ e^{-\psi} - A_- e^{\psi}] = Z_3 A_\tau, \tag{2.5.7}$$

and

$$A_+ e^{-\psi} + A_- e^{\psi} = A_\tau, \tag{2.5.8}$$

where $\psi \equiv \Gamma_1^{(2)} d$. From (2.5.5)–(2.5.8) the reflection (ρ) and transmission (τ) coefficients are determined base on the radial electric field and are given by

$$\begin{aligned}
\rho &\equiv \frac{Z_1 A_\rho}{Z_1 A_{\text{in}}} = \frac{\sinh(\psi)(Z_1 Z_3 - Z_2^2) + \cosh(\psi)(Z_1 Z_2 - Z_2 Z_3)}{\sinh(\psi)(Z_1 Z_3 + Z_2^2) + \cosh(\psi)(Z_1 Z_2 + Z_2 Z_3)}, \\
\tau &\equiv \frac{Z_3 A_\tau}{Z_1 A_{\text{in}}} = \frac{Z_3}{Z_1} \frac{2 Z_1 Z_2}{\sinh(\psi)(Z_1 Z_3 + Z_2^2) + \cosh(\psi)(Z_1 Z_2 + Z_2 Z_3)}.
\end{aligned} \tag{2.5.9}$$

After we have established the amplitudes of the magnetic vector potential it is possible to determine the electromagnetic field in each one of the regions, thus we can investigate the power flow in the system. Using Poynting's theorem the power conservation implies that

$$\text{Re}(Z_1)\left[|A_{\text{in}}|^2 - |A_\rho|^2\right] = \text{Re}(Z_3)|A_\tau|^2. \tag{2.5.10}$$

This expression relates the power in the first region to that in the third. It does not depend explicitly on the second region; if, for example, in the third region the wave is below cutoff, the characteristic impedance is imaginary and the right-hand side is zero. Consequently, the absolute value of the reflection coefficient is unity, regardless of what happens in the second region. On the other hand, if in regions 1 and 3 the wave is above cutoff, and in region 2 the wave is below cutoff, we still expect power to be transferred. However, the transmission coefficient decays exponentially with $\psi = \Gamma_1^{(2)} d$

$$\tau \sim \frac{4Z_3Z_2}{(Z_1Z_3 + Z_2^2) + Z_2(Z_1 + Z_3)} e^{-\psi}. \tag{2.5.11}$$

In spite of the discontinuities there can be frequencies at which the reflection coefficient (ρ) is zero if we design the structure such that

$$Z_1Z_3 = Z_2^2 \quad \text{and} \quad \psi = j\pi/2, \tag{2.5.12}$$

as one can conclude by examining the numerator of ρ. The expression in (2.5.12) defines the conditions for the so-called *quarter-λ transformer*. Figure 2.10 shows a typical picture of the transmission coefficient. Note that the peaks in the transmission correspond to constructive interference of the two waves in the central section; the valleys correspond to destructive interference of the same waves. Zero reflections also occur when

$$Z_1 = Z_3 \quad \text{and} \quad \psi = j\pi. \tag{2.5.13}$$

If in the first and third region the wave's frequency is *below cutoff* but in the middle region a wave can propagate, then the system will determine a set of discrete frequencies at which the wave can bounce between the two sections. These eigen-frequencies are determined by the geometric parameters and the dielectric coefficients. We can calculate these frequencies from the poles of the transmission or reflection coefficient, namely from the condition that its denominator is zero:

$$\sinh(\psi)(Z_1Z_3 + Z_2^2) + Z_2\cosh(\psi)(Z_1 + Z_3) = 0. \tag{2.5.14}$$

Equivalently, one can write equations (2.5.5)–(2.5.8) in a matrix form, set the input term to zero ($A_{\text{in}} = 0$) and look for the non-trivial solution by requiring that

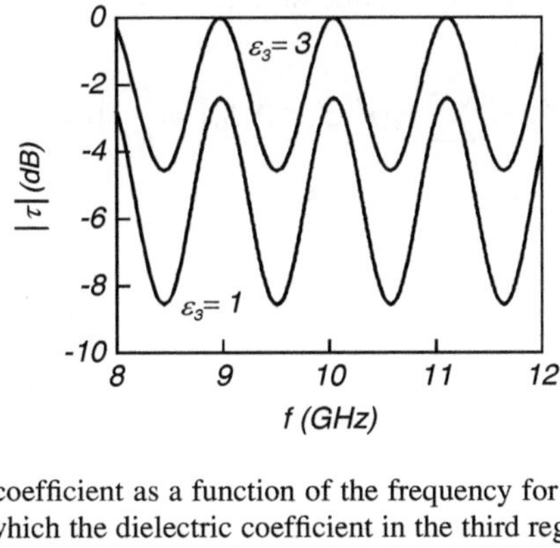

Fig. 2.10 Transmission coefficient as a function of the frequency for two cases: the upper trace represents a situation in which the dielectric coefficient in the third region equals that in the first, therefore at certain frequencies all the power is transferred – see (2.5.13). In the lower trace the two are different and the relation in (2.5.12) is not satisfied, therefore always a fraction of the energy is reflected

the determinant of the matrix is zero – the result is identical with (2.5.14). The reader is encouraged to determine Green's function of the configuration described in this section – see Exercise 2.10.

2.5.2 Geometric Discontinuity

Another source of reflected waves is a *geometric* discontinuity. In a sense these can be conceived as impedance discontinuities but of a more complex character since geometric variations *couple* between the different modes in the waveguide. The simplest configuration which can be considered quasi-analytically consists of a waveguide of radius R_1 and another of radius $R_2 < R_1$; the discontinuity occurs at $z = 0$ as illustrated in Fig. 2.11. A detailed analysis when a single mode impinges upon a discontinuity was reported in the literature e.g., Mittra and Lee (1971) or Lewin (1975).

Step I: We examine first the case when the source term is in the *left-hand side* ($z < 0$), therefore Green's function in the left-hand side has two components

$$G(z<0,r|z'<0,r') = \sum_{s=1}^{\infty} \frac{J_0(p_s r/R_1) J_0(p_s r'/R_1)}{\frac{1}{2} R_1^2 J_1^2(p_s)} \frac{\exp(-\Gamma_s^{(1)}|z-z'|)}{4\pi\Gamma_s^{(1)}}$$
$$+ \sum_{s=1}^{\infty} \rho_s(r', z'<0) \mathrm{J}_0\left(p_s \frac{r}{R_1}\right) \exp\left(\Gamma_s^{(1)} z\right), \qquad (2.5.15)$$

the non-homogeneous solution, which corresponds to an infinite waveguide and the homogeneous solution which is due to the discontinuity; $\Gamma_s^{(1)} = \sqrt{(p_s/R_1)^2 - (\omega/c)^2}$. In the right-hand side ($z > 0$),

$$G(z>0,r|z'<0,r') = \sum_{s=1}^{\infty} \tau_s(r', z'<0) \mathrm{J}_0\left(p_s \frac{r}{R_2}\right) \exp\left(-\Gamma_s^{(2)} z\right), \qquad (2.5.16)$$

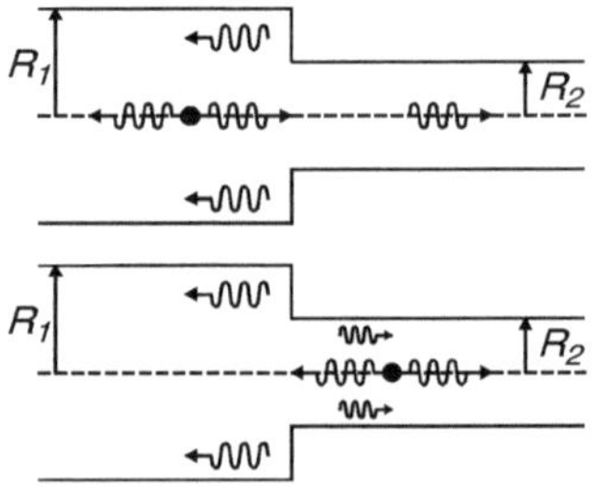

Fig. 2.11 Green's function calculation for one discontinuity in the geometry of a waveguide. In the upper figure the source is in the left and in the lower it is in the right

where $\Gamma_s^{(2)} = \sqrt{(p_s/R_2)^2 - (\omega/c)^2}$. Continuity of the radial electric field at $z = 0$ entails

$$\frac{\partial^2}{\partial z \partial r} G(r, z{=}\,0^- | r', z' < 0) = \begin{cases} \dfrac{\partial^2}{\partial z \partial r} G(r, z{=}\,0^+ | r', z' < 0) & \text{for} \quad 0 \le r < R_2, \\ 0 & \text{for} \quad R_1 \ge r \ge R_2 \end{cases} \tag{2.5.17}$$

In order to determine the amplitudes ρ_s and τ_s the last equation is multiplied by $\mathrm{J}_1(p_s r/R_1)$, the product is integrated from 0 to R_1 and using the orthogonality of the Bessel function [similar to (2.3.17) but for first order Bessel function] we obtain

$$g_s^{(1)}(r', z') - \rho_s(r', z') = \sum_{\sigma=1}^{\infty} Z_{s,\sigma} \tau_\sigma(r', z'), \tag{2.5.18}$$

where

$$g_s^{(1)}(r', z') = \frac{\mathrm{J}_0(p_s r'/R_1)}{\frac{1}{2} R_1^2 \mathrm{J}_1^2(p_s)} \frac{\exp\left(\Gamma_s^{(1)} z'\right)}{4\pi \Gamma_s^{(1)}}, \tag{2.5.19}$$

and

$$Z_{s,\sigma} \equiv \frac{\Gamma_\sigma^{(2)}}{\Gamma_s^{(1)}} \frac{p_\sigma}{p_s} \frac{R_1}{R_2} \frac{1}{\mathrm{J}_1^2(p_s)} \frac{2}{R_1^2} \int_0^{R_2} \mathrm{d}r r \mathrm{J}_1\left(p_s \frac{r}{R_1}\right) \mathrm{J}_1\left(p_\sigma \frac{r}{R_2}\right). \tag{2.5.20}$$

Continuity of the azimuthal magnetic field in the domain $0 < r < R_2$ implies

$$\frac{\partial}{\partial r} G(r, z = 0^+ | r', z' < 0) = \frac{\partial}{\partial r} G(r, z = 0^- | r', z' < 0). \tag{2.5.21}$$

As above, we use the fact that in the domain of interest, $\mathrm{J}_1(p_s r/R_2)$ form a complete orthogonal set of functions hence

$$\tau_\sigma(r', z') = \sum_{s=1}^{\infty} Y_{\sigma,s} \left[g_s^{(1)}(r', z') + \rho_s(r', z') \right], \tag{2.5.22}$$

where

$$Y_{\sigma,s} \equiv \frac{2}{R_1{}^2} \int_0^{R_2} \mathrm{d}r r \mathrm{J}_1\left(p_s \frac{r}{R_1}\right) \mathrm{J}_1\left(p_\sigma \frac{r}{R_2}\right). \tag{2.5.23}$$

The integral in both expressions for Z and Y can be calculated analytically (Abramowitz and Stegun 1968, p. 484) and it is given by

$$\int_0^1 d\xi \xi J_1(p_n\xi) J_1(p_m u\xi) = \begin{cases} \frac{1}{2} J_1^2(p_n) & \text{for} \quad p_n = p_m u \\ p_m u \left[p_n^2 - p_m^2 u^2\right]^{-1} J_1(p_n) J_0(p_m u) & \text{otherwise.} \end{cases} \tag{2.5.24}$$

From (2.5.18), (2.5.22) one can determine the amplitudes of the reflected and transmitted waves. Adopting a vector notation, i.e., $\rho_s(r', z' < 0) \rightarrow \mathbf{R}^{(-)}$, $\tau_s(r', z' < 0) \rightarrow \mathbf{T}^{(-)}$ and $g_s^{(1)}(r', z' < 0) \rightarrow \mathbf{g}^{(1)}$, these amplitudes can be formally written as

$$\mathbf{R}^{(-)} = (I + ZY)^{-1}(I - ZY)\mathbf{g}^{(1)} \tag{2.5.25}$$

and

$$\mathbf{T}^{(-)} = Y\left[I + (I + ZY)^{-1}(I - ZY)\right]\mathbf{g}^{(1)}. \tag{2.5.26}$$

Step II: In a similar way, if the source is in the *right-hand side* $(z' > 0)$ then Green's function in the left-hand side can be written as

$$G(z < 0, r|z' > 0, r') = \sum_{s=1}^{\infty} \rho_s(r', z' > 0) J_0\left(p_s \frac{r}{R_1}\right) \exp\left(\Gamma_s^{(1)} z\right), \tag{2.5.27}$$

and

$$\begin{aligned} G(z > 0, r|z' > 0, r') = &\sum_{s=1}^{\infty} \frac{J_0(p_s r/R_2) J_0(p_s r'/R_2)}{\frac{1}{2} R_2^2 J_1^2(p_s)} \frac{\exp\left(-\Gamma_s^{(2)}|z - z'|\right)}{4\pi\Gamma_s^{(2)}} \\ &+ \sum_{s=1}^{\infty} \tau_s(r', z' > 0) J_0\left(p_s \frac{r}{R_2}\right) \exp\left(-\Gamma_s^{(2)} z\right). \end{aligned} \tag{2.5.28}$$

Continuity of E_r at $z = 0$ implies

$$\rho_s(r', z') = \sum_{\sigma=1}^{\infty} Z_{s,\sigma}\left[g_\sigma^{(2)}(r', z') - \tau_\sigma(r', z')\right], \tag{2.5.29}$$

where

$$g_s^{(2)}(r', z') = \frac{J_0(p_s r'/R_2)}{\frac{1}{2} R_2^2 J_1^2(p_s)} \frac{\exp\left(-\Gamma_s^{(2)} z'\right)}{4\pi\Gamma_s^{(2)}}, \tag{2.5.30}$$

and the continuity of H_ϕ can be simplified to read

$$\tau_\sigma(r',z') + g_\sigma^{(2)}(r',z') = \sum_{s=1}^{\infty} Y_{\sigma,s}\rho_s(r',z'). \tag{2.5.31}$$

Again, adopting a vector notation $\tau_s(r',z'>0) \to \mathbf{T}^{(+)}$, $g_s^{(2)}(r',z'<0) \to \mathbf{g}^{(2)}$ and $\rho_s(r',z'>0) \to \mathbf{R}^{(+)}$ we can write for the reflected and transmitted waves the following expressions

$$\mathbf{T}^{(+)} = -(I+YZ)^{-1}(I-YZ)\mathbf{g}^{(2)}, \tag{2.5.32}$$

and

$$\mathbf{R}^{(+)} = Z\left[I + (I+YZ)^{-1}(I-YZ)\right]\mathbf{g}^{(2)}. \tag{2.5.33}$$

With Green's function established, we calculate now the *energy* emitted by a particle with a charge e as it traverses the discontinuity. Assuming a constant velocity v_0, the current distribution is given by (2.4.9) and the electric field which acts on the particle due to the discontinuity is given by

$$E_z(r,z,\omega) = \frac{e\mathrm{v}_0}{j\omega\varepsilon_0}\left[\frac{\omega^2}{c^2} + \frac{\partial^2}{\partial z^2}\right]\int_{-\infty}^{\infty} \mathrm{d}z' G(r,z|0,z')\exp\left(-j\frac{\omega}{\mathrm{v}_0}z'\right). \tag{2.5.34}$$

With this field component we can examine the total power transferred by the particle i.e.,

$$P(t) = -2\pi\int_{-\infty}^{\infty} \mathrm{d}z \int_0^{R(z)} \mathrm{d}r r J_z(r,z,t)E_z(r,z,t), \tag{2.5.35}$$

and also the total energy defined by

$$W = \int_{-\infty}^{\infty} dt P(t), \tag{2.5.36}$$

which explicitly reads

$$\begin{aligned}
W = &-\frac{e^2\mathrm{v}_0}{2\pi\varepsilon_0 R_1^2}\int_{-\infty}^{\infty}\mathrm{d}\omega\frac{1}{j\omega}\sum_{s=1}^{\infty}p_s^2\int_{-\infty}^{0}\mathrm{d}z'\exp\left(-j\frac{\omega}{\mathrm{v}_0}z'\right)\rho_s(0,z')\\
&\times\int_{-\infty}^{0} dt \exp\left[t\left(j\omega + \Gamma_s^{(1)}\mathrm{v}_0\right)\right]\\
&-\frac{e^2\mathrm{v}_0}{2\pi\varepsilon_0 R_2^2}\int_{-\infty}^{\infty}\mathrm{d}\omega\frac{1}{j\omega}\sum_{s=1}^{\infty}p_s^2\int_0^{\infty}\mathrm{d}z'\exp\left(-j\frac{\omega}{\mathrm{v}_0}z'\right)\tau_s(0,z')\\
&\times\int_0^{\infty} dt \exp\left[t\left(j\omega + \Gamma_s^{(1)}\mathrm{v}_0\right)\right]
\end{aligned} \tag{2.5.37}$$

According to (2.5.25), (2.5.32) and the definitions of $g^{(1)}$ and $g^{(2)}$, we can write

$$\rho_s(0, z' < 0) \equiv \sum_{s'} \alpha_{s,s'} \exp\left(\Gamma_{s'}^{(1)} z'\right) \quad (2.5.38)$$

and

$$\tau_s(0, z' > 0) \equiv \sum_{s'} \chi_{s,s'} \exp\left(-\Gamma_{s'}^{(2)} z'\right). \quad (2.5.39)$$

Consequently, the expression for the total energy reads

$$\begin{aligned} W = & \frac{e^2 \mathrm{v}_0^2}{2\pi\varepsilon_0 R_1^2} \int_{-\infty}^{\infty} \mathrm{d}\omega \frac{1}{j\omega} \sum_{s,s'=1}^{\infty} p_s^2 \frac{\alpha_{s,s'}}{j\omega - \mathrm{v}_0 \Gamma_{s'}^{(1)}} \frac{1}{j\omega + \mathrm{v}_0 \Gamma_s^{(1)}} \\ & - \frac{e^2 \mathrm{v}_0^2}{2\pi\varepsilon_0 R_2^2} \int_{-\infty}^{\infty} \mathrm{d}\omega \frac{1}{j\omega} \sum_{s,s'=1}^{\infty} p_s^2 \frac{\chi_{s,s'}}{j\omega + \mathrm{v}_0 \Gamma_{s'}^{(2)}} \frac{1}{j\omega - \mathrm{v}_0 \Gamma_s^{(2)}}. \end{aligned} \quad (2.5.40)$$

The matrices α and χ are frequency dependent, therefore numerical methods have to be invoked in order to have a quantitative answer regarding the energy transfer. Nevertheless, the spectrum can be readily derived from these two expressions. The first term represents the energy emitted when the particle moves in the left-hand side and the second corresponds to the energy emitted when it moves in the right one. It should be pointed out that each one of the terms has two contributions: a fraction of the energy propagates to the left and the remainder to the right. In the next subsection we present a simpler configuration which allows one to trace analytically the way the electromagnetic field develops in time in the case of reflections. We recommend the reader to solve Exercise 2.11 at the end of the chapter in order to assess the emitted spectrum.

Before concluding, one question needs to be addressed. In principle, the number of modes required to represent the field exactly is infinite, but practically only a finite number of terms is taken into consideration because of the need to invert the matrices numerically. The question is what should be the number of Bessel harmonics necessary for the representation of a discontinuity as the one presented above and what is the error associated with the truncation. In order to answer this question, let us consider a simple function

$$f(r) = \begin{cases} 1 & \text{for} \quad 0 \leq r < R_2, \\ 0 & \text{for} \quad R_2 < r \leq R_1, \end{cases} \quad (2.5.41)$$

as illustrated in Fig. 2.12.

This function can also be represented by a superposition of Bessel functions:

$$f(r) = \sum_{s=1}^{\infty} f_s \mathrm{J}_0\left(p_s \frac{r}{R_1}\right), \quad (2.5.42)$$

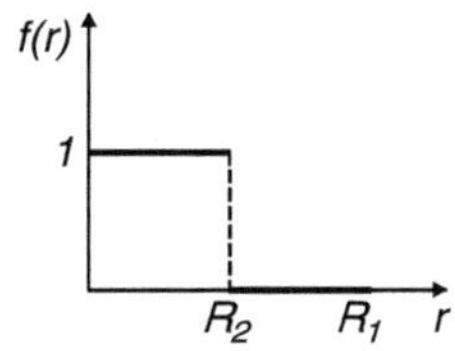

Fig. 2.12 Step function used to model the effect of truncation in a Bessel series representation

where

$$f_s = 2\frac{R_2}{R_1}\frac{\mathrm{J}_1(p_sR_2/R_1)}{p_s\mathrm{J}_1^2(p_s)}; \tag{2.5.43}$$

here we used the fact that the integral

$$\int_0^x \mathrm{d}\xi\xi\mathrm{J}_0(\xi) = x\mathrm{J}_1(x), \tag{2.5.44}$$

can be evaluated analytically (Abramowitz and Stegun 1968, p. 484). We now define the *relative error* made when representing the function only with a finite number of Bessel harmonics as the

$$\mathrm{Error}(N) \equiv \frac{\int_0^{R_1}\mathrm{d}rr\left[f(r) - \sum_{s=1}^{N} f_s\mathrm{J}_0(p_sr/R_1)\right]^2}{\int_0^{R_1}\mathrm{d}rrf^2(r)}. \tag{2.5.45}$$

Using (2.5.43), (2.5.44) the last relation can be simplified to read

$$\mathrm{Error}(N) = 1 - 4\sum_{s=1}^{N}\left[\frac{\mathrm{J}_1(p_sR_2/R_1)}{p_s\mathrm{J}_1(p_s)}\right]^2. \tag{2.5.46}$$

Figure 2.13 illustrates this error. Taking a single mode the normalized error is 36% for $R_2/R_1 = 0.5$ and it drops to 2% for 20 modes. However, even with 20 modes the error can be significantly higher if the radii ratio is small and it is more than 15% for

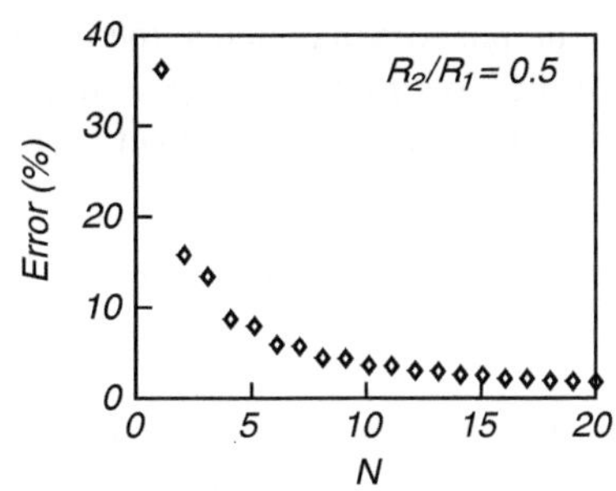

Fig. 2.13 Numerical error as a function of the number of terms

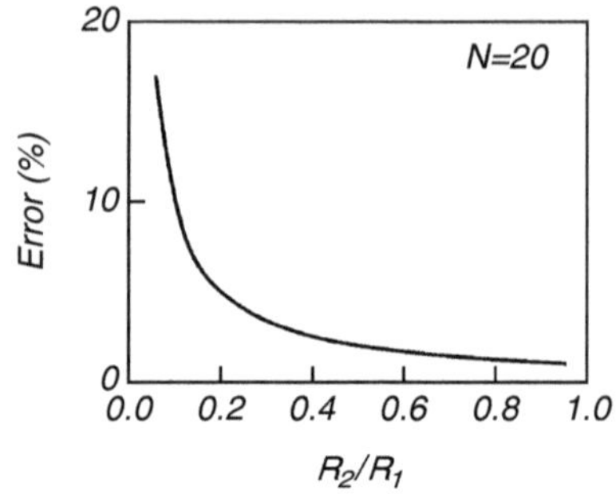

Fig. 2.14 Truncation error as a function of the radius ratio for a constant number of Bessel harmonics

$N = 20$ and $R_2/R_1 \simeq 0.1$ – see Fig. 2.14. These facts become crucial when an accurate solution with multiple discontinuities is necessary.

2.5.3 *Wake-Field in a Cavity*

In order to examine *transient phenomena* associated with reflected waves we calculate the electromagnetic energy in a cavity as a single point-charge traverses the structure. Consider a *lossless* cylindrical cavity of radius R and length d. A charged particle (e) moves along the axis at a constant velocity v_0. Consequently, the longitudinal component of the current density is the only non-zero term, thus

$$J_z(\mathbf{r},t) = -e\mathrm{v}_0 \frac{1}{2\pi r}\delta(r)\delta(z - \mathrm{v}_0 t). \tag{2.5.47}$$

It excites the longitudinal magnetic vector potential $A_z(\mathbf{r},t)$, which for an azimuthally symmetric system, satisfies

$$\left[\frac{1}{r}\frac{\partial}{\partial r}r\frac{1}{\partial r} + \frac{\partial^2}{\partial z^2} - \frac{1}{c^2}\frac{\partial^2}{\partial t^2}\right]A_z(r,z,t) = -\mu_0 J_z(r,z,t). \tag{2.5.48}$$

In this section, we consider only the internal problem, ignoring the electromagnetic phenomena outside the cavity. The boundary conditions on the internal walls of the cavity impose $E_z(r = R, z, t) = 0$, $E_r(r, z = 0, t) = 0$ and $E_r(r, z = d, t) = 0$ therefore, the magnetic vector potential reads

$$A_z(r,z,t) = \sum_{s=1,n=0}^{\infty} A_{s,n}(t)\mathrm{J}_0\left(p_s\frac{r}{R}\right)\cos\left(\frac{\pi n}{d}z\right). \tag{2.5.49}$$

Using the orthogonality of the trigonometric and Bessel functions, we find that the amplitude $A_{s,n}(t)$ satisfies

$$\left[\frac{d^2}{dt^2}+\Omega_{s,n}^2\right]A_{s,n}(t)=-\frac{e\mathrm{v}_0}{2\pi\varepsilon_0}\frac{1}{\frac{1}{2}R^2J_1^2(p_s)}\frac{1}{g_n d}\cos\left(\frac{\pi n}{d}\mathrm{v}_0 t\right)\left[h(t)-h\left(t-\frac{\mathrm{v}_0}{d}\right)\right],\tag{2.5.50}$$

where

$$g_n=\begin{cases}1 & \text{for}\quad n=0,\\ 0.5 & \text{otherwise,}\end{cases}\tag{2.5.51}$$

and

$$\Omega_{s,n}=c\sqrt{\left(\frac{p_s}{R}\right)^2+\left(\frac{\pi n}{d}\right)^2},\tag{2.5.52}$$

are the eigen-frequencies of the cavity. Before the particle enters the cavity ($t<0$), no field exists, therefore

$$A_{s,n}(t<0)=0.\tag{2.5.53}$$

For the time the particle is in the cavity namely, $0<t<d/\mathrm{v}_0$, the solution of (2.5.50) consists of the homogeneous and the excitation term:

$$A_{s,n}\left(0<t<\frac{d}{\mathrm{v}_0}\right)=B_1\cos(\Omega_{s,n}t)+B_2\sin(\Omega_{s,n}t)+\alpha_{s,n}\cos(\omega_n t),\tag{2.5.54}$$

where

$$\alpha_{s,n}=-\frac{e\mathrm{v}_0}{2\pi\varepsilon_0}\frac{1}{\frac{1}{2}R^2J_1^2(p_s)}\frac{1}{g_n d}\frac{1}{\Omega_{s,n}^2-\omega_n^2},\tag{2.5.55}$$

and

$$\omega_n=\frac{\pi n}{d}\mathrm{v}_0.\tag{2.5.56}$$

Since both the magnetic and the electric field are zero at $t=0$, the function $A_{s,n}(t)$ and its first derivative are zero at $t=0$ hence

$$B_1+\alpha_{s,n}=0,\tag{2.5.57}$$

and

$$B_2=0.\tag{2.5.58}$$

Consequently, the amplitude of the magnetic vector potential $[A_{s,n}(t)]$ reads

$$A_{s,n}(t) = \alpha_{s,n}\left[\cos(\omega_n t) - \cos(\Omega_{s,n} t)\right]. \quad (2.5.59)$$

Beyond $t = d/\mathrm{v}_0$, the particle is out of the structure thus the source term in (2.5.50) is zero and the solution reads

$$A_{s,n}\left(t > \frac{d}{\mathrm{v}_0}\right) = C_1 \cos\left[\Omega_{s,n}\left(t - \frac{d}{\mathrm{v}_0}\right)\right] + C_2 \sin\left[\Omega_{s,n}\left(t - \frac{d}{\mathrm{v}_0}\right)\right]. \quad (2.5.60)$$

As in the previous case, at $t = d/\mathrm{v}_0$ both $A_{s,n}(t > d/\mathrm{v}_0)$ and its derivative, have to be continuous:

$$\alpha_{s,n}\left[(-1)^n - \cos\left(\Omega_{s,n}\frac{d}{\mathrm{v}_0}\right)\right] = C_1, \quad (2.5.61)$$

and

$$\alpha_{s,n}\Omega_{s,n} \sin\left(\Omega_{s,n}\frac{d}{\mathrm{v}_0}\right) = C_2\Omega_{s,n}. \quad (2.5.62)$$

For this time-period, the explicit expression for the magnetic vector potential is

$$\begin{aligned} A_{s,n}\left(t > \frac{d}{\mathrm{v}_0}\right) &= \alpha_{s,n}\left[(-1)^n - \cos\left(\Omega_{s,n}\frac{d}{\mathrm{v}_0}\right)\right] \cos\left[\Omega_{s,n}\left(t - \frac{d}{\mathrm{v}_0}\right)\right] \\ &+ \alpha_{s,n} \sin\left(\Omega_{s,n}\frac{d}{\mathrm{v}_0}\right) \sin\left[\Omega_{s,n}\left(t - \frac{d}{\mathrm{v}_0}\right)\right], \end{aligned} \quad (2.5.63)$$

The expressions in (2.5.53), (2.5.59), (2.5.63) describe the magnetic vector potential in the cavity at all times. Figure 2.15 illustrates schematically this solution.

During the period the electron spends in the cavity, there are two frequencies which are excited: the eigen-frequency of the cavity $\Omega_{s,n}$ and the "resonances" associated with the motion of the particle, ω_n. The latter set corresponds to the case when the phase velocity, $\mathrm{v}_{\mathrm{ph}} = \omega/k$, equals the velocity L/R. Since the boundary conditions impose $k = \pi n/d$ and the resonance implies

$$\mathrm{v}_0 = \mathrm{v}_{\mathrm{ph}} = c\left(\frac{\omega}{c}\frac{d}{\pi n}\right), \quad (2.5.64)$$

thus we can immediately deduce the resonance frequencies ω_n as given in (2.5.56).

Now that the magnetic vector potential has been determined, we consider the effect of the field generated in the cavity on the moving particle. The relevant component is the longitudinal one

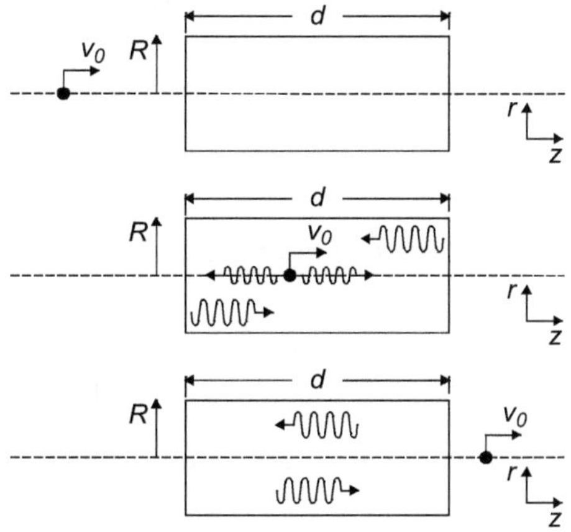

Fig. 2.15 Schematics of the field distribution generated by a particle as it traverse a cavity. Prior to its entrance, no field exists in the cavity. When in the cavity the field has two contributions: directly from the source (non-homogeneous) and reflections from the walls (homogeneous). After the particle leaves the cavity only the homogeneous contribution remains

$$A_z\left(r,z,0<t<\frac{d}{v_0}\right)=\sum_{s=1,n=0}\alpha_{s,n}J_0\left(p_s\frac{r}{R}\right)\cos\left(\frac{\pi n}{d}z\right)\left[\cos(\omega_n t)-\cos(\Omega_{s,n}t)\right]. \tag{2.5.65}$$

Note that we omitted the upper limit in the double summation since in practice, the actual dimensions of the particle, which so far was considered infinitesimally small, determines this limit. In order to quantify this statement we realize that the summation is over all eigenmodes which have a wavenumber much longer than the particle's dimension i.e., $\Omega_{s,n}\Delta_z/c<1$ and $p_sR_b/R<1$wherein Δ_z is the bunch length whereas R_b represents its radius.

According to Maxwell's equations, the longitudinal electric field is

$$\varepsilon_0\frac{\partial}{\partial t}E_z(\mathbf{r},t)=-J_z(\mathbf{r},t)+\frac{1}{r}\frac{\partial}{\partial r}rH_\phi(\mathbf{r},t). \tag{2.5.66}$$

Furthermore, the field that acts on the particle does not include the *self-field*, therefore we omit the current density term. Using the expression for the magnetic vector potential (2.1.32), we have

$$E_z(\mathbf{r},t)=-c^2\int dt\frac{1}{r}\frac{\partial}{\partial r}r\frac{\partial}{\partial r}A_z(\mathbf{r},t), \tag{2.5.67}$$

or explicitly,

$$E_z\left(r,z,0<t<\frac{d}{v_0}\right)=\sum_{s=1,n=0}\alpha_{s,n}\left(\frac{cp_s}{R}\right)^2J_0\left(\frac{p_sr}{R}\right)\cos\left(\frac{\pi n}{d}z\right)\left[\frac{\sin(\omega_n t)}{\omega_n}-\frac{\sin(\Omega_{s,n}t)}{\Omega_{s,n}}\right]. \tag{2.5.68}$$

In a lossless and closed cavity the total power flow is zero, therefore Poynting's theorem in its integral form reads

$$\frac{\mathrm{d}W}{\mathrm{d}t} = -2\pi \int_0^R \mathrm{d}rr \int_0^d \mathrm{d}zE_z(r,z,t)\mathrm{J}_z(r,z,t). \tag{2.5.69}$$

Thus substituting the current density (2.5.47) we obtain

$$W = e\mathrm{v}_0 \int_0^{d/\mathrm{v}_0} \mathrm{d}tE_z(r,z=\mathrm{v}_0 t,t), \tag{2.5.70}$$

which has the following explicit form

$$W = e\mathrm{v}_0 \sum_{s=1,n=0} \alpha_{s,n}\left(\frac{cp_s}{R}\right)^2 \int_0^{d/\mathrm{v}_0} \mathrm{d}t\cos(\omega_n t)\left[\frac{\sin(\omega_n t)}{\omega_n} - \frac{\sin(\Omega_{s,n}t)}{\Omega_{s,n}}\right]. \tag{2.5.71}$$

We can evaluate analytically the time integral in this expression. As can be readily deduced, the first term represents the *non-homogeneous* part of the solution and its contribution is identically zero whereas the second's reads

$$W = -e\mathrm{v}_0 \sum_{s=1,n=0} \alpha_{s,n}\left(\frac{cp_s}{R}\right)^2 \frac{1-(-1)^n\cos(\Omega_{s,n}d/\mathrm{v}_0)}{\Omega_{s,n}^2 - \omega_n^2}. \tag{2.5.72}$$

Substituting the explicit expression for $\alpha_{s,n}$ we have

$$\bar{W} \equiv W\left(\frac{e^2}{4\pi\varepsilon_0 d}\right)^{-1} = \sum_{s=1,n=0}\left(\frac{2p_s}{J_1(p_s)}\right)^2 \frac{1}{g_n}$$
$$\times \frac{1}{[p_s^2 + (\pi nR/d\gamma)]^2}\left[1-(-1)^n\cos\left(\frac{\Omega_{s,n}}{\mathrm{v}_0}d\right)\right]. \tag{2.5.73}$$

Figure 2.16 illustrates the normalized energy excited by a 10 MeV in the first frequencies $\omega < 10\,\Omega_{1,0}$. In this range the spectrum is virtually independent of the particles energy $(\gamma \gg 1)$

$$\bar{W}_{s,n}(\gamma \gg 1) = \left(\frac{2}{J_1(p_s)p_s}\right)^2 \frac{1}{g_n}\left(1-(-1)^n\cos\frac{\Omega_{s,n}}{c}d\right). \tag{2.5.74}$$

The impact of the homogeneous solution (reflections) on the interaction with electrons will be discussed in detail in Chap. 4 in the context of high power traveling wave tubes. Recently, Sotnikov et al. (2009) recognized that the homogeneous solution (quenching wave) may be of the same order of magnitude as the wake generated in high-gradient dielectric wakefield accelerator.

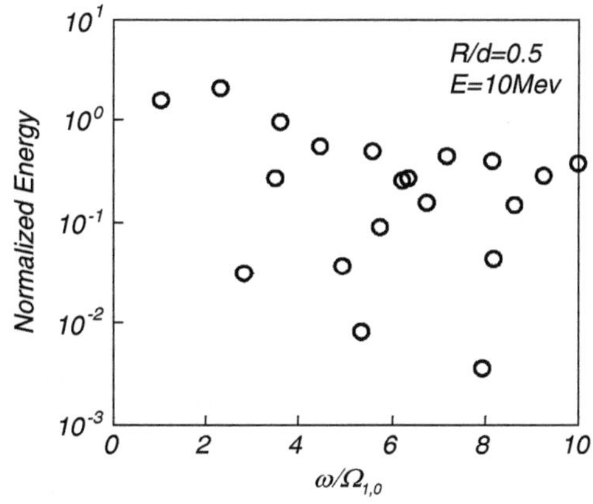

Fig. 2.16 Normalized energy deposited by a 10 MeV point charge in the first modes $(\omega < 10\,\Omega_{1,0})$ of a cylindrical cavity

2.6 Scattered Waves Phenomena

As an electromagnetic wave impinges upon an obstacle, it is scattered. This reflected energy can be harnessed for interaction with charged particles or for measurement purposes. In this section, we consider several cases chosen due to their relative simplicity.

2.6.1 *Plane Wave Scattered by a Dielectric Cylinder*

As a starting point, let us consider a plane wave that propagates in the x direction and it impinges upon a dielectric (ε_{cyl}) cylinder of radius R whose axis is parallel to the magnetic field component of the incident wave

$$H_z^{(\text{inc})}(x) = H_0 \exp\left(-j\frac{\omega}{c}\sqrt{\varepsilon_{\text{bg}}}x\right); \tag{2.6.1}$$

tacitly assuming a steady state regime $\exp(j\omega t)$ and the background medium is characterized by a dielectric coefficient ε_{bg} – see Fig. 2.17. Based on the generating Bessel function (Abramowitz and Stegun 1968, p. 361)

$$\exp\left[\frac{1}{2}u\left(v - \frac{1}{v}\right)\right] = \sum_{n=-\infty}^{\infty} v^n \mathrm{J}_n(u) \tag{2.6.2}$$

this incident component may be written in cylindrical coordinates ($x = r\cos\phi$) as

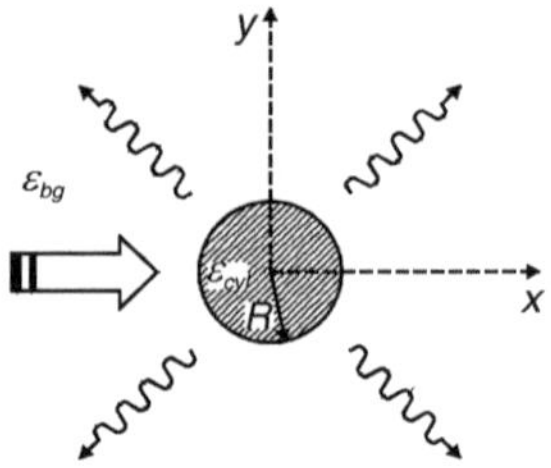

Fig. 2.17 A plane wave scattered by a dielectric cylinder

$$\begin{aligned} H_z^{(\mathrm{inc})}(r,\phi) &= H_0 \exp\left(-j\frac{\omega}{c}\sqrt{\varepsilon_{\mathrm{bg}}}r\cos\phi\right) \\ &= H_0 \sum_{n=-\infty}^{\infty} \exp\left[jn\left(\phi-\frac{\pi}{2}\right)\right] \mathrm{J}_n\left(\frac{\omega}{c}\sqrt{\varepsilon_{\mathrm{bg}}}r\right). \end{aligned} \tag{2.6.3}$$

The presence of the cylinder alters the electromagnetic field thus the secondary field is given by

$$H_z^{(\mathrm{sec})}(r,\phi) = H_0 \sum_{n=-\infty}^{\infty} \exp\left[jn\left(\phi-\frac{\pi}{2}\right)\right] \begin{cases} \tau_n \mathrm{J}_n\left(\frac{\omega}{c}\sqrt{\varepsilon_{\mathrm{cyl}}}r\right) & r \le R \\ \rho_n \mathrm{H}_n^{(2)}\left(\frac{\omega}{c}\sqrt{\varepsilon_{\mathrm{bg}}}r\right) & r \ge R \end{cases} \tag{2.6.4}$$

For imposing the boundary conditions, it is necessary to specify the azimuthal electric field

$$\begin{aligned} E_\phi^{(\mathrm{inc})}(r,\phi) &= \eta_0 H_0 \sum_{n=-\infty}^{\infty} \exp\left[jn\left(\phi-\frac{\pi}{2}\right)\right] \frac{j}{\sqrt{\varepsilon_{\mathrm{bg}}}} \dot{\mathrm{J}}_n\left(\frac{\omega}{c}\sqrt{\varepsilon_{\mathrm{bg}}}r\right) \\ E_\phi^{(\mathrm{sec})}(r,\phi) &= \eta_0 H_0 \sum_{n=-\infty}^{\infty} \exp\left[jn\left(\phi-\frac{\pi}{2}\right)\right] \begin{cases} \tau_n \frac{j}{\sqrt{\varepsilon_{\mathrm{cyl}}}} \dot{\mathrm{J}}_n\left(\frac{\omega}{c}\sqrt{\varepsilon_{\mathrm{cyl}}}r\right) & r \le R \\ \rho_n \frac{j}{\sqrt{\varepsilon_{\mathrm{bg}}}} \dot{\mathrm{H}}_n^{(2)}\left(\frac{\omega}{c}\sqrt{\varepsilon_{\mathrm{bg}}}r\right) & r > R \end{cases} \end{aligned} \tag{2.6.5}$$

Continuity of the two components facilitate to determine the amplitudes

$$\begin{aligned} \rho_n &= \frac{b\dot{\mathrm{J}}_n(b)\dot{\mathrm{J}}_n(a) - a\dot{\mathrm{J}}_n(b)\mathrm{J}_n(a)}{a\dot{\mathrm{J}}_n(b)\mathrm{H}_n^{(2)}(a) - b\mathrm{J}_n(b)\dot{\mathrm{H}}_n^{(2)}(a)}, \\ \tau_n &= b\frac{\dot{\mathrm{J}}_n(a)\mathrm{H}_n^{(2)}(a) - \mathrm{J}_n(a)\dot{\mathrm{H}}_n^{(2)}(a)}{a\dot{\mathrm{J}}_n(b)\mathrm{H}_n^{(2)}(a) - b\mathrm{J}_n(b)\dot{\mathrm{H}}_n^{(2)}(a)} \end{aligned} \tag{2.6.6}$$

where $a = \sqrt{\varepsilon_{\mathrm{bg}}}\omega R/c$ and $b = \sqrt{\varepsilon_{\mathrm{cyl}}}\,\omega R/c$. With the amplitudes established, two measures need to be considered. The first is the extent the cylinder scatters the wave namely, the *scattering cross-section*. For this purpose we determine the total power scattered in the cylindrical envelope of radius $r \gg R$ and height Δ_z

$$\begin{aligned}P_{\text{scatt}} &= \Delta_z r \int_0^{2\pi} d\phi \text{Re}\left\{\frac{1}{2} E_\phi^{(\text{sec})}(r,\phi)\left[H_z^{(\text{sec})}(r,\phi)\right]^*\right\}\\ &= \Delta_z 2\pi r\left(\frac{\eta_0 |H_0|^2}{2\sqrt{\varepsilon_{\text{bg}}}}\right)\sum_{n=-\infty}^{\infty}|\rho_n|^2 \text{Re}\left\{j\left[\text{H}_n^{(2)}\left(\frac{\omega}{c}\sqrt{\varepsilon_{\text{bg}}}r\right)\right]^* \dot{\text{H}}_n^{(2)}\left(\frac{\omega}{c}\sqrt{\varepsilon_{\text{bg}}}r\right)\right\}\\ &= \Delta_z 2\pi\left(\frac{\eta_0|H_0|^2}{2\varepsilon_{\text{bg}}}\right)\frac{2}{\pi\frac{\omega}{c}}\sum_{n=-\infty}^{\infty}|\rho_n|^2. \end{aligned} \tag{2.6.7}$$

The scattering cross section is defined by the ratio of the scattered power and the impinging energy flux, $S_x = \frac{1}{2}\frac{\eta_0}{\sqrt{\varepsilon_{\text{bg}}}}|H_0|^2$,

$$\sigma_{\text{scatt}} \equiv \frac{P_{\text{scatt}}}{S_x} = (2R\Delta_z)\frac{2}{a}\sum_{n=-\infty}^{\infty}|\rho_n|^2. \tag{2.6.8}$$

Figure 2.18 illustrates the normalized cross-section ($\bar{\sigma} = \sigma_{\text{scatt}}/2R\Delta_z$) as a function of the frequency.

It reveals the evident resonant character of the cross-section: for a dielectric coefficient $\varepsilon_{\text{cyl}} = 3.3$ and using $N = 100$ azimuthal harmonics, the cross section is almost 4 for $R \simeq 0.4\lambda_0$ and close to unity if the radius is $R \simeq 0.72\lambda_0$. Moreover, if due to dielectric loss, part of the reflected power is absorbed, the effective cross-section is systematically *smaller* than the lossless case.

The opposite is the case if the medium is active, as illustrated in Fig. 2.19. Since multiple reflections in the cylinder may enhance significantly the scattered power, the cross-section may be *larger*. In the figure the range between $0.9 < R/\lambda_0 < 1.2$ has been magnified, and for the specific parameters, the normalized cross section at

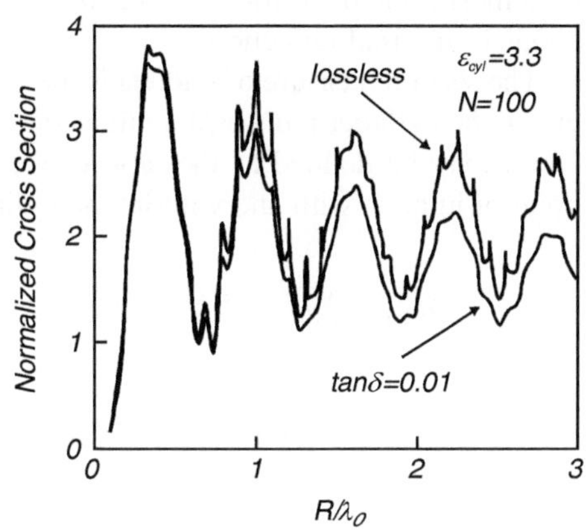

Fig. 2.18 Normalized cross section as a function of the radius normalized to the wavelength in vacuum. Comparing to the lossless case, the cross section in case of a lossy cylinder is systematically smaller since part of the power is absorbed. As a rough estimate $\bar{\sigma}_{\text{lossy}} \sim \bar{\sigma}_{\text{lossless}} \exp[-2\pi(R/\lambda_0)\sqrt{\varepsilon_{\text{cyl}}}\tan\delta]$

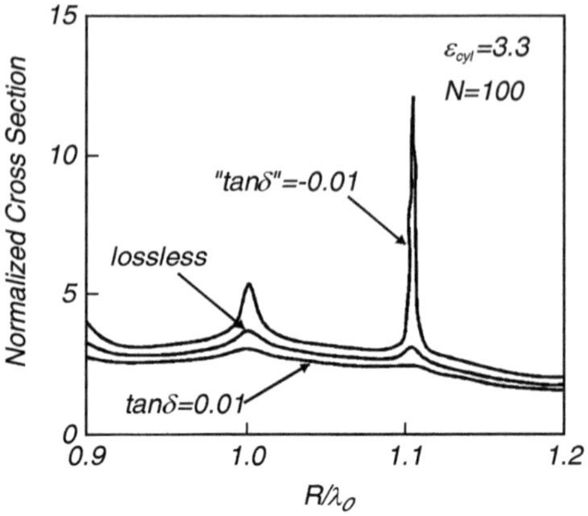

Fig. 2.19 Normalized cross section as a function of the radius normalized to the wavelength in vacuum – focusing in the range $0.9 < R/\lambda_0 < 1.2$. If the medium is active, the cross section can be significantly larger comparing to the lossless case

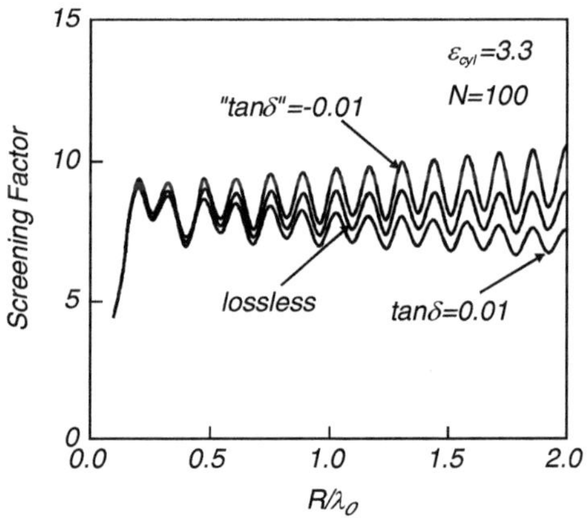

Fig. 2.20 Screening factor as a function of the radius normalized to the wavelength in vacuum. If the medium is active, the cross section can be significantly larger comparing to the lossless case

$R \simeq 1.1\lambda_0$ has dropped from 3.1 to 2.5 due to the dielectric loss but it has increased by almost a factor of four $\bar{\sigma} \sim 12$ in case of an active medium – corresponding to about 11 internal reflections.

The second measure of interest is the *screening factor*, which is indicative of the extent the cylinder reduces/magnifies the electromagnetic energy in its center. This factor may be defined as the ratio of the electromagnetic energy densities at the point of interest with and without the cylinder namely,

$$S \equiv 10\log\left[\frac{w_E^{(\mathrm{tr})}(r=0) + w_M^{(\mathrm{tr})}(r=0)}{w_E^{(\mathrm{inc})}(r=0) + w_M^{(\mathrm{inc})}(r=0)}\right] = 10\log\left(|\tau_0|^2 + |\tau_1|^2\right) \qquad (2.6.9)$$

For the parameters mentioned above, the screening factor is illustrated in Fig. 2.20 and evidently, the fiber tends to focus the electromagnetic energy. As may be expected, this focusing is suppressed by lossy material and it is amplified by active medium. In addition, we observe that lossy or gain medium do not alter the resonant pattern (peaks) associated with azimuthally propagating modes.

2.6.2 Evanescent Waves Scattered by a Dielectric Cylinder

In many cases of interest, waves attached to moving charges are scattered by various obstacles and these radiating modes may be used for the characterization of bunches of electrons as well as of the obstacle. We now exploit the relatively simple configuration of the dielectric cylinder in order to examine the scattering of evanescent waves attached to a charged line (Q/Δ_z) moving with a velocity v at a height $h > R$ – see Fig. 2.21. Near the cylinder the incident field is given by

$$\begin{aligned}
H_z^{(\text{inc})}(x, y < h, t) &= -\frac{Q}{4\pi\Delta_z}\int_{-\infty}^{\infty} d\omega \exp\left[j\omega\left(t - \frac{x}{\text{v}}\right) - \frac{|\omega|}{\gamma \text{v}}(h - y)\right] \\
&= -\frac{Q}{4\pi\Delta_z}\int_{-\infty}^{\infty} d\omega \exp(j\omega t)\sum_{n=-\infty}^{\infty}\mathfrak{I}_n(r, \omega, \text{v})\exp(jn\phi) \\
\mathfrak{I}_n(r, \omega, \text{v}) &\equiv \frac{1}{2\pi}\int_{-\pi}^{\pi} d\phi \exp\left[-jn\phi - j\frac{\omega}{c}r\cos\phi - \frac{|\omega|}{\gamma \text{v}}(h - r\sin\phi)\right]
\end{aligned} \tag{2.6.10}$$

Similar to the approach from the above,

$$\begin{aligned}
H_z^{(\text{sec})}(r, \phi, t) = &-\frac{Q}{4\pi\Delta_z}\int_{-\infty}^{\infty} d\omega \exp(j\omega t) \\
&\times \sum_{n=-\infty}^{\infty}\exp(jn\phi)\begin{cases}\tau_n(\omega)\text{J}_n\left(\frac{\omega}{c}\sqrt{\varepsilon_{\text{cyl}}}r\right) & r \le R \\ \rho_n(\omega)\text{H}_n^{(2)}\left(\frac{\omega}{c}\sqrt{\varepsilon_{\text{bg}}}r\right) & r \ge R\end{cases}
\end{aligned} \tag{2.6.11}$$

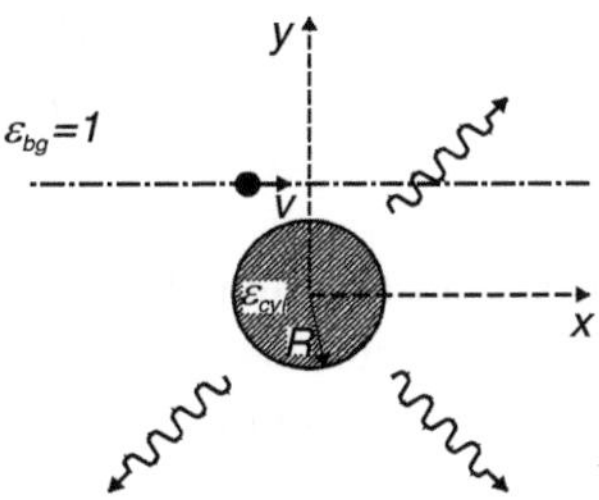

Fig. 2.21 Evanescent waves attached to a moving charged-line are scattered by a cylinder

and

$$E_{\phi}^{(\mathrm{inc})} = -\frac{Q}{4\pi\Delta_z}\int_{-\infty}^{\infty} d\omega \exp(j\omega t)\frac{-1}{j\omega\varepsilon_0\varepsilon_{\mathrm{bg}}}\sum_{n=-\infty}^{\infty}\partial_r\Im_n(r,\omega,\mathrm{v})\exp(jn\phi)$$

$$E_{\phi}^{(\mathrm{sec})} = -\frac{Q}{4\pi\Delta_z}\int_{-\infty}^{\infty} d\omega \exp(j\omega t)\frac{-1}{j\omega\varepsilon_0} \times \sum_{n=-\infty}^{\infty}\exp(jn\phi)\begin{cases}\frac{1}{\varepsilon_{\mathrm{cyl}}}\tau_n\left(\frac{\omega}{c}\sqrt{\varepsilon_{\mathrm{cyl}}}\right)\dot{\mathrm{J}}_n\left(\frac{\omega}{c}\sqrt{\varepsilon_{\mathrm{cyl}}}r\right) & r \le R\\ \frac{1}{\varepsilon_{\mathrm{bg}}}\rho_n\left(\frac{\omega}{c}\sqrt{\varepsilon_{\mathrm{bg}}}\right)\dot{\mathrm{H}}_n^{(2)}\left(\frac{\omega}{c}\sqrt{\varepsilon_{\mathrm{bg}}}r\right) & r \ge R\end{cases} \tag{2.6.12}$$

Thus imposing the boundary conditions we obtain the reflection and transmission coefficients

$$\rho_n = \frac{b\mathrm{J}_n(b)\dot{\Im}_n - a\dot{\mathrm{J}}_n(b)\Im_n}{a\dot{\mathrm{J}}_n(b)\mathrm{H}_n^{(2)}(a) - b\mathrm{J}_n(b)\dot{\mathrm{H}}_n^{(2)}(a)},$$
$$\tau_n = \frac{b\mathrm{H}_n^{(2)}(a)\dot{\Im}_n - b\dot{\mathrm{H}}_n^{(2)}(a)\Im_n}{a\dot{\mathrm{J}}_n(b)\mathrm{H}_n^{(2)}(a) - b\mathrm{J}_n(b)\dot{\mathrm{H}}_n^{(2)}(a)} \tag{2.6.13}$$

wherein $\dot{\Im}_n(R,\omega,\mathrm{v}) \equiv [R\partial_r\Im_n(r,\omega,\mathrm{v})]_{r=R}/a$, $a = \omega\sqrt{\varepsilon_{\mathrm{bg}}}R/c$, $b = \omega\sqrt{\varepsilon_{\mathrm{cyl}}}R/c$. With the field established, it is possible to determine the emitted energy during the passage of the charged-line near the cylinder

$$W = \int_{-\infty}^{\infty} dt P(t) = \Delta_z r\int_{-\pi}^{\pi} d\phi\int_{-\infty}^{\infty} dt\, E_{\phi}^{(\mathrm{sec})}(r,\phi,t)H_z^{(\mathrm{sec})}(r,\phi,t) = \frac{Q^2}{2\pi\varepsilon_0\Delta_z}\int_{-\infty}^{\infty} da\frac{1}{a}\sum_{n=-\infty}^{\infty}|\rho_n|^2 \tag{2.6.14}$$

which enables us to write the following expression for the normalized spectrum

$$\frac{d\bar{W}}{da} \equiv \frac{1}{\frac{Q^2}{2\pi\varepsilon_0\Delta_z}}\frac{dW}{da} = \frac{2}{a}\sum_{n=-\infty}^{\infty}|\rho_n|^2; \tag{2.6.15}$$

it should be pointed out that it has been assumed that the charged-line moves in vacuum thus, $\varepsilon_{\mathrm{bg}} = 1$. If the single line charge is replaced by train of M micro-bunches of length $\Delta_{\|}$ and thickness $\Delta_{\perp}$ with a spacing L between each two micro-bunches then the normalized spectrum is

$$\frac{d\bar{W}}{da} = \frac{2}{a}\sum_{n=-\infty}^{\infty}|\rho_n|^2\left[\mathrm{sinc}^2\left(\frac{1}{2}\frac{a}{\beta}\frac{\Delta_{\|}}{R}\right)\frac{\mathrm{sinc}^2\left(\frac{1}{2}M\frac{a}{\beta}\frac{L}{R}\right)}{\mathrm{sinc}^2\left(\frac{1}{2}\frac{a}{\beta}\frac{L}{R}\right)}\mathrm{sinhc}^2\left(\frac{1}{2}\frac{a}{\gamma\beta}\frac{\Delta_{\perp}}{R}\right)\right]. \tag{2.6.16}$$

Note that the total amount of charge remains Q. For simplicity sake, we assume for what follows that $\varepsilon_{\mathrm{cyl}} \to \infty$ namely, the field does not penetrate in the cylinder therefore $\rho_n = -\dot{\mathfrak{J}}_n/\dot{\mathrm{H}}_n^{(2)}(a)$ and also that the bunch is ultra-relativistic ($\gamma \to \infty$) or if to be more accurate, $\frac{a}{\gamma}\frac{\Delta_{\perp}}{R}, \frac{a}{\gamma}\frac{h}{R} \ll 1$. Since the last two transverse geometric parameters $\frac{\Delta_{\perp}}{R}, \frac{h}{R}$ are expected to be at the most of the order of unity, the previous condition limits the spectrum of our approximation to $a < a_{\mathrm{co}} \equiv 0.01\gamma$. Another implication is a significantly simpler expression for

$$\dot{\mathfrak{J}}_n \equiv \frac{d}{da}\left[\frac{1}{2\pi}\int_{-\pi}^{\pi} d\phi \exp(-jn\phi - ja\cos\phi)\right] = \dot{\mathrm{J}}_n(a)\exp\left(-jn\frac{\pi}{2}\right) \tag{2.6.17}$$

which finally implies

$$\begin{aligned}\frac{d\bar{W}}{da} &= \left[\frac{2}{a}\sum_{n=-\infty}^{\infty}\frac{\dot{\mathrm{J}}_n^2(a)}{\dot{\mathrm{J}}_n^2(a)+\dot{\mathrm{Y}}_n^2(a)}\right]\mathrm{sinc}^2\left(\frac{1}{2}\frac{\Delta_{\|}}{R}a\right)\frac{\mathrm{sinc}^2\left(\frac{M}{2}\frac{L}{R}a\right)}{\mathrm{sinc}^2\left(\frac{1}{2}\frac{L}{R}a\right)} \\ &\simeq 1.345[1-\exp(-2.5a)]\mathrm{sinc}^2\left(\frac{1}{2}\frac{\Delta_{\|}}{R}a\right)\frac{\mathrm{sinc}^2\left(\frac{M}{2}\frac{L}{R}a\right)}{\mathrm{sinc}^2\left(\frac{1}{2}\frac{L}{R}a\right)}.\end{aligned} \tag{2.6.18}$$

Evidently, the first term represents the contribution of the ideal line-charge whereas the two trailing terms represent the single bunch size effect and the multiple bunches impact respectively.

Before concluding this subsection, two comments are in place. First, the configuration considered above illustrates the coupling between the evanescent waves attached to the moving charged-line and the propagating waves scattered by cylinder. Due to the resemblance to regular diffraction, the emerging waves are also referred to as *diffraction radiation*. Second, we need to provide an alternative interpretation of the emitted spectral density as manifested in (2.6.16). From the way it has been developed, it obviously characterizes the *radiative* contribution far away from the charged-line. Based on Poynting theorem, the source of this radiation is the effect of the secondary field on the charged-line

$$
\begin{aligned}
W &= \int_{-\infty}^{\infty} dt P(t) = \int_{-\infty}^{\infty} dt \Delta_z \int_{-\infty}^{\infty} dx \int_{-\infty}^{\infty} dy E_x^{(\mathrm{sec})}(x, y, t) J_x(x, y, t) \\
&= -Q \int_{-\infty}^{\infty} dx \int_{-\infty}^{\infty} d\omega \exp\left(j \frac{\omega}{\mathrm{v}} x\right) E_x^{(\mathrm{sec})}(x, h; \omega)
\end{aligned}
\tag{2.6.19}
$$

Using the normalization employed above we get

$$
\frac{d\bar{W}}{da} = \bar{Z}_{\|} \equiv -\frac{2\pi}{R\eta_0} \left[\frac{1}{Q/\Delta_z} \int_{-\infty}^{\infty} dx \exp\left(j \frac{\omega}{\mathrm{v}} x\right) E_x^{(\mathrm{sec})}(x, h; \omega) \right] \tag{2.6.20}
$$

which clearly reveals that the spectral density of the emitted energy is proportional to the spatial Fourier transform of the electric field *as experienced by the charged-line*. It can be readily checked that the square brackets have units of ohm-meter and consequently, the normalized spectral density equals the, so called, (normalized) *longitudinal impedance* experienced in this case by the charged-line.

2.6.3 Evanescent Waves Scattered by a Metallic Wedge

Diffraction radiation is commonly employed by the particle accelerator community for characterizing the location and to some extent the shape of a charged bunch. This is generally done with thin metallic foils. A model for describing the system, as in the cylinder case, consists of a charged-line (Q/Δ_z) moving at a constant velocity v at a height h above the tip of an ideal wedge ($2\pi - \alpha_2 < \phi < 2\pi - \alpha_1$) – see Fig. 2.22. In the frequency domain, the magnetic field is a solution of

$$
\left[\frac{1}{r} \frac{\partial}{\partial r} r \frac{\partial}{\partial r} + \frac{1}{r^2} \frac{\partial^2}{\partial \phi^2} + \frac{\omega^2}{c^2} \right] H_z(r, \phi, \omega) = \left[\frac{\partial J_x}{\partial y} \right]_{\substack{y = r \sin\phi \\ x = r\cos\phi}} \tag{2.6.21}
$$

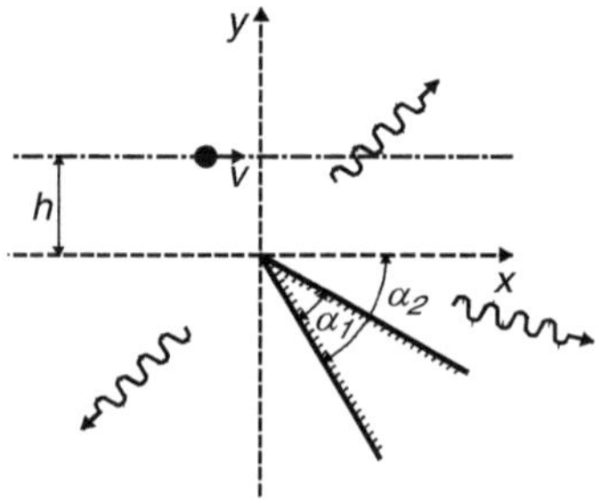

Fig. 2.22 Evanescent waves attached to a moving charged-line are scattered by a wedge

Obviously the radial electric field is zero on both sides of the wedge therefore it is natural to employ the orthogonality of the corresponding azimuthal eigenfunctions namely,

$$H_z(r,\phi,\omega)=\sum_{n=0}^{\infty}H_n(r)\cos[n\pi(\phi+\alpha_1)/(2\pi-\alpha_2+\alpha_1)], \tag{2.6.22}$$

to get

$$\left[\frac{1}{r}\frac{d}{dr}r\frac{d}{dr}-\frac{v^2}{r^2}+\frac{\omega^2}{c^2}\right]H_n(r)=S_n(r)\equiv\frac{1}{g_n}\frac{1}{2\pi-\alpha_2+\alpha_1}\int_{-\alpha_1}^{2\pi-\alpha_2}d\phi\frac{\partial J_x}{\partial y}\cos[v(\phi+\alpha_1)] \tag{2.6.23}$$

with $v=n\pi/(2\pi-\alpha_2+\alpha_1)$. The source term may be simplified

$$S_n(r)=\frac{Q}{2\pi\Delta_z}\frac{1}{g_n}\frac{(-v)}{2\pi-\alpha_2+\alpha_1}\frac{1}{r}\int_{-\alpha_1}^{2\pi-\alpha_2}d\phi\exp\left(j\frac{\omega}{\mathrm{v}}r\cos\phi\right)$$
$$\times\sin[v(\phi+\alpha_1)]\cos\phi\,\delta(y-h) \tag{2.6.24}$$

Next, we define $\phi_0=\arcsin(h/r)$ and take advantage of the Dirac delta function

$$\delta(y-h)=\frac{u(r-h)}{\sqrt{r^2-h^2}}[\delta(\phi-\phi_0)+\delta(\phi-\pi+\phi_0)] \tag{2.6.25}$$

Wherein $u(x)$ denotes the Heaviside step function thus

$$S_n(r)=\frac{-Q}{2\pi\Delta_z}\frac{v}{g_n}\frac{u(r-h)}{2\pi-\alpha_2+\alpha_1}\frac{1}{r^2}\left\{\exp\left(j\frac{\omega}{\mathrm{v}}\sqrt{r^2-h^2}\right)\sin[v(\phi_0+\alpha_1)]\right.$$
$$\left.-\exp\left(-j\frac{\omega}{\mathrm{v}}\sqrt{r^2-h^2}\right)\sin[v(\pi-\phi_0+\alpha_1)]\right\}. \tag{2.6.26}$$

For a solution of (2.6.23) we employ the corresponding Green's function

$$\mathrm{G}_v(r,r')=j\frac{\pi}{2}\begin{cases}\mathrm{J}_v\left(\frac{\omega}{c}r'\right)\mathrm{H}_v^{(2)}\left(\frac{\omega}{c}r\right) & r>r'\\ \mathrm{H}_v^{(2)}\left(\frac{\omega}{c}r'\right)\mathrm{J}_v\left(\frac{\omega}{c}r\right) & r<r'\end{cases} \tag{2.6.27}$$

and can formally determine the magnetic field

$$H_z(r,\phi,t)=\int_0^{\infty}d\omega\sum_{n=0}^{\infty}\cos[v(\phi+\alpha_1)]\,2\mathrm{Re}\left\{\exp(j\omega t)\int_0^{\infty}dr'\,r'\,\mathrm{G}_v(r,r')S_n(r')\right\} \tag{2.6.28}$$

With this field component established we may proceed and evaluate the *radiated energy* at $r \to \infty$

$$W = \Delta_z r \int_{-\infty}^{\infty} dt \int_{-\alpha_1}^{2\pi-\alpha_2} d\phi E_\phi(r,\phi,t) H_z(r,\phi,t)$$
$$= \eta_0 2\pi\left(\frac{\pi}{2}\right)(2\pi-\alpha_2+\alpha_1)\Delta_z 2\int_0^\infty d\omega \sum_{n=0}^{\infty} g_n \frac{1}{\frac{\omega}{c}}\left|\int_0^\infty dr' r' \mathrm{J}_\nu\left(\frac{\omega}{c}r'\right) S_n(r')\right|^2 \tag{2.6.29}$$

Defining $\Omega = \omega h/c$, $\bar{W} = W[Q^2/2(2\pi-\alpha_2+\alpha_1)\varepsilon_0\Delta_z]^{-1}$ and $\xi = r/h$ the normalized spectral density is given by

$$\frac{d\bar{W}}{d\Omega} = \frac{2}{\Omega}\sum_{n=1}^{\infty}\left|\int_1^\infty d\xi \frac{\nu \mathrm{J}_\nu(\Omega\xi)}{\xi}\left\{\begin{array}{l}\exp\left(j\frac{\Omega}{\beta}\sqrt{\xi^2-1}\right)\sin[\nu(\arcsin(1/\xi)+\alpha_1)] \\ -\exp\left(-j\frac{\Omega}{\beta}\sqrt{\xi^2-1}\right)\sin[\nu(\pi-\arcsin(1/\xi)+\alpha_1)]\end{array}\right\}\right|^2 \tag{2.6.30}$$

or finally, after defining $\xi = 1/\sin\psi$ we obtain

$$\frac{d\bar{W}}{d\Omega} = \frac{2}{\Omega}\sum_{n=1}^{\infty}\left|\int_0^{\pi/2} d\psi \frac{\nu \mathrm{J}_\nu\left(\frac{\Omega}{\sin\psi}\right)}{\tan\psi}\left\{\begin{array}{l}\exp\left(j\frac{\Omega}{\beta}\cot\psi\right)\sin[\nu(\psi+\alpha_1)] \\ -\exp\left(-j\frac{\Omega}{\beta}\cot\psi\right)\sin[\nu(\pi-\psi+\alpha_1)]\end{array}\right\}\right|^2. \tag{2.6.31}$$

Several observations are in place now that we have an explicit expression for the energy's spectral density. First, note that if the velocity is reversed $\mathrm{v} \to -\mathrm{v}$, this is equivalent to taking the complex conjugate of the term in the curled brackets, the spectral density is *invariant*. At the limit of very low frequencies, the spectral density emitted by a relativistic bunch ($\beta \sim 1$) is *inversely proportional to the frequency* as can be concluded from Fig. 2.23 where we plot this quantity as a function of the normalized frequency(Ω). It should be pointed out that the first twenty harmonics were considered and for fast convergence, the integration was performed in the range $\varepsilon < \psi < \pi/2 - \varepsilon$ where $\varepsilon = 0.0005\pi$. As reference, we also present the first two harmonics ($n = 1, 2$).

Examining the same quantity as a function of the normalized momentum of the bunch we conclude that – see Fig. 2.24 – for $\gamma\beta < 1$ the spectral density is proportional to β^4 (for $\Omega = 1$) however, this dependence is strongly dependent on the normalized frequency. For example, for $\Omega = 0.2$ the spectral density is proportional to $\gamma\beta\left[1 + (\gamma\beta/15)^4\right]^{-1/4}$.

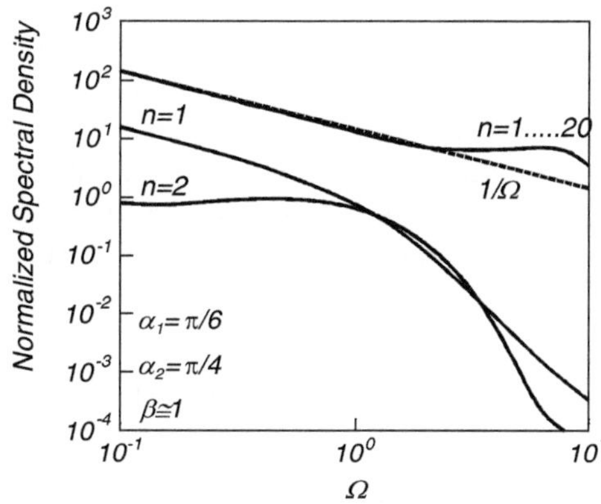

Fig. 2.23 Contribution of the first 20 harmonics to the normalized spectral density as a function of the normalized frequency; $\alpha_1 = \pi/6$, $\alpha_2 = \pi/4$ and $\beta = 1$. The dashed line clearly reveals that this quantity is inversely proportionnal to the frequency for $\Omega < 1$. As reference the first two harmonics are also presented

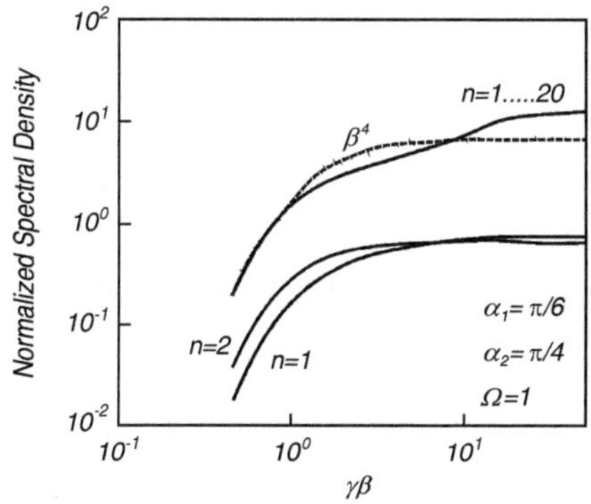

Fig. 2.24 Contribution of the first 20 harmonics to the normalized spectral density as a function of the normalized momentum; $\alpha_1 = \pi/6$, $\alpha_2 = \pi/4$ and $\Omega = 1$. For a relativistic bunch each harmonic reaches it asymptotic value at a different momentum. As reference, the first two harmonics are also plotted. For $\Omega = 1$ the dashed line clearly reveals that the spectral density is proportional to β^4; this dependence may be quite different at other frequencies

Although only the $n = 1$ and $n = 2$ harmonics are illustrated, we found that all harmonics reach an asymptotic value. In fact, this conclusion can be readily deduced from the fact that the energy spectral density depends on β and the latter approaches unity at high kinetic energy. The bending point depends both on the normalized frequency the geometry of the wedge as well as on the harmonic's index.

Exercises

2.1 Determine the boundary condition associated with charge conservation. How it relates to (2.1.12)–(2.1.15)?

2.2 In the context of Sect. 2.2.4, calculate the electromagnetic field associated with the moving charge (2.2.21)–(2.2.22). Calculate the

(continued)

Poynting vector associated with this field. With this result, calculate the total power. Is there a force acting on the moving particle?

2.3 Show that the power radiated in free space by the current distribution in (2.3.2) is given by $P = \eta_0 I^2 (\omega d/c)^2 / 12\pi$.

2.4 By virtue of the superposition principle, show that in case of multiple "wires" carrying currents I_ν located at (x_ν, y_ν) between the two plates of a radial transmission line the magnetic vector potential is given by

$$A_z = \sum_\nu I_\nu \frac{1}{4j} H_0^{(2)} \left[\frac{\omega}{c} \sqrt{(x - x_\nu)^2 + (y - y_\nu)^2} \right]$$

2.5 Calculate the energy velocity (Sect. 2.3.3) assuming two modes TM_{01} and TM_{02} above cut-off. Plot the energy velocity as a function of the ratio of the two modes $0.3 < \rho = |A_{01}/A_{02}| < 3.0$.

2.6 Calculate the radiation impedance of the TM_{01} in a circular waveguide of radius R. Assume a current distribution

$$J_r(r, z, \omega) = I \Delta_z \mathrm{J}_1(p_1 r/R) \delta(z) p_1 / 2\pi r R.$$

2.7 The expression that determines the magnetic vector potential of N electrons moving in a dielectric medium ε_r at a velocity v

$$A_z(r, z, \omega) = -\frac{e\mu_0}{(2\pi)^2} \mathrm{K}_0 \left(\frac{\omega}{c} r \sqrt{\frac{c^2}{\mathrm{v}^2} - \varepsilon_r} \right) \sum_{i=1}^{N} \exp\left[-j \frac{\omega}{\mathrm{v}} (z - z_i) \right]$$

Assuming that the electrons are uniformly distributed $|z_i| \leqslant \Delta/2$, calculate the power emitted by this bunch. Show that the power emitted in the range $\lambda > \Delta$ is proportional to N^2. What happens in the range $\lambda < \Delta$? Repeat the exercise for a Gaussian distribution.

2.8 In Sect. 2.4 we have demonstrated that

$$\frac{P_{\mathrm{av}}}{\dfrac{(eN)^2 \mathrm{v}}{2\pi\varepsilon_0 R^2}} = \frac{1/\varepsilon_r}{\varepsilon_r \beta^2 - 1} \sum_s \left[\frac{2\mathrm{J}_1(p_s R_b/R)}{(p_s R_b/R)\mathrm{J}_1(p_s)} \operatorname{sinc}\left(\frac{p_s}{2} \frac{\Delta/R}{\sqrt{\varepsilon_r \beta^2 - 1}} \right) \right]^2$$

Determine the condition(s) necessary to suppress the radiation excited in specific mode(s). Can you ensure zero power in many modes? Hint: for large values of s, $p_{s+1} - p_s \sim \pi$.

2.9 Draw the normalized average power (2.4.58) for $M = 100$ normalized to $M = 1$ as a function of L/R as in Fig. 2.7. Show that for specific values of L/R, this quantity is of order of unity. Analyze the spectrum in a few of these cases.

(continued)

2.10 Calculate Green's function associated with the system described in Sect. 2.5.1. Begin with case when the source is located to the left of the discontinuity and continue by solving the problem when the source is between the discontinuities. Can you deduce the Green's function for the third case, when the source is after the third discontinuity, from the first one. For the second case (point-source between the discontinuities) can you design the system such that the source emits zero power?

2.11 Based on (2.5.40), analyze the spectrum of the emitted radiation as a point-charge traverses a geometric discontinuity. Keep the ratio of the number of modes in each region proportional to the radii ratio.

2.12 Based on the formulation of the wake generated by a point charge in a loss-less cavity (Sect. 2.5.3), determine the spectrum of train of N point charges generated in the same cavity. Assume that the spacing between two adjacent charges is L.

2.13 Repeat the steps in Sect. 2.6.1 for the orthogonal polarization, $E_z = E_0 \exp\left(-j\frac{\omega}{c}\sqrt{\varepsilon_{bg}}x\right)$.

2.14 Analyze the normalized spectrum density in (2.6.16) as a function of the various parameters.

2.15 Plot the contour of constant far-field emitted energy density from a wedge (Sect. 2.6.3) for several values of the kinetic energy; $\alpha_1 = \pi/6$, $\alpha_2 = \pi/4$, $\Omega = 1$ and $\gamma = 2, 11, 21, 31$.

2.16 Extend (2.6.31) to the case of a train of M micro-bunches of spacing L and the length of each one is Δ_x whereas the thickness is Δ_y.

Chapter 3
Elementary Electron Dynamics

There are numerous topics regarding electron's dynamics, which can and probably should be discussed as background to the investigation of distributed electron-wave interaction. Among these, a fraction will be presented here with emphasis on basic concepts that are of relevance to the principles to be elaborated in the next chapters. All topics considered throughout the text rely on classical mechanics (Sect. 3.1) and without exception they are consistent with the special theory of relativity (Sect. 3.2), therefore the fundamentals of these two theories are summarized.

Beyond reviewing the fundamental concepts of relativistic classical dynamics, we consider in Sect. 3.3 some of the methods of electron generation and discuss the Child-Langmuir law which draws a limit on the maximum current achievable when applying a voltage on a cathode. After electrons are generated, they are typically guided by magnetic fields and waveguides to the interaction region. In Sect. 3.4 we present some basics of beam propagation in free space with uniform or periodic magnetic field. The section concludes with the basic measures of beam quality: emittance and brightness.

Section 3.5 is dedicated to space-charge waves. After introducing the basic concepts of fast and slow space-charge waves, we consider two instabilities that can develop when these waves are present. One is the resistive wall instability and the other is the two-beam instability. Interference of two space-charge waves is shown to play the crucial role in relativistic klystrons. Chapter 3 concludes with a brief discussion on radiation from moving charges (Sect. 3.6).

3.1 Classical Dynamics

In a substantial fraction of the interaction schemes, it is sufficient to describe the electron-wave interaction in the framework of classical mechanics and for this reason, we shall not discuss here quantum mechanic effects. The classical approach includes either the Newtonian equation of motion, or Lagrangian or Hamiltonian formalism. In all cases, the relativistic framework is considered. Furthermore, in all

L. Schächter, *Beam-Wave Interaction in Periodic and Quasi-Periodic Structures*, Particle Acceleration and Detection, DOI 10.1007/978-3-642-19848-9_3,

cases of interest, many particles are involved and consequently, statistical methods are invoked and for this purpose, we present very briefly the kinetic and the fluid approximations, which are used throughout the text.

3.1.1 Newtonian Equations of Motion

The elementary equations, which describe the dynamics of a particle at the classical level, are given by

$$\frac{\mathrm{d}}{\mathrm{d}t}[m\gamma(t)\mathbf{v}(t)] = \mathbf{F}(t), \tag{3.1.1}$$

where $\mathbf{F}(t)$ is the force acting on the particle and if an electromagnetic field is present then the force is given by the Lorentz force which reads

$$\mathbf{F}(t) = -e\{\mathbf{E}[\mathbf{r}(t), t] + \mathbf{v}(t) \times \mathbf{B}[\mathbf{r}(t), t]\}; \tag{3.1.2}$$

e and m represent the charge and the rest mass of the electron respectively, $\mathbf{v}(t)$ is its velocity vector at any point in time and

$$\gamma(t) = \frac{1}{\sqrt{1 - \mathbf{v}(t) \cdot \mathbf{v}(t)/c^2}}. \tag{3.1.3}$$

The electromagnetic field, $\mathbf{E}(\mathbf{r}(t), t)$ and $\mathbf{B}(\mathbf{r}(t), t)$ is the field at the particle's location.

A full description of the particle's dynamics requires to determine also the location of the particle at each point in time; this is given by

$$\frac{\mathrm{d}}{\mathrm{d}t}\mathbf{r}(t) = \mathbf{v}(t). \tag{3.1.4}$$

The state-vector of such a particle is a 6D vector and it consists of the relative location of the particle $\mathbf{r}(t)$ and its momentum i.e. $[\mathbf{r}(t), m\gamma(t)\mathbf{v}(t)]$.

As in the case of Maxwell's equations, the energy conservation can be deduced from these equations. For this purpose (3.1.1) is multiplied scalarly by $\mathbf{v}(t)$. After substituting (3.1.2) in the right-hand side, we can find that the second term contribution is identically zero since the product $\mathbf{v} \times \mathbf{B}$ is orthogonal to both the velocity vector and the magnetic induction. In the left-hand side we have

$$\begin{aligned} m\mathbf{v}(t) \cdot \frac{\mathrm{d}}{\mathrm{d}t}[\gamma(t)\mathbf{v}(t)] \\ = m\frac{1}{2}\gamma(t)\frac{\mathrm{d}}{\mathrm{d}t}[\mathbf{v}(t) \cdot \mathbf{v}(t)] + m[\mathbf{v}(t) \cdot \mathbf{v}(t)]\frac{\mathrm{d}}{\mathrm{d}t}\gamma(t), \end{aligned} \tag{3.1.5}$$

which can be simplified if we now use the definition of $\gamma(t)$ in (3.1.3) to express $\mathbf{v}\cdot\mathbf{v}$ as $c^2[1-\gamma^{-2}(t)]$ which yields

$$mc^2\frac{\mathrm{d}}{\mathrm{d}t}\gamma(t) = -e\mathbf{v}(t)\cdot\mathbf{E}[\mathbf{r}(t),t]. \tag{3.1.6}$$

This is the expression for single particle energy conservation.

3.1.2 Lagrangian Formalism

It is convenient in many cases to use a different approach when formulating the dynamics of a single particle. The basic idea is to introduce a *scalar* function L, called Lagrangian, from which the *vector* equation of motion can be derived. This function depends on the velocity and location of the particle and in general, it may also depend on time. Without loss of generality, we can define the *action* as

$$I = \int_{t_1}^{t_2} \mathrm{d}t L(\mathbf{v},\mathbf{r};t), \tag{3.1.7}$$

and require that the motion of the particle from time t_1 to time t_2 is such that the line integral is an extremum for the *path of motion*. To formulate this statement mathematically it implies to require that this action is at an extremum with respect to a virtual change δr, hence

$$\begin{aligned}
\delta I &= \delta\int_{t_1}^{t_2} \mathrm{d}t L(\mathbf{v},\mathbf{r};t) = 0,\\
&= \int_{t_1}^{t_2} \mathrm{d}t\delta L(\mathbf{v},\mathbf{r};t) = 0,\\
&= \int_{t_1}^{t_2} \mathrm{d}t\left[\frac{\partial L}{\partial \mathbf{r}}\delta\mathbf{r} + \frac{\partial L}{\partial \mathbf{v}}\frac{\mathrm{d}}{\mathrm{d}t}\delta\mathbf{r}\right] = 0,\\
&= \int_{t_1}^{t_2} \mathrm{d}t\delta\mathbf{r}\left[\frac{\partial L}{\partial \mathbf{r}} - \frac{\mathrm{d}}{\mathrm{d}t}\frac{\partial L}{\partial \mathbf{v}}\right] = 0,
\end{aligned} \tag{3.1.8}$$

In this context by "virtual" we mean an infinitesimal change in the configuration space due to an infinitesimal change of the coordinates system, $\delta\mathbf{r}$, consistent with the forces imposed on the particle at the given time. In the last line of the expression from the above, we used the fact that after the integration by parts, the variation at t_1 and t_2 is identically zero. Thus, in order to satisfy $\delta I = 0$, the Lagrangian has to be a solution of the following differential equation

$$\frac{\mathrm{d}}{\mathrm{d}t}\left(\frac{\partial L}{\partial \mathbf{v}}\right) - \frac{\partial L}{\partial \mathbf{r}} = 0. \tag{3.1.9}$$

This is called Lagrange's equation and it is identical with the (relativistic) equations of motion, provided that L is chosen to be

$$L = mc^2\sqrt{1-\mathbf{v}\cdot\mathbf{v}/c^2} + e(\Phi - \mathbf{v}\cdot\mathbf{A}), \tag{3.1.10}$$

where Φ is the scalar electric potential and $\mathbf{A}$ is the magnetic vector potential.

At this point, we are in position to define in a systematic way the momentum of a particle in the presence of an electromagnetic field. This will be referred to as the *canonical momentum* associated with the coordinate $\mathbf{r}$ and it is defined by

$$\mathbf{p} = \frac{\partial L}{\partial \mathbf{v}} = m\frac{\mathbf{v}}{\sqrt{1-\mathbf{v}\cdot\mathbf{v}/c^2}} - e\mathbf{A} = m\gamma\mathbf{v} - e\mathbf{A}. \tag{3.1.11}$$

With this definition in mind and Lagrange equation, we can already point out one of the advantages of the Lagrangian formalism. If L is not an explicit function of one of the coordinates (say x) then the second term in (3.1.9) vanishes. This, in conjunction with the last definition, implies that the corresponding component of the canonical momentum (in this case p_x) is a constant of motion. Therefore, the constants of motion can be deduced from the *symmetry* of the system.

3.1.3 Hamiltonian Formalism

In particular, if L does not depend explicitly on time then by its differentiation by parts and using Lagrange's equation we obtain

$$\left.\begin{aligned} \frac{dL}{dt} &= \frac{\partial L}{\partial \mathbf{r}}\cdot\frac{d\mathbf{r}}{dt} + \frac{\partial L}{\partial \mathbf{v}}\cdot\frac{d\mathbf{v}}{dt} = \frac{d}{dt}\left(\frac{\partial L}{\partial \mathbf{v}}\right)\cdot\mathbf{v} + \frac{\partial L}{\partial \mathbf{v}}\cdot\frac{d\mathbf{v}}{dt} \\ \frac{dL}{dt} &= \frac{d}{dt}\left(\mathbf{v}\cdot\frac{\partial L}{\partial \mathbf{v}}\right) \end{aligned}\right\} \Rightarrow \frac{d}{dt}\left(L - \mathbf{v}\cdot\frac{\partial L}{\partial \mathbf{v}}\right) = 0. \tag{3.1.12}$$

Subject to this condition, the expression in brackets is a constant and is proportional to the total energy in the system. Based on this last result it is convenient to define the so-called Hamiltonian of the system as

$$H = \mathbf{v}\cdot\mathbf{p} - L. \tag{3.1.13}$$

For a relativistic particle it reads

$$\begin{aligned} H &= \mathbf{v}\cdot(m\gamma\mathbf{v} - e\mathbf{A}) - \left[-mc^2\sqrt{1-\mathbf{v}\cdot\mathbf{v}/c^2} + e(\Phi - \mathbf{v}\cdot\mathbf{A})\right] \\ &= mc^2\gamma - e\Phi. \end{aligned} \tag{3.1.14}$$

According to the last expression and comparing it to the free particle case, the energy, in the presence of an electromagnetic field, is given by $E = H + e\Phi$. Bearing in mind that, according to the special theory of relativity, the energy and the momentum are related by $E^2 = \mathbf{p}^2c^2 + m^2c^4$, we conclude that the Hamiltonian of a relativistic particle expressed in terms of momentum $\mathbf{p}$ [using (3.1.11)] is given by

$$H = \sqrt{(\mathbf{p} + e\mathbf{A})^2 c^2 + (mc^2)^2} - e\Phi. \tag{3.1.15}$$

This is also a scalar function and as in the case of the Lagrange's function, there are a set of equations, which describe the motion of the system. These read

$$\frac{\mathrm{d}\mathbf{r}}{\mathrm{d}t} = \frac{\partial H}{\partial \mathbf{p}}, \tag{3.1.16}$$

and

$$\frac{\mathrm{d}\mathbf{p}}{\mathrm{d}t} = -\frac{\partial H}{\partial \mathbf{r}}. \tag{3.1.17}$$

Neither the Lagrangian nor the Hamiltonian formalisms include more information about the system than that provided by the Newtonian equations of motion; however as indicated in the case of the Lagrange function, the constants of motion can be determined in an easier and more systematic way. In addition, the formulation of the dynamics of more complex variables can be "naturally" formulated. Consider, for example, a dynamic variable $\rho(\mathbf{p}, \mathbf{r}, t)$ and suppose it is required to determine its equation of motion. At first glance the vector equations of motion [(3.1.1)–(3.1.2)] give us a limited hint as of how to proceed however, the Hamiltonian formalism is very helpful since firstly we can write

$$\frac{\mathrm{d}\rho}{\mathrm{d}t} = \frac{\partial \rho}{\partial t} + \frac{\partial \rho}{\partial \mathbf{r}}\frac{\partial \mathbf{r}}{\partial t} + \frac{\partial \rho}{\partial \mathbf{p}}\frac{\partial \mathbf{p}}{\partial t} = \frac{\partial \rho}{\partial t} + \frac{\partial \rho}{\partial \mathbf{r}}\frac{\mathrm{d}\mathbf{r}}{\mathrm{d}t} + \frac{\partial \rho}{\partial \mathbf{p}}\frac{\mathrm{d}\mathbf{p}}{\mathrm{d}t} \tag{3.1.18}$$

and secondly substitute Hamilton's equations. The result is

$$\frac{\mathrm{d}\rho}{\mathrm{d}t} = \frac{\partial \rho}{\partial t} + \frac{\partial \rho}{\partial \mathbf{r}}\frac{\partial H}{\partial \mathbf{p}} - \frac{\partial \rho}{\partial \mathbf{p}}\frac{\partial H}{\partial \mathbf{r}} \equiv \frac{\partial \rho}{\partial t} + \{\rho, H\}. \tag{3.1.19}$$

The latter definition is also known as the *Poisson brackets*. Hamiltonian formulation and a generalization of Poisson brackets provide the basis for the quantum formulation of microscopic electron's dynamic. A more detailed discussion on classical mechanics can be found in textbooks by Goldstein (1950) and Landau and Lifshitz (1960).

3.1.4 Kinetic Approximation: Liouville's Theorem

The formulation presented above relies on a *single* particle interacting with the electromagnetic field and no direct interaction is considered other than through this field. Even in the absence of an external electromagnetic field there are numerous electrons in any system and it is not possible to solve instantaneously the equations of motion for all electrons, therefore statistical methods are invoked. Instead of information regarding each particle, we consider the probability density, $f(\mathbf{r},\mathbf{p},t)$, to find a particle at a given time t in the 6-dimensional phase-space element $\mathbf{rp} \to (\mathbf{r}+\delta\mathbf{r})(\mathbf{p}+\delta\mathbf{p})$; this probability density satisfies

$$\int_{-\infty}^{\infty} \mathrm{d}\mathbf{r} \int_{-\infty}^{\infty} \mathrm{d}\mathbf{p} f(\mathbf{r},\mathbf{p},t) \equiv 1. \tag{3.1.20}$$

Although the notation is the same, it is important to realize the difference between $(\mathbf{r},\mathbf{p})$ in this sub-section and the previous one: previously, $(\mathbf{r},\mathbf{p})$ were the *coordinates* of a given particle in a 6D phase-space whereas here, we do not know the location of any of the particles. In (3.1.20), $(\mathbf{r},\mathbf{p})$ are the *variables* of the probability density. Assuming that we know this probability density function, the charge density is

$$\rho(\mathbf{r},t) \equiv -en(\mathbf{r},t) = -en_0 \int_{-\infty}^{\infty} \mathrm{d}\mathbf{p} f(\mathbf{r},\mathbf{p},t), \tag{3.1.21}$$

where n_0 is the average particle density and the current density is

$$\mathbf{J}(\mathbf{r},t) = -en_0 \int_{-\infty}^{\infty} \mathrm{d}\mathbf{p}\mathbf{v} f(\mathbf{r},\mathbf{p},t). \tag{3.1.22}$$

These two expressions indicate that, in principle, if we know this function we should be able to calculate the electromagnetic field. Motivated by this fact, we proceed and determine next the dynamics of this probability density function. According to Liouville's theorem, *the distribution function is a constant along any trajectory in the phase-space*. This is valid, for non-interacting particles and closed system; however the formulation can be generalized to include collisions and external effects. For a collisionless ensemble, the Liouville theorem can be mathematically formulated as

$$\frac{\mathrm{d}}{\mathrm{d}t} f(\mathbf{r},\mathbf{p},t) = 0. \tag{3.1.23}$$

Using the Hamiltonian dynamics in terms of Poisson brackets as formulated in (3.1.19) we have

$$
\begin{aligned}
\frac{\mathrm{d}}{\mathrm{d}t}f(\mathbf{r},\mathbf{p},t) &= 0 \\
&= \frac{\partial f(\mathbf{r},\mathbf{p},t)}{\partial t} + \{f(\mathbf{r},\mathbf{p},t),H\} \\
&= \frac{\partial f(\mathbf{r},\mathbf{p},t)}{\partial t} + \frac{\mathrm{d}\mathbf{r}}{\mathrm{d}t}\cdot\frac{\partial f(\mathbf{r},\mathbf{p},t)}{\partial \mathbf{r}} + \frac{\mathrm{d}\mathbf{p}}{\mathrm{d}t}\cdot\frac{\partial f(\mathbf{r},\mathbf{p},t)}{\partial \mathbf{p}} \\
&= \frac{\partial f(\mathbf{r},\mathbf{p},t)}{\partial t} + \mathrm{v}\cdot\frac{\partial f(\mathbf{r},\mathbf{p},t)}{\partial \mathbf{r}} \\
&\quad - e[E(\mathbf{r},t) + \mathbf{v}\times\mathbf{B}(\mathbf{r},t)]\cdot\frac{\partial f(\mathbf{r},\mathbf{p},t)}{\partial \mathbf{p}} = 0.
\end{aligned}
\tag{3.1.24}
$$

Let us now present a very simple solution of this equation, which also reflects on the character of the interaction of charges in plasma. In contrast to the case of a gas where each individual atom interacts only with its nearest neighbor due to the short range character of a neutral atom, in the case of charged particles the range of the Coulomb force is long and consequently many particles, in its vicinity, might be affected. We consider a static electric field ($\mathbf{E} = -\nabla\Phi$) which develops in a neutral system due to a local perturbation in the neutrality of the system. The Hamiltonian in this case reads

$$H = \frac{\mathbf{p}^2}{2m} - e\Phi, \tag{3.1.25}$$

and the solution of (3.1.24) can be checked to read

$$f(\mathbf{r},\mathbf{p},t) = f_0 \exp(-H/k_{\mathrm{B}}T), \tag{3.1.26}$$

where $k_{\mathrm{B}} = 1.38066 \times 10^{-23} \mathrm{J}K^{-1}$ is the Boltzman constant and T is the absolute temperature of the particles; f_0 is determined using (3.1.20). Integration over the momentum can be performed analytically thus the density of the particles, according to (3.1.21), reads

$$n(\mathbf{r}) = n_0 \exp\left[\frac{e\Phi(\mathbf{r})}{k_{\mathrm{B}}T}\right]. \tag{3.1.27}$$

A potential, which develops in the distribution, causes a change $n(\mathbf{r}) - n_0$ in the particle density; n_0 is the average density of the particles. The latter affects in turn the electric scalar potential Φ hence

$$\nabla^2\Phi = \frac{e}{\varepsilon_0} n_0 \left[\exp\left(\frac{e\Phi}{k_B T}\right) - 1\right]. \tag{3.1.28}$$

Assuming $e|\Phi|/k_BT \ll 1$, we can expand the right-hand side in Taylor series of which we keep only the first term. If we further assume spherical symmetry, we can readily solve (3.1.28) and the electrostatic potential is given by

$$\Phi(r) = -\frac{e}{4\pi\varepsilon_0}\frac{1}{r}\exp\left(-\frac{r}{\lambda_D}\right), \tag{3.1.29}$$

where λ_D is the Debye length, defined by

$$\lambda_D^2 = \frac{\varepsilon_0 k_B T}{n_0 e^2}. \tag{3.1.30}$$

The solution in (3.1.29) indicates that the potential generated by this perturbation is screened on a scale of the *Debye length* and beyond this radius its effect is vanishingly small. With this characteristic length parameter we can define the typical (Debye) sphere whose volume is $4\pi\lambda_D^3/3$. In this range, the charge has a non-negligible effect on adjacent particles. The number of the particles affected by the perturbation mentioned above, is proportional to the product of the averaged particles' density, n_0, and the volume of the Debye sphere. In order to avoid effects of such fluctuations it will be reasonable to require that no particles (other than the source) will be in this sphere i.e.,

$$n_0\left(4\pi\frac{1}{3}\lambda_D^3\right) < 1, \tag{3.1.31}$$

which also means that the density has to be smaller than a critical value n_c given by

$$n_0 < n_c \equiv \left(\frac{4\pi}{3}\right)^2\left(\frac{\varepsilon_0 k_B T}{e}\right)^3. \tag{3.1.32}$$

Whenever the kinetic approximation will be employed, it will be assumed that this condition is locally satisfied.

3.1.5 Hydrodynamic Approximation

In the framework of the kinetic approximation presented above, at a given location there is a finite probability to find particles of different velocities. This is to say that in an infinitesimal volume the energy spread of the electrons (represented by the temperature) is significant. In many cases of interest, this spread is virtually zero and as a result, we can attribute to all particles in an infinitesimal volume, a certain *velocity* and *density*. For determining the dynamics of the velocity, it is assumed that the reaction of the particles to an external force is identical to that of a single particle which satisfies

$$\frac{\mathrm{d}}{\mathrm{d}t}[m\gamma(\mathbf{r},t)\mathbf{v}(\mathbf{r},t)] = -e[\mathbf{E}(\mathbf{r},t) + \mathbf{v}(\mathbf{r},t) \times \mathbf{B}(\mathbf{r},t)]. \tag{3.1.33}$$

and γ satisfies

$$\frac{\mathrm{d}}{\mathrm{d}t}\left[mc^2\gamma(\mathbf{r},t)\right] = -e\mathbf{v}(\mathbf{r},t) \cdot \mathbf{E}(\mathbf{r},t). \tag{3.1.34}$$

The particles' density in the infinitesimal volume at the given time, $n(\mathbf{r},t)$, satisfies the *continuity equation*

$$\nabla \cdot [n(\mathbf{r},t)\mathbf{v}(\mathbf{r},t)] + \frac{\partial}{\partial t}n(\mathbf{r},t) = 0, \tag{3.1.35}$$

which is equivalent to the charge conservation introduced in context of Maxwell's equation (see Sect. 2.1.1). In the context of the equations above, the derivative $\mathrm{d}/\mathrm{d}t$ is given by

$$\frac{\mathrm{d}}{\mathrm{d}t} = \frac{\partial}{\partial t} + \mathbf{v}(\mathbf{r},t) \cdot \nabla. \tag{3.1.36}$$

Assuming that the velocity and density fields were established, the charge and current densities read

$$\rho(\mathbf{r},t) = -en(\mathbf{r},t), \tag{3.1.37}$$

and

$$\mathbf{J}(\mathbf{r},t) = -en(\mathbf{r},t)\mathbf{v}(\mathbf{r},t) \tag{3.1.38}$$

thus the connection to Maxwell's equations is completed. The three equations, (3.1.33)–(3.1.35) together with Maxwell's equations consist a self-consistent set of electrodynamic equations in the framework of the *hydrodynamic approximation*. Note that contrary to Sect. 3.1.1, the velocity $\mathbf{v}(\mathbf{r},t)$represents a field rather than a coordinate of a particle in the phase-space.

In order to quantify this approximation, we can state that any variations of the velocity (or density) field on the scale of an infinitesimal volume are negligible on the scale of the distance between any two particles in this volume. If the density in the mentioned volume is n then the characteristic distance between each two particles is $l \simeq n^{-1/3}$ thus

$$\frac{|\nabla \cdot \mathbf{v}|}{|\mathbf{v}|} \ll \frac{1}{l}. \tag{3.1.39}$$

If, for example, the largest spatial variation is determined by the radiation field i.e. λ, then the condition above implies that $\lambda \times n^{1/3} \ll 1$; for a relativistic beam of radius 3 mm carrying a current of 1 kA, the density is $7 \times 10^{11}\mathrm{m}^{-3}$, the characteristic length l is $l \simeq 110\ \mu$m therefore as long as the radiation wavelength is larger than $1,000\ \mu$m the approximation is fully justified. However, for a strongly bunched beam there will be regions in space where the density is orders of magnitude smaller than the (initial) average density and consequently the validity of the fluid approximation has to be properly re-examined. An additional perspective relies on the energy spread (temperature) which was assumed to be negligibly low. In terms of Debye critical density, defined in (3.1.32), low temperature entails low critical density n_c therefore, the hydrodynamic approximation is valid as long as $n \gg n_{\mathrm{c}}$.

3.1.6 Global Energy Conservation

In Chap. 2 we developed the Poynting's theorem from Maxwell's equations and it was indicated that it is associated with the power and energy conservation of the electromagnetic field. It was formulated as

$$\nabla \cdot \mathbf{S} + \frac{\partial}{\partial t}\left[\frac{1}{2}\varepsilon_0 \mathbf{E} \cdot \mathbf{E} + \frac{1}{2}\mu_0 \mathbf{H} \cdot \mathbf{H}\right] = -\mathbf{J} \cdot \mathbf{E}, \tag{3.1.40}$$

where $\mathbf{S} = \mathbf{E} \times \mathbf{H}$ is the Poynting vector and $\mathbf{J}$ was assumed to be given. At this stage we can release this constraint since the current density was determined (3.1.38) in terms of the density and velocity fields. Our goal now is to formulate the global energy conservation of the electromagnetic, velocity and density fields as one conservation law e.g., charge conservation in (2.1.12). For this purpose, we substitute the current density definition in (3.1.40). In addition, we use the definition of the total electromagnetic energy density W from (2.1.19) to write

$$\nabla \cdot \mathbf{S} + \frac{\partial}{\partial t} W = en\mathbf{v} \cdot \mathbf{E}. \tag{3.1.41}$$

The scalar product on the right-hand side is identical to that in (3.1.34), therefore the last equation yields

$$\nabla \cdot \mathbf{S} + \frac{\partial}{\partial t} W = -mc^2 n \frac{\mathrm{d}}{\mathrm{d}t}\gamma; \tag{3.1.42}$$

using the definition in (3.1.36) we have

$$\nabla \cdot \mathbf{S} + \frac{\partial}{\partial t} W = -mc^2 \left[n \frac{\partial}{\partial t} \gamma + n\mathbf{v} \cdot \nabla \gamma \right],$$
$$= -mc^2 \left[\frac{\partial}{\partial t} (n\gamma) - \gamma \frac{\partial}{\partial t} n + \nabla \cdot (n\mathbf{v}\gamma) - \gamma \nabla \cdot (n\mathbf{v}) \right]. \quad (3.1.43)$$

The continuity equation in (3.1.35) further simplifies this expression since the sum of the second and the fourth terms on the right-hand side is zero, hence

$$\nabla \cdot \left[\mathbf{S} + mc^2 n\gamma \mathbf{v} \right] + \frac{\partial}{\partial t} \left[W + mc^2 n\gamma \right] = 0. \quad (3.1.44)$$

This expression is the global energy conservation of the electromagnetic, velocity and density fields. The total energy flux is given by the first term and it is the sum of the electromagnetic Poynting vector and the kinetic energy flux: $\mathbf{S} + mc^2 n(\gamma - 1)\mathbf{v}$. The total energy density stored in the system is the sum of the electromagnetic energy density W and the kinetic energy density $W + mc^2 n(\gamma - 1)$. For these interpretations we have subtracted from (3.1.44) the continuity equation

$$\nabla \cdot \left[\mathbf{S} + mc^2 n(\gamma - 1)\mathbf{v} \right] + \frac{\partial}{\partial t} \left[W + mc^2 n(\gamma - 1) \right] = 0, \quad (3.1.45)$$

multiplied by the rest energy of the electron i.e., mc^2.

3.2 Special Theory of Relativity

Modern high power radiation sources and accelerators rely on the interaction of electromagnetic waves with electrons whose velocity is very close to c. In these conditions, one has to invoke relativistic dynamics.

3.2.1 Basic Principles

The dynamics of the electrons as formulated so far is consistent with what is known as the *Special Theory of Relativity* as formulated by Albert Einstein in 1905. For an adequate formulation of its principles, we have to introduce the concept of the system of reference also referred to as frame of reference. It consists of a set of rulers to measure the distance and enabling us to determine the location of an event in space and in addition, a series of clocks which monitor the time. An inertial frame of reference can be conceived as being attached to a free particle i.e., a particle which no forces act on, thus it moves with a constant velocity. Another frame of reference moving at a different but constant velocity is also inertial.

The first principle of the theory states that: (1) *The laws of nature are form-invariant with respect to the transformation from one frame of reference to another.* In other words, the laws of nature can be written in the same form in all inertial systems of reference. As an example, let us consider a motionless frame of reference $R(x, y, z, ct)$ and the law of nature to be examined is Maxwell's equations

$$\begin{aligned} &\nabla \times \mathbf{E} + \frac{\partial}{\partial t}\mathbf{B} = 0, \\ &\nabla \times \mathbf{H} - \frac{\partial}{\partial t}\mathbf{D} = \mathbf{J}, \\ &\nabla \cdot \mathbf{D} = \rho, \\ &\nabla \cdot \mathbf{B} = 0. \end{aligned} \tag{3.2.1}$$

Now, in another frame of reference, $R'(x', y', z', ct')$, Maxwell's equations have an identical form. "Primed" notation indicates that the numbers which indicate the location and the time of the event under consideration are different than these measured in laboratory frame. Similarly, primed field components or the source terms are those measured by an observer in the moving frame and according to (1) they satisfy

$$\begin{aligned} &\nabla' \times \mathbf{E}' + \frac{\partial}{\partial t'}\mathbf{B}' = 0, \\ &\nabla' \times \mathbf{H}' - \frac{\partial}{\partial t'}\mathbf{D}' = \mathbf{J}', \\ &\nabla' \cdot \mathbf{D}' = \rho', \\ &\nabla' \cdot \mathbf{B}' = 0. \end{aligned} \tag{3.2.2}$$

The two observers, one in the laboratory and the other in the moving frame of reference, intend to "compare notes" regarding data each one has measured. In this process they have to take into consideration the finite time it takes information to traverse the distance between two points. This brings us to the second principle of the special theory of relativity which states that (2) *The phase velocity of a plane electromagnetic wave in vacuum is the same in all inertial frames of reference.*

3.2.2 Lorentz Transformation

In contrast with Newtonian mechanics, where the spatial coordinates are variables and the time is a *parameter*, according to special theory of relativity the time a coordinate as the other three spatial coordinates. Therefore, in order to describe the motion of a wave in vacuum, we denote by $\mathrm{d}r$ the space interval it traverses in a time-interval $\mathrm{d}t$ hence, $\mathrm{d}r = c\mathrm{d}t$. The last expression can also be written as

$$\mathrm{d}s^2 = \mathrm{d}r^2 - c^2\mathrm{d}t^2 = \mathrm{d}x^2 + \mathrm{d}y^2 + \mathrm{d}z^2 - c^2\mathrm{d}t^2 = 0, \tag{3.2.3}$$

which is a generalization of the concept of distance in a regular three dimensional space. As clearly indicated, in the particular case of a plane wave, the distance it traverses in the four dimensional space (space-time) is zero. By virtue of the invariance of the phase-velocity, the moving observer can write the same statement as

$$(\mathrm{d}s')^2 = (\mathrm{d}x')^2 + (\mathrm{d}y')^2 + (\mathrm{d}z')^2 - (c\mathrm{d}t')^2 = 0. \tag{3.2.4}$$

Without loss of generality, we can assume that the relative motion of the two frames of reference is along the z axis and that at a certain point in space-time the two frames overlap; thus, we assume the following general transformation

$$\begin{aligned} \mathrm{d}x' &= \mathrm{d}x, \\ \mathrm{d}y' &= \mathrm{d}y, \\ \mathrm{d}z' &= a_{11}\mathrm{d}z - a_{12}c\mathrm{d}t, \\ c\mathrm{d}t' &= a_{22}c\mathrm{d}t - a_{21}\mathrm{d}z. \end{aligned} \tag{3.2.5}$$

Substituting these relations in (3.2.3)–(3.2.4) and comparing coefficients we find the following relations

$$\begin{aligned} a_{11}^2 - a_{21}^2 &= 1, \\ a_{22}^2 - a_{12}^2 &= 1, \\ a_{11}a_{12} - a_{22}a_{21} &= 0. \end{aligned} \tag{3.2.6}$$

At the origin ($z' = 0$), we must have $z = \mathrm{v}_0 t$ where v_0 is the relative velocity between the two frames, therefore $a_{12}/a_{11} = \beta \equiv \mathrm{v}_0/c$. With this observation we can now determine the coefficients of (3.2.5). Firstly, we substitute $a_{12} = \beta a_{11}$ in the second and third equation. Secondly, we substitute a_{22} from one of the resulting equations. The equation obtained for a_{21} can be solved and the result is $a_{21} = \gamma\beta$ where $\gamma = [1 - \beta^2]^{-1/2}$. The other two coefficients can be readily determined and they are given by $a_{12} = \gamma\beta$, $a_{22} = \gamma$ and $a_{11} = \gamma$. These coefficients define the so-called Lorentz transformation which for the 4-vector of the coordinates $(\mathbf{r}, ct)$ can be formulated as

$$\begin{aligned} x' &= x, \\ y' &= y, \\ z' &= \gamma(z - \beta ct), \\ ct' &= \gamma(ct - \beta z). \end{aligned} \tag{3.2.7}$$

The transformation from the laboratory frame of reference to the moving one is determined by reversing the sign of β and replacing the prime and unprimed variables namely,

$$\begin{aligned} x &= x', \\ y &= y', \\ z &= \gamma(z' + \beta ct'), \\ ct &= \gamma(ct' + \beta z'). \end{aligned} \tag{3.2.8}$$

The same transformation relates the components of *any* 4-vector in the moving frame and the laboratory.

3.2.3 Kinetic and Dynamic Quantities

The phase velocity of a plane wave was defined as the velocity an observer has to move in order to measure the same phase – e.g. to be on the crest of the wave. If according to the special theory of relativity this velocity is the same in all frames of reference, then we may expect that the *phase* itself is also invariant otherwise, the observer will measure a phase which varies. Hence,

$$\omega t - \mathbf{k} \cdot \mathbf{r} = \omega' t' - \mathbf{k}' \cdot \mathbf{r}' = \text{const.}, \tag{3.2.9}$$

where ω is the angular frequency and $\mathbf{k}$ is the wave-number vector. Substituting Lorentz transformations, we obtain the following transformation for the frequency and wave-numbers:

$$\begin{aligned} k'_x &= k_x, \\ k'_y &= k_y, \\ k'_z &= \gamma\left(k_z - \beta\frac{\omega}{c}\right), \\ \frac{\omega'}{c} &= \gamma\left(\frac{\omega}{c} - \beta k_z\right). \end{aligned} \tag{3.2.10}$$

As in the space-time transformation, the inverse is obtained by reversing the sign of β and replacing the prime and unprimed variables i.e.,

$$\begin{aligned} k_x &= k'_x, \\ k_y &= k'_y, \\ k_z &= \gamma\left(k'_z + \beta\frac{\omega'}{c}\right), \\ \frac{\omega}{c} &= \gamma\left(\frac{\omega'}{c} + \beta k'_z\right). \end{aligned} \tag{3.2.11}$$

Similar to the 4-vector of the coordinates $(\mathbf{r}, ct)$ which describes an event in an inertial frame of reference, the set $(\mathbf{k}, \omega/c)$ also forms a 4-vector which describes the propagation of an electromagnetic wave in vacuum.

Another 4-vector is that of the source densities $(\mathbf{J}, c\rho)$ as can be concluded after applying invariance principle on the charge conservation law as expressed in (2.1.12). The detailed proof is left as an exercise to the reader. In addition, the potentials $(\mathbf{A}, \Phi/c)$ form a 4-vector provided that the Lorentz gauge is imposed.

In order to prove the last statement one can check, by analogy to (3.2.3), that the wave equation operator

$$\frac{\partial^2}{\partial x^2}+\frac{\partial^2}{\partial y^2}+\frac{\partial^2}{\partial z^2}-\frac{1}{c^2}\frac{\partial^2}{\partial t^2}, \tag{3.2.12}$$

is relativistically invariant. Subject to the Lorentz gauge, we can write the wave equations for the potentials

$$\left[\frac{\partial^2}{\partial x^2}+\frac{\partial^2}{\partial y^2}+\frac{\partial^2}{\partial z^2}-\frac{1}{c^2}\frac{\partial^2}{\partial t^2}\right]\left(A,\frac{\Phi}{c}\right)=\mu_0(\mathbf{J},c\rho). \tag{3.2.13}$$

Since the right-hand side is relativistically invariant, as is the wave-equation operator, we conclude that the three components of $\mathbf{A}$ and Φ, i.e. $(\mathbf{A},\Phi/c)$, form a relativistically invariant 4-vector.

Comment 3.1. While all the quantities specified so far form a 4-vector and the *transverse* components are relativistically invariant, in the case of the electromagnetic field components, the *longitudinal* components are invariant ($E'_z=E_z$, $B'_z=B_z$). The transverse components satisfy

$$\begin{aligned} E'_x &= \gamma(E_x-\beta cB_y) & cB'_x &= \gamma(cB_x+\beta E_y)\\ E'_y &= \gamma(E_y+\beta cB_x) & cB'_y &= \gamma(cB_y-\beta E_x)\end{aligned}$$

and all six components form an anti-symmetric 2nd rank *field-strength tensor*

$$\begin{pmatrix} 0 & cB_z & -cB_y & -E_x\\ -cB_z & 0 & cB_x & -E_y\\ cB_y & -cB_x & 0 & -E_z\\ E_x & E_y & E_z & 0\end{pmatrix}.$$

It is instructive to examine a few of the principles presented above in a free electron laser – see Sect. 1.2.4. This consists of a static periodic magnetic field (wiggler) and an electron beam, which is injected in this field with a velocity v_0 from $z\to-\infty$ to ∞ as illustrated in Fig. 3.1. If the period of the wiggler is L then the magneto-static ($\omega=0$) "wave-number" associated with this field is $k_w=2\pi/L$ i.e., $B_x=B_0\cos(k_w z)$. This magneto-static field is seen by the moving electron as a wave since if we substitute the third equation of (3.2.8) we obtain $\cos[k_w\gamma(z'+\beta ct')]$. In the frame of reference attached to the electron this wave propagates from $z'\to\infty$ to $-\infty$. Based on the argument of the trigonometric function we can readily identify the characteristic frequency the electron oscillates as $\omega'=ck_w\gamma\beta$ and the wave number as $k'_z=-k_w\gamma$. The same result is achieved by employing the transformations in (3.2.10): firstly, we observe that the wiggler's wave-number in the laboratory frame of reference is $k_z=-k_w$ and as indicated

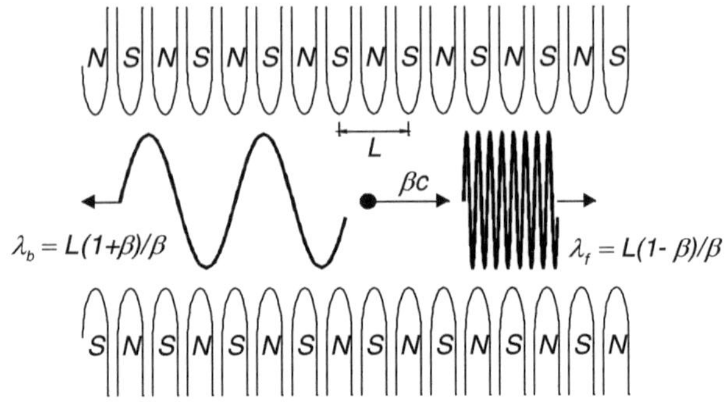

Fig. 3.1 Radiation emitted by an electron moving in a periodic magnetic field as measured in the laboratory frame of reference

previously, the field is static, therefore $\omega = 0$. Secondly, we substitute these two quantities in (3.2.10) and obtain:

$$\begin{aligned} k'_z &= \gamma[(-k_\mathrm{w}) - 0] = -\gamma k_\mathrm{w}, \\ \frac{\omega'}{c} &= \gamma[0 - \beta(-k_\mathrm{w})] = \gamma\beta k_\mathrm{w}. \end{aligned} \tag{3.2.14}$$

Again, this corresponds to a backward propagating wave since

$$k'_z = -\frac{1}{\beta}\frac{\omega'}{c}. \tag{3.2.15}$$

Under the influence of this wiggler the electron oscillates around its initial location ($z' = 0$) thus, the force in its frame is expected to have the form $\cos(k_\mathrm{w}\gamma\beta ct')$. As it oscillates, it emits radiation in all direction however, for simplicity sake we consider only the waves emitted along the z' axis. In the positive direction it emits a wave which oscillates at ω' and its wave-number is $k'_z = \omega'/c$. Substituting in (3.2.11) we translate the parameters of this wave into the laboratory frame of reference, the result being

$$\frac{\omega}{c} = \gamma\left(\frac{\omega'}{c} + \beta\frac{\omega'}{c}\right) = k_\mathrm{w}\gamma^2\beta(1+\beta). \tag{3.2.16}$$

For relativistic particles ($\beta \to 1$) the radiation wavelength is

$$\lambda \simeq \frac{1}{2\gamma^2}L, \tag{3.2.17}$$

and if $L = 4\,\mathrm{cm}$, the radiation wavelength for a 100 MeV electron is 0.5 μm, therefore this scheme has the potential of generating tunable radiation at frequencies which are not achievable with atomic or molecular lasers.

The wave emitted in the negative direction of the z' axis has a wave number $k'_z = -\omega'/c$ and in the laboratory frame of reference its frequency is

$$\frac{\omega}{c} = \gamma\left[\frac{\omega'}{c} + \beta\left(-\frac{\omega'}{c}\right)\right] = k_{\mathrm{w}}\frac{\beta}{1+\beta}. \tag{3.2.18}$$

Clearly in this direction (anti-parallel to the electron) the frequency is much lower and even for a relativistic particle it reaches the $\omega \simeq ck_{\mathrm{w}}/2$ level. The reader is encouraged to calculate the frequency emitted at an arbitrary angle θ.
Many other analytic examples employing the concepts of Einstein's Special Theory of Relativity can be found in books by Pauli (1958), Van Bladel (1984) and Schieber (1986).

3.3 Electron Generation

In all interaction mechanisms to be discussed in this text, the electrons are free. But in nature electrons are attached to atoms or molecules, therefore they have to be extracted from the material in order to utilize them for conversion of energy. In this section, we consider a few topics associated with free electrons' generation. There are several ways to free electrons from the bulk material: they can be extracted from metals by applying an electric field perpendicular to the metal-vacuum interface – this is called *field-emission*. This concept gained a renewed interest due to the possibility of building small tips (of sub-microns scale) using micro-electronics technology. With this technique, each tip emits small currents but since many such tips can be made on one centimeter square, the current density can become quite significant. The applied voltage is also low comparing to usual field-emitters. A second method to extract electrons is the *thermionic emission*. In this case, the emitting surface (cathode) is heated and a fraction of the electrons in the material can overcome the work function and they become free. A third mechanism relies on *photo-emission*. In this case, a laser beam illuminates the cathode its photons providing the electrons with sufficient energy to overcome the work function. A fourth mechanism relies on the emission of electrons from metallic tips adjacent to dielectric materials. A fifth possible mechanism relies on what is called *secondary emission* in which case one electron hits a surface and releases more electrons. In the subsections that follow, we consider several topics related to electrons generation.

3.3.1 Field Emission from a Rough Surface

In the framework of a crude model, the electrons in a metal can be conceived to behave as an ideal gas confined by a potential well – of characteristic height determined by the work-function. If an external field is applied, a fraction of these electrons tunnel through this potential well. While in vacuum, these electrons are pushed away by the extracting field and pulled back by the corresponding image-charges. In 1928 Fowler and Nordheim (1928) have employed these concepts to demonstrate that the current density emitted is proportional to $E^2 \exp[-(...)/E]$. A revised version of the original formula is accepted to have the form (Miller 1982)

$$J[\mathrm{A/cm^2}] = \frac{1.54 \cdot 10^{-6} E^2}{\varphi\, t^2\left(3.79 \cdot 10^{-4}\dfrac{\sqrt{E}}{\varphi}\right)} \exp\left[-6.83 \cdot 10^7 \frac{\varphi^{3/2}}{E}\, \mathrm{v}\left(3.79 \cdot 10^{-4}\frac{\sqrt{E}}{\varphi}\right)\right] \tag{3.3.1}$$

where $E[\mathrm{V/cm}]$ denotes the electric field, the work function is denoted by $\varphi[\mathrm{e}V]$ and the two other functions can be approximated by $\mathrm{v}(u) \simeq 1 - u^{1.6897}$, $t(u) \simeq 1 + 0.1107u^{1.3343}$. This represents an ideal situation whereby the emitting surface is of zero roughness. In practice, the surface has a *finite roughness* and as a result the *local* field is enhanced by a geometric representing the micro-protrusion therefore locally, we may expect current density enhancement. At the macroscopic level, namely after averaging over all microscopic deviations from ideal flatness, the emission may be corrected by replacing the local electric field (E) by an enhancement factor β or explicitly, $E \to \beta E$.

In this subsection, we consider an idealized model to describe the random micro-protrusions in terms of a corrugated surface for establishing the effect of such a geometry on the *global* character of the emission. Several researchers (Miller 1967, 1984; Lewis 1955; Chatterton 1966) have considered the effect of various geometric configurations on the emission of electrons with the main emphasis on evaluating the geometric enhancement factor β of a *single* emitter. Even when multiple emitters are involved, as is the case of arrays of field emitters, each tip is isolated and controlled individually thus, in zero order, the *coupling effect* between adjacent tips is neglected. When the coupling is considered, some grating parameters facilitate generation of maximum current and others lead to maximum current density. In order to envision the reason for such a maximum to occur we have to bear in mind that once the tips are apart, the electric field may be large and consequently, the current density at its peak is high but the total current generated in a unit length of the structure is small since the emitting area is miniscule. At the other extreme, if the tips are very close to each other, they affect each other, the electric field (at the tip) is relatively small, and again the total current generated in a unit length is small although the emitting area is considerably larger. Between these two low values, the current is expected to achieve a maximum value.

In order to examine in detail this effect let us consider a two-dimensional system as illustrated in the top frame of Fig. 3.2. It consists of a uniform anode and a corrugated cathode with rectangular grooves and teeth. The period of the system is denoted by L, the anode-cathode gap is denoted by g, the tooth's width is d, its height is h and a voltage V_0 is applied to the anode. For this relatively simple geometry, we can calculate the electrostatic potential distribution in the entire

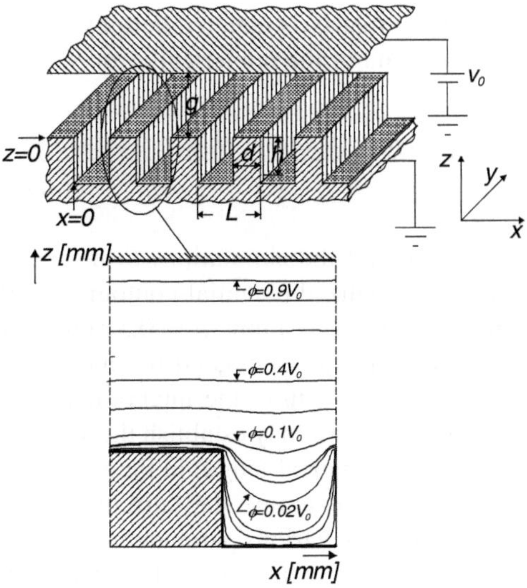

Fig. 3.2 Schematic of the system is presented in the top-frame. In the lower frame typical potential-contours are illustrated (for $L = 1$ mm, $h = 1$ mm, $d = 0.5$ mm and $g = 1$ mm)

volume. The boundary conditions on the anode ($z = g$) and on the grooves specify the following solution of Laplace's equation:

$$\Phi(x,z) = V_0 \begin{cases} \dfrac{z}{g} + \sum\limits_{n=-\infty}^{\infty} A_n \exp(-jk_n x) \dfrac{\sinh[k_n(z-g)]}{\sinh(k_n g)} & \begin{matrix} 0<x<L \\ 0<z<g \end{matrix} \\ \sum\limits_{m=1}^{\infty} B_m \sin\left(\dfrac{\pi m x}{L-d}\right) \sinh\left[\dfrac{\pi m}{L-d}(z+h)\right] & \begin{matrix} 0<x<L-d \\ -h<z<0 \end{matrix} \end{cases} \tag{3.3.2}$$

where $k_n = 2\pi n/L$. It is convenient to define,

$$\begin{aligned} \zeta_{nm} &\equiv \frac{1}{L-d} \int_0^{L-d} dx \exp(-jk_n x) \sin\left[\frac{\pi m x}{L-d}\right], \\ U_m &\equiv \frac{1}{g} \frac{L-d}{\pi m} [1-(-1)^m], \\ V_{m'm} &\equiv \frac{2}{\pi m} \left(\frac{L-d}{L}\right)^2 \tanh\left[\frac{\pi m h}{L-d}\right] \sum_{n=-\infty}^{\infty} (k_n L) \zeta_{nm}^* \zeta_{nm'} \coth(k_n g). \end{aligned} \tag{3.3.3}$$

With these definitions we may express the relations obtained when imposing the boundary conditions, at $z = 0$, in terms of $\underline{\widehat{B}} = \left[\underline{\underline{I}} + \underline{\underline{V}}\right]^{-1} \underline{U}$ thus the two sets of amplitudes read

$$\begin{aligned} B_m &= \widehat{B}_m \frac{2}{\pi m \cosh\left(\dfrac{\pi m g}{L-d}\right)}, \\ A_n &= -\frac{L-d}{L} \sum_{m=1}^{\infty} \zeta_{nm}^* B_m \sinh\left(\frac{\pi m g}{L-d}\right); \end{aligned} \tag{3.3.4}$$

the identity matrix is denoted here by $(\underline{\underline{I}})$. This quasi-analytic formulation enables good estimate of the numerical error, defined by

$$\text{Error} \equiv \frac{\int_0^{L-d} \left[\Phi(x,z=0^+) - \Phi(x,z=0^-)\right]^2 dx + \int_{L-d}^{L} \left[\Phi(x,z=0^+)\right]^2 dx}{\int_0^{L-d} \left[\Phi(x,z=0^-)\right]^2 dx} \tag{3.3.5}$$

and throughout the simulations to be presented this error was kept below 0.1%. Typical normalized potential-contours are illustrated in the lower frame of Fig. 3.2 (for $L = 1$ mm, $g = 1$ mm, $d = 0.5$ mm and $h = 1$ mm). The total current (I) due to field emission was calculated by integrating (3.3.1) over all four surfaces of *one period* of the structure; the integration was performed numerically. Based on the calculation of I we may establish the average current density $J_{\text{av}} = I/L\Delta_y$ where Δ_y represents the typical length in the y direction.

Let us now examine the current (I) and the average current density (J_{av}) as a function of the various parameters taking $\Delta_y = 1$ cm, $\varphi = 2$ eV and $V_0 = 200$ kV. When choosing the last parameter it has been tacitly assumed that the surface of the material that forms the grating is smooth ($\beta = 1$).

Figure 3.3 illustrates the average current density J_{av} as a function of the fraction d/L for various values of h/g. The plot reveals the existence of an optimal value close to $d/L = 0.008$. According to this plot this optimal value is virtually independent of the ratio h/g; in this case $L = g = 1$ mm. Furthermore, bearing in mind that the anode-cathode spacing (g) is constant, this plot indicates that the current increases as the groove's height increases. This increase continues as long as $h \leq L - d$; beyond this value there is no further increase in the peak current density. No optimal value (d/L) exists if we examine the current rather than the average current density. However, if we do plot the current in *one period* (I) as a function of L (with d/L is a parameter) we find that there is an optimal period L as clearly revealed in Fig. 3.4.

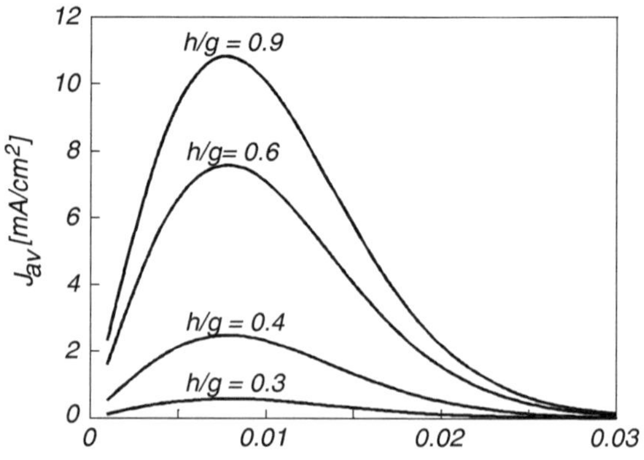

Fig. 3.3 The variation of the average current density J_{av} as a function of the ratio d/L, for various values of g/h. The other parameters are $L = 1$ mm and $g = 1$ mm

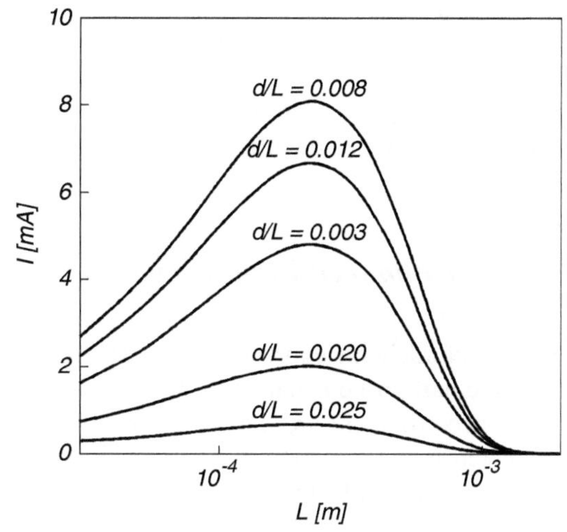

Fig. 3.4 The current emitted in *one period* as a function of the period length L (logarithmic scale), where d/L is a parameter; $h = 0.5\,\text{mm}$ and $g = 1\,\text{mm}$

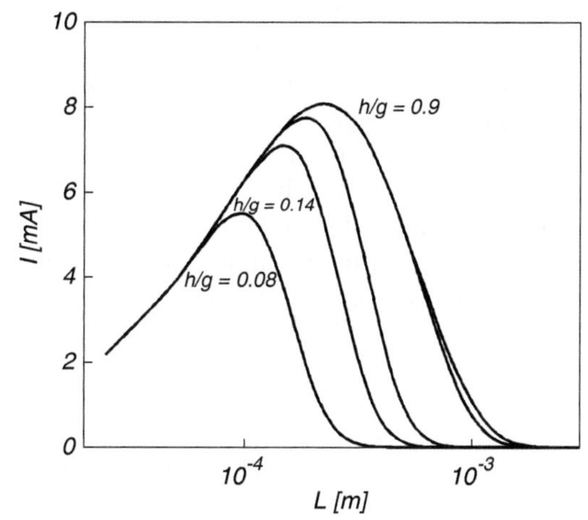

Fig. 3.5 The variation of the current emitted in one period as a function of the period length L for various values of h/g. The other parameters are $g = 1\,\text{mm}$ and $d/L = 0.008$

In this particular case the maximum occurs for $L = 0.25\,\text{mm}$ where the other parameters are: $h = 0.5\,\text{mm}$ and $g = 1\,\text{mm}$. Note that the location of the peak is almost independent of the ratio d/L. Finally, if we keep the ratio d/L constant and allow h to vary ($g = 1\,\text{mm}$ and $d/L = 0.008$), the peak of the current is strongly dependent on h – see Fig. 3.5. At high h values $(h \geq L - d)$ the current peak value reaches a constant value. No maximum occurs when the current in one period is plotted as a function of d/L.

In conclusion, for an efficient field emitter it is necessary to determine the optimization criterion: either we choose a grating geometry for generation of maximum current in one period or maximum average current-density. A different geometry meets the specification in either case. In case of a grating of a given period, we may expect to obtain maximum average current-density as a function of

d/L. On the other hand, if manufacturing constraints impose the ratio d/L, it is reasonable to expect an optimal grating's period in order to achieve maximum current. This approach may be employed for establishing the effective β of the surface, for example, the average current density illustrated in Fig. 3.3 at the given voltage, determines the effective β for a given grating configuration.

3.3.2 Enhanced Field Emission by a Dielectric Medium

Fowler-Nordeheim expression was developed subject to the assumption that the metallic half-space is flat and it faces vacuum. In many cases of interest, dielectric substances (oxides) are deposited on the metals either intentionally or not. Such a dielectric material may insulate the emitting surface and suppress the emission, or enhances locally the emission because of effective "focusing" of the electric field – as is the case of light rays and a lens. While this effect may be devastating in certain cases when the surface needs to sustain intense electric fields, it might be of great benefit in cases when controlled field emission is required e.g. electron emission using ferro-electric based cathodes. In this sub-section we present a crude model revealing the effect of a dielectric material on the field-emission from a metallic edge.

Consider a sharp edge of angle α attached to a dielectric (ε_r) half-space as illustrated in Fig. 3.6. It is assumed that the radius of curvature of the edge is much smaller than the longitudinal dimension (Δ_z) of the edge such that the system may be assumed to be infinite in the z direction. Furthermore, on the edge the electrostatic potential is assumed to be zero therefore, we may write the following solution for the Laplace equation

$$\Phi(r,\phi) = \begin{cases} A_1 \sin[\nu(\phi - \pi + \alpha)] r^{\nu} & 0<\phi<\pi-\alpha \\ A_2 \sin[\nu(\phi + \pi)] r^{\nu} & -\pi<\phi<0. \end{cases} \tag{3.3.6}$$

This solution satisfies the boundary conditions at $\phi = \pi - \alpha$ and $\phi = -\pi$. The curvature (ν) of the field is determined by imposing the boundary conditions

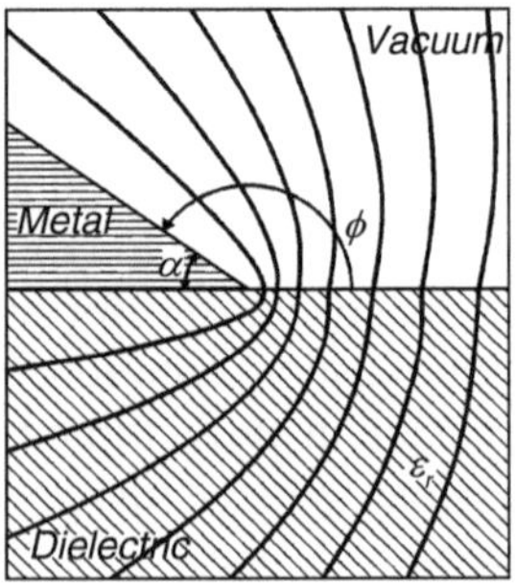

Fig. 3.6 Schematic of the triple-point system and the associated constant potential lines

for Φ and D_ϕ at $\phi = 0$. From the two resulting equations we can establish the expression which determines the variation of curvature parameter,ν, as a function of α and ε_r, it reads

$$\varepsilon_r \tan[\nu(\pi - \alpha)] = -\tan(\pi\nu). \tag{3.3.7}$$

Solution of this expression determines the general behavior of the potential near the edge. A typical distribution is illustrated in Fig. 3.6 for $\alpha = \pi/6$ and $\varepsilon_r = 300$.

At the limit of very high dielectric coefficient ($\varepsilon_r \to \infty$) the curvature approaches the value of $\nu \to 0.5$. For the other extreme ($\varepsilon_r = 1$), the solution of (2) has an analytic form which reads

$$\nu = \frac{1}{2}\left(1 - \frac{\alpha}{2\pi}\right)^{-1}. \tag{3.3.8}$$

Figure 3.7 illustrates the variation of the curvature (ν) as a function of the dielectric coefficient (ε_r) between the two limits mentioned above.

The curvature parameter determines the electric field, which in turn controls the surface charge distribution on the metallic edge. Let us now calculate the charge stored on both sides of the edge on a strip which is Δ_z long and its width is R. The choice of radius (R) is arbitrary but it is tacitly assumed that it is much larger than the radius of curvature of the edge and much smaller than the distance to the anode. On the top strip of the edge the total charge is given by $Q_{\text{top}} = -\Delta_z \int_0^R \mathrm{d}r D_\phi(r, \phi = \pi - \alpha) = \Delta_z \varepsilon_0 A_1 R^\nu$ whereas on the bottom strip $Q_{\text{bottom}} = -\Delta_z \varepsilon_0 \varepsilon_r A_2 R^\nu$. The electrostatic potential also enables us to calculate the total electrostatic energy (W_E) stored in a partial ($\pi - \alpha > \phi > -\pi$) cylinder of radius R and length Δ_z. This energy enables us to define the capacitance of the system according to $C \equiv Q_{total}^2/2W_E$ or explicitly

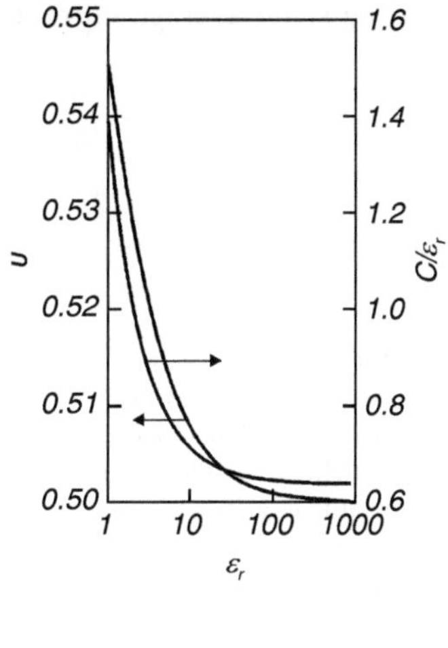

Fig. 3.7 The "dispersion" relation between the curvature parameter ν and the dielectric coefficient ε_r. The second curve illustrates the correlation between the nonlinear characteristic of the capacitance and ν; $\alpha = \pi/6$

$$C = \varepsilon_0 \Delta_z \frac{2\{\sin(\nu\pi) + \varepsilon_r \sin[\nu(\pi - \alpha)]\}^2}{(\pi - \alpha)\sin^2(\nu\pi) + \pi\varepsilon_r \sin^2[\nu(\pi - \alpha)]}, \tag{3.3.9}$$

where we used the definition of the total charge on both sides of the edge namely, $Q_{\text{total}} = Q_{\text{top}} + Q_{\text{bottom}}$. In zero order this capacitance is *linear* in ε_r however if we plot the ratio $C' \equiv C/\varepsilon_r\varepsilon_0\Delta_z$ (see Fig. 3.7) we find that the curvature parameter (ν) determines also the deviation from linearity of the capacitance. With the capacitance established we further define an *effective* voltage as $V_{\text{eff}} \equiv Q_{\text{total}}/C$ hence

$$V_{\text{eff}} = \frac{1}{2} A_1 R^\nu \frac{(\pi - \alpha)\sin^2(\nu\pi) + \pi\varepsilon_r \sin^2[\nu(\pi - \alpha)]}{\sin(\nu\pi)\{\sin(\nu\pi) + \varepsilon_r \sin[\nu(\pi - \alpha)]\}}. \tag{3.3.10}$$

We next examine the current emitted from the top strip of the edge assuming either a constant voltage (V_{eff}) or a constant charge (Q_{top}). For simplicity sake we adopt here a simple version of the Fowler-Nordheim expression, $J \simeq k_1 E^2 e^{-k_2/E}$ wherein $k_1 = 1.54 \times 10^{-6}/W[\text{A/V}^2]$, $k_2 = 6.83 \times 10^7 W^{1.5}[\text{V/m}]$ and W is the work function of the metal. Using the analytic expressions for the electric field derived based on (3.3.6), we can determine the current emitted from the top surface

$$I = I_0 \frac{\nu^2}{2\nu - 1}\left[e^{-a_0/\nu} + \frac{a_0}{\nu}\int_0^1 dy y^{(2\nu-1)/(1-\nu)} e^{-a_0 y/\nu}\right] \tag{3.3.11}$$

where $I_0 = \Delta_z R k_1 E_{\text{eff}}^2$, and $a_0 = k_2/E_{\text{eff}}$, $E_{\text{eff}} = Q_{\text{top}}/\varepsilon_0 R\Delta_z$. This expression indicates that the emitted current is inversely proportional to $2\nu - 1$ and bearing in mind that the curvature parameter (ν) approaches 0.5 as ε_r tends to infinity, we may expect the current to increase as ε_r increases. This result is confirmed in Fig. 3.8 where we plot the average current density defined as $I/\Delta_z R$. In fact (3.3.11) reveals that within a reasonable approximation $(a_0 \ll 1)$ the current is *linear* in ε_r.

A closer look at the variation of the current is revealed in Fig. 3.9 where we plot the same average current density but divided by ε_r and, as in the case of the

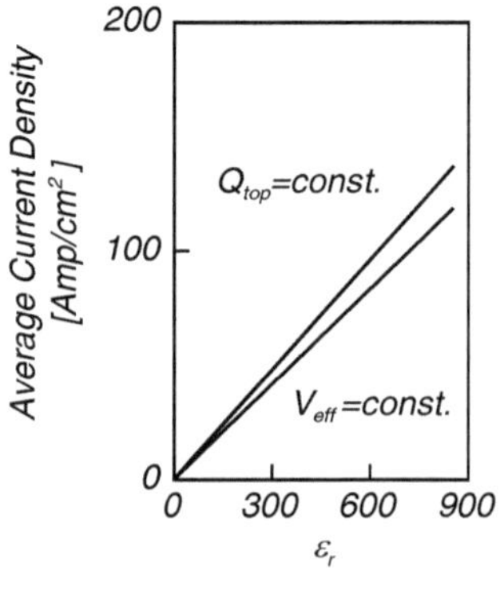

Fig. 3.8 Average current density as a function of the dielectric coefficient ε_r for constant charge on the top of the metallic edge Q_{top} and constant voltage V_{eff}; $\alpha = \pi/6$

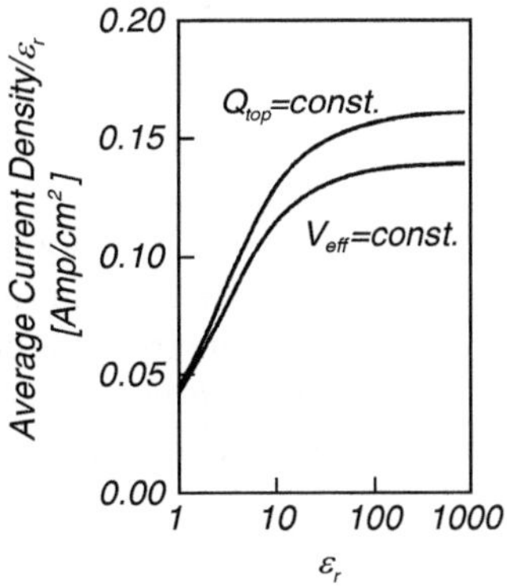

Fig. 3.9 Average current density divided by ε_r as a function of the dielectric coefficient ε_r for constant charge on the top of the metallic edge Q_{top} and constant voltage V_{eff}; $\alpha = \pi/6$

capacitance, we observe that the curvature parameter (v) affects the total current emitted by the metal; the parameters are as follows: $\Delta_z = 1\,\text{cm}$, $R = 1\,\text{mm}$, $\alpha = \pi/6$, $W = 4.5\,\text{eV}$ (Tungsten) and $Q_{\text{top}} = 0.5 \times 10^{-6}\Delta_z R$.

A similar behavior is observed when V_{eff} is assumed to be constant; the current in this case is given by

$$I(v) = I_1 \frac{\chi^2(v)}{2v-1}\left\{\exp\left[-\frac{b_0}{\chi(v)}\right] + \frac{b_0}{\chi(v)}\int_0^1 \mathrm{d}y y^{(2v-1)/(1-v)} \exp\left[-\frac{b_0 y}{\chi(v)}\right]\right\} \quad (3.3.12)$$

where $I_1 = \Delta_z R k_1 (V_{\text{eff}}/R)^2$, $b_0 = k_2/(V_{\text{eff}}/R)$ and

$$\chi(v) = \frac{2v\sin(v\pi)\{\sin(v\pi) + \varepsilon_r \sin[v(\pi-\alpha)]\}}{(\pi-\alpha)\sin^2(v\pi) + \pi\varepsilon_r \sin^2[v(\pi-\alpha)]}. \quad (3.3.13)$$

Figures 3.8 and 3.9 illustrate the same aspects discussed previously for the case when $V_{\text{eff}} = 80\,\text{V}$.

In either one of the cases, if the electric field is sufficiently high ($k_2 << E$) such that $e^{-k_2/E} \approx 1$, then the expression for the current has a simple form

$$\begin{aligned} I_Q &\equiv I_{Q_{\text{top}}=\text{const.}}(v) = I_0 \frac{v^2}{2v-1}, \\ I_V &\equiv I_{V_{\text{eff}}=\text{const.}}(v) = I_1 \frac{\chi^2(v)}{2v-1}. \end{aligned} \quad (3.3.14)$$

At the two extremes ($\varepsilon_r = 1$ and $\varepsilon_r \gg 1$) and assuming $\alpha \ll 2\pi$, these expressions can be further simplified to read

$$\begin{aligned} \varepsilon_r = 1: I_Q &= I_0\left(\frac{\pi}{2\alpha}\right), & I_V &= I_1\left(\frac{2}{\pi\alpha}\right) \\ \varepsilon_r \gg 1: I_Q &\simeq I_0\left(\frac{\pi}{2\alpha}\right)\left(\frac{\varepsilon_r}{2}\right), & I_V &\simeq I_1\left(\frac{2}{\pi\alpha}\right)\frac{\varepsilon_r}{2} \end{aligned} \quad (3.3.15)$$

thus the field-emission enhancement factor for $\varepsilon_r \gg 1$ is given by

$$\beta \simeq \sqrt{\frac{\varepsilon_r}{2}}. \tag{3.3.16}$$

The result presented above is subject to at least one tacit assumption: we have used the Fowler-Nordheim formula near the *edge* although it has been developed for a *planar* geometry. This clearly sets limits on the radius of curvature of the tip – it should be much larger than a quantum level radius of curvature ($\sim 1-5$ nm) and much smaller than macroscopic curvature $\sim 1-5\ \mu$m.

3.3.3 Child-Langmuir Limiting Current: Planar Diode

The microscopic details of the field emission from metal were considered in literature and a short review can be found in Miller's (1982) book. It is beyond the scope of our presentation to investigate the dynamics of electrons at the microscopic level in the metal and in what follows we assume that whatever field is applied normal to the metallic cathode, electrons are being emitted. In fact, the discussion to follow is independent of the mechanism electrons are extracted from the cathode and the question to be addressed is: what is the *limiting current* one can extract from a cathode by applying a voltage on the anode? The logic behind this question is simple: As a voltage is applied, electrons leave the cathode and move towards the anode. Since they traverse the anode-cathode gap in a finite time, the cathode is screened by these electrons and the field it experiences is weaker – the situation is illustrated in Fig. 3.10. We expect to reach the maximum current limit as the electric field on the anode is zero that is to say, the cathode is screened.

A typical voltage pulse on the anode is 100 ns or longer and for comparison, the time an electron moving at 0.5 c traverses a 3 cm gap is on the order of 0.2 ns. The three orders of magnitude difference between these two time scales justifies the static approximation used next. In a one-dimension system, the electric scalar potential is a solution of the Poisson equation

$$\frac{d^2\Phi}{dz^2} = \frac{e}{\varepsilon_0} n(z), \tag{3.3.17}$$

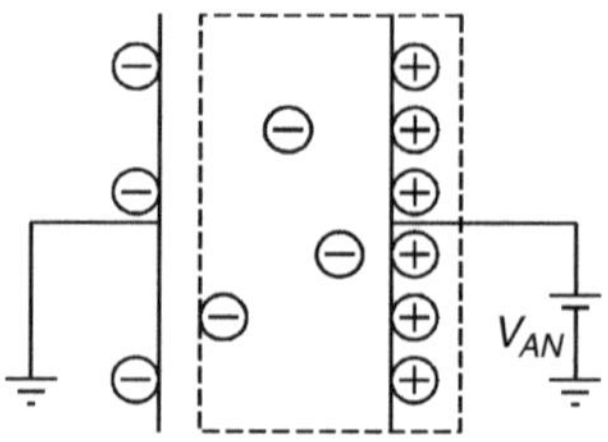

Fig. 3.10 Child-Langmuir limiting current: electrons in the gap screen the cathode

where $n(z)$ is the particle's density in the anode-cathode gap and it is yet to be determined; e is the charge of one electron. The dynamics of the particles' density is determined, within the framework of the hydro-dynamic approximation, by the continuity equation which in the case of a static problem reads

$$\frac{\mathrm{d}}{\mathrm{d}z}[n(z)\mathrm{v}(z)] = 0. \tag{3.3.18}$$

The velocity $\mathrm{v}(z)$, is governed by the equations of motion but in this particular case it is more convenient to use the single particle energy conservation (3.1.34) which in conjunction with (3.1.36) and the static case considered here ($\partial/\partial t = 0$) reads

$$\frac{\mathrm{d}}{\mathrm{d}z}\left[\gamma(z) - \frac{e\Phi(z)}{mc^2}\right] = 0. \tag{3.3.19}$$

At the cathode, the potential is zero and the initial kinetic energy of the electrons is $mc^2(\gamma(0) - 1)$ implying,

$$\gamma(z) = \gamma(0) + \frac{e\Phi(z)}{mc^2}. \tag{3.3.20}$$

Expression (3.3.18) indicates that the current density J is constant in space and it reads

$$J = -en(z)\mathrm{v}(z) = \text{const}. \tag{3.3.21}$$

Equations (3.3.17)–(3.3.19) govern the dynamics of the electron in a static potential. In order to proceed to a solution of these equations it is convenient to substitute (3.3.20) in (3.3.17); in the resulting expression, we substitute the density from (3.3.21) and obtain

$$\frac{\mathrm{d}^2}{\mathrm{d}z^2}\gamma = \frac{eJ\eta_0}{mc^2}\frac{\gamma}{\sqrt{\gamma^2 - 1}}, \tag{3.3.22}$$

where we also used the fact that $\eta_0 \equiv 1/c\varepsilon_0$. The coefficient on the right-hand side has units of 1/length^2, therefore we firstly define

$$K^2 = \frac{eJ\eta_0}{mc^2}, \tag{3.3.23}$$

and secondly, employ this definition to normalize coordinate $\zeta = Kz$. Assuming that the electric field is not zero over the entire domain, the next step is to multiply (3.3.22) by $\mathrm{d}\gamma/\mathrm{d}\zeta$ and get

$$\frac{1}{2}\frac{\mathrm{d}}{\mathrm{d}\zeta}\left(\frac{\mathrm{d}\gamma}{\mathrm{d}\zeta}\right)^2 - \frac{\mathrm{d}}{\mathrm{d}\zeta}\sqrt{\gamma^2 - 1} = 0, \tag{3.3.24}$$

which implies that

$$\frac{1}{2}\left(\frac{\mathrm{d}\gamma}{\mathrm{d}\zeta}\right)^2 - \sqrt{\gamma^2 - 1} = \text{const.} \tag{3.3.25}$$

As indicated above, we consider the limit when the cathode is completely screened by the space-charge in the gap, therefore at $\zeta = 0$ the electric field vanishes and according to (3.3.19) $\mathrm{d}\gamma/\mathrm{d}\zeta = 0$. However, for the sake of generality, let us assume that the electrons enter the diode gap with virtually zero kinetic energy i.e., $\gamma(\zeta = 0) = 1$ implying

$$\frac{1}{\sqrt{2}}\frac{\mathrm{d}\gamma}{\mathrm{d}\zeta} = \left(\gamma^2 - 1\right)^{1/4}. \tag{3.3.26}$$

The last expression can be integrated and the formal result is

$$\begin{aligned} Kg\sqrt{2} &= \int_1^{\gamma_{AN}} \frac{\mathrm{d}\gamma}{\left(\gamma^2 - 1\right)^{1/4}} = \begin{cases} 2^{-\frac{1}{4}}\frac{4}{3}\left(\frac{eV_{AN}}{mc^2}\right)^{\frac{3}{4}} & \frac{eV_{AN}}{mc^2} < 1 \\ -1.6 + 2\left(\frac{eV_{AN}}{mc^2}\right)^{\frac{1}{2}} & \frac{eV_{AN}}{mc^2} \gg 1 \end{cases} \\ &\simeq \frac{2^{-\frac{1}{4}}\frac{4}{3}\left(\frac{eV_{AN}}{mc^2}\right)^{\frac{3}{4}}}{\left[1 + \left(\frac{8}{81}\frac{eV_{AN}}{mc^2}\right)^{\frac{3}{4}}\right]^{\frac{1}{3}}} \end{aligned} \tag{3.3.27}$$

where $\gamma_{AN} = 1 + eV_{AN}/mc^2$, V_{AN} is the voltage applied on the anode and g is the anode-cathode gap; the last term is an approximation within less than 3% of the exact expression. With the result in (3.3.27) it is possible to determine the explicit expression for the Child-Langmuir *limiting current*

$$J = \frac{16}{18\sqrt{2}}\frac{mc^2}{e\eta_0}\frac{1}{g^2}\frac{\left(\frac{eV_{AN}}{mc^2}\right)^{\frac{3}{2}}}{\left[1 + \left(\frac{8}{81}\frac{eV_{AN}}{mc^2}\right)^{\frac{3}{4}}\right]^{\frac{2}{3}}} \tag{3.3.28}$$

or more conveniently

$$J[\mathrm{kA/cm^2}] = \frac{2.33}{g^2[\mathrm{cm}]}\frac{V_{AN}^{3/2}[\mathrm{MV}]}{\left[1 + 0.291 \times V_{AN}^{3/4}[\mathrm{MV}]\right]^{2/3}}. \tag{3.3.29}$$

Although this expression was developed for a planar diode, the scaling of the current with the voltage remains the same in cylindrical and spherical geometries, therefore in analogy to the conductance in a metallic resistor, one can define the *perveance* as, $p \equiv I/V_{AN}^{3/2}$, where I is the total current which flows in the diode.

In order to have a feeling about this relation (3.3.29) we can calculate the limiting current for 1 MV voltage applied on a 3 cm gap. The result is 260 A/cm^2. It is important to realize that this is a supremum. For example, if the applied voltage is only 100 kV, then the limiting current is 8.2 A/cm^2 but a field $E \simeq V_{AN}/g \simeq 3\,\text{MV/m}$ is not sufficient to extract virtually any current, via field emission. In the case of thermionic emission, the current J emitted when a metallic cathode whose work function Φ_w is heated to an absolute temperature T is given by

$$J[\text{kA/cm}^2] \simeq 0.12\,T^2 \exp\left(-\frac{e\Phi_w}{k_B T}\right); \tag{3.3.30}$$

for $T = 1{,}400^\circ$K and $\Phi_w = 2\,\text{V}$ the current density is 15 A/cm^2. If we go back to the example above, we realize that for the first case ($V_{AN} = 1$ MV) we can not reach the space-charge limiting current since the thermionic cathode can generate only one third of this limit. On the other hand, applying $V_{AN} = 100$ kV the system is space-charge limited and we do not utilize the entire potential of the thermionic emission. A similar situation occurs in the case of photo-emission.

3.3.4 Child-Langmuir Limiting Current: Emitting Edge

In Sect. 3.3.2 we have examined emission from a metallic edge in the vicinity of a dielectric half-space and we have shown that for sufficiently large dielectric coefficient the emitted current is proportional to the dielectric coefficient. In the framework of this analysis, we have ignored space-charge effects. In the present sub-section, the space-charge will be shown to "smooth" the edge effect.

Consider a dielectric (ε_r) medium occupying the space $\pi < \phi < 2\pi + \beta$ and a metallic edge of an angle α that may emit electrons at any electric field; see Fig. 3.11. Assuming a uniform system in the z direction, the potential (Φ) satisfies the 2D Poisson equation and it is assumed to be driven by a density of electrons denoted by n. No time variations occur in this system and the motion of the charges is described by a velocity field $\mathbf{v}$. Consequently, the current density, defined as $\mathbf{J} \equiv -en\mathbf{v}$, satisfies the continuity equation that in the stationary case reads, $\nabla \cdot (n\mathbf{v}) = 0$. Furthermore, the two relevant equations of motion, $(\mathbf{v} \cdot \vec{\nabla})\mathbf{v} = (e/m)\nabla\Phi$, are equivalent to energy conservation $mv^2/2 = e\Phi$ and $\vec{\nabla} \times \mathbf{v} = 0$; the last two relations will be used in what follows. Bearing in mind that the boundary conditions are conveniently expressed in a cylindrical coordinate system, we may summarize the governing equations as follows

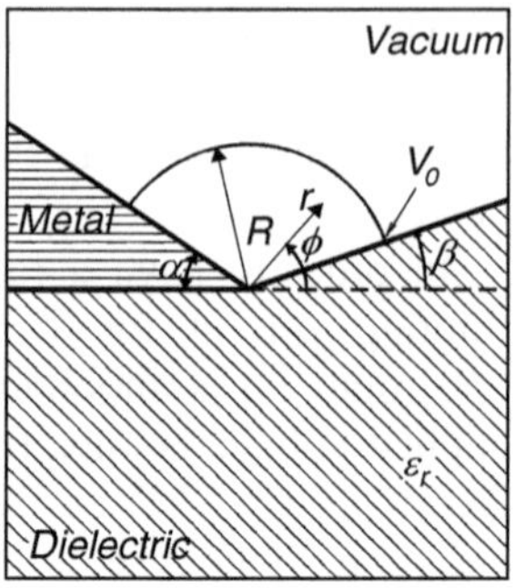

Fig. 3.11 The schematics of metallic (α) and dielectric ($\pi + \beta$) edges. At an arbitrary radius R at the interface between the vacuum and the dielectric, the potential is assumed to be known and is denoted by V_0

$$\begin{aligned}
&\left[\frac{1}{r}\frac{\partial}{\partial r}r\frac{\partial}{\partial r}+\frac{1}{r^2}\frac{\partial^2}{\partial\phi^2}\right]\Phi=\frac{e}{\varepsilon_0}n,\\
&\frac{\partial}{\partial r}(rn\mathrm{v}_r)+\frac{\partial}{\partial\phi}(n\mathrm{v}_\phi)=0,\\
&\mathrm{v}_r^2+\mathrm{v}_\phi^2=\frac{2e}{m}\Phi,\\
&\frac{\partial}{\partial r}(r\mathrm{v}_\phi)-\frac{\partial}{\partial\phi}\mathrm{v}_r=0.
\end{aligned} \tag{3.3.31}$$

The last expression corresponds to the z component of $\vec{\nabla}\times\mathbf{v}=0$ and it is satisfied by choosing $\mathbf{v}$ to be the gradient of a scalar flow function Ψ i.e. $\mathbf{v}=\nabla\Psi$. In the dielectric region, where no free electrons are present, the solution of the potential satisfies Laplace's equation and it may be written as

$$\Phi(r,-\pi\le\phi\le\beta)=A\left(\frac{r}{R}\right)^{\nu}\sin[\nu(\phi+\pi)] \tag{3.3.32}$$

where the *radial* curvature parameter ν is yet to be determined and, as can be readily checked, this solution satisfies the boundary condition on the metal $\phi=-\pi$. Now, since the potential has to be continuous at $\phi=\beta$ we may assume that in the space-charge region the potential may be assumed to have the form

$$\Phi(r,\beta\le\phi\le\pi-\alpha)=\Phi_0(\phi)\left(\frac{r}{R}\right)^{\nu}. \tag{3.3.33}$$

With this type of solution in mind we conclude, based on Poisson equation and energy conservation, that

$$\begin{aligned}
n(r,\phi)&=N(\phi)\left(\frac{r}{R}\right)^{\nu-2},\\
\Psi(r,\phi)&=\Psi_0(\phi)\left(\frac{r}{R}\right)^{1+\frac{1}{2}\nu}.
\end{aligned} \tag{3.3.34}$$

Substituting these expressions in the Poisson equation, the continuity equation and energy conservation equation, we obtain

$$\begin{aligned}
&\left[\frac{d^2}{d\phi^2}+v^2\right]\Phi_0=\frac{eR^2}{\varepsilon_0}N,\\
&\frac{d}{d\phi}\left[N\frac{d\Psi_0}{d\phi}\right]+\left(1+\frac{1}{2}v\right)\left(\frac{3v}{2}-1\right)N\Psi_0=0,\\
&\left(1+\frac{1}{2}v\right)^2\Psi_0^2+\left[\frac{d\Psi_0}{d\phi}\right]^2=\frac{2eR^2}{m}\Phi_0.
\end{aligned}\tag{3.3.35}$$

These equations describe the dynamics of the electrons and the field in the vacuum region.

Associated with these equations there is a set of boundary conditions that will be formulated next. Specifically, the potential at $\phi=\beta$ is continuous $A\sin[v(\pi+\beta)]=\Phi_0(\phi=\beta)$ and so is D_ϕ therefore, $v\varepsilon_r A\cos[v(\pi+\beta)]=[d\Phi_0/d\phi]_{\phi=\beta}$; from these two conditions we find that

$$\cot[v(\pi+\beta)]=\frac{1}{v\varepsilon_r}\left[\frac{d}{d\phi}\ln(\Phi_0)\right]_{\phi=\beta}.\tag{3.3.36}$$

If we assume that $\Phi_0(\phi)$ is known, then this relation enables us to evaluate the radial curvature parameter v. In addition to the conditions above, the potential vanishes on the edge, it equals V_0 at the interface with the dielectric [$r=R$ and $\phi=\beta$] and the electric field normal to the edge at a radius R is zero since the electrons emitted in the diode gap screen the cathode [Child-Langmuir condition, $E_\phi(\phi=\pi-\alpha)=0$] or explicitly,

$$\Phi_0(\phi=\pi-\alpha)=0,\qquad \Phi_0(\phi=\beta)=V_0,\quad \text{and}\quad \left[\frac{d\Phi_0}{d\phi}\right]_{\phi=\pi-\alpha}=0.\tag{3.3.37}$$

It is important to emphasize that in principle V_0 is determined in conjunction with the configuration away from the edge and in the framework of this analysis, it is assumed to be known.

In addition to the boundary conditions associated with the potential, there are three other conditions to be imposed on the velocity field: (1) its *radial* component has to vanish on the edge $v_r(r,\phi=\pi-\alpha)=0$. (2) Its *azimuthal* velocity at the interface with the dielectric has to be zero $v_\phi(r,\phi=\beta)=0$ and (3) the byproduct of the last two relations, in conjunction with energy conservation, implies that $v_\phi(\phi=\pi-\alpha)=0$. Explicitly these three conditions read

$$\Psi_0(\phi=\pi-\alpha)=0,\qquad \left[\frac{d\Psi_0}{d\phi}\right]_{\phi=\beta}=0\quad\text{and}\quad \left[\frac{d\Psi_0}{d\phi}\right]_{\phi=\pi-\alpha}=0.\tag{3.3.38}$$

As already stated, our goal is to determine the limiting current emitted from the metallic edge. The current density from the edge is given by $J_\phi(r, \phi = \pi - \alpha) = -e[n(r,\phi)\mathrm{v}_\phi(r,\phi)]_{\phi=\pi-\alpha}$ therefore the current in a unit area $\Delta_z \times R$ reads

$$\begin{aligned} I &= e\Delta_z\left[N(\phi)\frac{\mathrm{d}\Psi_0}{\mathrm{d}\phi}\right]_{\phi=\pi-\alpha}\int_0^R \mathrm{d}r\frac{1}{r}\left(\frac{r}{R}\right)^{\nu-2}\left(\frac{r}{R}\right)^{1+\frac{1}{2}\nu} \\ &= e\Delta_z\left[N(\phi)\frac{\mathrm{d}\Psi_0}{\mathrm{d}\phi}\right]_{\phi=\pi-\alpha}\left(\frac{3\nu}{2}-1\right)^{-1} \end{aligned} \tag{3.3.39}$$

where it was tacitly assumed that $\nu > 2/3$. According to this result it is necessary to evaluate all the quantities of interest near $\phi \simeq \pi - \alpha$. In particular, note that based on the continuity equation in (3.3.35) we have to determine N and Ψ_0 since

$$\left[N\frac{\mathrm{d}\Psi_0}{\mathrm{d}\phi}\right]_{\phi=\pi-\alpha} = -\left(1+\frac{\nu}{2}\right)\left(\frac{3\nu}{2}-1\right)\int_\beta^{\pi-\alpha}\mathrm{d}\phi' N(\phi')\Psi_0(\phi'). \tag{3.3.40}$$

The starting point is the electrostatic potential. From the three boundary conditions in (3.3.38), we may "construct" a simple solution for the potential that has the following form

$$\Phi_0(\phi \simeq \pi - \alpha) = V_0\left(\frac{\pi-\alpha-\phi}{\pi-\alpha-\beta}\right)^p; \tag{3.3.41}$$

where p is the *azimuthal* curvature parameter and it is yet to be determined. However, from the boundary conditions we may conclude that it has to be larger than unity $(p > 1)$. Based on Poisson's equation and assuming that $p < 2$, then in the vicinity of $\phi \simeq \pi - \alpha$

$$N(\phi \simeq \pi - \alpha) = \frac{\varepsilon_0 V_0}{eR^2}\frac{p(p-1)}{(\pi-\alpha-\beta)^2}\left(\frac{\pi-\alpha-\phi}{\pi-\alpha-\beta}\right)^{p-2}. \tag{3.3.42}$$

In the framework of the approximation associated with (3.3.41) and relying on the expression for energy conservation, the flow function Ψ_0 is given by

$$\Psi_0(\phi \simeq \pi - \alpha) = \sqrt{V_0\frac{2eR^2}{m}}\frac{1}{1+p/2}\frac{(\pi-\alpha-\phi)^{1+p/2}}{(\pi-\alpha-\beta)^{p/2}}. \tag{3.3.43}$$

Since the current density has to be finite at $\phi = \pi - \alpha$ we conclude that $p - 2 + p/2 = 0$ implying $p = 4/3$; this is exactly the characteristic spatial variation of the potential in a planar diode at the limiting current $[\Phi = V_0(z/g)^{4/3}$, where g is the diode gap].

Substituting (3.3.42) and (3.3.43) in (3.3.40) and (3.3.39) the average limiting current density is given by

$$J_{\mathrm{av}} \equiv \frac{I}{\Delta_z R} = \left[\frac{3}{10}\left(1+\frac{v}{2}\right)\right]\left[\frac{4}{9}\varepsilon_0\sqrt{\frac{2e}{m}}\frac{V_0^{3/2}}{R^2}\right]. \tag{3.3.44}$$

It is evident that the right term in the brackets has the form of Child-Langmuir (CL) law. Moreover, it is important to point out that at the denominator the length parameter, R, is not the same as in regular planar CL where it represents the diode gap. Here this parameter represents the typical height of a *micro-protrusion* on the cathode. Consequently, the limiting current may exceed the "regular" CL current determined based on the anode-cathode spacing, by many orders of magnitude. The right term indicates that the average limiting current density is linear with the *radial* curvature parameter, v, which according to (3.3.36) and subject to the approximation in (3.3.41) is a solution of

$$\cot[v(\pi+\beta)] \simeq -\frac{p}{v\varepsilon_{\mathrm{r}}}\frac{1}{\pi-\alpha-\beta}. \tag{3.3.45}$$

According to this last expression the radial curvature parameter, v, depends on the geometry of the system (α and β) and on the electrical property (ε_{r}) of the medium. Its dependence is particularly simple at the limit of very large dielectric coefficient, $\varepsilon_{\mathrm{r}} \gg 1$, in which case $v(\pi+\beta) = \pi(1/2+i)$ where $i = 0, \pm1, \pm2$..... Subject to this condition, the average current density is not explicitly dependent on the angle of the metallic edge (α) since $v = (3/2)(1+\beta/\pi)^{-1}$. For a given geometry ($\alpha$ and β) the average current density decreases as a function of ε_{r} until it reaches the asymptotic level mentioned above; a typical behavior is illustrated in Fig. 3.12.

Another interesting aspect is revealed when comparing the radial curvature of the field in vacuum (v_{V}) with that in a space-charge dominated region (v_{CL}). The former is a solution of $\varepsilon_{\mathrm{r}} = -\tan[v_{\mathrm{V}}(\pi+\beta)]\cot[v_{\mathrm{V}}(\pi-\alpha-\beta)]$ and it can be checked that for $\varepsilon_{\mathrm{r}} = 1$, $v_{\mathrm{V}} = \pi/(2\pi-\alpha)$ thus for $\alpha = \pi/6$ the vacuum curvature

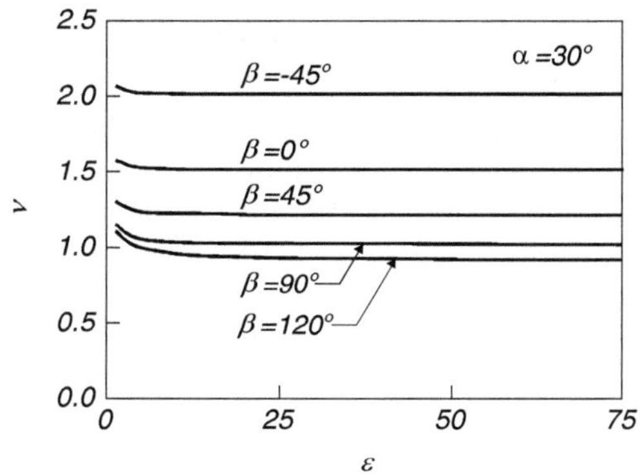

Fig. 3.12 For a given geometry (α and β) the average current density decreases as a function of ε_{r} until it reaches the asymptotic level

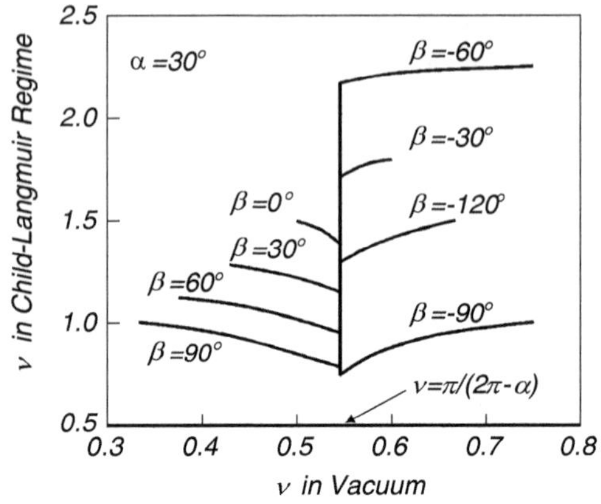

Fig. 3.13 Comparison of the radial curvature of the field in vacuum (ν_V) and a space-charge dominated region (ν_{CL}). It is evident that in many cases the space-charge leads to elimination of singularity i.e. ν rather than being smaller than unity, becomes larger than unity under the effect of the space-charge

is $\nu_V = 6/11$. This point and the dependence of ν_V and ν_{CL} are illustrated in Fig. 3.13. It is evident that there are cases whereby the space-charge leads to *elimination of singularity* i.e. ν rather than being smaller than unity, it becomes larger than unity.

3.4 Beam Propagation

Once the electron beam has been generated, it has to be confined to a small region in space and guided towards the interaction region. Naturally, beams tend to diverge under the repelling effect of the electrostatic force and while this divergence is somewhat reduced by the inherent magnetic force there is still need for external means in order to preserve the beam shape. The most common way to guide electron beams is to apply a static magnetic field, which can be either uniform or periodic. In this section, we consider the propagation of a cylindrical beam i.e., assuming an azimuthally symmetric system.

3.4.1 Beam Propagation in Free Space: Uniform B-Field

In the absence of a guiding magnetic field, a non-neutral beam will diverge under the influence of the repelling electrostatic force. If the applied magnetic field is too low, the beam will also diverge but after a longer distance. As the guiding field exceeds a certain value, to be determined next, the trajectories are stable. In order to investigate the electron motion let us consider the two transverse components of the equations of motion. The radial component reads

$$\frac{\mathrm{d}}{\mathrm{d}t}\mathrm{v}_r + \frac{1}{\gamma}\frac{\mathrm{d}\gamma}{\mathrm{d}t}\mathrm{v}_r - \frac{1}{r}\mathrm{v}_\phi^2 = -\frac{e}{m}\frac{1}{\gamma}\left(E_r + \mathrm{v}_\phi B_z - \mathrm{v}_z B_\phi\right), \qquad (3.4.1)$$

whereas the azimuthal component is given by

$$\frac{1}{r}\frac{\mathrm{d}}{\mathrm{d}t}(\gamma r \mathrm{v}_\phi) = -\frac{e}{m}(E_\phi - \mathrm{v}_r B_z + \mathrm{v}_z B_r). \tag{3.4.2}$$

Before we actually proceed to the stability analysis, it will be instructive to examine the last equation more thoroughly. For this purpose, we define the magnetic flux encompassed by an electron moving at a distance r from the axis by

$$\Psi(r,z) = \int_0^r \mathrm{d}r' 2\pi r' B_z(r',z). \tag{3.4.3}$$

The absolute time variation of this quantity is given by

$$\frac{\mathrm{d}\Psi}{\mathrm{d}t} = \frac{\partial\Psi}{\partial r}\frac{\mathrm{d}r}{\mathrm{d}t} + \frac{\partial\Psi}{\partial z}\frac{\mathrm{d}z}{\mathrm{d}t}, \tag{3.4.4}$$

and it can be simplified by using the definition of the magnetic flux and also the fact that the divergence of the magnetic inductance $\mathbf{B}$ vanishes i.e., $\nabla \cdot \mathbf{B} = 0$:

$$\frac{\mathrm{d}\Psi}{\mathrm{d}t} = 2\pi r B_z \mathrm{v}_r - 2\pi r B_r \mathrm{v}_z. \tag{3.4.5}$$

If we now compare the right-hand side of the last equation with the right-hand side of (3.4.2) we observe that in the absence of azimuthal electric field, they are virtually identical and therefore, by substituting (3.4.2) in (3.4.5) we obtain

$$\frac{\mathrm{d}}{\mathrm{d}t}\left[\Psi - \frac{2\pi}{e} r(m\gamma \mathrm{v}_\phi)\right] = 0. \tag{3.4.6}$$

This expression indicates that for a particle "born" with zero azimuthal motion in a magnetic field, the azimuthal motion is determined entirely by the local flux, its radial location and the energy. A different interpretation of the same result can be achieved by noting that we can represent the magnetic field in terms of the azimuthal magnetic vector potential as $B_z = [\partial(rA_\phi)/\partial r]/r$ therefore substituting in the definition of the flux we obtain

$$r(m\gamma \mathrm{v}_\phi - eA_\phi) = rp_\phi = \text{const.}, \tag{3.4.7}$$

which is the longitudinal component of the *canonical angular momentum*.

With this last result we can now investigate the radial component of the equation of motion. For this purpose, it is assumed that the guiding magnetic field is uniform (B_0), thus the azimuthal component of the magnetic vector potential is $A_\phi = rB_0/2$. Consequently, the conservation of the canonical angular momentum implies

$$rp_\phi = r(m\gamma \mathrm{v}_\phi - eA_\phi) = r\left(m\gamma \mathrm{v}_\phi - e\frac{1}{2}rB_0\right) = rm\left(\gamma \mathrm{v}_\phi - \frac{1}{2}\Omega_\mathrm{c} r\right). \tag{3.4.8}$$

where Ω_c is the (non-relativistic) cyclotron angular frequency defined by

$$\Omega_c \equiv \frac{eB_0}{m}. \tag{3.4.9}$$

We can use the last expression in (3.4.8) and substitute v_ϕ in (3.4.1); the result is

$$\frac{d^2}{dt^2}r - \frac{\Omega_c^2}{4\gamma^2 r^3}(r^2 - r_0^2)^2 + \frac{1}{2}\frac{\Omega_c^2}{r\gamma}(r^2 - r_0^2) = -\frac{e}{m\gamma}(E_r - v_z B_\phi). \tag{3.4.10}$$

In this expression, we neglected the energy variation and r_0 is the radius where the electron was "born" – it is tacitly assumed that at this location the azimuthal velocity vanishes ($v_\phi = 0$).

For further simplification of the equation which describes the radial motion we next evaluate the two field components in the right-hand side of (3.4.10). The radial component of the electric field represents the field generated by electrons located at radii smaller than that of the particle therefore, using (2.1.3) which in our case reads

$$\frac{1}{r}\frac{d}{dr}rE_r(r) = -\frac{e}{\varepsilon_0}n(r), \tag{3.4.11}$$

and assuming uniform distribution of particles, we find that

$$E_r(r) = -\frac{en}{2\varepsilon_0}r. \tag{3.4.12}$$

In a similar way, we consider (2.1.2) to calculate the azimuthal component of the magnetic field; the relevant component reads

$$\frac{1}{r}\frac{d}{dr}\left[rH_\phi(r)\right] = -en(r)v_z(r). \tag{3.4.13}$$

As in the previous case, we assume that both the density and the velocity are uniform in space hence

$$H_\phi(r) = -\frac{env_z}{2}r. \tag{3.4.14}$$

These two expressions (3.4.12), (3.4.14) can be substituted in the right-hand side of the radial component of the equation of motion (3.4.10) which then reads

$$\frac{d^2}{dt^2}r - \frac{\Omega_c^2}{4\gamma^2 r^3}(r^2 - r_0^2)^2 + \frac{1}{2}\frac{\Omega_c^2}{r\gamma}(r^2 - r_0^2) = \frac{1}{2}\frac{e^2 n}{m\varepsilon_0\gamma^3}r, \tag{3.4.15}$$

In equilibrium, there are no time variations, therefore the beam radius (R_b) is a solution of

$$-\frac{\Omega_c^2}{4\gamma^2 R_b^3}(R_b^2 - r_0^2)^2 + \frac{1}{2}\frac{\Omega_c^2}{R_b\gamma}(R_b^2 - r_0^2) = \frac{1}{2\gamma^3}\omega_p^2 R_b. \tag{3.4.16}$$

which is

$$R_b^2 = \frac{r_0^2}{\sqrt{1 - 2\omega_p^2/\Omega_c^2\gamma}}; \tag{3.4.17}$$

in these expressions ω_p is the non-relativistic plasma frequency

$$\omega_p^2 = \frac{e^2 n}{m\varepsilon_0}. \tag{3.4.18}$$

From the expression in (3.4.17) we conclude that a stable solution exists only if

$$\Omega_c^2 \geq 2\omega_p^2 \frac{1}{\gamma}. \tag{3.4.19}$$

Note that from (3.4.15) we can have a clear measure of the way the beam diverges in the absence of a guiding magnetic field. When $\Omega_c = 0$ it is convenient to define $\bar{r} \equiv r(t)/r_0$ and $\tau \equiv t\omega_p/\gamma^{3/2}\sqrt{2}$. With this notation (3.4.15) reads

$$\left[\frac{d^2}{d\tau^2} - 1\right]\bar{r} = 0. \tag{3.4.20}$$

A general solution of this equation is a superposition of $\sinh(\tau)$ and $\cosh(\tau)$. If, for simplicity, we assume that the radial velocity at $\tau = 0$ is zero then $\bar{r} = \cosh(\tau)$ which clearly indicates that the beam diverges.

Note that there are two characteristic length parameters which determine the beam radius: the first (r_0) is trivial and denotes the radius where the particle was "born". In order to establish the second, we use the definition of the total current $I = env_0\pi R_b^2$ and substitute it in the expression for the plasma frequency, to rewrite (3.4.17) as

$$R_b^2 = \frac{r_0^2}{\sqrt{1 - L_0^2/R_b^2}}, \tag{3.4.21}$$

where

$$L_0^2 \equiv \frac{2}{\pi}\frac{1}{\gamma\beta}\frac{e\eta_0 I}{ecB_0}\frac{mc^2}{ecB_0} \tag{3.4.22}$$

and finally the beam radius is determined by

$$R_{\rm b}^2 = \frac{1}{2}\left(L_0^2 + \sqrt{L_0^4 + 4r_0^4}\right). \tag{3.4.23}$$

From this expression, we conclude that the beam radius increases monotonically with L_0, therefore for a given r_0, the beam is compressed when the guiding field or the kinetic energy is increased. The radius increases when the current is raised. In addition, it is possible to deduce the equilibrium radius of the beam corresponding to electrons "born" in a zero magnetic field and on axis ($r_0 \simeq 0$), its value ($R_{\rm b} \simeq L_0$) is determined entirely by the guiding field, current and kinetic energy.

3.4.2 Beam Propagation in Free Space: Periodic B-Field

Generation of a uniform magnetic field for guiding an intense relativistic beam may require a substantial amount of energy. This magnetic field is generated by discharging a large bank of capacitors in solenoids and it can become quite energy consuming when a high repetition rate is required. In the latter case the main alternative is to consider the use of *permanent periodic magnets* (PPM) for guiding the beam. The magnetic field configuration is

$$\begin{aligned} B_z(r,z) &= B_0 \cos(k_{\rm w} z) {\rm I}_0(k_{\rm w} r), \\ B_r(r,z) &= B_0 \sin(k_{\rm w} z) {\rm I}_1(k_{\rm w} r), \end{aligned} \tag{3.4.24}$$

where $k_{\rm w} = 2\pi/L$ and L is the period of the permanent magnetic field. In order to present the stability condition, the analysis will be limited to a narrow pencil beam of maximum radius $R_{\rm b}$, which is much smaller than the periodicity of the field namely, $k_{\rm w} R_{\rm b} \ll 1$ hence

$$\begin{aligned} B_z(r,z) &= B_0 \cos(k_{\rm w} z), \\ B_r(r,z) &= B_0 \sin(k_{\rm w} z) \frac{1}{2} k_{\rm w} r. \end{aligned} \tag{3.4.25}$$

This field can be derived from the azimuthal component of the magnetic vector potential:

$$A_\phi = B_0 \cos(k_{\rm w} z) \frac{1}{2} r. \tag{3.4.26}$$

For simplicity we further assume that the electrons are "born" in a region of zero magnetic field and their initial azimuthal motion is zero; therefore according to the conservation of the canonical angular momentum (3.4.7) we conclude that

$$v_\phi = \frac{1}{2\gamma} r\Omega_c \cos(k_w z). \tag{3.4.27}$$

Neglecting energy variations and substituting the last expression in the radial component of the equation of motion (3.4.1), we obtain

$$\frac{d^2}{dt^2}r + r\left[\frac{1}{2\gamma}\Omega_c \cos(k_w z)\right]^2 = -\frac{e}{m\gamma}\left[E_r - v_z B_\phi\right]. \tag{3.4.28}$$

Using the expressions for the electric and magnetic field developed in the previous section (3.4.12), (3.4.14) and assuming that the longitudinal motion is dominant i.e., $|v_z| \gg |v_r|, |v_\phi|$ we find

$$\frac{d^2}{dt^2}r + \left\{\left[\frac{\Omega_c}{2\gamma}\cos(\Omega_w t)\right]^2 - \frac{1}{2\gamma^3}\omega_p^2\right\} r = 0, \tag{3.4.29}$$

where $\Omega_w = k_w v_0$ and v_0 is the velocity of the electron. For a *zero order* stability estimate, one can average the square brackets over time, thus the resulting coefficient of $r(t)$ has to be positive for the electrons to follow confined trajectories i.e.,

$$\Omega_c^2 \geq \frac{4}{\gamma}\omega_p^2. \tag{3.4.30}$$

Comparing this expression with the condition for a uniform field (3.4.19) it is evident that in the periodic case, on axis, the amplitude has to be by a factor of $\sqrt{2}$ larger. In Sect. 7.6.2 we further elaborate on the self-focusing of an electron beam in a wiggler field.

3.4.3 Beam Propagation in a Waveguide

In the previous sub-section we determined the necessary condition for the propagation of an electron beam in free-space when guided by a static magnetic field. The effect of the *vacuum chamber* was ignored. In this sub-section, we investigate the propagation of the beam in a waveguide assuming that an infinite magnetic field is applied such that only longitudinal motion is permitted – the basic configuration is illustrated schematically in Fig. 3.14. The region under investigation is far away

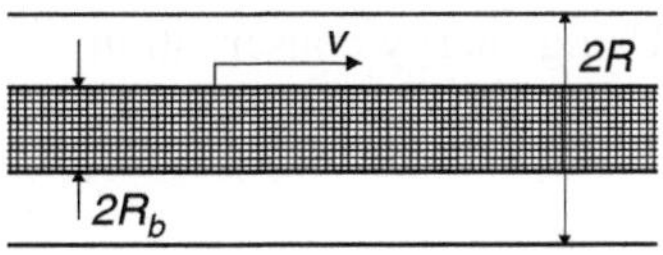

Fig. 3.14 Electron beam in a circular waveguide

from the entrance to the waveguide, therefore longitudinal variations are neglected and only radial variations are of interest. The presence of the metallic boundary at $r = R$ changes the potential experienced by the beam and consequently since the total energy is the sum of kinetic and potential energy, the former varies across the beam.

In order to envision the effect, consider a beam which at the input (free space) has a uniform spatial distribution and, more important, all the electrons have the same kinetic energy $mc^2(\gamma_0 - 1)$. Since the potential varies in the waveguide's space, electrons, which are at a different distance from the wall, experience different potential and therefore, their kinetic energy differs. Obviously, the potential energy comes at the expense of the kinetic energy. That is to say, that if electrons injected into a waveguide have at the input the same kinetic energy, then by increasing their number (N) the potential is elevated and while their velocity (v) is reduced. In terms of the injected current, which is proportional to the product Nv, it reaches a *maximum value* and beyond it, even if we increase the number of electrons injected, the current remains unchanged since electrons bounce back. Let us now investigate this phenomenon in a systematic way. In the waveguide the electrostatic potential satisfies the Poisson equation:

$$\frac{1}{r}\frac{\mathrm{d}}{\mathrm{d}r}r\frac{\mathrm{d}}{\mathrm{d}r}\Phi(r) = \frac{1}{\varepsilon 0}en(r). \tag{3.4.31}$$

Green's function associated with this equation is a solution of

$$\frac{1}{r}\frac{\mathrm{d}}{\mathrm{d}r}r\frac{\mathrm{d}}{\mathrm{d}r}G(r|r') = -\frac{1}{2\pi r}\delta(r - r'), \tag{3.4.32}$$

and it reads

$$G(r|r') = -\frac{1}{2\pi}\begin{cases} \ln(r/R) & \text{for} \quad 0 \le r' \le r < R, \\ \ln(r'/R) & \text{for} \quad 0 < r \le r' < R. \end{cases} \tag{3.4.33}$$

Based on Green's scalar theorem the potential reads

$$\Phi(r) = \frac{-e}{\varepsilon\ 0}\left[\left[\int_0^r \mathrm{d}r'r' \ln\left(\frac{r}{R}\right)n(r') + \int_r^R \mathrm{d}r'r' \ln\left(\frac{r'}{R}\right)n(r')\right]\right], \tag{3.4.34}$$

and in particular at the edge of the beam ($r = R_\mathrm{b}$) it reads

$$\Phi(R_\mathrm{b}) = -\frac{e}{\varepsilon 0}\ \ln\left(\frac{R_\mathrm{b}}{R}\right)\int_0^{R_\mathrm{b}} \mathrm{d}r'r'n(r'). \tag{3.4.35}$$

Using energy conservation

$$\gamma - \frac{e\Phi}{mc^2} = \text{const.}, = \gamma_0, \tag{3.4.36}$$

and charge conservation

$$J = -env = \text{const.}, \tag{3.4.37}$$

we can write

$$\Phi(R_\mathrm{b}) = -\frac{J}{c\varepsilon_0}\ln\left(\frac{R}{R_\mathrm{b}}\right)\int_0^{R_\mathrm{b}} \mathrm{d}r\, r\left[1-\left(\gamma_0+\frac{e\Phi(r)}{mc^2}\right)^{-2}\right]^{-1/2}. \tag{3.4.38}$$

The expression $mc^2(\gamma_0 - 1)$ represents the kinetic energy *before* the particles entered the waveguide. For a narrow pencil beam, the potential is not expected to vary significantly on the beam cross-section, therefore we can replace $\Phi(r)$ with $\Phi(R_\mathrm{b})$ thus

$$\begin{aligned}\xi\left(\frac{e\Phi(R_\mathrm{b})}{mc^2}\right) &\equiv \left(\frac{e\Phi(R_\mathrm{b})}{mc^2}\right)\sqrt{1-\left(\gamma_0+\frac{e\Phi(R_\mathrm{b})}{mc^2}\right)^{-2}},\\ &= -\frac{1}{2\pi}\frac{eI\eta_0}{mc^2}\ln\left(\frac{R}{R_\mathrm{b}}\right).\end{aligned} \tag{3.4.39}$$

In Fig. 3.15 the function $\xi(x)$ is plotted for $\gamma_0 = 3$ and it clearly shows that it has a minimum. This minimum occurs for $x_0 = -\gamma_0 + \gamma_0^{1/3}$ implying that $\xi(x_0) = -\left(\gamma_0^{2/3} - 1\right)^{3/2}$ and therefore, the maximum current which can flow in the waveguide is given by

$$I_\mathrm{max} = \frac{mc^2}{e\eta_0}\frac{2\pi}{\ln(R/R_\mathrm{b})}\left(\gamma_0^{2/3} - 1\right)^{3/2}. \tag{3.4.40}$$

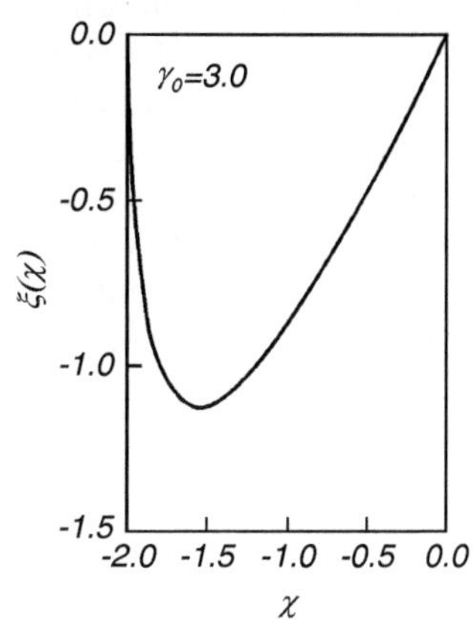

Fig. 3.15 ξas defined in (3.4.39) is plotted as a function of the normalized potential ($x \equiv e\Phi(R_\mathrm{b}/mc^2)$) on the beam envelope

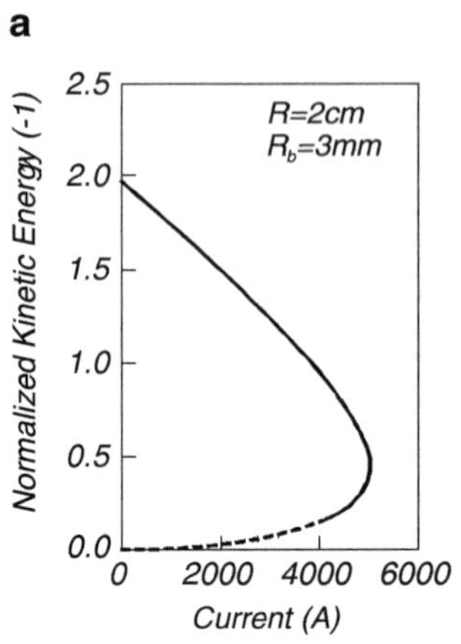

Fig. 3.16 (**a**) Normalized kinetic energy of the electrons in the waveguide as a function of the injected current. The initial energy is always the same. (**b**) Maximum potential efficiency as a function of the perveance

As already mentioned, any attempt to inject a higher current in the waveguide will cause electrons to bounce back to the diode and a *virtual cathode* may develop.

Another implication of this phenomenon refers to the remnant kinetic energy of the beam. For a given *initial* energy at the entrance to the waveguide, the kinetic energy decreases with the increasing current due to the potential energy associated with the space-charge effect. Figure 3.16a illustrates the effective kinetic energy of the electrons as a function of the current, for $R = 2\,\text{cm}$, $R_\text{b} = 3\,\text{mm}$ and $\gamma_0 = 3$. For this particular set of parameters the effective kinetic energy drops to $\gamma = 2.0$ when 3.9 kA are injected; the limiting current in this case is 5 kA in which case $\gamma = 1.56$. In other words, consider a situation whereby in the diode the electrons are accelerated to the 1 MeV level. For the electrons to propagate in the waveguide, the electrons give up part of their kinetic energy to "build" the potential associated with the space-charge – this potential barrier is almost 500 keV. This effect inflicts limitations on the maximum efficiency of a device. If, for example, the diode had generated a $1\,\text{MeV} \times 3.9\,\text{kA} \simeq 3.9\,\text{GW}$ beam, in the waveguide, the maximum power available for radiation generation is $0.5\,\text{MeV} \times 3.9\,\text{kA} \simeq 2\,\text{GW}$ and the maximum radiation conversion efficiency is 50% since what counts is the *available* kinetic energy in the waveguide. Figure 3.16b illustrates the maximum theoretical efficiency, defined as $(\gamma - 1)/(\gamma_0 - 1)$, as a function of the perveance.

The limiting current which can propagate in the waveguide is directly related to the Child-Langmuir limiting current in a diode as it becomes evident in particular at low voltages $(eV_\text{AN}/mc^2 \ll 1)$ since $\gamma_0 = 1 + eV_\text{AN}/mc^2$ and

$$I_\text{max} = \frac{mc^2}{e\eta_0} \frac{2\pi}{\ln(R/R_\text{b})} \left(\frac{2}{3} V_{AN}\right)^{3/2}, \tag{3.4.41}$$

which has an identical form to (3.3.28). For this reason, a high efficiency device has always to be designed based on a low perveance diode.

It was indicated that space-charge effects cause a transverse spatial variation of the electron's kinetic energy, which clearly may alter the interaction with electromagnetic waves. However, in the analysis above, it was assumed that the beam is sufficiently narrow such that variations across its section could be neglected. A parameter of importance is the *location* of the beam relative to the waveguide's wall as it becomes evident from (3.4.40). Roughly, by increasing the distance between the external wall and the beam, the limiting current becomes smaller. In order to examine this effect more accurately, one can examine the limiting current of an annular beam of radius R_b and δr thickness (much smaller than R_b). The approach is similar to the above and the result is similar except for the different meaning R_b has in this case. The closer the beam is to the waveguide's wall, the higher the limiting current and therefore the lower the potential depression. To emphasize this effect even further, one can consider two thin annular beams of two different radii but at the entrance of the pipe, they have the same kinetic energy. It can be shown that the kinetic energy of the electrons in the outer beam is actually larger than that of the electrons in the inner beam.

For a corrugated wall, as the one illustrated in Fig. 3.17, the limiting current is given by

$$I_{\max} = \frac{1}{F_b}\frac{mc^2}{e\eta_0}\frac{2\pi}{\ln(R_{\text{int}}/R_{\text{b}})}\left(\gamma_0^{2/3} - 1\right)^{3/2}, \tag{3.4.42}$$

where F_b is boundary form factor illustrated in Fig. 3.18. It shows the dependence of this factor on the external radius R_{ext}, disk width d and periodicity of the structure for $R_b = 3\,\text{mm}$ and $R_{\text{int}} = 9\,\text{mm}$. In the left frame we observe that, for $L = 7.7\,\text{mm}$ and $d = 1\,\text{mm}$, F_b reaches a maximum value of 1.1 when $R_{\text{ext}} = 1.5R_{\text{int}}$ and any further increase in this ratio does not change the boundary factor. In the central frame, this factor is plotted as a function of the disk thickness and we observe that it decreases with the increasing d. Finally, the right frame indicates that the boundary factor increases with the increasing periodicity ($d = L/2$).

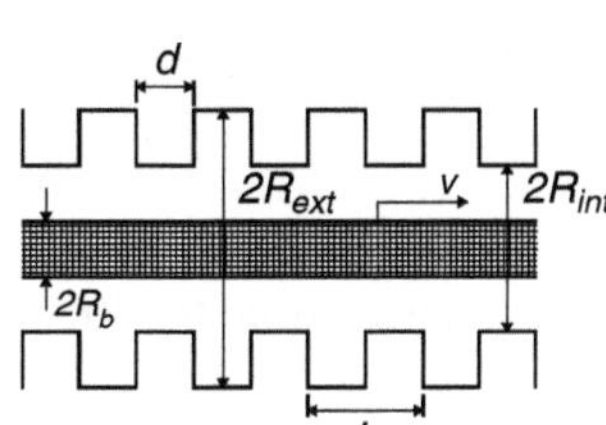

Fig. 3.17 Beam propagation in a corrugated waveguide

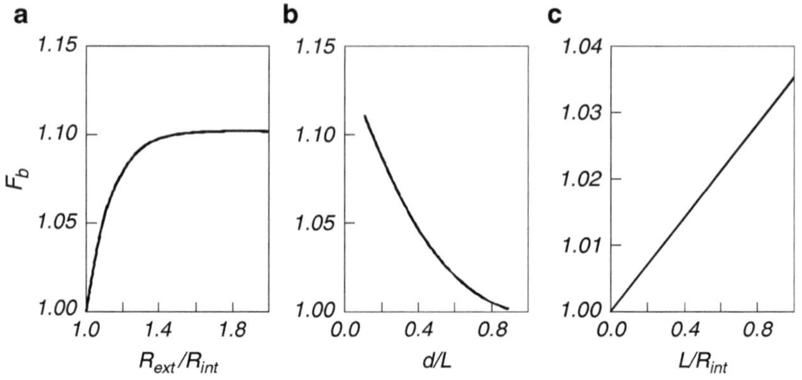

Fig. 3.18 Boundary factor, F_b, as a function of the (**a**) external radius, (**b**) disk thickness and (**c**) period

3.4.4 Beam Emittance and Brightness

So far, we considered in this Section a few equilibrium phenomena associated with beam propagation. Before concluding, it is important to introduce two important notions that quantify the *transverse* fluctuations of the beam around a specified equilibrium. It is explicitly assumed that these fluctuations (δx_i) are stable and they represent small deviations from the stable trajectory, which in the present context is assumed to be co-linear. This is the case for virtually all radiation sources and in acceleration structures. Further assuming that the longitudinal motion is predominant such that $v_z \gg |v_x|, |v_y|$, the transverse motion is approximated to be harmonic

$$\frac{d^2}{dz^2}\delta x_i + K_0^2 \delta x_i = 0, \tag{3.4.43}$$

wherein K_0 represents the attracting transverse force associated with the lattice in the case of an accelerator or with the guiding of the beam in a radiation source. Consequently, assuming the transverse force is z independent then the trajectory is described by

$$\delta x_i(z) = A_i \cos(K_0 z + \phi_i), \tag{3.4.44}$$

and its first derivative, denoted by $\delta \dot{x}_i(z)$, is given by

$$\delta \dot{x}_i(z) = -A_i K_0 \sin(K_0 z + \phi_i). \tag{3.4.45}$$

In the phase-space the locus of each trajectory is an ellipse

$$\left(\frac{\delta x_i}{A_i}\right)^2 + \left(\frac{\delta \dot{x}_i}{K_0 A_i}\right)^2 = 1. \tag{3.4.46}$$

If we denote by $a \equiv \max(A_i)$ the largest amplitude among all particles, then the area of the ellipse is

$$\pi\varepsilon \equiv \pi a(aK_0) = \pi K_0 a^2, \tag{3.4.47}$$

and it defines the *emittance* (ε) of an ideal beam. This corresponds to the area in the δx, $\delta\dot{x}$ space occupied by the entire ensemble. It is natural to define the emittance in the "normal" phase-space i.e., x, p_x. For this purpose the $\delta\dot{x}$ is multiplied by $\gamma\beta$ since $p_x = mc\gamma\beta\delta\dot{x}$. Consequently, the *normalized emittance*, ε_n, is defined as

$$\varepsilon_n = \gamma\beta\varepsilon, \tag{3.4.48}$$

omitting the (constant) mc term from the momentum definition. Note that occasionally, the emittance is defined with the π included. In this text, we prefer the definition from the above.

An ideal periodic motion is only a convenient model which provides us with an intuitive interpretation of the emittance. In all practical cases the system is not uniform and we now extend the model to include space dependent, $K_0(z)$, effects

$$\frac{d^2}{dz^2}\delta x_i + K_0^2(z)\delta x_i = 0. \tag{3.4.49}$$

The solution in this case has the form

$$\delta x_i(z) = A_i(z)\cos[\psi(z) + \phi_i(z)], \tag{3.4.50}$$

where for a constant K_0 we have $\psi(z) = \int dz K_0(z)$. Substituting in (3.4.49) and using the orthogonality of the trigonometric functions, we obtain two equations

$$\begin{aligned} \cos[..]&: \frac{d^2A_i}{dz^2} + K_0^2(z)A_i - A_i\left(\frac{d\psi}{dz}\right)^2 = 0, \\ \sin[..]&: 2\frac{dA_i}{dz}\frac{d\psi}{dz} + A_i\frac{d^2\psi}{dz^2} = 0. \end{aligned} \tag{3.4.51}$$

The second equation can be multiplied by A_i and written as

$$\frac{d}{dz}\left[A_i^2\frac{d\psi}{dz}\right] = 0 \rightarrow A_i^2\frac{d\psi}{dz} = C_i. \tag{3.4.52}$$

In particular, a particle, which oscillates with maximum amplitude, a satisfies

$$a^2\frac{d\psi}{dz} = C. \tag{3.4.53}$$

For a constant K_0 and comparing with (3.4.47) we have

$$C = \varepsilon. \tag{3.4.54}$$

With this relation we can determine the equation for the envelope a by substituting (3.4.53) in the first equation of (3.4.51); the result is

$$\frac{d^2}{dz^2}a + K_0^2(z)a - \varepsilon^2\frac{1}{a^3} = 0. \tag{3.4.55}$$

Next stage is to introduce a general definition of the emittance, which is not dependent on the solutions presented above but rather as a way to characterize the transverse phase-space. Lapostolle (1971) used an alternative definition,

$$\bar{\varepsilon} \equiv 4\left[\langle \delta x^2\rangle\langle(\delta\dot{x})^2\rangle - \langle\delta x\delta\dot{x}\rangle^2\right]^{1/2}. \tag{3.4.56}$$

This definition can be tested against the trivial solution in (3.4.44)–(3.4.45). Assuming that the phases ϕ_i and the amplitudes A_i are (statistically) independent, it can be readily shown that the second term is identically zero for a uniform distribution of phases and

$$\left\langle(\delta x)^2\right\rangle = \frac{1}{2}\langle A_i^2\rangle, \quad \left\langle(\delta\dot{x})^2\right\rangle = \frac{1}{2}K_0^2\langle A_i^2\rangle, \tag{3.4.57}$$

therefore

$$\bar{\varepsilon} = 2K_0\langle A_i^2\rangle. \tag{3.4.58}$$

If the quantity $\langle A_i^2\rangle$ corresponds to the mean square value of the amplitudes $\langle A_i^2\rangle = a^2/2$ the emittance obtained is identical with that determined in (3.4.47).

In the discussion, so far we have considered only one out of the two transverse dimensions; in order to attribute the emittance to a specific dimension we denote the emittance associated with the motion in the x direction by ε_x and in a similar way, we define ε_y as the emittance associated with the motion in the y direction. With these two definitions, it is convenient to introduce another quantity, which provides a figure of merit regarding the beam quality. This is the *brightness*:

$$B \equiv \frac{I}{(\pi\varepsilon_x)(\pi\varepsilon_y)}F_{\mathrm{f}}. \tag{3.4.59}$$

F_{f} is a geometrical form factor on the order of unity. Similar to the emittance, the normalized brightness can be defined as $B_n \equiv B/\beta^2\gamma^2$. Both brightness and emittance provide information regarding the beam quality. However, while the emittance addresses primarily the transverse geometrical characteristics of a beam, the brightness accounts also for the beam intensity. Lawson (1988) provides the reader a detailed discussion on emittance.

3.5 Space-Charge Waves

All the effects considered so far were either static or quasi-static. In this section, we introduce some elementary concepts of waves, which propagate along electron beams. For this purpose consider a beam whose unperturbed beam density is n_0 and its zero order velocity is v_0 (the effect of potential depression is already included); the beam is guided by a very strong magnetic field and as a result, the motion is confined to the longitudinal direction. Consider an electric field E_z excited in the system and its form being

$$E_z = E\exp(-jkz). \tag{3.5.1}$$

The z component of the linearized equation of motion implies that the linear perturbation in the velocity field denoted by $\delta\mathrm{v}$ is

$$m\gamma^3 j(\omega - \mathrm{v}_0 k)\delta\mathrm{v} = -eE_z; \tag{3.5.2}$$

here m and e are the mass and the charge of an electron respectively; $\gamma = 1/\sqrt{1-(\mathrm{v}_0/c)^2}$. Next, we can determine the perturbation in density (δn) using the continuity equation, the result is

$$\delta n = n_0 \frac{k}{\omega - \mathrm{v}_0 k}\delta\mathrm{v}. \tag{3.5.3}$$

The current density defined in (3.1.38) is linearized in the perturbation terms i.e., $J_z = -e(\delta n \mathrm{v}_0 + n_0\delta\mathrm{v})$ and it reads

$$J_z = -j\omega\varepsilon_0 \frac{\omega_\mathrm{p}^2}{\gamma^3(\omega - \mathrm{v}_0 k)^2} E_z, \tag{3.5.4}$$

where ω_p is the plasma frequency as defined in (3.4.18). Next, we use (2.1.38) and substitute the result in the wave equation for the magnetic vector potential (2.1.36) which then reads

$$\left[\frac{1}{r}\frac{\mathrm{d}}{\mathrm{d}r}r\frac{\mathrm{d}}{\mathrm{d}r} + \left(\frac{\omega^2}{c^2} - k^2\right)\left(1 - \frac{\omega_\mathrm{p}^2}{\gamma^3(\omega - \mathrm{v}_0 k)^2}\right)\right]A_z(r,k;\omega) = 0. \tag{3.5.5}$$

In an *infinite* system $\partial/\partial r = 0$, there are two sets of solutions

$$k^2 - \frac{\omega^2}{c^2} = 0, \qquad \left(k - \frac{\omega}{\mathrm{v}_0}\right)^2 - \frac{\omega_\mathrm{p}^2}{\mathrm{v}_0^2\gamma^3} = 0. \tag{3.5.6}$$

The first set corresponds to a pure electromagnetic wave and the second represents the dispersion relation of the so-called *space-charge waves*. There are two such waves, both propagating parallel to the beam with a phase velocity close to the average velocity of the beam

$$k_\pm = \frac{\omega}{\mathrm{v}_0} \pm K_\mathrm{p} \tag{3.5.7}$$

with $K_\mathrm{p}^2 \equiv \omega_\mathrm{p}^2/\gamma^3\mathrm{v}_0^2$. In contrast to regular electromagnetic waves, in the 1D case considered momentarily, the space-charge wave has only an electric field and its magnetic component, is identically zero even if the time variations are very rapid.

For a 2D case, we consider a beam in a radial waveguide of radius R. The boundary condition $[E_z(r = R) = 0]$ imposes the following dispersion relation

$$\left(\frac{\omega^2}{c^2} - k^2\right)\left[1 - \frac{\omega_\mathrm{p}^2}{\gamma^3(\omega - \mathrm{v}_0 k)^2}\right] = \frac{p_s^2}{R^2} \tag{3.5.8}$$

and as above there are two groups of solutions: the electromagnetic modes group whose asymptotic behavior can be determined from the limit when no beam is present $(\omega_\mathrm{p}^2 = 0)$ namely, $(\omega/c)^2 - k^2 = (p_s/R)^2$. We discussed this group in the context of electromagnetic TM modes in Chap. 2. The second group represents waves that propagate along the beam and they can be approximated by

$$k_s^{(\pm)} \simeq \frac{\omega}{\mathrm{v}_0} \pm \delta k_s \tag{3.5.9}$$

$\left[\delta k_s^2 = K_p^2\xi_s^2,\ \xi_s^{-2} = 1 + (\gamma\beta p_s c/\omega R)^2\right]$ provided that the plasma wave-number, K_p, is much smaller than ω/v_0. Note that the factor ξ_s is always smaller than unity such that each mode sees a *reduced plasma frequency*. An extensive discussion on space-charge wave can be found in a book by Beck (1958). In the remainder of this section we review only a few instructive topics. Note that in case of finite transverse extent, the magnetic field associated with this wave is not zero.

3.5.1 Slow and Fast Space-Charge Waves

The waves that correspond to $k_s^{(+)}$ have a phase velocity

$$\mathrm{v}_\mathrm{slow} = \frac{\omega}{k_s^{(+)}} = \frac{\mathrm{v}_0}{1 + \mathrm{v}_0\delta k_s/\omega} < \mathrm{v}_0 \tag{3.5.10}$$

which is slower than the beam average velocity (v_0) therefore they are called *slow space-charge waves*. The waves that correspond to $k_s^{(-)}$ have a phase velocity

$$v_{\text{fast}} = \frac{\omega}{k_s^{(-)}} = \frac{v_0}{1 - v_0 \delta k_s/\omega} > v_0, \tag{3.5.11}$$

which is greater than the average velocity of the beam and these are referred to as *fast space-charge waves.*

3.5.2 "Negative" and "Positive" Energy

The contribution of the space-charge wave to the average kinetic energy density is determined based on the global energy conservation in (3.1.45) which is given by

$$\delta E = mc^2 \frac{1}{4} [\delta n \delta \gamma^* + \delta n^* \delta \gamma]. \tag{3.5.12}$$

We can now express $\delta\gamma$ in terms of δv namely $\delta\gamma = \gamma^3 \beta \delta v/c$ and then use the expression in (3.5.3) to write

$$\delta E = \frac{1}{2} mc^2 |\delta n|^2 \gamma^3 \beta \frac{1}{n_0 c} \left(\frac{\omega}{k} - v_0 \right). \tag{3.5.13}$$

This result indicates that the slow space-charge waves have a total kinetic energy density which is *smaller* than the average kinetic energy of the beam i.e.,

$$\delta E_{\text{slow}} = \frac{1}{2} mc^2 |\delta n|^2 \gamma^3 \beta \frac{1}{n_0 c} (v_{\text{slow}} - v_0) < 0. \tag{3.5.14}$$

For this reason, these waves are also referred to as "negative" energy waves. Fast space-charge waves have "positive" energy since

$$\delta E_{\text{fast}} = \frac{1}{2} mc^2 |\delta n|^2 \gamma^3 \beta \frac{1}{n_0 c} (v_{\text{fast}} - v_0) > 0. \tag{3.5.15}$$

When *distributed* interaction between electrons and electromagnetic waves is possible we will see that, the slow space-charge waves play an important role in the process. Both fast and slow space-charge waves play a very important role in klystrons where the interaction is limited to the close vicinity of a cavity gap.

If we examine the average power carried by each one of the modes we may readily conclude that this is *identically zero* as is readily seen by examining the current density term in (3.5.4) and

$$\mathrm{Re}\left(\frac{1}{2}E_zJ_z^*\right) = \mathrm{Re}\left[j\omega\varepsilon_0\frac{\omega_p^2}{\gamma^3(\omega - v_0k)^2}\frac{1}{2}|E_z|^2\right]. \tag{3.5.16}$$

Since all the k's are real then clearly the right hand side is zero. However, in a superposition of two space-charge waves the real power may be non-zero. This is the basis for the operation of relativistic klystrons.

3.5.3 Resistive Wall Instability

When electro-magnetic waves propagate in a lossy wall waveguide, a fraction of the power is absorbed in the wall. Since the process is linear, namely the power absorbed per unit length is proportional to the local power flow, the wave decays exponentially in space. In the case of space-charge waves, the situation is different – it is shown next that the slow space-charge wave can actually be amplified due to the complex impedance at the metallic surface.

Before examining a realistic case, it is instructive to investigate a simplified model. For this purpose, we assume that the beam propagates in a lossy medium that is characterized by $\varepsilon_r = 1 + \sigma/j\omega\varepsilon_0$ – where σ is the conductivity of the medium. It can be shown that the dispersion relation of the space-charge waves as formulated in (3.5.6) should be updated by replacing $\varepsilon_0 \to \varepsilon_0\varepsilon_r$. This implies that $\omega_p^2 \to \omega_p^2/\varepsilon_r$. In the right-hand side the expression is complex and since the solution for k has the form $k_\pm = \omega/v_0 \pm K_p/\sqrt{\varepsilon_r}$ we clearly see that slow space-charge wave grows in space since $\mathrm{Im}(k_+) = K_p\sqrt{\omega\varepsilon_0/2\sigma} > 0$.

Let us now examine this process in a more realistic system. Consider a waveguide of radius R made of a material of *finite conductivity* ($\sigma \gg \omega\varepsilon_0$). The beam which propagates is electromagnetically characterized by (3.5.4) and for simplicity, we assume that it fills the entire waveguide. The magnetic vector potential for, $r < R$, is a solution of (3.5.5) and its solution reads

$$A_z(r, k; \omega) = A\mathrm{I}_0(\Lambda r), \tag{3.5.17}$$

where

$$\Lambda^2 \equiv \left(k^2 - \frac{\omega^2}{c^2}\right)\left[1 - \frac{\omega_p^2}{\gamma^3(\omega - v_0k)^2}\right]. \tag{3.5.18}$$

The impedance at $r = R$ is the ratio

$$Z_{\mathrm{beam}} \equiv \frac{E_z(r = R)}{H_\phi(r = R)} = j\eta_0\frac{(\omega/c)^2 - k^2}{\Lambda\omega/c}\frac{\mathrm{I}_0(\Lambda R)}{\mathrm{I}_1(\Lambda R)}. \tag{3.5.19}$$

In the metallic wall, the magnetic vector potential satisfies

$$\left[\frac{1}{r}\frac{\mathrm{d}}{\mathrm{d}r}r\frac{\mathrm{d}}{\mathrm{d}r} - k^2 - j\omega\mu_0\sigma\right]A_z(r,k;\omega) = 0, \tag{3.5.20}$$

and the solution for $r \geq R$ reads

$$A_z(r,k;\omega) = B\mathrm{K}_0(\Xi r), \tag{3.5.21}$$

where $\Xi^2 = k^2 + j\omega\mu_0\sigma$. The impedance at the discontinuity is

$$Z_{\text{wall}} \equiv \frac{E_z(r=R)}{H_\phi(r=R)} = -j\eta_0\frac{(\omega/c)^2 - k^2}{\Xi\omega/c}\frac{\mathrm{K}_0(\Xi R)}{\mathrm{K}_1(\Xi R)}. \tag{3.5.22}$$

Imposing the boundary condition, as we have done before, is equivalent to the requirement that the two impedances match i.e.,

$$Z_{\text{wall}} = Z_{\text{beam}}, \tag{3.5.23}$$

and this determines the dispersion relation in a lossy waveguide. In order to illustrate the effect of the lossy material on the propagation of a space-charge wave, we consider a solution which has the form

$$k = \frac{\omega}{\mathrm{v}_0} + \delta k, \tag{3.5.24}$$

and $\omega/\mathrm{v}_0 \gg |\delta k|$. Furthermore, for sufficiently high conductivity we have $\sigma \gg \omega\varepsilon_0$ and for simplicity we consider the limit $\omega R/c \gg 1$. Consequently, the dispersion relation can be simplified to read

$$\sqrt{1 - \frac{\omega_{\mathrm{p}}^2}{\gamma^3\mathrm{v}_0^2\delta k^2}} = -\gamma\beta\sqrt{j\frac{\sigma}{\omega\varepsilon_0}}. \tag{3.5.25}$$

According to the last term on the right, for a relativistic beam the right-hand side is much larger than unity, therefore on the left-hand side the second term has to satisfy $|\delta k|^2 \ll \omega_{\mathrm{p}}^2/\mathrm{v}_0^2\gamma^3$, hence

$$\sqrt{\frac{\omega_{\mathrm{p}}^2}{\gamma^3\mathrm{v}_0^2\delta k^2}} = j\gamma\beta\frac{c}{\omega}\sqrt{j\omega\mu_0\sigma}. \tag{3.5.26}$$

The expression for the dispersion relation can be simplified by defining

$$q_0 \equiv \sqrt{\frac{1}{2}\frac{eJ\eta_0}{mc^2}\frac{\omega\varepsilon_0}{\sigma}\frac{1}{(\gamma\beta)^5}}, \tag{3.5.27}$$

therefore, the solution of interest: $\delta k = (1 + j)q_0$ represents a slow space-charge wave which grows exponentially in space

$$|\exp(-jk_+ z)| = \exp(q_0 z), \tag{3.5.28}$$

whose imaginary part determines the spatial growth rate of what is called the *resistive wall instability*. Note that the *"negative" energy wave* is the one which grows. This is the case in all the schemes of collective beam-wave interaction. The second solution, $\delta k = -(1 + j)q_0$, describes a fast space-charge wave which decays exponentially in space. The similarity to the simple model presented at the beginning of this sub-section is evident.

Four characteristics of the growth warrant special attention: (1) the spatial growth q_0 is proportional to the square root of the current $q_0 \propto \sqrt{I}$, (2) it is inversely proportional to the normalized momentum $q_0 \propto (\gamma\beta)^{-5/2}$, (3) it is proportional to the square root of the frequency $q_0 \propto \sqrt{\omega}$ and (4) it is inversely proportional to the square root of the conductivity $q_0 \propto 1/\sqrt{\sigma}$.

Resistive wall instability occurs as space-charge waves propagate in *vacuum* but in the vicinity of a metallic wall of finite conductivity. A similar phenomenon occurs if the space-charge wave traverses a lossy plasma or neutral gas characterized by a series of resonances $\omega_{0,\mu}$ of width $\omega_{1,\mu}$; the density of atoms that have these resonances is denoted by n_μ. Consequently, it is convenient to define the electronic plasma frequency associated with each resonance by $\omega_{\mathrm{p},\mu}^2 = e^2 n_\mu / m\varepsilon_0$ and the dielectric coefficient of this system is

$$\varepsilon(\omega) = 1 + \sum_\mu \frac{\omega_{\mathrm{p},\mu}^2}{\omega_{0,\mu}^2 - \omega^2 + j\omega\omega_{1,\mu}}. \tag{3.5.29}$$

This configuration combines two well known effects: resistive wall instability and resonant absorption, in order to amplify radiation at selected frequencies. In other words, we use *microscopic* cavities i.e. atoms or molecules, for the generation of an instability that causes space-charge waves to grow in space. For simplicity we ignore the scattering process of these electrons with the constituents of the resonant medium and the possible ionization of the medium by beam's front. The dispersion relation of the electromagnetic and space-charge waves ignoring any transverse variations

$$\left[\varepsilon(\omega)\frac{\omega^2}{c^2} - k^2\right]\left[1 - \frac{\omega_\mathrm{p}^2}{(\omega - kv_0)^2}\frac{1}{\varepsilon(\omega)}\right] = 0. \tag{3.5.30}$$

Clearly the first term, $k = (\omega/c)\sqrt{\varepsilon(\omega)}$, represents the TEM mode therefore the spatial decay is $\mathrm{Im}(k_{\mathrm{EM}}) = (\omega/c)\mathrm{Im}\left[\sqrt{\varepsilon(\omega)}\right]$ and in a similar way the spatial growth of the space-charge wave is given by

$$\mathrm{Im}(k_{\mathrm{SC}}) = \frac{\omega_\mathrm{p}}{c\beta\gamma^{3/2}}\mathrm{Im}\left[\frac{1}{\sqrt{\varepsilon(\omega)}}\right]. \tag{3.5.31}$$

this result indicates that the imaginary part of the dielectric function is responsible to the spatial growth of the space-charge wave. In other words, the same term that causes the opaqueness of the medium, is responsible to the spatial growth of the space-charge wave. There are three distinct differences between the last result and the regular resistive wall instability: (1) the "conductivity" in this case is *narrow-band*. (2) The beam propagates through the medium rather than near the medium and (3) this mechanism has particular appeal in the millimeter, sub-millimeter or even in the optical range where resonant absorption is a significant effect.

For a given current, the spatial growth in (3.5.31) is inversely proportional to the momentum to the power of 3/2 [i.e. $\propto (\gamma\beta)^{-3/2}$] implying that there must be an optimal momentum since at very low momentum (corresponding to a few hundreds eV) the process will diminish due to ionization of the medium.

In order to envision the potential of this mechanism let us consider the analysis by Gorgy et al. (1966) that indicates that peak absorption coefficient $[\alpha_{max} = 2\text{Im}(k_{EM})]$ of a pure rotational line at room temperature is $\alpha_{max} \simeq 55 \times 10^{-12}$ $\mu^2 f^3/\Delta f$. The peak frequency (in GHz) is denoted by f, Δf represents the linewidth in GHz and μ is the dipole moment of the atom or molecule expressed in Debye; the typical values of μ vary between 1–4 Debye (e.g. $\mu_{\text{Ammonia}} = 1.44$ Debye, $\mu_{\text{Water}} = 1.84$ Debye). The line-width depends on the characteristics of the molecule as well as the mechanism of energy loss such as collisions between molecules. Typically, the higher the pressure, the shorter the mean free path and therefore the larger the line-width. In fact, for sufficiently high pressure, the line-width is proportional to the pressure. Accordingly, the line-width may vary quite dramatically from a few kHz in the case of the 24 GHz resonance of Ammonia to MHz and higher. In the example to follow we assume $\Delta f = 100$ MHz and $\mu = 2.5$ Debye and operation at 125 GHz thus $\alpha_{\max} = 6.72\,\text{cm}^{-1}$ $[\text{Im}(k_{\text{EM}}) = 3.36\,\text{cm}^{-1}$.

A relativistic beam $(\gamma \gg 1)$ is injected in this gas and for a beam plasma frequency of 2 GHz, the growth rate of the system is $0.054\ \text{cm}^{-1}$. This corresponds to 0.47 dB/cm thus, we may expect a 50 dB gain in about 120 cm of interaction length. Consequently, assuming an initial modulation that corresponds to 1 kW at the input-end, after 120 cm a total power of 100 MW can be expected. This gain calculation assumes an equal excitation of both space-charge waves. Moreover, we ignored the possibility of ionization of the medium by the intense microwave radiation and non-linear effects.

3.5.4 Two-Beam Instability

The phase velocity of space-charge waves is close to the average velocity of the beam. If two beams move close to each other at two different but close velocities, the space-charge waves that may develop along these beams can convert energy to rf. In order to investigate the effect of two different velocities we examine two beams which have two different velocities v_1 and v_2; for simplicity sake, it is assumed that both occupy the same volume and the electrons density is n_1 and n_2 respectively. By virtue of the linearity we may readily conclude that the contribution of the two current density to the magnetic vector potential implies

$$\left[\frac{1}{r}\frac{\mathrm{d}}{\mathrm{d}r}r\frac{\mathrm{d}}{\mathrm{d}r}+\left(\frac{\omega^2}{c^2}-k^2\right)\left(1-\frac{\omega_{\mathrm{p},1}^2}{\gamma_1^3(\omega-\mathrm{v}_1k)^2}-\frac{\omega_{\mathrm{p},2}^2}{\gamma_2^3(\omega-\mathrm{v}_2k)^2}\right)\right]A_z(r,k;\omega)=0. \tag{3.5.32}$$

As before, for simplicity sake we ignore the transverse radiation thus the dispersion of the space-charge waves is

$$D(k)\equiv\frac{\omega_{\mathrm{p},1}^2}{\gamma_1^3(\omega-\mathrm{v}_1k)^2}+\frac{\omega_{\mathrm{p},2}^2}{\gamma_2^3(\omega-\mathrm{v}_2k)^2}-1=0. \tag{3.5.33}$$

Figure 3.19 illustrates this dispersion function. It clearly reveals two regimes: in one case all four solutions are real (dashed line) whereas in the other case, two of the solutions are real and two others are complex. For a *rough* estimate regarding the necessary condition for the latter to occur, we can expand the dispersion function in Taylor series in the vicinity of $k\simeq k_0\equiv\omega(1/2\mathrm{v}_1+1/2\mathrm{v}_2)$. Defining,

$$\begin{aligned}
k&=\frac{\omega}{2}\left(\frac{1}{\mathrm{v}_1}+\frac{1}{\mathrm{v}_2}\right)+\delta k,\\
\frac{1}{\delta\mathrm{v}}&=\frac{1}{\mathrm{v}_1}-\frac{1}{\mathrm{v}_2},\\
\kappa_1^2&=4\frac{\omega_{\mathrm{p},1}^2}{\gamma_1^3\mathrm{v}_1^2}\frac{\delta\mathrm{v}^2}{\omega^2},\\
\kappa_2^2&=4\frac{\omega_{\mathrm{p},2}^2}{\gamma_2^3\mathrm{v}_2^2}\frac{\delta\mathrm{v}^2}{\omega^2},
\end{aligned} \tag{3.5.34}$$

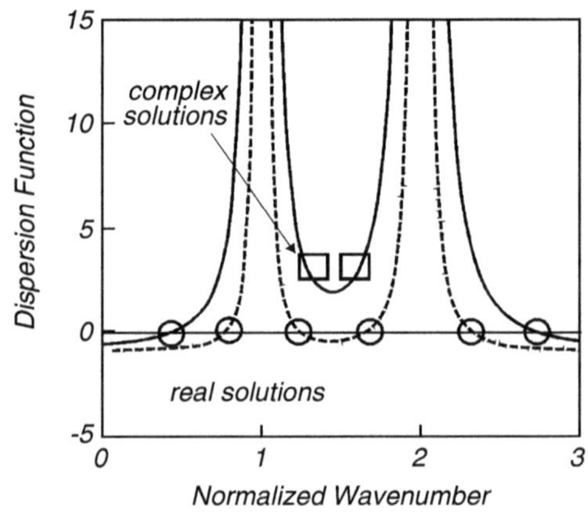

Fig. 3.19 Function that its zeros determine the real or complex wave-numbers

the dispersion function is given by

$$12\left(\kappa_1^2+\kappa_2^2\right)\left(\delta \mathrm{v}\frac{\delta k}{\omega}\right)^2-4\left(\kappa_1^2-\kappa_2^2\right)\left(\delta \mathrm{v}\frac{\delta k}{\omega}\right)-1+\kappa_1^2+\kappa_2^2=0 \tag{3.5.35}$$

and the condition for complex solutions to occur may be readily deduced from the two relevant solutions

$$\delta k_{\pm}=\frac{\omega/\delta \mathrm{v}}{6\left(\kappa_1^2+\kappa_2^2\right)}\left[\kappa_1^2-\kappa_2^2\pm\sqrt{\left(\kappa_1^2-\kappa_2^2\right)^2-3\left(\kappa_1^2+\kappa_2^2\right)\left(\kappa_1^2+\kappa_2^2-1\right)}\right] \tag{3.5.36}$$

namely,

$$\left(\kappa_1^2-\kappa_2^2\right)^2-3\left(\kappa_1^2+\kappa_2^2\right)\left(\kappa_1^2+\kappa_2^2-1\right)<0. \tag{3.5.37}$$

Assuming that both κ's are of the same order of magnitude, the last condition simplifies to $\kappa^2>1/2$ or

$$\delta \mathrm{v}^2>\mathrm{v}_1^2\frac{1}{8}\frac{\omega^2}{\omega_{\mathrm{p}}^2}\gamma^3 \tag{3.5.38}$$

implying that there is a minimal velocity difference that must be met for the two-beam instability to occur.

Subject to the condition in (3.5.37), the spatial growth is

$$\mathrm{Im}(\delta k)=\frac{\omega}{|\delta \mathrm{v}|}\frac{1}{6\left(\kappa_1^2+\kappa_2^2\right)}\left[\sqrt{3\left(\kappa_1^2+\kappa_2^2\right)\left(\kappa_1^2+\kappa_2^2-1\right)-\left(\kappa_1^2-\kappa_2^2\right)^2}\right]. \tag{3.5.39}$$

Case 1: *Motionless Background Plasma*. The simplest manifestation of the two beam instability is the case when one of the beams is actually motionless say $\mathrm{v}_2=0$ and the operating frequency is *below* the plasma frequency $\omega<\omega_{\mathrm{p},2}$. The dispersion relation needs no approximation and its solution is

$$k_{\pm}=\frac{\omega}{\mathrm{v}_1}\pm\sqrt{\frac{\omega_{\mathrm{p},1}^2}{\mathrm{v}_1^2\gamma_1^3}\frac{\omega^2}{\omega^2-\omega_{\mathrm{p},2}^2}} \tag{3.5.40}$$

indicating that the imaginary part is non-zero and it is proportional to the square-root of the current and $1/(\gamma_1\beta_1)^{3/2}$.

Case 2: *Ultra-Relativistic Beam*. Another regime of interest is when both beams are relativistic but one is ultra-relativistic $(\gamma_1\gg\gamma_2\gg1)$ thus

$$\frac{\omega_{\mathrm{p},1}^2\gamma_1^{-3}}{\left[\omega-kc+\omega/2\gamma_1^2\right]^2}+\frac{\omega_{\mathrm{p},2}^2\gamma_2^{-3}}{\left[\omega-kc+\omega/2\gamma_2^2\right]^2}=1 \tag{3.5.41}$$

Defining $\Delta\omega = \omega - kc$, assuming that $\omega/2\gamma_2^2 \gg |\Delta\omega| \gg \omega/2\gamma_1^2$ we get a second order polynomial

$$\frac{\omega_{p,1}^2 \gamma_1^{-3}}{\Delta\omega^2} + 4\gamma_2 \frac{\omega_{p,2}^2}{\omega^2} = 1 \tag{3.5.42}$$

which has complex solutions, subject to the assumption $4\gamma_2\omega_{p,2}^2 > \omega^2$. They are given by

$$k_{\pm} = \frac{\omega}{c} \pm \frac{1}{c}\sqrt{\frac{\omega_{p,1}^2}{\gamma_1^3} \frac{\omega^2}{\omega^2 - 4\gamma_2\omega_{p,2}^2}}. \tag{3.5.43}$$

These three expressions, [(3.5.39), (3.5.40) and (3.5.43)], are the simplest manifestations of the *two-beam instability*.

Case 3: *Two Thin Annular Beams*. In practice, the two beams occupy different volumes and in what follows we consider two annular beams of radius $R_{b,1}$ and $R_{b,2}$. Their thickness is much smaller than the wavelength and it is equal in both cases ($\Delta \ll \lambda$). The inner beam (subscript 1) carries a current I_1, and I_2 is the current carried by the outer beam – subscript 2. Both beams propagate along a very strong magnetic field, in a metallic cylindrical (lossless) waveguide of radius R – see Fig. 3.20.

For formulating the dispersion relation of the space-charge and the electromagnetic waves we define: $\alpha_\nu = \Gamma R_{b,\nu}$, $\Gamma^2 = k^2 - \omega^2/c^2$,

$$\begin{aligned}\theta_1 &= \frac{\mathrm{I}_1(\alpha_1)}{\mathrm{I}_0(\alpha_1)} - \frac{\omega_{p,1}^2}{\gamma_1^3(\omega - k\mathrm{v}_1)^2}\Gamma\Delta,\\ \theta_2 &= \frac{\mathrm{I}_1(\alpha_2)K_0(\Gamma R) + \mathrm{I}_0(\Gamma R)K_1(\alpha_2)}{\mathrm{I}_0(\alpha_2)K_0(\Gamma R) - \mathrm{I}_0(\Gamma R)K_0(\alpha_2)} + \frac{\omega_{p,2}^2}{\gamma_2^3(\omega - k\mathrm{v}_2)^2}\Gamma\Delta.\end{aligned} \tag{3.5.44}$$

With these definitions the dispersion relation reads

$$\frac{\theta_1 \mathrm{I}_0(\alpha_1) - \mathrm{I}_1(\alpha_1)}{\theta_2 \mathrm{I}_0(\alpha_2) - \mathrm{I}_1(\alpha_2)} = \frac{\theta_1 \mathrm{K}_0(\alpha_1) + \mathrm{K}_1(\alpha_1)}{\theta_2 \mathrm{K}_0(\alpha_2) + \mathrm{K}_1(\alpha_2)}. \tag{3.5.45}$$

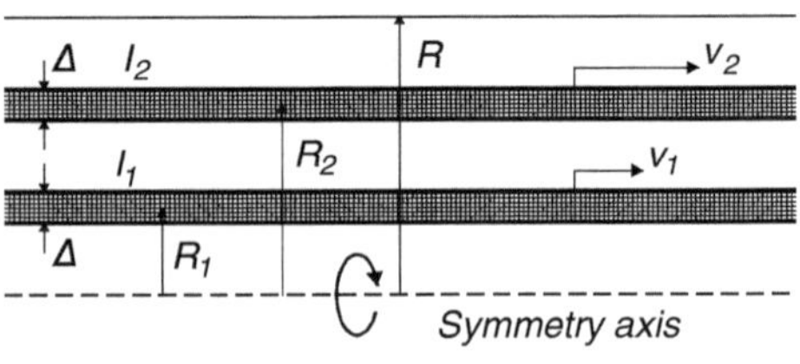

Fig. 3.20 Two-beam instability: two beams of different currents and different velocities move in a cylindrical waveguide. Under certain circumstances dc energy is converted into rf

A more familiar shape is obtained by assuming that the beam radius is large on the scale of the wavelength i.e., $|\alpha_\nu| \gg 1$ implying

$$\begin{aligned}\theta_1 &= 1 - \frac{\omega_{\mathrm{p},1}^2}{\gamma_1^3(\omega - k\mathrm{v}_1)^2}\Gamma\Delta,\\ \theta_2 &= -\coth(\alpha - \alpha_2) + \frac{\omega_{\mathrm{p},2}^2}{\gamma_2^3(\omega - k\mathrm{v}_2)^2}\Gamma\Delta.\end{aligned} \tag{3.5.46}$$

and

$$\begin{aligned}&\frac{\omega_{\mathrm{p},2}^2\Gamma\Delta}{\gamma_2^3(\omega - k\mathrm{v}_2)^2} + \frac{\omega_{\mathrm{p},1}^2\Gamma\Delta}{\gamma_1^3(\omega - k\mathrm{v}_1)^2} = \frac{\exp(\alpha - \alpha_2)}{\sinh(\alpha - \alpha_2)} + \tanh(\alpha_2 - \alpha_1)\\ &\times\left\{1 + \left[1 - \frac{\omega_{\mathrm{p},1}^2\Gamma\Delta}{\gamma_1^3(\omega - k\mathrm{v}_1)^2}\right]\left[\coth(\alpha - \alpha_2) - \frac{\omega_{\mathrm{p},2}^2\Gamma\Delta}{\gamma_2^3(\omega - k\mathrm{v}_2)^2}\right]\right\}.\end{aligned} \tag{3.5.47}$$

Keeping only terms linear in the plasma frequency and assuming that $\alpha \equiv \Gamma R \gg \alpha_2$, the dispersion relation of space-charge waves in two planar beams of thickness Δ is virtually identical to (3.5.33) up to a form factor

$$\frac{\omega_{\mathrm{p},2}^2}{\gamma_2^3(\omega - k\mathrm{v}_2)^2} + \frac{\omega_{\mathrm{p},1}^2}{\gamma_1^3(\omega - k\mathrm{v}_1)^2} = \frac{2}{\Gamma\Delta}. \tag{3.5.48}$$

Another case that warrants special attention is when the outer beam is very close to the metallic wall $\alpha \simeq \alpha_2$. As one may expect, there is no instability since the electric field is "short-circuited" by the metallic wall. In fact, it can be readily shown that the space-charge wave propagates only along the inner beam. Its dispersion relation being determined by

$$\frac{\omega_{\mathrm{p},1}^2\Gamma\Delta}{\gamma_1^3(\omega - k\mathrm{v}_1)^2} = \frac{\exp(\alpha - \alpha_1)}{\sinh(\alpha - \alpha_1)}. \tag{3.5.49}$$

Comment 3.2. Note that the two annular beams may be generated on the same cathode provided the potential's suppression is significant.

3.5.5 Interference of Space-Charge Waves

After establishing the existence of space-charge waves, let us examine next how they carry a small perturbation. For this purpose, it is convenient to adopt a transmission-line notation whereby the longitudinal electric field is associated with the voltage

$$V(z) = V_{+}\exp\left[-j\left(k_e + K_p\right)z\right] + V_{-}\exp\left[-j\left(k_e - K_p\right)z\right], \qquad (3.5.50)$$

and the azimuthal magnetic field is associated with the current

$$I(z) = \frac{V_{+}}{Z_{+}}\exp\left[-j\left(k_e + K_p\right)z\right] + \frac{V_{-}}{Z_{-}}\exp\left[-j\left(k_e - K_p\right)z\right]. \qquad (3.5.51)$$

In these expressions $k_e = \omega/\mathrm{v}_0$, $K_\mathrm{p} = \omega_\mathrm{p}/\mathrm{v}_0\gamma^{3/2}$,

$$Z_{\pm} = \eta_0\left(\frac{1}{\beta} \pm \frac{cK_\mathrm{p}}{\omega}\right), \qquad (3.5.52)$$

$\beta = \mathrm{v}_0/c$ and $\gamma = [1 - \beta^2]^{-1/2}$. Note that, in contrast to transmission lines where the two possible waves propagate in opposite directions, here both waves propagate in the same direction but with two different phase velocities. Let us assume now that both the voltage and the current are known at $z = 0$,

$$\begin{aligned} V(0) &= U_0, \\ I(0) &= I_0. \end{aligned} \qquad (3.5.53)$$

Subject to these two conditions the amplitudes $V_{\pm}$ can be calculated and consequently, the voltage and current modulation on the beam may be determined at any location

$$\begin{aligned} V_{+} &= \frac{Z_{+}}{Z_{+} - Z_{-}}(U_0 - Z_{-}I_0), \\ V_{-} &= -\frac{Z_{-}}{Z_{+} - Z_{-}}(U_0 - Z_{+}I_0), \\ V(z) &= \frac{Z_{+}}{Z_{+} - Z_{-}}(U_0 - Z_{-}I_0)\exp\left[-j\left(k_e + K_p\right)z\right] \\ &\quad + \frac{-Z_{-}}{Z_{+} - Z_{-}}(U_0 - Z_{+}I_0)\exp\left[-j\left(k_e - K_p\right)z\right], \\ I(z) &= \frac{1}{Z_{+} - Z_{-}}(U_0 - Z_{-}I_0)\exp\left[-j\left(k_e + K_p\right)z\right] \\ &\quad + \frac{-1}{Z_{+} - Z_{-}}(U_0 - Z_{+}I_0)\exp\left[-j\left(k_e - K_p\right)z\right],. \end{aligned} \qquad (3.5.54)$$

When the initial current modulation is zero ($I_0 = 0$) these equations read

$$\begin{aligned} V(z) &= \frac{U_0}{Z_{+} - Z_{-}}\exp(-jk_e z)\left[Z_{+}\exp\left(-jK_p z\right) - Z_{-}\exp\left(jK_p z\right)\right] \\ I(z) &= \frac{U_0}{Z_{+} - Z_{-}}\exp(-jk_e z)\left[\exp\left(-jK_p z\right) - \exp\left(jK_p z\right)\right]. \end{aligned} \qquad (3.5.55)$$

and associated with this modulation the average real power which develops along the beam oscillates in space and it is given by

$$P(z) = \tfrac{1}{2}\mathrm{Re}[V(z)I^*(z)] \\ = \frac{U_0^2}{2\eta_0\beta}\left(\frac{\omega}{cK_p}\right)^2 \sin^2(K_p z). \tag{3.5.56}$$

It is important to point out that along an ideal space-charge wave the Poynting vector is identically zero and the term above is proportional to $\int dV E_z J_z^*$. Note that both the current and the power is zero at $z = 0$. Down the stream, they both grow monotonically until it peaks after a quarter of the plasma wavelength $(\lambda_p \equiv 2\pi/K_p)$ namely $z = \lambda_p/4$ – similar to the interference of two regular electromagnetic waves. This fact is utilized in relativistic klystrons for further amplification of the modulation by placing a cavity at $z = \lambda_p/4$. Let us now examine this effect.

In the framework of the current model the cavity is represented by an RLC circuit whose impedance is

$$Z_{\mathrm{cav}} = Z_0 \frac{j\omega\omega_0}{\omega_0^2 - \omega^2 + 2j\omega\omega_0/Q}, \tag{3.5.57}$$

where Q is known as the quality factor of the cavity and together with Z_0 they determine the impedance at resonance i.e., $Z_{\mathrm{cav}} = Z_0 Q/2$ as illustrated in Fig. 3.21. The incident waves are given by (3.5.50)–(3.5.51) and the transmitted ones by

$$V_{\mathrm{tr}}(z) = V_{\mathrm{tr},+}\exp\left[-j(k_e + K_p)\left(z - \frac{\lambda_p}{4}\right)\right] + V_{\mathrm{tr},-}\exp\left[-j(k_e - K_p)\left(z - \frac{\lambda_p}{4}\right)\right],$$
$$I_{\mathrm{tr}}(z) = \frac{V_{\mathrm{tr},+}}{Z_+}\exp\left[-j(k_e + K_p)\left(z - \frac{\lambda_p}{4}\right)\right] + \frac{V_{\mathrm{tr},-}}{Z_-}\exp\left[-j(k_e - K_p)\left(z - \frac{\lambda_p}{4}\right)\right]. \tag{3.5.58}$$

The boundary condition at $z = \lambda_p/4$ can be determined from the fact that the current associated with the incident and transmitted waves has to be continuous i.e.

$$I\left(\frac{\lambda_p}{4}\right) = I_{\mathrm{tr}}\left(\frac{\lambda_p}{4}\right), \tag{3.5.59}$$

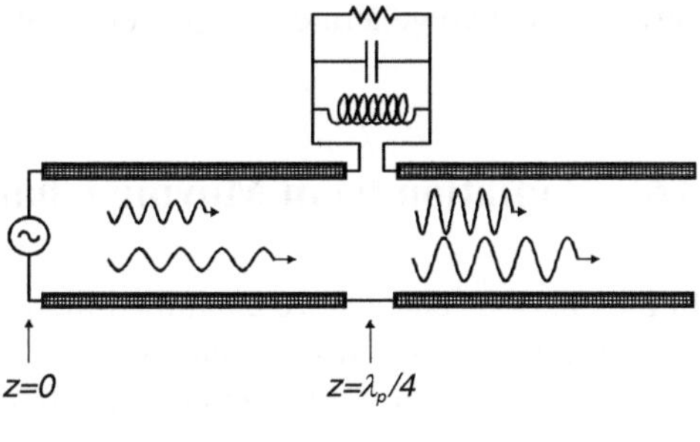

Fig. 3.21 Schematic of beam-cavity interaction in the framework of the transmission line formulation. The amplitude of the space-charge waves is significantly affected by the cavity

and the voltage associated with the incident waves is the sum of the transmitted and the voltage on the cavity aperture:

$$V\left(\frac{\lambda_{\rm p}}{4}\right) = V_{\rm tr}\left(\frac{\lambda_{\rm p}}{4}\right) + V_{\rm cav}. \tag{3.5.60}$$

Subject to these boundary conditions, the transmitted waves are given by

$$\begin{aligned} V_{\rm tr}(z) &= V(z) + Z_{\rm cav} I\left(\frac{\lambda_{\rm p}}{4}\right)\exp\left[-jk_e\left(z-\frac{1}{4}\lambda_p\right)\right] \\ &\quad \times\left\{-\cos\left[K_{\rm p}\left(z-\frac{1}{4}\lambda_p\right)\right] + j\frac{Z_+ + Z_-}{Z_+ - Z_-}\sin\left[K_{\rm p}\left(z-\frac{1}{4}\lambda_p\right)\right]\right\} \\ I_{\rm tr}(z) &= I(z) + 2jI\left(\frac{\lambda_{\rm p}}{4}\right)\exp\left[-jk_e\left(z-\frac{1}{4}\lambda_p\right)\right] \\ &\quad \times \sin\left[K_{\rm p}\left(z-\frac{1}{4}\lambda_p\right)\right]\frac{Z_{\rm cav}}{Z_+ - Z_-}. \end{aligned} \tag{3.5.61}$$

With these relations we realize that a small voltage modulation evolves, after quarter of the plasma wavelength $z = \lambda_p/4$ into a relatively low power at the cavity location is $P(z = \lambda_p/4) = (U_0^2/2\eta_0\beta)(\omega/cK_p)^2$ whereas after another quarter of the plasma wavelength

$$\frac{P\left(\frac{\lambda_{\rm p}}{2}\right)}{P\left(\frac{\lambda_{\rm p}}{4}\right)} \simeq \left(\frac{|Z_{\rm cav}|}{\eta_0}\frac{\omega}{cK_p}\right)^2 = \left(\frac{Z_0}{2\eta_0}Q\right)^2\frac{\omega^2}{\beta\omega_p^2}(\gamma\beta)^3 \gg 1. \tag{3.5.62}$$

the available power is several orders of magnitude larger than that available at the location of the cavity. Three main terms contribute to this ratio: (1) the quality factor of the cavity, (2) the ratio of the operating frequency and the plasma frequency and (3) the momentum of the electrons.

Before we conclude it is important to emphasize here that in contrast to the case of the two instabilities introduced in the last two sections, the amplitudes of each one of the space-charge waves are *uniform* in space and only the interference between the two generates the power mentioned above.

3.6 Radiation from Moving Charges

In the previous chapter we examined the process of radiation generation when the charge moves at a constant velocity near a uniform dielectric medium (Cerenkov radiation) or near a finite-size body (diffraction radiation). Before concluding this

Chapter, we briefly examine the process of radiation generation due to non-uniform velocity of charged particles.

3.6.1 *Radiation from an Oscillating Dipole*

Among the simplest and at the same time widespread manifestation of radiation emission is the one occurring when a charge (q) oscillates at a given angular frequency (ω) and in this subsection we consider the power generated by this moving charge. Denoting by $d/2$ the amplitude of oscillation, the current density associated with this particle is $\bar{J}_z = j\omega(d/2)(q/2\pi r)\delta(r)\delta(z)$. Further defining the dipole moment $p = qd$, the vector magnetic potential and the electric scalar counterpart read

$$\begin{aligned} \bar{A}_z &= j\frac{\omega}{c^2}\frac{p}{4\pi\varepsilon_0}\frac{1}{r}\exp\left(-j\frac{\omega}{c}r\right), \\ \bar{\Phi} &= \frac{p}{4\pi\varepsilon_0 r^2}\left(1+j\frac{\omega}{c}r\right)\cos\theta\exp\left(-j\frac{\omega}{c}r\right). \end{aligned} \tag{3.6.1}$$

In what follows we consider only terms that contribute to the *far-field* $r \gg \lambda$ namely, field components that are inversely proportional to r such that the radial component of the Poynting vector

$$\bar{S}_r = \frac{1}{2}\left[\bar{E}_\theta \bar{H}_\varphi^* - \bar{E}_\varphi \bar{H}_\theta^*\right] \tag{3.6.2}$$

is inversely proportional to r^2 and as a result, the total power emitted in vacuum does not depend on the radius where the measurement is performed. Moreover, it can be readily shown that the azimuthal component of the electric far-field is zero thus $\bar{S}_r = \bar{E}_\theta \bar{H}_\varphi^*/2$. Consequently, the relevant far-field components are

$$\begin{aligned} \bar{H}_\varphi &\simeq -\frac{1}{\mu_0}\frac{\omega^2}{C^3}\frac{P}{4\pi\varepsilon_0}\left[\frac{1}{r}\exp\left(-j\frac{\omega}{c}r\right)\right]\sin\theta, \\ \bar{E}_\theta &\simeq -\frac{\omega^2}{C^2}\frac{P}{4\pi\varepsilon_0}\left[\frac{1}{r}\exp\left(-j\frac{\omega}{c}r\right)\right]\sin\theta, \end{aligned} \tag{3.6.3}$$

implying

$$\bar{S}_r = \frac{1}{2\mu_0 c}\left(\frac{\omega}{c}\right)^4\left(\frac{p}{4\pi\varepsilon_0}\right)^2\frac{\sin^2\theta}{r^2} \tag{3.6.4}$$

and finally the *average* power reads

$$\bar{P} = r^2\int_0^{2\pi} d\varphi \int_0^{\pi} d\theta \sin\theta\, \bar{S}_r = \frac{1}{2\eta_o}\left(\frac{p}{4\pi\varepsilon_0}\frac{\omega^2}{c^2}\right)^2\frac{8\pi}{3}. \tag{3.6.5}$$

Several comments are in place at this stage:

Comment 3.3. With the average power established, we may deduce the instantaneous power by replacing the harmonic oscillation $\zeta(t) = (d/2)\cos(\omega t)$ with an arbitrary acceleration

$$P(t) = \frac{2}{3} mc^2 \frac{r_e}{c^3} \ddot{\zeta}^2(t), \tag{3.6.6}$$

wherein

$$r_e \equiv \frac{e^2}{4\pi\varepsilon_0} \frac{1}{mc^2} \simeq 2.8 \times 10^{-15}[\mathrm{m}] \tag{3.6.7}$$

is the *classical radius of the electron.*

Comment 3.4. Another interesting observation relies on the fact that the average energy emitted during one period of the radiation field is $W \equiv PT = (\eta_0/2)(p\omega/c)^2\omega$. Bearing in mind that the energy of a single photon is expressed in terms of Planck's constant ($\hbar = 1.054 \times 10^{-34}\,\mathrm{J\,s}$) is given by $W_{\mathrm{ph}} = \hbar\omega$, then the number of emitted photons by the dipole is

$$N_{\mathrm{ph}} = \frac{2\pi}{3}\alpha\left(\frac{p}{e}\frac{\omega}{c}\right)^2 \tag{3.6.8}$$

wherein

$$\alpha \equiv \frac{e^2}{4\pi\varepsilon_0} \frac{1}{\hbar c} \simeq \frac{1}{137}, \tag{3.6.9}$$

is the so-called *fine-structure constant*. Examining this relation (3.6.8) for a single dipole reveals that the number of photons emitted per dipole is very small bearing in mind that typically the size of the dipole is much smaller than the wavelength.

Comment 3.5. Subject to the far-field approximation, we have concluded that the amplitude of the oscillating dipole is not constant. From the perspective of the oscillating charge this implies that in addition to the "binding" force there is another force associated with the radiation field. A rough evaluation of this force (F_{rad}) is possible by pursuing the following approach. In a lossless medium, the electromagnetic energy emitted comes at the expense of the kinetic energy of the particle namely,

$$\int dt \dot{\zeta}(t) F_{\mathrm{rad}}(t) = \frac{2}{3} mc^2 \frac{r_e}{c^3} \int dt \ddot{\zeta}^2(t). \tag{3.6.10}$$

Bearing in mind that

$$\int dt \ddot{\zeta}^2(t) = \int dt \left[\frac{d}{dt}\left(\dot{\zeta}\ddot{\zeta}\right) - \dddot{\zeta}\ \dot{\zeta}\right] \tag{3.6.11}$$

and assuming that the contribution of the second term vanishes in the relevant time interval we conclude that the *radiation reaction force* is

$$F_{\text{rad}}(t) = \frac{2}{3} mc^2 \frac{r_e}{c^3} \dddot{\zeta}(t). \tag{3.6.12}$$

With this force term, the equation of motion may be generalized when an external force is applied

$$m\left(\frac{d^2\zeta}{dt^2} - \tau_{\text{rad}} \frac{d^3\zeta}{dt^3}\right) = F_{\text{ext}}(t) \tag{3.6.13}$$

wherein the *radiative damping time* is

$$\tau_{\text{rad}} \equiv \frac{2}{3}\frac{r_e}{c} = 0.52 \times 10^{-23}[\text{sec}]. \tag{3.6.14}$$

3.6.2 Radiation from a Moving Charge

The power emitted by an oscillating dipole is independent of whether it moves with a constant velocity or it is motionless. In this subsection, we consider a dipole moving with a constant velocity $\mathbf{v}$ and for the sake of simplicity, it is first assumed to be *parallel* to the oscillation. In the frame of reference where the dipole is stationary,

$$P(t') = \frac{2}{3} mc^2 \frac{r_e}{c^3} \left[\frac{d^2\zeta'(t')}{dt'^2}\right]^2. \tag{3.6.15}$$

In terms of the laboratory frame of reference variables $dt' = dt/\gamma$, $d\zeta' = d\zeta\,\gamma$ we get

$$\begin{aligned} P(t) &= \frac{2}{3} mc^2 \frac{r_e}{c^3} \left[\gamma^3 \frac{d^2\zeta}{dt^2}\right]^2, \\ &= \frac{2}{3}\frac{r_e}{c} mc^2 \left[\frac{1}{mc}\frac{d}{dt}\left(m\gamma \frac{d\zeta}{dt}\right)\right]^2, \\ &= \frac{2}{3}\frac{r_e}{c} mc^2 \left[\frac{F_{\|}}{mc}\right]^2. \end{aligned} \tag{3.6.16}$$

For a longitudinal gradient of the order of $100[\text{GeV/m}]$ which has been reported in recent plasma laser wake-field acceleration experiment, the power emitted by a single electron is of the order of nWatt which is completely negligible on the scale of the electromagnetic power required to generate such a gradient.

In case of an oscillation *perpendicular* to the motion $dt' = dt/\gamma$, $d\zeta' = d\zeta$ therefore,

$$\begin{aligned} P(t) &= \frac{2}{3}mc^2\frac{r_e}{c^3}\left[\gamma^2\frac{d^2\zeta}{dt^2}\right]^2, \\ &= \frac{2}{3}\frac{r_e}{c}mc^2\left[\frac{\gamma}{mc}\frac{d}{dt}\left(m\gamma\frac{d\zeta}{dt}\right)\right]^2, \\ &= \frac{2}{3}\frac{r_e}{c}mc^2\left[\frac{\gamma F_\perp}{mc}\right]^2. \end{aligned} \tag{3.6.17}$$

As an example, one may consider a relativistic electron (5 GeV) bent by a dipole ($B_d = 0.2[T]$) in a damping ring. The vertical force is $F_\perp \simeq eB_d c$, and the corresponding radius of curvature is about $\rho \simeq 80[\mathrm{m}]$ and consequently, the energy lost per unit length is $\Delta E/2\pi\rho = P/c \simeq 1.4[\mathrm{keV/m}]$. Bearing in mind that the electrons may make billions of revolutions this gradient is significant and the essence of the operation of a *damping ring* relies on this phenomenon: electrons emitting *synchrotron radiation* are decelerated proportional to γ^4 in all three dimensions and reaccelerated only in the longitudinal direction, reducing in the process the transverse phase-space.

For completeness, it may be demonstrated (Jackson 1962) that the power emitted in case of an arbitrary angle between the acceleration and the velocity is given by

$$P(t) = \frac{2}{3}mc^2\frac{r_e}{c}\gamma^6\left[\dot{\boldsymbol{\beta}}\cdot\dot{\boldsymbol{\beta}} - (\boldsymbol{\beta}\times\dot{\boldsymbol{\beta}})\cdot(\boldsymbol{\beta}\times\dot{\boldsymbol{\beta}})\right]. \tag{3.6.18}$$

3.6.3 Ensemble of Radiating Sources

In the previous two subsections, we considered a single, point-like radiator. In this subsection we consider an ensemble of N radiators located at (X_ν, Y_ν, Z_ν) where $\nu = 1, 2...N$. For simplicity sake, the size of each source is assumed to be finite in each one of the dimensions and its size is denoted by $(\Delta_x, \Delta_y, \Delta_z)$. Our starting point is the expressions for the far-field determined in (3.6.3) explicitly, the exponent may be replaced by

$$\begin{aligned} \exp\left(-j\frac{\omega}{c}r\right) \to{}& \exp\left(-j\frac{\omega}{c}r\right)\mathrm{sinc}\left(\frac{1}{2}\Delta_x\frac{\omega}{c}\sin\theta\cos\phi\right) \\ &\times \mathrm{sinc}\left(\frac{1}{2}\Delta_y\frac{\omega}{c}\sin\theta\sin\phi\right)\mathrm{sinc}\left(\frac{1}{2}\Delta_z\frac{\omega}{c}\cos\phi\right) \\ &\times \sum_\nu \exp\left[j\frac{\omega}{c}(X_\nu\sin\theta\cos\phi + Y_\nu\sin\theta\sin\phi + Z_\nu\cos\theta)\right]. \end{aligned} \tag{3.6.19}$$

With this observation, we may deduce that the power emitted by the N radiators is given by

$$\begin{aligned}\bar{P}_N = &\frac{1}{2\eta_o}\left(\frac{qNd}{4\pi\varepsilon_0}\frac{\omega^2}{c^2}\right)^2\frac{8\pi}{3}\\ &\times\frac{1}{2\pi}\int_0^{2\pi}d\varphi\frac{3}{4}\int_0^{\pi}d\theta\sin^3\theta\,\text{sinc}^2\left(\frac{1}{2}\Delta_x\frac{\omega}{c}\sin\theta\cos\phi\right)\\ &\times\text{sinc}^2\left(\frac{1}{2}\Delta_y\frac{\omega}{c}\sin\theta\sin\phi\right)\text{sinc}^2\left(\frac{1}{2}\Delta_z\frac{\omega}{c}\cos\phi\right)\\ &\times\frac{1}{N^2}\sum_{\nu,\mu}\exp\left\{j\frac{\omega}{c}\left[(X_\nu - X_\mu)\sin\theta\cos\phi + (Y_\nu - Y_\mu)\sin\theta\sin\phi\right]\right\}\\ &\times\exp\left[j\frac{\omega}{c}(Z_\nu - Z_\mu)\cos\theta\right].\end{aligned} \tag{3.6.20}$$

A relatively simple expression may be developed at the limit of a point-source $\Delta_x = \Delta_y = \Delta_z \to 0$ and subject to the assumption that they all are in the $Z_\nu = 0$ plane thus

$$\begin{aligned}\bar{P}_N = &N^2\bar{P}_1\\ &\times\frac{1}{N^2}\sum_{\nu,\mu}\frac{3}{4}\int_0^{\pi}d\theta\sin^3\theta\,J_0\left[\frac{\omega}{c}\sqrt{(X_\nu - X_\mu)^2 + (Y_\nu - Y_\mu)^2}\sin\theta\right].\end{aligned} \tag{3.6.21}$$

Of specific interest is the case when all emitters form a line therefore without loss of generality we consider $X_\nu = \nu a$, $Y_\nu = 0$ thus

$$\bar{P}_N = N^2\bar{P}_1\left\{\frac{1}{N^2}\sum_{\nu,\mu}\frac{3}{4}\int_0^{\pi}d\theta\sin^3\theta\,J_0\left[\frac{\omega}{c}a|\nu - \mu|\sin\theta\right]\right\}. \tag{3.6.22}$$

The curled brackets represent a form factor that is always smaller than unity: for small arguments ($\omega a/c \ll 1/N$) it is unity and for large arguments ($\omega a/c \gg 1$) it asymptotically approaches $1/N$ as illustrated in Fig. 3.22 and even better in Fig. 3.23.

A reasonable approximation (error <5%) of the form factor reveals that the power is given by

$$\begin{aligned}\bar{P}_N &= N^2\bar{P}_1\left\{\frac{1}{N} + \left(1 - \frac{1}{N}\right)\left[1 + \left(\frac{N}{4.5}\frac{\omega}{c}a\right)^2\right]^{-1}\right\},\\ &= \bar{P}_1\left\{N + N(N-1)\left[1 + \left(\frac{N}{4.5}\frac{\omega}{c}a\right)^2\right]^{-1}\right\},\end{aligned} \tag{3.6.23}$$

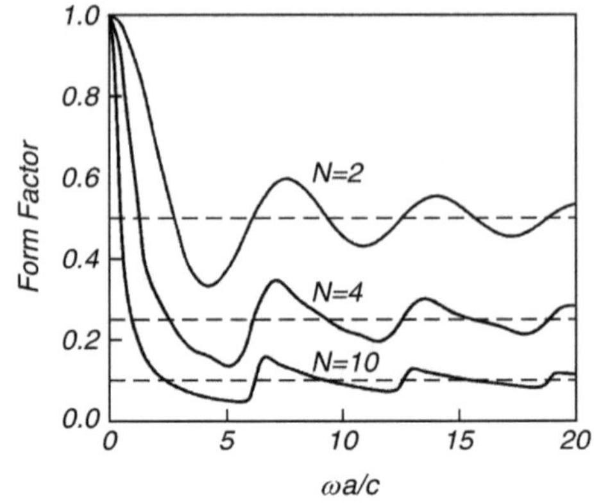

Fig. 3.22 Form factor in (3.6.22) as a function of the normalized frequency. For large values, the form factor approaches the value 1/N

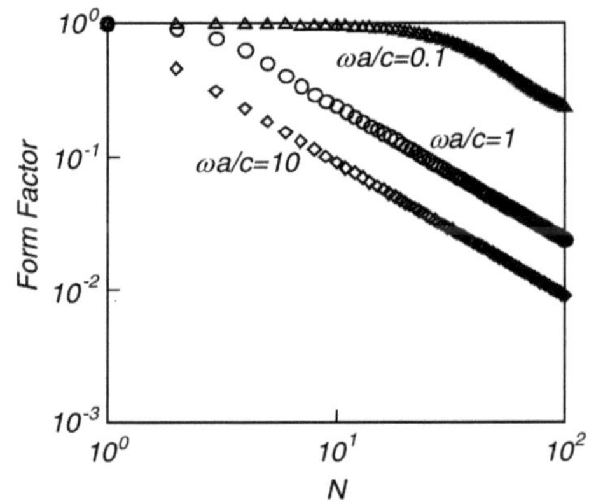

Fig. 3.23 Form factor in (3.6.22) as a function of the number of sources. For large values, the form factor approaches the value 1/N

which in turn indicates that if the radiated wavelength is much longer than the spacing between the sources ($\omega a/c \ll 4.5/N$), the radiation is emitted coherently (recall that all dipoles have the same phase) being proportional to N^2 whereas in the opposite case ($\omega a/c \gg 4.5/N$), the emitted power is not coherent being proportional to the number of emitters ($\propto N$).

3.6.4 Synchrotron Radiation of an Ensemble of Electrons

Electrons organized in bunches may generate coherent Cerenkov radiation according to the size of the bunch – see Exercise 2.6 in the previous chapter. In Sect. 3.6.2 we have evaluated based on simple arguments the total power of synchrotron radiation emitted by a single relativistic ($\gamma \gg 1$) electron following a circular (R) trajectory during one revolution;

$$P_1 = \frac{2}{3}\gamma^4 \frac{e^2 c}{4\pi\varepsilon_0 R^2}\ . \tag{3.6.24}$$

Our goal in this subsection is to formulate and examine some of the simple scaling laws regarding the *spectrum* of coherent and non-coherent synchrotron radiation generated by some simple electrons configuration. With this purpose in mind, we consider an ensemble of relativistic electrons following a circular trajectory with an angular frequency Ω; the latter is contemplated to be due to a uniform magnetic field B thus ($\Omega = eB/m\gamma$). The first stage is to determine the electromagnetic field in the entire space. Associated with this ensemble there is an azimuthal current density

$$J_\phi(r,\phi,z,t) = -e(\Omega R)\frac{1}{r}\,\delta(r-R)\,\delta(z)\,\sum\nolimits_{\nu=1}^{N}\delta[\phi-\Omega(t-t_\nu)]; \qquad (3.6.25)$$

N is the total number of electrons and t_ν is the time the νth electron reaches the point $\phi = 0$. This current density generates an electromagnetic field that may be derived from the two z-components of the field

$$H_z(r,\phi,z,t) = \int_{-\infty}^{\infty} dk\,\exp(-jkz)\sum_{n=-\infty}^{\infty}\exp(-jn\phi)\sum_{m=-\infty}^{\infty}\exp(jm\Omega t)\,\mathcal{H}_{n,m}(k,r)$$
$$\mathcal{H}_{n,m}(k,r) = \begin{cases} A_{n,m}\,\mathrm{J}_n(\Lambda_m r) & r \le R \\ B_{n,m}\,\mathrm{H}_n^{(2)}(\Lambda_m r) & r \ge R \end{cases}$$
$$E_z(r,\phi,z,t) = \int_{-\infty}^{\infty} dk\,\exp(-jkz)\sum_{n=-\infty}^{\infty}\exp(-jn\phi)\sum_{m=-\infty}^{\infty}\exp(jm\Omega t)\,\mathcal{E}_{n,m}(k,r)$$
$$\mathcal{E}_{n,m}(k,r) = \begin{cases} C_{n,m}\,\mathrm{J}_n(\Lambda_m r) & r \le R \\ D_{n,m}\,\mathrm{H}_n^{(2)}(\Lambda_m r) & r \ge R \end{cases} \qquad (3.6.26)$$

with $\Lambda_m^2 = (m\Omega/c)^2 - k^2$. In this notation, we tacitly assumed that the spectrum is discrete

$$\omega = m\Omega. \qquad (3.6.27)$$

For evaluation of the four coefficients (A, B, C and D) it is necessary to impose the continuity of E_ϕ, E_z and H_ϕ at $r = R$ whereas the discontinuity on H_z is determined by the surface current density derived from J_ϕ. Based on these constraints the amplitudes are

$$A_{n,m} = \left[\frac{\pi}{2j}\psi_n\,\overset{\circ}{\mathrm{H}}{}_n^{(2)}(\psi_n)\right]S_n\delta_{n,m} \quad B_{n,m} = \left[\frac{\pi}{2j}\psi_n\,\overset{\circ}{\mathrm{J}}{}_n(\psi_n)\right]S_n\delta_{n,m}$$
$$C_{n,m} = \left[\frac{\pi}{2j}\,\mathrm{H}_n^{(2)}(\psi_n)\left(\frac{-jk}{\Omega\varepsilon_0}\right)\right]S_n\delta_{n,m} \quad D_{n,m} = \left[\frac{\pi}{2j}\,\mathrm{J}_n(\psi_n)\left(\frac{-jk}{\Omega\varepsilon_0}\right)\right]S_n\delta_{n,m} \qquad (3.6.28)$$

wherein $\psi_n = \Lambda_n R,\ S_n = e\Omega g_n/(2\pi)^2,\quad g_n = \sum_{\nu=1}^{N}\exp(jn\Omega t_\nu)$. The latter is the form factor determined by the spatial distribution of the ensemble.

With the field established, we may proceed to the next stage to determine and simplify the power emitted by the ensemble. Formally, the emitted power is

$$\begin{aligned} P(t) &= -\int_{-\infty}^{\infty} dz \int_0^{\infty} dr\, r \int_0^{2\pi} d\phi\, J_\phi(r,\ \phi, z, t) E_\phi(r,\ \phi, z, t) \\ &= e\Omega R \sum_{\nu=1}^{N} E_\phi[r = R,\ \phi = \Omega(t - t_\nu), z = 0, t] \end{aligned} \tag{3.6.29}$$

and since the contribution to the power stems only from the "radiative" part of the spectrum namely, $k^2 \le (n\Omega/c)^2$, it reads

$$\begin{aligned} \bar{P} = 2\beta \sum_{n=1}^{\infty} n\, |g_n|^2 \Bigg\{ &\int_0^{n\beta} d\xi\, \xi^2 \left[(n\beta)^2 - \xi^2\right]^{-1} \mathrm{J}_n^2\left(\sqrt{(n\beta)^2 - \xi^2}\right) \\ &+ \beta^2 \int_0^{n\beta} d\xi\, \mathring{\mathrm{J}}_n^2\left(\sqrt{(n\beta)^2 - \xi^2}\right)\Bigg\}; \end{aligned} \tag{3.6.30}$$

$\beta = \Omega R/c$ representing the normalized local velocity of the electron and the normalized power is $\bar{P} = P/(e^2 c/4\pi\varepsilon_0 R^2)$. Further simplification of this result may be achieved by defining a new variable $\xi = n\beta\cos\theta$, with the observation that for relativistic particles $\beta \simeq 1 - 1/2\gamma^2$as well as the fact that most of the radiation is emitted in a narrow angle $\delta\theta$ in the plane defined by $\theta = \pi/2$, it simplifies to be

$$\bar{P} = 2\sum_{n=1}^{\infty} n^2\, |g_n|^2 \int d\delta\theta \left\{ \mathrm{J}_n^2\left[n\left(1 - \frac{1}{2\gamma^2} - \frac{\delta\theta^2}{2}\right)\right] + \delta\theta^2\, \mathring{\mathrm{J}}_n^2\left[n\left(1 - \frac{1}{2\gamma^2} - \frac{\delta\theta^2}{2}\right)\right]\right\}. \tag{3.6.31}$$

This result makes the use of asymptotic Bessel functions

$$\begin{aligned} \mathrm{J}_n\left(n\frac{1}{\cosh\psi}\right) &\simeq \frac{\exp[-n(\psi - \tanh\psi)]}{\sqrt{2\pi n\, \tanh\psi}} \simeq \frac{1}{\sqrt{2\pi n\,\psi}} \exp\left[-n\frac{\psi^3}{3}\right], \\ \mathring{\mathrm{J}}_n\left(n\frac{1}{\cosh\psi}\right) &\simeq \sqrt{\frac{\sinh 2\psi}{4\pi n}} \exp[-n(\psi - \tanh\psi)] \simeq \sqrt{\frac{\psi}{2\pi n}} \exp\left[-n\frac{\psi^3}{3}\right], \end{aligned} \tag{3.6.32}$$

rather natural and therefore, identifying $\psi \simeq \sqrt{\delta\theta^2 + \gamma^{-2}}$ we have

$$\bar{P} \simeq \frac{1}{\pi}\sum_{n=1}^{\infty} n\, |g_n|^2 \int d\delta\theta \left(\psi + \delta\theta^2 \frac{1}{\psi}\right) \exp\left(-n\frac{2}{3}\psi^3\right). \tag{3.6.33}$$

This, in turn, may be further simplified by defining $\gamma\delta\theta = \sinh u$ leading to

$$\bar{P} \simeq \frac{1}{\pi\gamma^2}\sum_{n=1}^{\infty} n\,|g_n|^2 \int_0^{\infty} du\ \cosh 2u\ \exp\left[-n\frac{2}{3}\frac{1}{\gamma^3}\cosh^3 u\right]. \tag{3.6.34}$$

The last integral can be evaluated numerically the result is

$$\int_0^{\infty} du\ \cosh 2u\ \exp\left[-x\cosh^3 u\right] \simeq \frac{0.883}{x^{\frac{2}{3}}}\exp(-x), \tag{3.6.35}$$

and finally implying

$$P = \frac{e^2 c}{4\pi\varepsilon_0 R^2}\frac{1.157}{\pi}\sum_{n=1}^{\infty} n^{1/3}\,|g_n|^2 \exp\left(-\frac{2n}{3\gamma^3}\right). \tag{3.6.36}$$

This expression determines the *synchrotron radiation* power emitted by an ensemble of electrons during one revolution.

In order to assess the *numerical error* associated with the approximations so far we replace the sum for the single electron case ($g_n = 1$) with an integral that can be evaluated analytically

$$\sum_{n=1}^{\infty} n^{1/3}\exp\left(-\frac{2n}{3\gamma^3}\right) \to \int_0^{\infty} dn\, n^{1/3}\exp(-2n/3\gamma^3) = 0.893\left(\frac{2}{3\gamma^3}\right)^{-4/3} \tag{3.6.37}$$

leading to

$$P = 0.57\gamma^4\frac{e^2 c}{4\pi\varepsilon_0 R^2}. \tag{3.6.38}$$

This is within less than 20% error of the exact expression introduced in (3.6.24).

Since the expression in (3.6.36) provides us with the accurate scaling of the emitted power on γ during the period of one revolution, we proceed now to the third stage and this is: analysis of the spectrum as well as the power for several simple configurations.

Comment 3.6. The spontaneous power emitted by N un-correlated electrons is obviously given by

$$P_{\rm sp} = \frac{2}{3}\gamma^4\frac{e^2 c}{4\pi\varepsilon_0 R^2}N. \tag{3.6.39}$$

Comment 3.7. A *single bunch* of N electrons of Gaussian distribution $(\Delta_{\rm b})$, namely a probability density $f(\zeta) = \left(1/\sqrt{2\pi}\Delta_b\right)\exp\left(-\zeta^2/2\Delta_{\rm b}^2\right)$ leads to

$$\frac{1}{N^2}|g_n|^2 = \left|\int_{-\infty}^{\infty} d\zeta f(\zeta)\exp\left(jn\frac{\Omega}{c}\zeta\right)\right|^2 = \exp\left[-\left(n\frac{\Omega}{c}\Delta_b\right)^2\right], \tag{3.6.40}$$

the corresponding spectrum is proportional to

$$n^{1/3}\exp\left[-\frac{2n}{3\gamma^3} - \left(n\frac{\Delta_b}{R}\right)^2\right] \tag{3.6.41}$$

and consequently, the emitted power is

$$\begin{aligned} P_b &= \frac{2}{3}\gamma^4\frac{e^2c}{4\pi\varepsilon_0 R^2}\left\{N + N(N-1)\frac{\int_0^\infty dx\,x^{1/3}\exp\left[-x - x^2\left(\frac{\Delta_b}{R}\frac{3\gamma^3}{2}\right)^2\right]}{\int_0^\infty dx\,x^{1/3}\exp(-x)}\right\} \\ &\simeq \frac{2}{3}\gamma^4\frac{e^2c}{4\pi\varepsilon_0 R^2}\left\{N + N(N-1)\left[1 + \left(\frac{\Delta_b}{R}\frac{3\gamma^3}{2}\right)^{3/2}\right]^{-1}\right\}. \end{aligned} \tag{3.6.42}$$

Evidently, the second term is the contribution of the *coherent synchrotron radiation*. In order to have an estimate regarding the orders of magnitude involved, let us consider $N = 10^{10}$, 5 GeVelectrons forming a $\Delta_b = 1$ cm long bunch, following a circular trajectory of radius $R = 100$ m. The coherent term is 0.5% of the total power. On the other hand, if the number of electrons is increased to $N = 10^{11}$, the energy reduced to 1 GeV and keeping the other parameters the same, the coherent term is almost two orders of magnitude larger than the spontaneous term. Based on this result we may develop a relatively simple criterion for the coherent synchrotron radiation from a bunch to become dominant namely,

$$\frac{2R}{3\gamma^3}N^{2/3} \gg \Delta_b. \tag{3.6.43}$$

Comment 3.8. The same approach may be employed for a *train of micro-bunches*. Each micro-bunch has a length Δ_{mb} (Gaussian distribution), there are M micro-bunches and the distance between any two bunches is $\Delta_T \gg \Delta_{mb}$. In each micro-bunch there are N_{el}electrons and the total number of electrons is assumed to be as above $N = N_{el}M$. In this case

$$\frac{1}{N^2}|g_n|^2 = \exp\left[-\left(n\frac{\Delta_{mb}}{R}\right)^2\right]\left[\frac{\operatorname{sinc}\left(\frac{1}{2}n\frac{\Delta_T}{R}M\right)}{\operatorname{sinc}\left(\frac{1}{2}n\frac{\Delta_T}{R}\right)}\right]^2 \tag{3.6.44}$$

and the emitted power is

$$P_{\text{train}} = \frac{2}{3}\gamma^4 \frac{e^2 c}{4\pi\varepsilon_0 R^2}\left\{ N + N(N-1)\frac{\int_0^\infty dx\, x^{1/3}\exp\left(-x - x^2\bar{\Delta}_{\text{mb}}^2\right)\left[\frac{\operatorname{sinc}\left(x\bar{\Delta}_T M\right)}{\operatorname{sinc}\left(x\bar{\Delta}_T\right)}\right]^2}{\int_0^\infty dx\, x^{1/3}\exp(-x)} \right\},$$

$$\bar{\Delta}_{\text{mb}} \equiv \frac{\Delta_{\text{mb}}}{R}\frac{3\gamma^3}{2}, \quad \bar{\Delta}_T \equiv \frac{1}{2}\frac{\Delta_T}{R}\frac{3\gamma^3}{2}. \tag{3.6.45}$$

However, for the sake of simplicity, we limit the discussion to a "zero" size micro-bunch such that

$$P_{\text{train}} = \frac{2}{3}\gamma^4 \frac{e^2 c}{4\pi\varepsilon_0 R^2}\left\{ N + N(N-1)\frac{\int_0^\infty dx x^{1/3}\exp(-x)\left[\frac{\operatorname{sinc}\left(x\bar{\Delta}_T M\right)}{\operatorname{sinc}\left(x\bar{\Delta}_T\right)}\right]^2}{\int_0^\infty dx x^{1/3}\exp(-x)} \right\} \tag{3.6.46}$$

and it may be approximated by

$$P_{\text{train}} = \frac{2}{3}\gamma^4 \frac{e^2 c}{4\pi\varepsilon_0 R^2}\left[\left[N + \frac{N(N-1)}{M}\left[1 + \frac{M-1}{1+\left(M\bar{\Delta}_T/2\right)^2}\right]\right]\right]. \tag{3.6.47}$$

Assuming the same macro-bunch length (1 cm), a bunch spacing corresponding to optical wavelength (1 μm) and correspondingly, $M \sim 10^4$, the coherent term is inversely proportional to M and, in fact, numerical analysis of this term reveals that

$$P_{\text{train}} \simeq \frac{2}{3}\gamma^4 \frac{e^2 c}{4\pi\varepsilon_0 R^2}\left[N_{\text{el}}M + N_{\text{el}}(N_{\text{el}}M - 1)\right]. \tag{3.6.48}$$

However, while simple, this result has limited practical relevance since the size of the micro-bunch plays a dominant role. For considering the latter, we need to evaluate (3.6.46) numerically. As before, let us consider $N = 10^{10}$, 5 GeV electrons forming a $\Delta_b = 1$ cm long bunch, following a circular trajectory of radius $R = 100$ m. The spacing between two bunches is 1 μm corresponding to $M = 10^4$ micro-bunches and the size of one micro-bunch is assumed to be $\Delta_{\text{mb}} \simeq 0.1$ μm. The coherent term increases from 0.5% to 13.6% of the total emitted power. Reducing the energy to 1 GeV and increasing the number of electrons in the macro-bunch to $N = 10^{11}$cause an increase of three orders of magnitude (950) in the coherent component relative to the spontaneous term – this is an increase of more than one order of magnitude above coherent power emitted by a uniform bunch.

In the configuration presented here the bunches were assumed to be pre-formed. In practice, a uniform bunch may become bunched – this process has been observed experimentally by Byrd et al. (2002) and discussed by Venturini et al. (2005) but it is beyond the scope of this chapter.

Comment 3.9. It is instructive at this stage to determine some of the characteristics of the spectrum emitted by a *point-charge* specified in (3.6.37). Obviously, it increases monotonically to a peak value which occurs at $n_{\max} = \gamma^3/2$ and correspondingly, the peak value is proportional to γ. Beyond this peak value, the spectrum decreases exponentially. In terms of frequencies, according to (3.6.27),

$$\omega_{\max} = \Omega\frac{\gamma^3}{2} = \left(\frac{eB}{m\gamma}\right)\frac{\gamma^3}{2} = \frac{c}{R}\frac{\gamma^3}{2} \tag{3.6.49}$$

reflecting the fact that for a given bending magnetic field $\omega_{\max} \propto \gamma^2$ whereas for a given radius of curvature $\omega_{\max} \propto \gamma^3$; here we used the fact that for a relativistic particle $\Omega R/c \simeq 1$. The average frequency emitted is proportional to $\omega_{\max}$ or explicitly,

$$\begin{aligned}\langle\omega\rangle = \Omega\langle n\rangle &= \Omega\frac{\sum\limits_{n=1}^{\infty} n^{4/3}\exp\left(-\frac{2n}{3\gamma^3}\right)}{\sum\limits_{n=1}^{\infty} n^{1/3}\exp\left(-\frac{2n}{3\gamma^3}\right)} = \Omega\frac{\int\limits_0^{\infty} dn\, n^{4/3}\exp\left(-\frac{2n}{3\gamma^3}\right)}{\int\limits_0^{\infty} dn\, n^{1/3}\exp\left(-\frac{2n}{3\gamma^3}\right)} \\ &= \Omega\frac{3\gamma^3}{2}\frac{\int\limits_0^{\infty} dx\, x^{4/3}\exp(-x)}{\int\limits_0^{\infty} dx\, x^{1/3}\exp(-x)} = \Omega\frac{3\gamma^3}{2}\frac{\Gamma(7/3)}{\Gamma(4/3)} = 2\gamma^3\Omega = 4\omega_{\max}\end{aligned} \tag{3.6.50}$$

and in a similar way,

$$\begin{aligned}\langle\omega^2\rangle = \Omega^2\langle n^2\rangle &= \Omega^2\left(\frac{3\gamma^3}{2}\right)^2\frac{\int\limits_0^{\infty} dx\, x^{7/3}\exp(-x)}{\int\limits_0^{\infty} dx\, x^{1/3}\exp(-x)} \\ &= \Omega^2\left(\frac{3\gamma^3}{2}\right)^2\frac{\Gamma(10/3)}{\Gamma(4/3)} = 7\Omega^2\gamma^6 = 28\,\omega_{\max}^2\end{aligned} \tag{3.6.51}$$

implying that $\Delta\omega = 2\sqrt{3}\omega_{\max}$ and $\Delta\omega/\langle\omega\rangle \simeq 0.866$.

Comment 3.10. Following a similar approach for a *finite length bunch*, we establish first the probability of obtaining a photon of energy $\hbar n\Omega$, accounting for both spontaneous and coherent spectrum,

$$p_n = \frac{Nn^{1/3}\exp\left(-\frac{2n}{3\gamma^3}\right) + N(N-1)n^{1/3}\exp\left[-\frac{2n}{3\gamma^3} - \left(n\frac{\Delta_b}{R}\right)^2\right]}{\sum\limits_{m=1}^{\infty}\left\{Nm^{1/3}\exp\left(-\frac{2m}{3\gamma^3}\right) + N(N-1)m^{1/3}\exp\left[-\frac{2m}{3\gamma^3} - \left(m\frac{\Delta_b}{R}\right)^2\right]\right\}}. \tag{3.6.52}$$

Assuming that the particle is ultra-relativistic namely,

$$\frac{\Delta_b}{R}\gamma^3 \gg 1, \tag{3.6.53}$$

the first two moments $\langle n\rangle = \sum_n np_n$ and $\langle n^2\rangle = \sum_n n^2 p_n$ are given by

$$\begin{aligned}\langle n\rangle &\simeq \frac{\sum\limits_{n=1}^{\infty}\left\{n^{4/3}\exp\left(-\frac{2n}{3\gamma^3}\right)+Nn^{4/3}\exp\left[-\left(n\frac{\Delta_b}{R}\right)^2\right]\right\}}{\sum\limits_{m=1}^{\infty}\left\{m^{1/3}\exp\left(-\frac{2m}{3\gamma^3}\right)+Nm^{1/3}\exp\left[-\left(m\frac{\Delta_b}{R}\right)^2\right]\right\}}\\ &\simeq \frac{\left(\frac{3\gamma^3}{2}\right)^{7/3}\int\limits_0^\infty dx x^{4/3}\exp(-x)+N\left(\frac{R}{\Delta_b}\right)^{7/3}\int\limits_0^\infty dx x^{4/3}\exp(-x^2)}{\left(\frac{3\gamma^3}{2}\right)^{4/3}\int\limits_0^\infty dx x^{1/3}\exp(-x)+N\left(\frac{R}{\Delta_b}\right)^{4/3}\int\limits_0^\infty dx x^{1/3}\exp(-x^2)}\\ &\simeq \frac{4}{3}\frac{3\gamma^3}{2}\frac{1+0.389N\left(\frac{2}{3\gamma^3}\frac{R}{\Delta_b}\right)^{7/3}}{1+0.758N\left(\frac{2}{3\gamma^3}\frac{R}{\Delta_b}\right)^{4/3}}\end{aligned} \tag{3.6.54}$$

and

$$\begin{aligned}\langle n^2\rangle &\simeq \frac{\sum\limits_{n=1}^{\infty}\left\{n^{7/3}\exp\left(-\frac{2n}{3\gamma^3}\right)+Nn^{7/3}\exp\left[-\left(n\frac{\Delta_b}{R}\right)^2\right]\right\}}{\sum\limits_{m=1}^{\infty}\left\{m^{1/3}\exp\left(-\frac{2m}{3\gamma^3}\right)+Nm^{1/3}\exp\left[-\left(m\frac{\Delta_b}{R}\right)^2\right]\right\}}\\ &\simeq \frac{\left(\frac{3\gamma^3}{2}\right)^{10/3}\int\limits_0^\infty dx x^{7/3}\exp(-x)+N\left(\frac{R}{\Delta_b}\right)^{10/3}\int\limits_0^\infty dx x^{7/3}\exp(-x^2)}{\left(\frac{3\gamma^3}{2}\right)^{4/3}\int\limits_0^\infty dx x^{1/3}\exp(-x)+N\left(\frac{R}{\Delta_b}\right)^{4/3}\int\limits_0^\infty dx x^{1/3}\exp(-x^2)}\\ &\simeq \frac{28}{9}\left(\frac{3\gamma^3}{2}\right)^2\frac{1+0.162N\left(\frac{2}{3\gamma^3}\frac{R}{\Delta_b}\right)^{10/3}}{1+0.758N\left(\frac{2}{3\gamma^3}\frac{R}{\Delta_b}\right)^{4/3}}.\end{aligned} \tag{3.6.55}$$

As an example, let us consider $N = 10^{10}$, 2 GeVelectrons forming a $\Delta_b = 1$ cm long bunch, following a circular trajectory of radius $R = 100$ m implying $\frac{2}{3\gamma^3}\frac{R}{\Delta_b} \simeq 10^{-6}$ and consequently,

$$\left.\begin{aligned}\langle n\rangle &\simeq \frac{2}{76.8}\gamma^3 \simeq 0.026\gamma^3\\ \Delta n &= \sqrt{\langle n^2\rangle - \langle n\rangle^2} = 0.3\gamma^3\end{aligned}\right\} \Rightarrow \frac{\Delta n}{\langle n\rangle} = \frac{\Delta\omega}{\langle\omega\rangle} \sim 11.56 \tag{3.6.56}$$

which is one order of magnitude higher than the case of a point-charge.

Exercises

3.1 Show by substituting (3.1.10) in Lagrange's equation of motion that they are identical to the Newtonian equations of motion. Repeat the exercise with Hamilton's equations of motion.

3.2 Show in a systematic way that if the charge of an electron is the same in all frames of reference then the set $(\mathbf{J}, c\rho)$ forms a 4-vector.

3.3 Show that the condition in (3.4.19) is sufficient to ensure stability of the beam. Hint: linearize the equation of motion and assume that in spite oscillation the total current is preserved.

3.4 Determine the frequency emitted by an electron wiggling in a magnetic field at any angle θ – see end of Sect. 3.2

3.5 Based on Sect. 3.3.1 determine the effective enhancement parameter as a function of the geometry of the grating. Analyze the effect of the various geometric parameters. Can you suggest a measurement such that based on the presented analysis it would be possible to assess, at least part of, the surface topology.

3.6 Determine the limiting current in a cylindrical and spherical diode for the non-relativistic regime. Show that like in the planar configuration, the current is proportional to $V_{AN}^{3/2}$. Hint: follow Langmuir (1913) and determine the corresponding proportionality factors.

3.7 Calculate the limiting currents of two thin annular beams of radii R_1 and R_2 moving in a waveguide of radius R $(R > R_2 > R_1)$; both beams are generated by the same diode, thus the initial kinetic energy $[mc^2(\gamma_0 - 1)]$ is the same. Determine the kinetic energy of the electrons in each one of the beams. Repeat the exercise for two beams of different initial energies and currents.

3.8 Show that in dielectric loaded waveguide the limiting current of a pencil beam is given by

$$I_{\max} = 2\pi\frac{mc^2}{e\eta_0}\left(\gamma_0^{2/3} - 1\right)^{3/2}\ln^{-1}\left[\frac{R_b}{R_d}\left(\frac{R_d}{R}\right)^{1/\varepsilon_r}\right].$$

The dielectric ($\varepsilon_r > 1$) fills the region $R \geq r \geq R_d$; the beam radius is R_b and it is smaller than the inner radius of the dielectric, R_d. Note that with the dielectric, for the same waveguide radius R, the limiting current is

(*continued*)

larger which means that the potential depression is smaller and consequently, the kinetic energy is larger.

3.9 Calculate the limiting current for an annular beam of radius R_b which carries a current I and its thickness is Δ and it moves *outside* a metallic waveguide of radius R.

3.10 Prove the dispersion relation specified by (3.5.45). Analyze (numerically) the condition for instability to occur. In particular examine the dependence on the geometric parameters. Examine also the spatial growth as a function of the frequency, geometry, momenta and the currents.

3.11 Calculate the two-beam instability for the case of a motionless plasma and a pencil beam moving through the plasma in a waveguide of radius R. The radius of the electron beam is R_b, its density is n_e and the density of the background plasma is denoted by n_{bg}.

3.12 Compare the Debye length with the plasma wavelength. For a temperature of $1,000°$K, which one is larger? What is your conclusion regarding the use of the hydrodynamic and kinetic approximations.

3.13 Develop the Child-Langmuir limiting current using the Lagrangian formulation. Hint: consult Chodorow and Susskind (1964) p. 125.

3.14 Use the integral $\oint \mathbf{E} \cdot dl = 0$ in order to calculate the potential depression of a pencil beam in a waveguide. Hints: (a) include the diode in your contour, (b) assume perfect conductors and (c) in the beam region the contour closes on axis.

3.15 3.15 Determine the effect of the vacuum chamber on the coherent synchrotron radiation. Hint: consult Schwinger(1949), Schiff (1946), Nodvick (Nodvick and Saxon 1954), Warnock and Morton (1990) or Ng (1990).

3.16 Consider a train of micro-bunches. Each micro-bunch has a length Δ_{mb}(Gaussian distribution), there are M microbunches and the distance between any two bunches is $\Delta_T \gg \Delta_{mb}$. In each micro-bunch there are N_{el} electrons. Relying on (3.6.44) determine the probability p_n for emission of a photon of energy $\hbar\Omega n$-accounting for both spontaneous as well as coherent radiation. Subject to the assumption $\Delta_b\gamma^3 \gg R$ analize the effect of the number of micro-bunches on $\Delta\omega/\langle\omega\rangle$. Use the parameters in Comment 3.10 in Sect. 3.6.4.

Chapter 4
Models of Beam–Wave Interaction in Slow-Wave Structures

So far we either assumed that the trajectory of the electrons is known and the electromagnetic field has been evaluated or vice versa, the electromagnetic field was known and the problem was to determine the trajectory of charged particles. In reality, both vary and at any point and instant, they need to be established in a self-consistent way. In this chapter, we investigate the fundamentals of distributed beam–wave interaction in a *slow-wave* structure. In Chap. 7, while discussing the interaction in a free electron laser, we will formulate the self-consistent interaction in a *fast-wave* device. Some additional aspects will be considered in Chap. 8 when discussing beam–wave interaction in accelerators.

A dielectric loaded waveguide and a pencil beam are chosen as the basic model in the first sections because it enables us to illustrate the essence of the interaction without the complications associated with complex boundary conditions. Throughout this chapter, the electron beam is assumed to be guided by a very strong magnetic field such that the electrons' motion is confined to the longitudinal direction. Furthermore, the kinetic energy of the electrons is assumed to take into consideration the potential depression associated with the injection of a beam into a metallic waveguide.

In the first section, we present part of Pierce's theory for the traveling-wave *amplifier* applied to dielectric loaded structure and extended to the *relativistic regime*. The interaction for a semi-infinitely long system is formulated in terms of the interaction impedance introduced in Chap. 2. Finite length effects are considered in the second section where we first examine the other extreme of the beam–wave interaction namely, the *oscillator*. In the context of an amplifier, it is shown that reflections affect the bandwidth and in addition, the beam shifts the frequency where maximum transmission occurs.

The macro-particle approach is described in Sect. 4.3 where the beam dynamics instead of being considered in the framework of the hydrodynamic approximation i.e. as a single fluid flow, is represented by a large number of clusters of electrons. Each one of the clusters is free to move at a different velocity according to the local field it experiences but the electrons that constitute the cluster are "glued" together. This formalism enables us to examine the interaction in phase-space either in the

L. Schächter, *Beam-Wave Interaction in Periodic and Quasi-Periodic Structures*, Particle Acceleration and Detection, DOI 10.1007/978-3-642-19848-9_4,

linear regime of operation or close to saturation. It also permits investigation of tapered structures and analysis of the interaction of pre-bunched beams in tapered structures.

The chapter concludes with a further extension of the macro-particle approach formalism to include the effect of reflections. This framework combines the formulations of an amplifier and an oscillator and permits us to quantify and illustrate the operation of a realistic device, which is neither an ideal amplifier nor an ideal oscillator.

With the exception of the first section that, as indicated, is a review of Pierce's TWT theory, most of the material presented in this chapter has been developed during the nineties as part of an effort to develop high power traveling-wave amplifiers. The first experiments on high power TWT performed at Cornell University and conducted by Professor John A. Nation (Shiffler et al. 1989) indicated that 100 MW of power at 8.76 GHz can be achieved before the system oscillates. Although no rf break-down was observed, the fact that the input is no longer isolated from the output, allows waves to be reflected backwards and this feedback could cause the system to oscillate. In order to isolate the input from the output the TWT was split in two sections separated by a sever (waveguide made of lossy material which operates below cut-off).

The second set of experiments on a two stage high power TWT indicated that power levels in excess of 400 MW are achievable with no indication of rf break-down (Shiffler et al. 1991). In this case, the spectrum of output frequencies was 300 MHz wide and a significant amount of power (up to 50%) was measured in asymmetric sidebands. The latter observation was investigated theoretically (Schächter et al. 1991) and it was concluded that it is a result of amplified noise at frequencies selected by the interference of the two waves bouncing between the ends of the last stage. In fact, we have shown (Schächter and Nation 1992) that what we call amplifier and oscillator are the two extreme of possible operation regimes and any practical device operates somewhere in between, according to the degree of control we have on the reflection process. One possible way to eliminate the problem is to use the transit-time isolation method. Its essence is to design a low group velocity structure such that by the time the reflected electromagnetic pulse reaches the input end, there are no longer electrons to interact with. This method was successfully demonstrated (Kuang et al. 1993) experimentally and power levels of 200 MW were achieved at 9 GHz. The spectrum of the output signal was less than 50 MHz wide and the pass-band of the periodic structure was less than 200 MHz.

4.1 Semi-Infinite Structure: Pierce-Like Theory

In the previous chapter it was justified to decouple the two groups of solutions described by the dispersion relation in (3.5.8) because the propagating electromagnetic modes have a phase velocity larger than c, whereas the space-charge waves

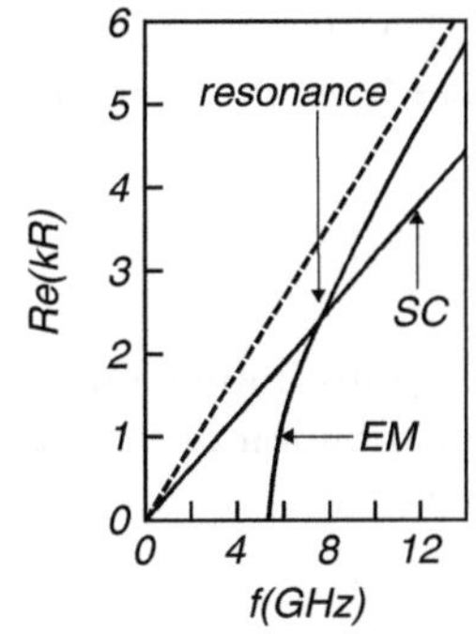

Fig. 4.1 Solution of the dispersion relation of the electromagnetic and space-charge waves. The resonance is at the intersection of the two curves. The *upper curve* is the asymptote of the pure electro-mangetic mode: $k = \omega\sqrt{\varepsilon_r}/c$

have a phase velocity which is of the order of v_0. In principle, it is possible to slow down the phase velocity of the electromagnetic wave, in the absence of the beam, below c and then, the waves may become coupled – it is there where resonance occurs. One possibility to slow down the phase velocity, which will be considered throughout this chapter, is to load the waveguide with a dielectric material. In Fig. 4.1 we illustrate schematically the dispersion curve of an electromagnetic wave which propagates in a dielectric loaded waveguide. The space-charge wave intersects the former curve at resonance.

4.1.1 Dielectric Filled Waveguide

As a first step, we assume full overlap between the beam and the dielectric. Although, in general no such overlap is permissible, this model will be used to explain in a simple way the quantities that describe the beam–wave interaction in a slow-wave structure. A realistic but somewhat more complex picture will be presented in the next sub-section. Based on *hydrodynamic* description of the beam-dynamics, as in Sect. 3.5, the dispersion relation of the $\mathrm{TM}_{0\mathrm{s}}$ modes in the presence of a dielectric material (ε_r) is given by

$$\left(\varepsilon_\mathrm{r}\frac{\omega^2}{c^2} - k_s^2\right)\left[1 - \frac{\omega_\mathrm{p}^2}{\varepsilon_\mathrm{r}\gamma^3(\omega - \mathrm{v}_0 k_s)^2}\right] = \frac{p_s^2}{R^2}, \tag{4.1.1}$$

k_s represents the wave-number of the sth mode and p_s is the zero of the zero-order Bessel function of the first kind i.e., $\mathrm{J}_0(p_s) \equiv 0$. For simplicity sake, it is assumed that only the first mode, $s = 1$, is excited and in the *absence* of the beam the solution of the electromagnetic wave is given by

$$k_1^{(0)} = \sqrt{\varepsilon_\mathrm{r}\frac{\omega^2}{c^2} - \frac{p_1^2}{R^2}}, \tag{4.1.2}$$

where R is the radius of the waveguide. The solution of the dispersion relation with the *beam present* is further assumed to have the form

$$k_1 = k_1^{(0)} + \delta k. \tag{4.1.3}$$

Substituting this relation in (4.1.1), assuming that the beam effect on the distribution of the electromagnetic field is small on the scale of one wavelength i.e.,

$$|\delta k| \ll k_1^{(0)}, \tag{4.1.4}$$

neglecting the beam effect on the wave which propagates anti-parallel to the beam and finally assuming that

$$\frac{p_1^2}{R^2} \gg 2k_1^{(0)}|\delta k|, \tag{4.1.5}$$

we obtain the following simplified version of the dispersion relation

$$\delta k(\delta k - \Delta k)^2 = -K_0^3 \equiv -\frac{1}{2}\frac{p_1{}^2}{R^2}\frac{\omega_{\mathrm{p}}^2}{\varepsilon_{\mathrm{r}} k_1 \mathrm{v}_0^2 \gamma^3}. \tag{4.1.6}$$

In this expression K_0 is the coupling wave-number and $\Delta k \equiv \omega/\mathrm{v}_0 - k_1^{(0)}$ represents the slip between the beam and the electromagnetic wave; this is more easily observed when dividing Δk by ω since

$$\frac{\Delta k}{\omega} = \frac{1}{\mathrm{v}_0} - \frac{1}{\mathrm{v}_{\mathrm{ph}}}. \tag{4.1.7}$$

The dispersion relation in (4.1.1) is a fourth order polynomial whereas (4.1.6) is a third order polynomial since the effect of the beam on the reflected wave was neglected. This type of dispersion relation was initially presented by Pierce (1947) in the context of beam–wave interaction in a helix – and it will be referred to hereafter as the Pierce's approach. Soon afterwards, Chu and Jackson (1948) presented the formulation based on full electromagnetic field analysis. In both cases, the dynamics of the beam was non-relativistic since the regime of operation at the time did not require relativistic analysis.

A third order polynomial has explicit analytic solution (see Abramowitz and Stegun 1968, p. 17). Two of its roots are complex provided that

$$q \equiv \Delta k + \frac{3}{4^{1/3}} K_0 > 0, \tag{4.1.8}$$

and then the imaginary part of δk is equal to

$$\mathrm{Im}(\delta k) = \frac{\sqrt{3}}{2}\left[-q^3 - \frac{1}{4}K_0^3 + (K_0 q)^{3/2}\right]^{1/3} - \frac{\sqrt{3}}{2}\left[-q^3 - \frac{1}{4}K_0^3 - (K_0 q)^{3/2}\right]^{1/3}, \tag{4.1.9}$$

which can be readily shown, by assuming that Δk and K_0 are *independent*, to have its maximum at

$$\Delta k \equiv \omega\left(\frac{1}{v_0} - \frac{1}{v_{ph}}\right) = 0. \tag{4.1.10}$$

This is also the resonance condition (see Fig. 4.1) and it can be formulated as

$$v_{ph} = v_0, \tag{4.1.11}$$

indicating that maximum growth rate occurs when the electron beam is synchronous with the wave. At resonance (4.1.6) has three solutions

$$\delta k_1 = -K_0, \quad \delta k_2 = K_0\left(\frac{1}{2} - j\frac{\sqrt{3}}{2}\right), \quad \delta k_3 = K_0\left(\frac{1}{2} + j\frac{\sqrt{3}}{2}\right), \tag{4.1.12}$$

corresponding to the three waves, which propagate in the forward direction. The first wave has a constant amplitude and its phase velocity is larger than v_0; the other two waves have a slower phase velocity and their amplitude vary in space. The third solution corresponds to a wave whose amplitude grows exponentially in space. The maximum spatial growth rate is therefore

$$\mathrm{Im}(\delta k)_{max} = \frac{\sqrt{3}}{2}K_0. \tag{4.1.13}$$

Further discussion of the interaction between electromagnetic waves and space-charge waves in slow-wave structures can be found in the early literature e.g., Pierce (1950), Slater (1950), Hutter (1960), Chodorow and Susskind (1964), or more recently in Gilmour (1986). At this point we wish to emphasize the difference between the operation of a traveling-wave tube in the relativistic and non-relativistic regime (Naqvi et al. 1996). We have emphasized, in the context of (4.1.9)–(4.1.10), that maximum gain occurs at resonance if we assume that K_0 and Δk are independent. But clearly this is not generally the case since both quantities are velocity dependent as revealed by their definition in (4.1.6)–(4.1.7). Furthermore, the validity of (4.1.8)–(4.1.9) is limited to the close vicinity of the expansion point, therefore we should solve (4.1.1) with no additional approximations. The result is illustrated in Fig. 4.2 for two cases. In one case the

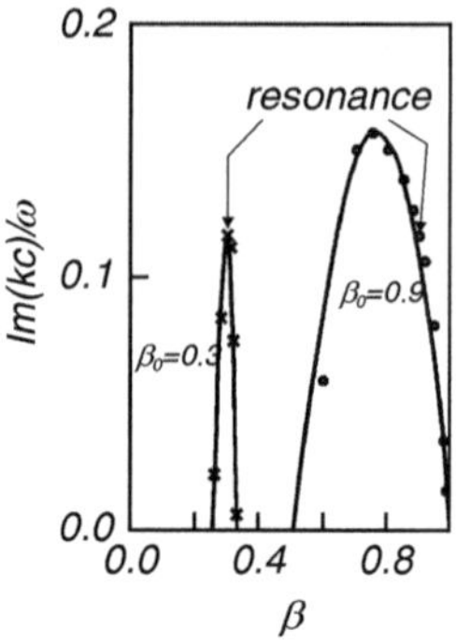

Fig. 4.2 Imaginary part of the wavenumber as a function of the beam velocity in two different cases. In one the wave is designed to propagate at low phase velocity ($\beta_{\text{ph}} = 0.3$) and the second at high phase velocity ($\beta_{\text{ph}} = 0.9$). In both cases the gain at resonance, $\beta = \beta_{\text{ph}}$, was designed to be the same

phase velocity of the wave is $\beta_{\text{ph}} = 0.3$ corresponding to a non-relativistic regime and in the other, $\beta_{\text{ph}} = 0.9$. In both cases at "resonance" ($\beta = \beta_{\text{ph}}$) the spatial growth rate was designed to be the same. We observe that at low velocities the predictions of the Pierce approach behave as expected but at higher energies the peak gain occurs at much lower velocity than anticipated by the model. This can be attributed to the $\gamma\beta$ dependence of the coupling coefficient (K_0). At low energies the coupling occurs in a relatively narrow range of velocities and as a result, the change in K_0 is small and the peak gain occurs at resonance. At higher energies, although the gain might have decreased if K_0 were constant, it actually increases because of the increase in K_0 due to its $1/\gamma\beta$ dependence. For a large deviation from resonance, the slip will ultimately take over and the gain drops. Now to some further comments regarding the *kinematics* of the interaction:

Comment 4.1. From the resonance condition (4.1.11) we conclude that maximum gain is achieved at a frequency which is related to the geometric and mechanical parameters by

$$\omega_{\text{r}} = \frac{p_1 \text{v}_0}{R} \frac{1}{\sqrt{\varepsilon_{\text{r}}(\text{v}_0/c)^2 - 1}}. \tag{4.1.14}$$

If we assume that at a frequency, ω_0, there is no longer growth, i.e., $q(\omega = \omega_0) = 0$, and ω_0 is only slightly apart from the resonance frequency ω_{r} ($\omega_0 = \omega_{\text{r}} + \delta\omega$ and $|\delta\omega| \ll \omega_{\text{r}}$) then we can make a crude estimate of the frequency range in which the ideal system under consideration will amplify. For this purpose we use the condition for an imaginary solution in (4.1.8) and the definition of Δk in (4.1.10): we substitute $\omega = \omega_{\text{r}} + \delta\omega$ and $k_1 = k_1^{(0)} + \text{v}_{\text{gr}}(\omega = \omega_{\text{r}})\delta\omega$ thus

$$\frac{\omega_{\text{r}} + \delta\omega}{\text{v}_0} - k_1^{(0)} - \delta\omega \text{v}_{gr} + \frac{3}{4^{1/3}} K_0 = 0. \tag{4.1.15}$$

We shall define the *interaction bandwidth*, $\delta\omega_i$, as equal $\delta\omega$:

$$\delta\omega_i \equiv \delta\omega = \frac{3}{4^{1/3}} K_0 \text{v}_{gr} \frac{\text{v}_0}{\text{v}_0 - \text{v}_{gr}}. \tag{4.1.16}$$

This result indicates that the interaction extends beyond the resonance frequency (ω_r) in an interval which is *linearly* proportional to the maximum spatial growth rate ($\mathrm{Im}(\delta k)_{\max}$):

$$\delta\omega_i = 3^{1/2}2^{1/3}\frac{v_0}{v_0 - v_{gr}} v_{gr}\mathrm{Im}(\delta k)_{\max}. \tag{4.1.17}$$

Furthermore, the closer the group velocity is to the beam velocity, the broader the interaction. Finally, this quantity is not dependent on the total interaction length.

Comment 4.2. The coupling coefficient K_0^3 can now be represented in terms of quantities that are more familiar. First we define the average current which flows along the waveguide as $I = en_0 v_0 \pi R^2$. Next we define the cross-section through which the wave propagates as $S_w = \pi R^2$. The interaction impedance defined in (2.3.29) was calculated for the present configuration in (2.3.30), reads $Z_{\mathrm{int}} = \eta_0 (p_1 c/\varepsilon_r \omega R)^2/\beta_{\mathrm{en}}$. With these quantities we can express K_0^3 as

$$K_0^3 = \frac{1}{2}\frac{1}{S_w}\frac{\omega}{c}\frac{eIZ_{\mathrm{int}}}{mc^2}\frac{1}{(\gamma\beta)^3}. \tag{4.1.18}$$

In the linear regime, the linearity of K_0^3 in the interaction impedance is a general feature whenever the electrons interact with a TM mode and their oscillation is longitudinal. And so is its dependence on the normalized momentum of the particle, $\gamma\beta$. The last expression can also be formulated in terms of the energy velocity using (2.3.33) as

$$K_0^3 = \frac{1}{2}\frac{1}{S_w}\frac{\omega}{c}\frac{eI\eta_0}{mc^2}\frac{1}{(\gamma\beta)^3}\frac{1}{\varepsilon_{\mathrm{int}}\beta_{\mathrm{en}}}, \tag{4.1.19}$$

which indicates that the growth rate is inversely proportional to the energy velocity. Since in most cases of (our) interest this is equal to the group velocity, we observe that if we substitute in the expression for the interaction bandwidth, the latter still decreases with the group velocity as $\delta\omega_i \propto {v_{gr}}^{2/3}$.

Comment 4.3. The entire approach relies on a *linearized* hydrodynamic approximation, which implies that the deviation from the initial *average* energy is small i.e.,

$$\gamma \gg |\delta\gamma|. \tag{4.1.20}$$

In Sect. 4.3 we adopt the macro-particle approach for description of the electron dynamics and it will be shown that an *individual* particle can have energy which is more than twice the *average* initial energy. Nevertheless, the average energy modulation of all particles can still be relatively small – namely the relation in (4.1.20) still holds.

So far we discussed the kinematics of the interaction, now we focus our attention onto the *dynamics* of the interaction. Let us consider a system of length d and assume that we know the value of the field at the input i.e., $E_z(r, z = 0) = E_0 \mathrm{J}_0(p_1 r/R)$. Furthermore, at this location the beam is not modulated yet thus, $\delta \mathrm{v}(z = 0) = 0$ and $\delta n(z = 0) = 0$. According to the three modes we found previously (4.1.12) we can write the solution for E_z as:

$$E_z(r, z, \omega) = \mathrm{J}_0\left(p_1 \frac{r}{R}\right) \mathrm{e}^{-jk_1^{(0)} z} \left[E_1 \mathrm{e}^{-j\delta k_1 z} + E_2 \mathrm{e}^{-j\delta k_2 z} + E_3 \mathrm{e}^{-j\delta k_3 z}\right]. \qquad (4.1.21)$$

The three conditions above determine three sets of algebraic equations:

$$\begin{aligned} &E_1 + E_2 + E_3 = E_0, \\ &\frac{E_1}{\omega/v_0 - k_1^{(0)} - \delta k_1} + \frac{E_2}{\omega/v_0 - k_1^{(0)} - \delta k_2} + \frac{E_3}{\omega/v_0 - k_1^{(0)} - \delta k_3} = 0, \\ &\frac{E_1(k_1^{(0)} + \delta k_1)}{\left(\omega/v_0 - k_1^{(0)} - \delta k_1\right)^2} + \frac{E_2(k_1^{(0)} + \delta k_2)}{\left(\omega/v_0 - k_1^{(0)} - \delta k_2\right)^2} + \frac{E_3(k_1^{(0)} + \delta k_3)}{\left(\omega/v_0 - k_1^{(0)} - \delta k_3\right)^2} = 0 \end{aligned} \qquad (4.1.22)$$

and in principle, we can now solve for E_1, E_2 and E_3 such that we can determine the total electromagnetic field at the output ($z = d$). For the sake of simplicity, we limit our discussion to the solution near resonance (where according to the Pierce approach the gain reaches its maximum). The three $\delta k'$s are of the same order of magnitude so we can estimate that $|E_1| \simeq |E_2| \simeq |E_3| \simeq E_0/3$. Therefore, the z component of the electric field is

$$E_z(r, z, \omega) \simeq \frac{E_0}{3} \mathrm{J}_0\left(p_1 \frac{r}{R}\right) \mathrm{e}^{-jk_1^{(0)} z} \left[\mathrm{e}^{-j\delta k_1 z} + \mathrm{e}^{-j\delta k_2 z} + \mathrm{e}^{-j\delta k_3 z}\right] \qquad (4.1.23)$$

According to the three solutions in (4.1.12) the first δk is always real therefore its amplitude is constant, the amplitude of the second decays exponentially, and the third grows in space. The local gain is the ratio between the local amplitude and the amplitude at the input is

$$\begin{aligned} G(z) &\equiv \frac{|E(z)|}{|E(0)|} \\ &= \frac{1}{3}\left|\mathrm{e}^{-j\delta k_1 z} + \mathrm{e}^{-j\delta k_2 z} + \mathrm{e}^{-j\delta k_3 z}\right| \\ &= \frac{1}{3}\left|\mathrm{e}^{3jK_0 z/2} + \mathrm{e}^{-\sqrt{3}K_0 z/2} + \mathrm{e}^{\sqrt{3}K_0 z/2}\right|. \end{aligned} \qquad (4.1.24)$$

The expression in (4.1.24) is illustrated in Fig. 4.3 where we present $G(z)$ in dB for several values of $K_0 d$. Although one of the solutions grows exponentially in the first part of the interaction region, the local gain is zero – this effect is referred to in literature as *spatial lethargy* since it takes some space for the exponentially growing

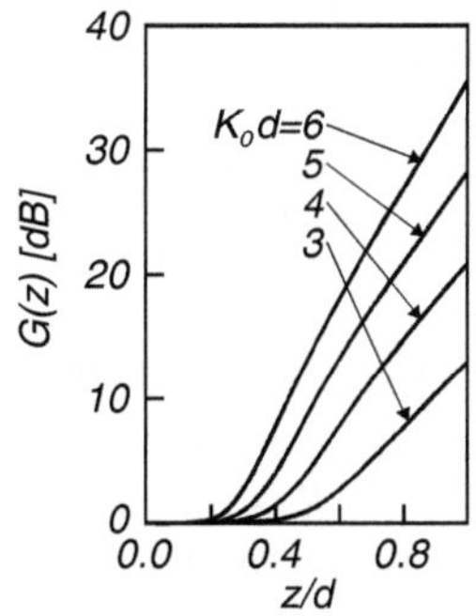

Fig. 4.3 The way the gain develops in space for different values of the coupling coefficient K_0

wave to become dominant. We can estimate this lethargy length d_L by determining the location where the amplitude of the first and third waves combined, reach the value at the input i.e., $|\exp(3jK_0d_L/2) + \exp(\sqrt{3}K_0d_L/2)|/3 = 1$. This equation can be solved numerically and the result is

$$d_L = \frac{1.412}{K_0}; \tag{4.1.25}$$

the second wave decays exponentially therefore it is neglected in this calculation. We shall further discuss this effect in the context of the macro-particle approach in Sect. 4.3. At the end of the interaction region we can also neglect the mode which has a constant amplitude (assuming that the gain is large enough) and the total gain (at $z = d$) is defined as

$$\text{gain}_{\text{dB}} = 20\log_{10}\left[\frac{1}{3}e^{\sqrt{3}K_0d/2}\right]. \tag{4.1.26}$$

The total gain and the lethargy length are related by

$$\frac{d_L}{d} = \frac{1.412\sqrt{3}}{2}\ln^{-1}\left[3 \times 10^{\text{gain(dB)}/20}\right]. \tag{4.1.27}$$

The effect is evident in Fig. 4.3 and using the relations above we found for $K_0d = 3$ the lethargy length is 0.47 d while the gain is 13 dB. For twice this growth ($K_0d = 6$) the lethargy length is 0.23 d and the gain is 35 dB.

4.1.2 Partially Filled Waveguide

Although in the previous model, the beam and the dielectric were occupying the entire space of the waveguide, has its tutorial merit, it is impractical concerning the generation of radiation. In general no beam-dielectric overlap is permitted namely, there has to be a significant distance between the electron beam and the structure that slows down the wave. In this sub-section, we consider the interaction between

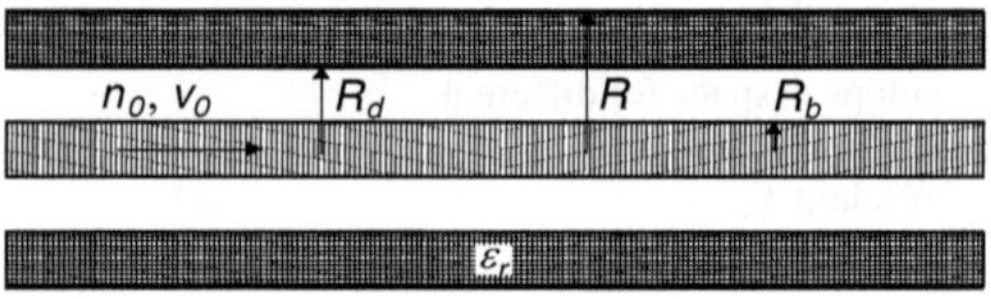

Fig. 4.4 A dielectric loaded waveguide. The radius of the guide is denoted by R. R_d and R_b stand for the dielectric inner radius and beam's radius, respectively

a beam of radius R_b and a wave which propagates in a waveguide partially filled with dielectric material (ε_r). The dielectric occupies the region between $R_d < r < R$ where R is the radius of the waveguide and $R_d > R_b$ – see Fig. 4.4. The system is semi-infinitely long, thus no reflections occur.

As a first stage, we consider the electromagnetic problem in the *absence* of the electrons. A TM_{01} mode is assumed to propagate along the waveguide and it is described by the z component of the magnetic vector potential, which in the vacuum gap ($0 < r < R_d$) reads:

$$A_z(r,z,\omega) = A_0 \mathrm{I}_0(\Gamma r)\mathrm{e}^{-jkz}, \tag{4.1.28}$$

where $\Gamma^2 = k^2 - (\omega/c)^2$. In the dielectric material ($R_d < r < R$) the magnetic vector potential is given by

$$A_z(r,z,\omega) = B_0 T_0(\kappa r)\mathrm{e}^{-jkz}, \tag{4.1.29}$$

where $\kappa^2 = \varepsilon_r(\omega/c)^2 - k^2$, $\mathrm{I}_0(\xi)$ is the zero order modified Bessel function of the first kind and

$$T_0(\kappa r) \equiv \mathrm{J}_0(\kappa r)\mathrm{Y}_0(\kappa R) - \mathrm{Y}_0(\kappa r)\mathrm{J}_0(\kappa R). \tag{4.1.30}$$

The electromagnetic field in the vacuum gap is

$$\begin{aligned} H_\phi(r,z,\omega) &= -A_0\frac{1}{\mu_0}\Gamma \mathrm{I}_1(\Gamma r)\mathrm{e}^{-jkz}, \\ E_r(r,z,\omega) &= -A_0\frac{c^2 k}{\omega}\Gamma \mathrm{I}_1(\Gamma r)\mathrm{e}^{-jkz}, \\ E_z(r,z,\omega) &= -A_0\frac{c^2}{j\omega}\Gamma^2 \mathrm{I}_0(\Gamma r)\mathrm{e}^{-jkz} \end{aligned} \tag{4.1.31}$$

In a similar way, in the dielectric material,

$$\begin{aligned} H_\phi(r,z,\omega) &= B_0\frac{1}{\mu_0}\kappa T_1(\kappa r)\mathrm{e}^{-jkz}, \\ E_r(r,z,\omega) &= B_0\frac{c^2 k}{\omega\varepsilon_r}\kappa T_1(\kappa r)\mathrm{e}^{-jkz}, \\ E_z(r,z,\omega) &= B_0\frac{c^2}{j\omega\varepsilon_r}\kappa^2 T_0(\kappa r)\mathrm{e}^{-jkz}, \end{aligned} \tag{4.1.32}$$

where

$$T_1(\kappa r) \equiv \mathrm{J}_1(\kappa r)\mathrm{Y}_0(\kappa R) - \mathrm{Y}_1(\kappa r)\mathrm{J}_0(\kappa R). \tag{4.1.33}$$

In order to determine the wave-number k we now impose the boundary conditions at $r = R_\mathrm{d}$: the continuity of H_ϕ implies

$$-A_0\Gamma\mathrm{I}_1(\Gamma R_\mathrm{d}) = B_0\kappa T_1(\kappa R_\mathrm{d}), \tag{4.1.34}$$

whereas the continuity of E_z results in

$$-A_0\Gamma^2\mathrm{I}_0(\Gamma R_\mathrm{d}) = B_0\frac{1}{\varepsilon_\mathrm{r}}\kappa^2 T_0(\kappa R_\mathrm{d}). \tag{4.1.35}$$

From these two equations the dispersion relation of the passive (subscript pa) device reads

$$D_\mathrm{pa}(\omega, k) \equiv \varepsilon_\mathrm{r}\mathrm{I}_0(\theta_d)T_1(\chi_d) - \frac{\chi_d}{\theta_d}\mathrm{I}_1(\theta_d)T_0(\chi_d) = 0, \tag{4.1.36}$$

where $\theta_d = \Gamma R_\mathrm{d}$ and $\chi_d = \kappa R_\mathrm{d}$. Figure 4.5 illustrates a solution of this dispersion relation [line *(a)*]. For comparison, two other dispersion relations are plotted: curve *(b)* represents the empty waveguide, whereas curve *(c)* corresponds to a waveguide filled with the same dielectric. From the dispersion relation we observe that at low frequencies (long wavelength) the mode behaves as if no dielectric exists. At high frequencies, the dielectric slab primarily confines the mode.

Next we consider the effect of the electron beam on the propagating waves. In this case, the magnetic vector potential in the beam region $0 < r < R_\mathrm{b}$ is given by

$$A_z(r, z, \omega) = A_0\mathrm{I}_0(\Lambda r)\mathrm{e}^{-jkz},$$
$$\Lambda^2 = \Gamma^2\left[1 - \frac{\omega_\mathrm{p}^2}{\gamma^3(\omega - \mathrm{v}_0 k)^2}\right], \tag{4.1.37}$$

with the plasma frequency, ω_p, defined in (3.4.18). Accordingly, the electromagnetic field reads

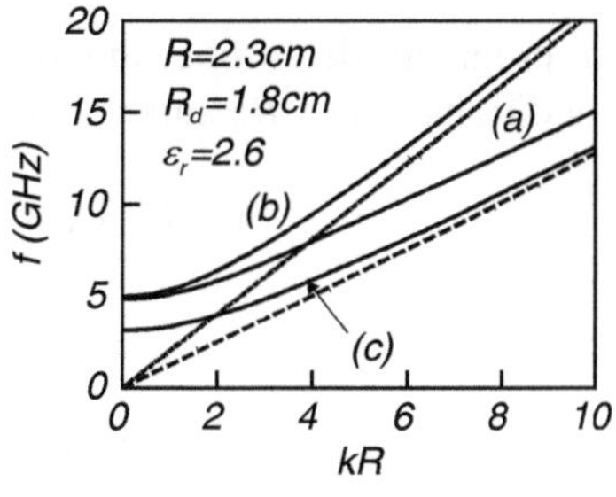

Fig. 4.5 (**a**) Solution of the dispersion relation in (4.1.36). (**b**) represents the empty waveguide, whereas (**c**) corresponds to a waveguide filled with dielectric. The upper dashed line represents the dispersion of a TEM mode in vacuum whereas the lower one represents the same mode in the dielectric medium

$$\begin{aligned} H_\phi(r,z,\omega) &= -A_0\frac{1}{\mu_0}\Lambda \mathrm{I}_1(\Lambda r)\mathrm{e}^{-jkz}, \\ E_r(r,z,\omega) &= -A_0\frac{c^2k}{\omega}\Lambda \mathrm{I}_1(\Lambda r)\mathrm{e}^{-jkz}, \\ E_z(r,z,\omega) &= -A_0\frac{c^2}{j\omega}\Gamma^2 \mathrm{I}_0(\Lambda r)\mathrm{e}^{-jkz}. \end{aligned} \tag{4.1.38}$$

In the vacuum gap between the beam and the dielectric ($R_\mathrm{b} < r < R_\mathrm{d}$) the potential is

$$A_z(r,z,\omega) = [B_0\mathrm{I}_0(\Gamma r) + C_0\mathrm{K}_0(\Gamma r)]\mathrm{e}^{-jkz}, \tag{4.1.39}$$

while the electromagnetic field is

$$\begin{aligned} H_\phi(r,z,\omega) &= -\frac{1}{\mu_0}\Gamma[B_0\mathrm{I}_1(\Gamma r) - C_0\mathrm{K}_1(\Gamma r)]\mathrm{e}^{-jkz}, \\ E_r(r,z,\omega) &= -\frac{c^2k}{\omega}\Gamma[B_0\mathrm{I}_1(\Gamma r) - C_0\mathrm{K}_1(\Gamma r)]\mathrm{e}^{-jkz}, \\ E_z(r,z,\omega) &= -\frac{c^2}{j\omega}\Gamma^2[B_0\mathrm{I}_0(\Gamma r) + C_0\mathrm{K}_0(\Gamma r)]\mathrm{e}^{-jkz}. \end{aligned} \tag{4.1.40}$$

In the dielectric the expression is similar to (4.1.29):

$$A_z(r,z,\omega) = D_0T_0(\kappa r)\mathrm{e}^{-jkz}, \tag{4.1.41}$$

and the electromagnetic field

$$\begin{aligned} H_\phi(r,z,\omega) &= D_0\frac{1}{\mu_0}\kappa T_1(\kappa r)\mathrm{e}^{-jkz}, \\ E_r(r,z,\omega) &= D_0\frac{c^2k}{\omega\varepsilon_\mathrm{r}}\kappa T_1(\kappa r)\mathrm{e}^{-jkz}, \\ E_z(r,z,\omega) &= D_0\frac{c^2}{j\omega\varepsilon_\mathrm{r}}\kappa^2 T_0(\kappa r)\mathrm{e}^{-jkz}. \end{aligned} \tag{4.1.42}$$

In order to determine the dispersion relation we now impose the boundary conditions at $r = R_\mathrm{b}$ and $r = R_\mathrm{d}$. Continuity of H_ϕ implies at $r = R_\mathrm{b}$:

$$A_0\Lambda\mathrm{I}_1(\Lambda R_\mathrm{b}) = \Gamma[B_0\mathrm{I}_1(\Gamma R_\mathrm{b}) - C_0\mathrm{K}_1(\Gamma R_\mathrm{b})], \tag{4.1.43}$$

while at $r = R_\mathrm{d}$:

$$-\Gamma[B_0\mathrm{I}_1(\Gamma R_\mathrm{d}) - C_0\mathrm{K}_1(\Gamma R_\mathrm{d})] = D_0\kappa T_1(\kappa R_\mathrm{d}). \tag{4.1.44}$$

Similarly the continuity of the longitudinal component of the electric field (E_z) implies at $r = R_b$:

$$A_0(-\Gamma^2)\mathrm{I}_0(\Lambda R_b) = (-\Gamma^2)[B_0\mathrm{I}_0(\Gamma R_b) + C_0\mathrm{K}_0(\Gamma R_b)], \quad (4.1.45)$$

and at $r = R_d$:

$$(-\Gamma^2)[B_0\mathrm{I}_0(\Gamma R_d) + C_0\mathrm{K}_0(\Gamma R_d)] = D_0\frac{1}{\varepsilon_r}\kappa^2 T_0(\kappa R_d). \quad (4.1.46)$$

These are four homogeneous equations and the non-trivial solution is determined from the condition that the determinant of the corresponding matrix is zero. Thus the dispersion equation of the active (subscript act) system is

$$D_{act}(\omega, k) \equiv D_{pa}(\omega, k) + D_{beam}(\omega, k) = 0. \quad (4.1.47)$$

The first term is the dispersion relation of the passive system and the second (D_{beam}) represents the beam effect:

$$D_{beam}(\omega, k) = \frac{\Gamma \mathrm{I}_1(\Gamma R_b)\mathrm{I}_0(\Lambda R_b) - \Lambda \mathrm{I}_0(\Gamma R_b)\mathrm{I}_1(\Lambda R_b)}{\Gamma \mathrm{K}_1(\Gamma R_b)\mathrm{I}_0(\Lambda R_b) + \Lambda \mathrm{K}_0(\Gamma R_b)\mathrm{I}_1(\Lambda R_b)} \times \left[\varepsilon_r \mathrm{K}_0(\theta_d)T_1(\chi_d) + \frac{\chi_d}{\theta_d}\mathrm{K}_1(\theta_d)T_0(\chi_d)\right]. \quad (4.1.48)$$

The solution of the dispersion relation in the beam absence is denoted by $k^{(0)}(\omega)$ and the effect of the beam will be represented by a deviation, δk, from this value namely, $k = k^{(0)} + \delta k$. If we now expand the first term in (4.1.47) around $k^{(0)}$, we find

$$D_{pa}(\omega, k^{(0)}) + \delta k\left[\frac{\partial D_{pa}(\omega, k)}{\partial k}\right]_{k=k^{(0)}} = \delta k\left[\frac{\partial D_{pa}(\omega, k)}{\partial k}\right]_{k=k^{(0)}}, \quad (4.1.49)$$

since by the definition of $k^{(0)}$, the dispersion function $D_{pa}(\omega, k^{(0)})$ is identically zero. Regarding the second term in the dispersion relation of the active system (4.1.47), we can readily check that for $\omega_p = 0$ the beam term D_{beam} is zero. We shall consider only the correction of the first order in the term $\omega_p^2/(\omega - v_0 k)^2\gamma^3$, therefore we can neglect the beam effect in the denominator of D_{beam} which after a first order Taylor expansion implies

$$D_{beam}(\omega, k^{(0)}) \simeq \frac{\omega_p^2}{2\gamma^3(\omega - v_0 k)^2}\left[\theta_b^2\left(\mathrm{I}_0^2(\theta_b) - \mathrm{I}_1^2(\theta_b)\right)\frac{\chi_d}{\theta_d^2}\frac{T_0(\chi_d)}{\mathrm{I}_0(\theta_d)}\right]_{k=k^{(0)}}. \quad (4.1.50)$$

In the process of evaluating this expression the following relation was used $\mathrm{I}_0(x)\mathrm{K}_1(x) + \mathrm{I}_1(x)\mathrm{K}_0(x) = 1/x$ and $\theta_b = \Gamma R_b$. Equations (4.1.49)–(4.1.50) can be substituted in the active dispersion relation (4.1.47) and the result written in an identical form as (4.1.6). The beam–wave coupling is determined by

$$K_0^3 = \frac{1}{2}\frac{eI\eta_0}{mc^2}\frac{1}{(\gamma\beta)^3}\frac{1}{\pi R^2}\left[\mathrm{I}_0^2(\theta_b) - \mathrm{I}_1^2(\theta_b)\right]_{k=k^{(0)}} \times \left[\left(\frac{R}{R_\mathrm{d}}\right)^2 \frac{\chi_d T_0(\chi_d)}{\mathrm{I}_0(\theta_d)}\left(\frac{\partial D_\mathrm{pa}}{\partial k}\right)^{-1}\right]_{k=k^{(0)}}. \tag{4.1.51}$$

The effect of the radius in the case of a pencil beam is revealed since the coupling coefficient K_0^3 is

$$K_0^3(R_\mathrm{b}) = K_0^3(0)[\mathrm{I}_0^2(\theta_b) - \mathrm{I}_1^2(\theta_b)]_{k=k^{(0)}}. \tag{4.1.52}$$

The modified Bessel function of the first kind is a monotonic function thus for $R_\mathrm{b} \to 0$ the coupling coefficient K_0^3 has its minimum and grows with increasing beam radius. This fact can be readily understood bearing in mind that the slow wave, which interacts with the electrons, decays exponentially from the dielectric surface inward. The larger the radius of the beam, the stronger the electric field it encounters, therefore the coupling is stronger. Walsh (1987) presented further discussion on the interaction in a dielectric loaded waveguide.

4.2 Finite Length Effects

All the examples of beam–wave interaction presented so far, such as resistive wall instability – Sect. 3.5.3, two-beam instability – Sect. 3.5.4 and traveling-wave interaction in the last two sub-sections, disregard the possibility of reflections from the output end. In practice, there are several causes for reflections to occur (1) the characteristic impedance of an electromagnetic wave in the interaction region differs from the impedance of the input and output waveguides. (2) Even if at a given frequency this impedance mismatch can be tuned, at other frequencies, reflections may dominate and control the interaction process. (3) In any interaction scheme eventually, the electromagnetic energy has to be decoupled from the beam. This decoupling process is always associated with some kind of discontinuity, therefore in this regard, reflections are an inherent part of the interaction.

In order to illustrate the effect of reflections, we first consider the extreme where the interaction is dominated by reflections, which is the case in an *oscillator*. It is shown that in zero order, the amplitude of the electromagnetic field in an oscillator is constant in space but varies in time whereas in an amplifier, it is constant in time but it varies in space. The oscillator analysis is followed by an investigation of the effect of reflections on an amplifier and the section concludes with some remarks on the interaction in an extended slow-wave cavity.

4.2.1 *Oscillator*

The simplest configuration we can conceive for an oscillator is a section of an amplifier with two reflecting mirrors at the two ends. There are two major differences between an amplifier and an oscillator (1) in an amplifier, the frequency is set externally whereas the wave-number is determined by the waveguide and the interaction. Consequently, (2) the amplitude of the wave varies in space but at a given location it is constant in time. The opposite holds for an oscillator; the wave-number is set by the cavity (mirrors) and the frequency is determined internally by both cavity and interaction. The amplitude is constant in space (at a given moment) and it varies in time. Therefore, if we assume that the distance between the two mirrors is d, then from the condition that $E_r(r,z=0,d)=0$ we conclude that

$$k = \nu\frac{\pi}{d}, \tag{4.2.1}$$

where ν is an integer which labels each longitudinal mode. In the remainder, we assume that the interaction is only with one of these modes ($\nu = 1$). In the beam absence the resonant frequency is

$$\omega_{\rm r} = \frac{c}{\sqrt{\varepsilon_{\rm r}}}\sqrt{\frac{{p_1}^2}{R^2}+\frac{\pi^2}{d^2}}, \tag{4.2.2}$$

and the beam introduces a small deviation $\delta\omega = \omega - \omega_{\rm r}$ which is a solution of the following dispersion relation

$$\delta\omega(\delta\omega - \Delta\omega)^2 = \Omega_0^3 \equiv \frac{1}{2}\left(\frac{p_1 c}{R}\right)^2 \frac{\omega_{\rm p}^2}{\omega_{\rm r}\varepsilon_{\rm r}^2\gamma^3}, \tag{4.2.3}$$

where similarly to the amplifier case Ω_0 is the coupling frequency and $\Delta\omega = {\rm v}_0\pi/d - \omega_{\rm r}$ is the slip. The main difference here is that in contrast to (4.1.6) the right-hand side is *positive* thus the same analysis of the third order polynomial can be applied for (4.2.3). We shall not repeat it but rather present the important results. The maximum temporal growth rate occurs at resonance and it is given by

$$\delta\omega_1 = \Omega_0, \quad \delta\omega_2 = -\frac{1}{2}\Omega_0(1-j\sqrt{3}), \quad \delta\omega_3 = -\frac{1}{2}\Omega_0(1+j\sqrt{3}). \tag{4.2.4}$$

The first solution corresponds to a wave whose phase velocity is larger than the average velocity of the electrons (${\rm v}_0$). The other two have phase velocities smaller than ${\rm v}_0$. The second solution represents a wave that is decaying in time whereas the third grows.

Comparing the dispersion relation for the amplifier (4.1.6) with the dispersion relation for the oscillator (4.2.3) we find that the relation between Ω_0 and K_0 is given by $\Omega_0^3 = K_0^3 v_{gr} v_0^2$, which implies that the connection between the maximum *temporal growth rate* $[\mathrm{Im}(\delta\omega)_{max}]$ in an oscillator and the maximum *spatial growth rate* in an amplifier $[\mathrm{Im}(\delta k)_{max}]$ is

$$\frac{\mathrm{Im}(\delta\omega)_{max}}{\mathrm{Im}(\delta k)_{max}}\frac{1}{c} = \left(\beta_{gr}\beta^2\right)^{1/3}. \tag{4.2.5}$$

As in the amplifier case, we can define the *temporal lethargy* as the period of time, τ_L, during which the electromagnetic energy in the oscillator starts to grow exponentially relative to the situation before the beam was launched (here we have assumed that initially the cavity was filled with electromagnetic energy):

$$\tau_L = \frac{1.412}{\Omega_0}. \tag{4.2.6}$$

If we consider a system which is characterized by $K_0 d = 4$, $d = 20\,\mathrm{cm}$, $\beta = 0.9$ and $\beta_{gr} = 0.2$ then the temporal lethargy is 0.4 ns, therefore on the scale of a pulse of hundreds of nanoseconds this is negligible. However, for $K_0 d = 0.04$ and assuming that all other parameters remain the same, the temporal lethargy is 40 ns which is significant.

In most cases of interest, the cavity is not initially filled with electromagnetic energy and the mirrors/walls are not ideal reflectors. As a result, the electromagnetic signal has to build up from noise and at the same time overcome ohm, dielectric or radiation loss through extraction ports/mirrors. All the combined loss mechanisms cause a decay in the electromagnetic field which, in zero order can be assumed to be proportional to the total amount of energy stored at a given time i.e.,

$$\frac{\mathrm{d}}{\mathrm{d}t}W + \frac{2}{\tau_{loss}}W = 0. \tag{4.2.7}$$

On the other hand, in Sect. 2.5.3 it was shown that a single particle excites a variety of electromagnetic waves in a cavity. When a uniform distribution of particles is injected into a cavity, waves of different frequencies and phases are generated and absorbed in the same time. This is *noise* induced by the beam in the cavity. For a *coherent* signal to develop from noise it is necessary that the beam–wave interaction exceed some threshold which can be expressed in terms of the injected current. This amount is determined from the condition that the growth due to the interaction at least cancels the decay due to loss mechanisms i.e.,

$$\mathrm{e}^{\sqrt{3}\Omega_0 t/2}\mathrm{e}^{-t/\tau_{loss}} \geq 1. \tag{4.2.8}$$

Using this expression we can explicitly write for the *threshold current*

$$
\begin{aligned}
I_{\mathrm{th}} &= \frac{16}{3\sqrt{3}} \frac{mc^2}{e\eta_0} \frac{S_w}{(c\tau_{\mathrm{loss}})^2} \frac{\gamma^3\beta}{\omega_{\mathrm{r}}\tau_{\mathrm{loss}}} \varepsilon_{\mathrm{int}} \\
&= 3.08 I_e \gamma^3 \beta \frac{S_w}{(c\tau_{\mathrm{loss}})^2 (\omega_{\mathrm{r}}\tau_{\mathrm{loss}})} \varepsilon_{\mathrm{int}}.
\end{aligned}
\tag{4.2.9}
$$

Note that this current is proportional to the dielectric coefficient of the interaction and it scales as γ^3.

4.2.2 Gain and Bandwidth Considerations

In the previous sub-section we examined the dramatic change in the characteristics of the beam–wave interaction as reflections are deliberately introduced causing temporal rather than spatial growth. Based on the (pure) electromagnetic analysis presented in Chap. 2 the assumption of zero reflections in an amplifier is not justified in general since the wavelengths of the electromagnetic wave (and thus the characteristic impedance) in the interaction region and the extraction region are different. If discontinuities in the characteristic impedance occur they generate reflected waves. The reflected waves interfere (constructively or destructively) with the incoming waves to generate transmission patterns that were discussed in Sect. 2.5.1. According to this picture, there are frequency ranges for which the transmission coefficient has a maximum or minimum. When the beam is present, the situation is somewhat more complex since, in addition to the regular electromagnetic modes that can propagate, there are also space-charge waves which carry energy. However, in the case of a sufficiently long system such that at the output end the exponentially growing mode is dominant we can still assume only two waves bouncing between the input and output ends.

The starting point is similar to what was presented in Sect. 2.5.1: consider a waveguide of radius R is filled with a dielectric material according to

$$
\varepsilon_{\mathrm{r}}(z) = \begin{cases} 1 & -\infty < z < 0, \\ \varepsilon_{\mathrm{r}} & 0 < z < d, \\ 1 & d < z < \infty. \end{cases}
\tag{4.2.10}
$$

A wave is launched from $z \to -\infty$ toward the discontinuity at $z = 0$ and for sake of simplicity we assume that this wave is composed of a single mode (TM_{01} i.e., $s = 1$). The z component of the magnetic vector potential in the first region $(-\infty < z < 0)$ is given by

$$
A_z(r, -\infty < z < 0, \omega) = \mathrm{J}_0\left(p_1 \frac{r}{R}\right) \left[A_{\mathrm{in}} \mathrm{e}^{-jk_1 z} + A_{\mathrm{ref}} \mathrm{e}^{jk_1 z}\right],
\tag{4.2.11}
$$

where A_{in} is the amplitude of the incoming wave and A_{ref} represents the amplitude of the reflected wave because of the discontinuity; $k_1 = \sqrt{(\omega/c)^2 - (p_1/R)^2}$. Between the two discontinuities at $z = 0$ and $z = d$ the solution has a similar form

$$A_z(r,\ 0 < z < d, \omega) = \mathrm{J}_0\left(p_1 \frac{r}{R}\right)[A\mathrm{e}^{-jKz} + B\mathrm{e}^{jk_2 z}], \qquad (4.2.12)$$

where $k_2 = \sqrt{\varepsilon_r(\omega/c)^2 - (p_1/R)^2}$. The term $A\mathrm{e}^{-jKz}$ is an "effective" wave which represents all three modes we discussed in Sect. 4.1. Finally, in the third region there is no reflected wave thus

$$A_z(r,\ d < z < \infty,\ \omega) = \mathrm{J}_0\left(p_1 \frac{r}{R}\right) A_{tr} \mathrm{e}^{-jk_1(z-d)}. \qquad (4.2.13)$$

The four, as yet unknown, amplitudes A_{ref}, A_{tr}, A and B are determined by imposing the boundary conditions at $z = 0, d$:

$$Z_1[A_{in} - A_{ref}] = Z_2[A - B], \qquad (4.2.14)$$

$$A_{in} + A_{ref} = A + B, \qquad (4.2.15)$$

$$Z_2[A\mathrm{e}^{-j\bar{\psi}_+} - B\mathrm{e}^{j\bar{\psi}_-}] = Z_1 A_{tr} \qquad (4.2.16)$$

and

$$A\mathrm{e}^{-j\bar{\psi}_+} + B\mathrm{e}^{j\bar{\psi}_-} = A_{tr}. \qquad (4.2.17)$$

In these expressions, the following definitions were used:

$$Z_1 \equiv \eta_0 \sqrt{1 - \left(\frac{p_1}{\omega R}\right)^2}, \quad Z_2 \equiv \eta_0 \frac{1}{\varepsilon_r} \sqrt{\varepsilon_r - \left(\frac{p_1}{\omega R}\right)^2}, \qquad (4.2.18)$$

$\bar{\psi}_+ \equiv Kd$ represents the phase and amplitude variation of the effective wave as it propagates from $z = 0$ to d; $\bar{\psi}_- = k_2 d$ represents the phase shift of the backward wave. Note that the effect of the beam was neglected in the impedance terms. The transmission coefficient is defined as $\tau = A_{tr}/A_{in}$ and is given by

$$\tau = \frac{4Z_1 Z_2 \mathrm{e}^{-j(\bar{\psi}_+ - \bar{\psi}_-)}}{\mathrm{e}^{j\bar{\psi}_-}(Z_1 + Z_2)^2 - \mathrm{e}^{-j\bar{\psi}_+}(Z_1 - Z_2)^2}. \qquad (4.2.19)$$

Before we consider the beam effect on the transmission coefficient let us examine the passive device namely when no beam is present. The wave-numbers in this case are the same $\bar{\psi}_+ = \bar{\psi}_- = \bar{\psi} = k_2 d$ and the transmission coefficient reads

$$\tau = \frac{4Z_1Z_2}{e^{j\bar{\psi}}(Z_1+Z_2)^2 - e^{-j\bar{\psi}}(Z_1-Z_2)^2}. \tag{4.2.20}$$

The peaks of the transmission coefficient occur when $2k_2^{(0)}d = 2n\pi$ and the valleys at $2k_2 d = n\pi$. According to the first relation, the "distance" (Δk_2) between two peaks is $(\Delta k_2)d = \pi$. Using the definition of the group velocity, v_{gr} in (2.3.24), the "distance" Δf, between two peaks is approximately given by

$$(\Delta f)_{\mathrm{pa}} \simeq \frac{\mathrm{v}_{\mathrm{gr}}}{2d}. \tag{4.2.21}$$

The subscript pa indicates that this is the bandwidth of a *passive* device. In general, the bandwidth of a given peak is the difference between the two frequencies for which the transmission coefficient is half (−3 dB) of its peak value. Since there are cases where the total height of the peak is less than 3 dB, we define in this case the bandwidth as the distance between two peaks (or two bottom points). If the impedance mismatch is much larger and the difference between the peak and bottom values is much larger then the bandwidth is explicitly dependent on both impedances. At least in what concerns the operation of a traveling-wave tube the trend is to work with minimum reflections, therefore the definition in (4.2.21) is sufficient for our purpose.

Our next step is to calculate the bandwidth of an *active* device. For this purpose we direct our attention back to the expression for the transmission coefficient in (4.2.19). It was mentioned above that the effect of the beam is effectively represented by $e^{-j\bar{\psi}_+}$. This is the ratio between the amplitude of the wave at the output and input. According to the simplified interaction model we developed in Sect. 4.1 we can represent this ratio as

$$e^{-j\bar{\psi}_+} \simeq \frac{1}{3} e^{-j(k_2+\delta k_3)d}, \tag{4.2.22}$$

where δk_3 is that solution of (4.1.6) which has a positive imaginary part as presented in (4.1.12) for the resonance case. In this expression, we tacitly assumed that this growing wave is dominant.

In this effective representation, we have two contributions: one is the real part of the wave-number and the other is its imaginary part. We consider first the effect of the *real part*. We already indicated that, without electrons, the peaks in the transmission coefficient are separated by $2k_2 d = 2\pi$ $(n = 0, \pm 1, \dots)$. In a similar way when the beam is injected they occur at $[2k_2 + \mathrm{Re}(\delta k_3)]d = 2\pi n$. Consequently, the frequency shift, δf, in the location of the peak due to the interaction is

$$\delta f \simeq -\frac{1}{4\pi} \mathrm{v}_{\mathrm{gr}} \mathrm{Re}(\delta k_3), \tag{4.2.23}$$

The maximum frequency shift is expected to occur at maximum gain, namely at resonance hence:

$$\delta f_{\max} \simeq -\frac{1}{4\pi\sqrt{3}} \mathrm{v}_{\mathrm{gr}} \mathrm{Im}(\delta k_3). \tag{4.2.24}$$

As an example, consider coupling coefficient $K_0 = 30\,\mathrm{m}^{-1}$ and group velocity of 0.5 c; the anticipated frequency shift is 180 MHz. For a group velocity ten times smaller the frequency shift drops to 18 MHz. Here it is important to emphasize two aspects (1) The effect of the beam on the reflection process at the end of the extended cavity was ignored. This is not always justified since the capacitive effects at the ends may become significant and consequently the impedances, which we assumed to be virtually frequency independent, may vary significantly causing an additional frequency shift. (2) When we mention here "frequency shift" what is meant is that the frequency where maximum gain occurs, shifts from one frequency to another but the system operates all along in a *linear* regime, namely the frequency at the output is identical with that at the input and only the frequency where maximum transmission occurs, varies because of the interaction.

Next we examine the effect of the *imaginary part* of the wave-number on the transmission coefficient. The transmission coefficient near the peak and close to resonance is given by

$$|\tau| \simeq \frac{4Z_1Z_2\mathrm{e}^{\sqrt{3}K_0d/2}}{(Z_1+Z_2)^2-(Z_2-Z_1)^2\mathrm{e}^{\sqrt{3}K_0d/2}}, \tag{4.2.25}$$

where K_0 has been defined in (4.1.6). In the relatively close vicinity of a peak, we may approximate the transmission coefficient for the case when no beam is present by:

$$|\tau_{\mathrm{pa}}| \simeq \frac{f_1}{\sqrt{(f-f_0)^2+f_1^2}}, \tag{4.2.26}$$

where f_0 is the frequency where the peak is located and, $2\sqrt{3}f_1$ is the bandwidth of the peak which according to the definition in (4.2.21) equals $(2\sqrt{3}f_1 =)\mathrm{v}_{\mathrm{gr}}/2d$. When the beam is present, the wave is amplified by a gain factor, $g(f)$ which at resonance reads $g(f_0) \simeq \frac{1}{3}\mathrm{e}^{\sqrt{3}K_0d/2}$. By analogy with the expression in (4.2.25) we can write the transmission coefficient for an active system as

$$|\tau_{\mathrm{act}}| \simeq \frac{f_1g(f)}{\sqrt{(f-f_0-\delta f_{\max})^2g^2(f)+f_1^2}}. \tag{4.2.27}$$

The bandwidth is the difference between the two frequencies at which $|\tau_{\rm act}|$ reaches 1/2 of its peak value. Ignoring the frequency shift, these two frequencies are a solution of

$$(f-f_0)^2 = 4f_1^2 g^{-2}(f_0) - f_1^2 g^{-2}(f). \tag{4.2.28}$$

Next we assume that the interaction bandwidth [see (4.1.17)] is much broader than f_1, therefore the right-hand side in the last expression can be approximated with $3f_1^2 g^{-2}(f_0)$. This result indicates that the bandwidth of an active (and high gain) system $(\Delta f)_{\rm act}$ is related to the gain and passive device bandwidth, $(\Delta f)_{\rm pa}$, by

$$(\Delta f)_{\rm act} = \frac{(\Delta f)_{\rm pa}}{g(f_0)}, \tag{4.2.29}$$

or

$$(\Delta f)_{\rm act} = (\Delta f)_{\rm pa} 10^{-\frac{\text{gain}[dB]}{20}}. \tag{4.2.30}$$

This result indicates that the product *bandwidth* × *gain* is constant. For example, a gain of 25 dB in a system whose passive bandwidth is 200 MHz causes the bandwidth of the active device to be 11 MHz.

4.2.3 Interaction in an Extended Cavity

If we examine the condition for the occurrence of the peaks in the last section and the resonance condition for cavity creation (see Sect. 4.2.1) – we find that the two are identical. In fact, we have indicated in Sect. 2.5.1 that the denominator of the transmission coefficient determines the resonance frequencies of the system. Whether these frequencies are real or imaginary depends on whether electromagnetic energy can leave the system either as a propagating wave (Davis et al. 1994) or via a dissipative (Ohm loss). An additional insight on the nature of the process can be achieved if the transmission coefficient of a system with two discontinuities (three characteristic impedances – see Fig. 2.9 in Sect. 2.5.1) is represented in terms of the local reflection and transmission coefficients. For this purpose let us define the transmission coefficient from the first region $(-\infty < z < 0)$ to the second $(0 < z < d)$ by τ_{12} as

$$\tau_{12} = \frac{2Z_1}{Z_1 + Z_2}, \tag{4.2.31}$$

where Z_1 and Z_2 are the characteristic impedances in each one of the regions. The reflection from the first region when a wave impinges from the second region is denoted by ρ_{21} and is given by

$$\rho_{21} = \frac{Z_2 - Z_1}{Z_1 + Z_2}, \tag{4.2.32}$$

and correspondingly the wave reflected from the third section when the wave impinges from the second is

$$\rho_{23} = \frac{Z_2 - Z_3}{Z_3 + Z_2}; \tag{4.2.33}$$

in a similar way, the wave which is transmitted into the third section in this case is

$$\tau_{23} = \frac{2Z_2}{Z_3 + Z_2}. \tag{4.2.34}$$

Using this notation in addition to the phase and amplitude advance as described in the previous sub-section ($e^{-j\bar{\psi}_+}$ and $e^{j\bar{\psi}_-}$), we find for the transmission coefficient of the active system (4.2.19) the following expression

$$\tau_{act} = \tau_{12} \frac{e^{-j\bar{\psi}_+}}{1 - \rho_{21}\rho_{23}e^{-j\bar{\psi}_+}e^{-j\bar{\psi}_-}} \tau_{23}. \tag{4.2.35}$$

Using this notation, we can now emphasize several aspects of the finite length effect:

Comment 4.4. The transmission coefficient of the active system depends on the ability to couple the power into the system (τ_{12}), the gain and the reflection process in the interaction region (the middle term) and on the ability to extract the power out of the system (τ_{23})

Comment 4.5. The middle term denominator includes all the information about the effect of reflections on the interaction process and in addition, it provides us with a criterion regarding transition to oscillation. As in the case of the empty cavity the eigen-frequencies of the system are determined by the zeros of the denominator i.e.,

$$1 - \rho_{21}\rho_{23}e^{-j\bar{\psi}_+}e^{-j\bar{\psi}_-} = 0. \tag{4.2.36}$$

From this expression we conclude that the necessary condition for oscillation is

$$g(f)|\rho_{23}||\rho_{21}| \geq 1. \tag{4.2.37}$$

The physical interpretation of this expression is the following: consider a wave of an amplitude 1 at the input end of the interaction region. As it traverses the system the wave is amplified according to the gain in the system, $g(f)$. At the output end it is partially reflected ($|\rho_{23}|$) and it undergoes an additional reflection ($|\rho_{21}|$) at the input. If the amplitude after this last reflection is larger than unity the amplitude will continue to grow in time after each round trip thus the system will oscillate.

In order to envision the effect let us consider two systems (1) $|\rho_{23}| = |\rho_{21}| = 0.1$, for which case the maximum gain before oscillation occurs is $20|\log_{10}(0.1 \times 0.1)| = 40\,\text{dB}$. (2) The other case of interest represents a situation in which one end (typically the input) is effectively short circuited thus $|\rho_{21}| = 1.0$ and the second has a reasonably good transition such that the effective reflection coefficient is $|\rho_{23}| = 0.05$; the maximum gain before oscillation in this case is $20|\log_{10}(0.05 \times 1.0)| = 26\,\text{dB}$.

Comment 4.6. If the system does not operate in a regime which is close to oscillation, it is possible to write the transmission coefficient of the active system, in the following form:

$$\begin{aligned}\tau_{\text{act}} &= \tau_{12}\mathrm{e}^{-j\bar{\psi}_+}\left[\sum_{n=0}^{\infty}\left(\rho_{21}\rho_{23}\mathrm{e}^{-j\bar{\psi}_+}\mathrm{e}^{-j\bar{\psi}_-}\right)^n\right]\tau_{23}\\ &= \tau_{12}\mathrm{e}^{-j\bar{\psi}_+}\left[1+\rho_{21}\rho_{23}\mathrm{e}^{-j\bar{\psi}_+}\mathrm{e}^{-j\bar{\psi}_-}+\left(\rho_{21}\rho_{23}\mathrm{e}^{-j\bar{\psi}_+}\mathrm{e}^{-j\bar{\psi}_-}\right)^2\cdots\right]\tau_{23} \end{aligned}\qquad(4.2.38)$$

In the framework of this notation, it is tacitly assumed that the electron pulse is infinitely long and there are an infinite number of reflections (as the number of terms in the sum). This is obviously not the case in practice and only a limited number of terms has to be considered according to the pulse length and the time it takes the signal to complete one round trip.

Comment 4.7. If there are fluctuations in the current or voltage the expression in (4.2.38) is more adequate for generalization purposes than (4.2.25). Let us denote the total round trip amplitude and phase shift by $\bar{R} = \rho_{21}\rho_{23}\mathrm{e}^{-j\bar{\psi}_+}\mathrm{e}^{-j\bar{\psi}_-}$; using this notation, the transmission coefficient of the active system reads

$$\tau_{act} = \tau_{12}\mathrm{e}^{-j\bar{\psi}_+}\left[1+\bar{R}+\bar{R}^2+\bar{R}^3\cdots\right]\tau_{23}. \qquad (4.2.39)$$

If the current varies along the pulse, then $\bar{R}(I)$ is a function of the current and the natural generalization will be

$$\tau_{\text{act}} = \tau_{12}\mathrm{e}^{-j\bar{\psi}_+}[1+\bar{R}(I_1)+\bar{R}(I_1)\bar{R}(I_2)+\bar{R}(I_1)\bar{R}(I_2)\bar{R}(I_3)\cdots]\tau_{23}, \qquad (4.2.40)$$

where I_ν indicates the average current in the course of the νth reflection. We shall return to this subject when we discuss the generalized formulation of an amplifier and an oscillator.

4.2.4 Backward-Wave Oscillator

In the type of structures on which we have based our model so far the wave and the energy, it carries flow both in the same direction. Therefore, if the input of an amplifier is at $z = 0$ and the beam flows in the *positive* direction, then power is converted from the beam to the wave. Consequently, comparing the power at a location d far enough from the input we will find that $P(z = d) > P(z = 0)$. Furthermore, the wave at $z = 0$ does not "know" that it is going to be amplified since there is no reflected wave or in other words, there is no feedback to provide this information. On the other hand, in an oscillator, the mirrors at both ends together with the structure itself provide a feedback that causes the amplitude at the input to follow the amplitude at the output such that approximately the amplitude of the wave is constant in space.

Imagine now a situation in which the wave propagates in the positive direction, but the energy flows in the opposite direction – this is exactly the case in periodic structures that will be discussed in the next chapter. The information regarding the interaction is carried by the wave opposite to the beam and in fact the input and the output trade places: the input in such a case is at $z = d$ and the output is at $z = 0$. In order to quantify our statements we start from the expression for the interaction wave-number as presented in (4.1.19) and since it was assumed that the energy velocity and the phase velocity are parallel we can now consider a situation in which β_{en} is negative and so is K_0^3. As a result we get, instead of (4.1.6),

$$\delta k(\delta k - \Delta k)^2 = K_0^3, \tag{4.2.41}$$

where the only difference is that the right-hand side is positive (as in the oscillator). The solution at resonance is different and it reads

$$\delta k_1 = K_0, \quad \delta k_2 = -K_0\left(\frac{1}{2} - j\frac{\sqrt{3}}{2}\right), \quad \delta k_3 = -K_0\left(\frac{1}{2} + j\frac{\sqrt{3}}{2}\right). \tag{4.2.42}$$

As in the traveling-wave amplifier the wave propagates in the forward direction, therefore similar to (4.1.21), we can write

$$\begin{aligned} E_z(r, z, \omega) &= \mathrm{J}_0\left(p_1 \frac{r}{R}\right) \mathrm{e}^{-jk_1^{(0)}(z-d)} \\ &\quad \times \left[E_1 \mathrm{e}^{-j\delta k_1(z-d)} + E_2 \mathrm{e}^{-j\delta k_2(z-d)} + E_3 \mathrm{e}^{-j\delta k_3(z-d)}\right]. \end{aligned} \tag{4.2.43}$$

Since at the input ($z = d$) the beam is assumed to be uniform (not bunched) and the initial amplitude is E_0, the boundary conditions imply

$$\begin{aligned} &E_1 + E_2 + E_3 = E_0, \\ &\frac{E_1}{\omega/\mathrm{v}_0 - k_1^{(0)} - \delta k_1} + \frac{E_2}{\omega/\mathrm{v}_0 - k_1^{(0)} - \delta k_2} + \frac{E_3}{\omega/\mathrm{v}_0 - k_1^{(0)} - \delta k_3} = 0, \\ &\frac{E_1(k_1^{(0)} + \delta k_1)}{\left(\omega/\mathrm{v}_0 - k_1^{(0)} - \delta k_1\right)^2} + \frac{E_2(k_1^{(0)} + \delta k_2)}{\left(\omega/\mathrm{v}_0 - k_1^{(0)} - \delta k_2\right)^2} + \frac{E_3(k_1^{(0)} + \delta k_3)}{\left(\omega/\mathrm{v}_0 - k_1^{(0)} - \delta k_3\right)^2} = 0; \end{aligned} \tag{4.2.44}$$

as in the traveling-wave tube case these three equations determine the amplitudes E_1, E_2 and E_3. At the output, the third solution is dominant; thus the gain is given by

$$G \equiv \frac{|E(0)|}{|E(d)|} \simeq \frac{1}{3} e^{\sqrt{3}K_0 d/2}. \tag{4.2.45}$$

Although the right-hand side is identical to the traveling-wave amplifier result, the fact that in this case the feedback is inherent in the interaction process and is not dependent on load impedance, makes the backward device substantially less sensitive to the load. Furthermore, Carmel et al. (1989) has shown experimentally that the presence of a stationary background plasma (gas) can improve substantially the efficiency of the system. In fact, several years before that, Carmel and Nation (1973) had shown that high power microwave radiation can be generated by a backward-wave oscillator driven by a relativistic, high current, electron beam.

4.3 Macro-particle Approach

The hydrodynamic approximation is adequate for the description of the interaction in the linear regime when we wish to consider the variation in the *average* dynamic variables – density and velocity fields. As we approach saturation, the spread in the velocity and density field becomes significant and the validity of the hydrodynamic approximation becomes questionable. In order to solve the problem we have to adopt a more fundamental approach, which is based on the solution of the single-particle equation of motion.

There are at least three ways to develop the simplified set of equations, which describes the interaction between electrons and an electromagnetic wave. All three have the one particle equation of motion in common and assume that the basic form of the solution of the electromagnetic field is preserved. The three methods differ in the way the equation that describes the amplitude and the phase of the electromagnetic field is developed. One possibility is to start from the non-homogeneous wave equation for the magnetic vector potential, the second method is to start from the wave equation for E_z and in the third method, the starting point is Poynting's theorem. Throughout this text we use either the first (Chap. 6) or the third. It is the latter that will be used in this section.

4.3.1 Simplified Set of Equations

The starting point is Poynting's theorem:

$$\nabla \cdot \mathbf{S} + \frac{\partial}{\partial t} W = -\mathbf{J} \cdot \mathbf{E}. \tag{4.3.1}$$

Assuming that the walls of the system are made of an ideal metal, then all the power flux flows in the z direction thus we can integrate over the cross-section (πR^2) of the system:

$$\frac{\partial}{\partial z}2\pi\int_0^R \mathrm{d}r r S_z(r,z,t) + \frac{\partial}{\partial t}2\pi\int_0^R \mathrm{d}r r W(r,z,t) = -2\pi\int_0^R \mathrm{d}r r J_z(r,z,t)E_z(r,z,t). \tag{4.3.2}$$

We assume that a very strong magnetic field confines the electron motion to the z direction therefore, the only non-zero component of the current density is longitudinal. Furthermore, the system is assumed to operate in the linear regime and it oscillates at a single frequency ω. For the present purposes we average out over one period of the wave $T = 2\pi/\omega$ and if we assume that there is no reflected wave and consequently, there is no change in the electromagnetic energy stored in the system, then (4.3.2) reads

$$\frac{\mathrm{d}}{\mathrm{d}z}\left[2\pi\int_0^R \mathrm{d}r r\frac{1}{T}\int_0^T \mathrm{d}t S_z(r,z,t)\right] = -2\pi\int_0^R \mathrm{d}r r\frac{1}{T}\int_0^T \mathrm{d}t J_z(r,z,t)E_z(r,z,t) \tag{4.3.3}$$

The first term is the total average power that propagates along the system:

$$P(z) = 2\pi\int_0^R \mathrm{d}r r\frac{1}{T}\int_0^T \mathrm{d}t S_z(r,z,t), \tag{4.3.4}$$

and according to the definition of the interaction impedance for a very thin pencil beam it is given by

$$P(z) = \frac{\pi R^2}{Z_{\mathrm{int}}}\frac{1}{T}\int_0^T \mathrm{d}t E_z^2(R_{\mathrm{b}},z,t). \tag{4.3.5}$$

The factor two difference between this equation and (2.3.29) is due to the fact that in the latter, the field has already been averaged on time. The principal assumption of the current approach is that at a given frequency and at a given location, the same interaction impedance, which relates the total average power to the longitudinal electric field in vacuum, relates the same quantities when the beam is also present. Consequently, Poynting's theorem now reads

$$\pi R^2\frac{\mathrm{d}}{\mathrm{d}z}\left[\frac{1}{Z_{\mathrm{int}}}\frac{1}{T}\int_0^T \mathrm{d}t E_z^2(R_{\mathrm{b}},z,t)\right] = -2\pi\frac{1}{T}\int_0^T \mathrm{d}t E_z(R_{\mathrm{b}},z,t)\int_0^{R_{\mathrm{b}}} \mathrm{d}r r J_z(r,z,t), \tag{4.3.6}$$

where again we used the thin beam approximation namely, the transverse variations of the electric field are negligible across the beam thickness, therefore the electric field was extracted from the integral in the right-hand side of the equation.

Within the framework of the single particle description, the current density of an azimuthally symmetric flow of electrons is given by

$$J_z(r,z,t) = -e\sum_i \mathrm{v}_i(t)\delta[z-z_i(t)]\frac{1}{2\pi r}\delta[r-r_i(t)]; \tag{4.3.7}$$

$z_i(t)$ and $r_i(t)$ are the longitudinal and the radial location of the ith electron at a time t. With this definition of the current density the radial integration is straightforward and it reads

$$\frac{\mathrm{d}}{\mathrm{d}z}\left[\frac{1}{T}\int_0^T \mathrm{d}tE_z^2(R_\mathrm{b},z,t)\right] = \frac{eZ_\mathrm{int}}{\pi R^2}\frac{1}{T}\int_0^T \mathrm{d}tE_z(R_\mathrm{b},z,t)\sum_{i=1}^N \mathrm{v}_i(t)\delta[z-z_i(t)] \tag{4.3.8}$$

The second main assumption in this approach is that the effect of the beam on the (single mode) distribution of the electric field is only longitudinal. In other words, if in the beam absence, the longitudinal electric field in the beam region was given by $E_z(R_\mathrm{b},z,t) = E_0\cos(\omega t - kz - \theta_0)$ where the amplitude (E_0) and the phase (θ_0) are constant, in the presence of the beam the same component reads

$$E_z(R_\mathrm{b},z,t) = E(z)\cos[\omega t - kz - \theta(z)] \tag{4.3.9}$$

and both the amplitude and the phase are allowed to vary in the longitudinal direction.

We proceed now by performing the time integration on both sides of (4.3.8). In the left-hand side, the integration over the trigonometric functions is straightforward whereas in the right-hand side, we take advantage of the Dirac delta function, thus

$$\frac{\mathrm{d}}{\mathrm{d}z}\left[\frac{1}{2}E^2(z)\right] = \frac{eZ_\mathrm{int}N}{\pi R^2 T}E(z)\langle\cos[\omega\tau_i(z) - kz - \theta(z)]\rangle, \tag{4.3.10}$$

where $\tau_i(z)$ is defined as

$$\tau_i(z) = \tau_i(0) + \int_0^z \mathrm{d}\zeta\frac{1}{\mathrm{v}_i(\zeta)}, \tag{4.3.11}$$

and it represents the time it takes the ith electron to reach the point z. $\mathrm{v}_i(z)$ is the velocity of the ith electron at z, N is the total number of electrons in one period (T) of the wave and $\langle\cdots\rangle \equiv N^{-1}\sum_{i=1}^N\cdots$. We can now identify eN/T as the average current during the period T namely, $I = eN/T$.

It is convenient at this point to adopt a complex notation namely, $\bar{E}(z) \equiv E(z)\mathrm{e}^{-j\theta(z)}$, which permits us to write (4.3.10) as

$$\begin{aligned}\frac{1}{2}\frac{\mathrm{d}}{\mathrm{d}z}E^2(z) &= \frac{1}{2}\frac{\mathrm{d}}{\mathrm{d}z}[\bar{E}(z)\bar{E}^*(z)] = \frac{1}{2}\left[\bar{E}(z)\frac{\mathrm{d}}{\mathrm{d}z}\bar{E}^*(z) + \bar{E}^*(z)\frac{\mathrm{d}}{\mathrm{d}z}\bar{E}(z)\right]\\ &= \frac{IZ_\mathrm{int}}{\pi R^2}\frac{1}{2}\left[\bar{E}(z)\left\langle\mathrm{e}^{j\chi_i(z)}\right\rangle + \bar{E}^*(z)\left\langle\mathrm{e}^{-j\chi_i(z)}\right\rangle +\right],\end{aligned} \tag{4.3.12}$$

wherein $\chi_i(z) = \omega\tau_i(z) - kz$. The last expression can also be written as

$$\bar{E}(z)\left[\frac{\mathrm{d}}{\mathrm{d}z}\bar{E}^*(z) - \frac{IZ_{\mathrm{int}}}{\pi R^2}\langle \mathrm{e}^{j\chi_i(z)}\rangle\right] + \bar{E}^*(z)\left[\frac{\mathrm{d}}{\mathrm{d}z}\bar{E}(z) - \frac{IZ_{\mathrm{int}}}{\pi R^2}\langle \mathrm{e}^{-j\chi_i(z)}\rangle\right] = 0 \quad (4.3.13)$$

and since it has to be satisfied for any $\bar{E}(z)$, we conclude that

$$\frac{\mathrm{d}}{\mathrm{d}z}\bar{E}(z) - \frac{IZ_{\mathrm{int}}}{\pi R^2}\langle \mathrm{e}^{-j\chi_i(z)}\rangle = 0, \quad (4.3.14)$$

which describes the dynamics of the amplitude and phase of the electromagnetic field and its dependence on the distribution of particles.

The next step is to simplify the equation of motion of the electrons. Since the motion is in one dimension, it is more convenient to use the single particle energy conservation as introduced in (3.1.6). Using the explicit expression for the electric field in (4.3.9) and following the motion of the electron in space we have

$$\frac{\mathrm{d}}{\mathrm{d}z}\gamma_i(z) = -\frac{e}{mc^2}\frac{1}{2}\left[\bar{E}(z)\mathrm{e}^{j\chi(z)} + \mathrm{c.c.}\right]. \quad (4.3.15)$$

It is more convenient to present these two equations (4.3.14) and (4.3.15) using a normalized notation. For this purpose, we normalize z to the length (d) of the interaction region and define $\zeta = z/d$ as a normalized coordinate. The normalized (complex) amplitude of the longitudinal component of the electric field in the region of the beam is

$$a(\zeta) = \frac{e\bar{E}(z)d}{mc^2}, \quad (4.3.16)$$

and the coupling coefficient α is

$$\alpha = \frac{eIZ_{\mathrm{int}}}{mc^2}\frac{d^2}{\pi R^2}. \quad (4.3.17)$$

Using this notation the variation in space of the normalized amplitude is given by

$$\frac{\mathrm{d}}{\mathrm{d}\zeta}a(\zeta) = \alpha\left\langle \mathrm{e}^{-j\chi_i(\zeta)}\right\rangle \quad (4.3.18)$$

and the single particle energy conservation reads

$$\frac{\mathrm{d}}{\mathrm{d}\zeta}\gamma_i(\zeta) = -\frac{1}{2}\left[a(\zeta)\mathrm{e}^{j\chi_i(\zeta)} + \mathrm{c.c.}\right]. \quad (4.3.19)$$

To complete the description of the particles' dynamics we have to determine the dynamics of the phase term χ_i. According to its definition and (4.3.11) we find

$$\frac{d}{d\zeta}\chi_i(\zeta) = \Omega\frac{1}{\beta_i} - K = \Omega\left(\frac{1}{\beta_i} - \frac{1}{\beta_{ph}}\right), \tag{4.3.20}$$

where $K = kd$, $\Omega = \omega d/c$ and β_{ph} is the normalized phase velocity. The last three equations form a closed set of equations, which describe the interaction.

Before we proceed to solutions of this set of equations for a practical system, we show that the approximations involved do not affect the global energy conservation. This is readily obtained by averaging the single particle energy conservation (4.3.19) and substituting the equation for the complex normalized amplitude (4.3.18):

$$\frac{d}{d\zeta}\left[\langle\gamma_i\rangle + \frac{1}{2\alpha}|a(\zeta)|^2\right] = 0. \tag{4.3.21}$$

In addition, we show how this set of first order differential equations leads to the same solution we found using the hydrodynamic approximation. To retrieve this limit we take twice the derivative of the amplitude equation in (4.3.18). After the first derivative we obtain

$$\frac{d^2}{d\zeta^2}a(\zeta) = -j\alpha\left\langle\left(\frac{\Omega}{\beta_i} - K\right)e^{-j\chi_i(\zeta)}\right\rangle, \tag{4.3.22}$$

and after the second

$$\frac{d^3}{d\zeta^3}a(\zeta) = -j\alpha\Omega\left\langle\frac{1}{(\beta_i\gamma_i)^3}e^{-j\chi_i(\zeta)}\frac{d}{d\zeta}\gamma_i\right\rangle - \alpha\left\langle\left(\frac{\Omega}{\beta_i} - K\right)^2 e^{-j\chi_i(\zeta)}\right\rangle. \tag{4.3.23}$$

Next we substitute the explicit expression for the single-particle energy conservation from (4.3.19) and consider only the slow varying term. The result is:

$$\left[\frac{d^3}{d\zeta^3} - j\frac{1}{2}\alpha\Omega\left\langle\frac{1}{(\beta_i\gamma_i)^3}\right\rangle\right]a(\zeta) = -\alpha\left\langle\left(\frac{\Omega}{\beta_i} - K\right)^2 e^{-j\chi_i(\zeta)}\right\rangle. \tag{4.3.24}$$

The differential equation on the left-hand side is equivalent to the third order polynomial obtained using the hydrodynamic approximation. According to this expression, if the variation in the momentum is small, the spatial growth rate is given by the imaginary part of the root of the characteristic polynomial:

$$\mathrm{Im}(k) = \frac{\sqrt{3}}{2}\left[\frac{1}{2}\frac{eIZ_{int}}{mc^2}\frac{1}{\pi R^2}\frac{\omega}{c}\left\langle\frac{1}{(\beta_i\gamma_i)^3}\right\rangle\right]^{1/3}, \tag{4.3.25}$$

and this is identical with the result in (4.1.13). The right-hand side term in (4.3.24) represents the driving term in the system. If at the input the phase and the velocity of the particles are completely uncorrelated then its contribution is zero. As the interaction progresses in space, the phase and the velocity of the particles become correlated and its contribution increases.

4.3.2 Phase-Space Distribution: Linear Regime

We now consider the beam–wave interaction using this set of simplified equations (4.3.18)–(4.3.20). The slow-wave structure consists of a dielectric loaded wave-guide which is 20 cm long. The system is driven by a 850 kV, 450 A electron pencil beam. In addition, a wave is launched at the input. The longitudinal component of the electric field at the beam location is assumed to be 1 MV/m. For a practical solution of the equations of motion, we divide the entire ensemble of electrons into 64 clusters equally populated with electrons. The internal distribution in each one of these clusters is assumed to remain unchanged along the interaction process. Figure 4.6 illustrates the way the gain and the efficiency

$$\eta(\%) \equiv \frac{\langle\gamma(z=0)\rangle - \langle\gamma(z)\rangle}{\langle\gamma(z=0)\rangle - 1} \times 100, \qquad (4.3.26)$$

vary along the system. As in the hydrodynamic model we observe first the "build-up" region where the gain is effectively zero, followed by a region where the gain (in dB) increases linearly. The efficiency in this case is less than 10% which means that the *average energy* has dropped by less than 10%. This is, on average, what one can expect from a single-stage traveling-wave tube (TWT) without special intervention. Both the lethargy and the linear gain section are in reasonable agreement with the regular Pierce approach.

Next step is to exploit the present formalism to investigate more systematically the interaction process. Figure 4.7 illustrates the way the phase-space distribution

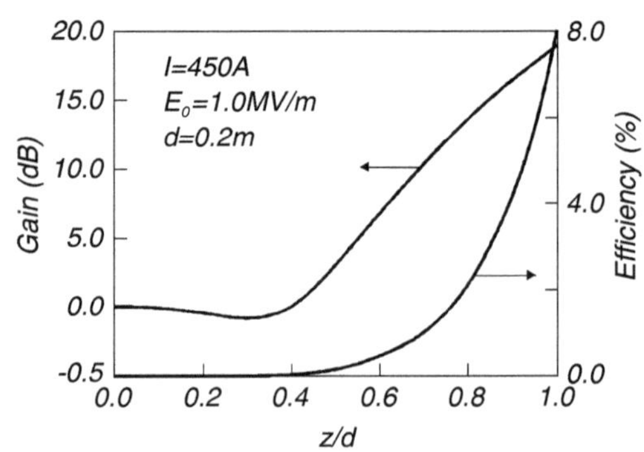

Fig. 4.6 Gain and efficiency along the interaction region. The *upper curve* represents the gain and the lower one shows the efficiency

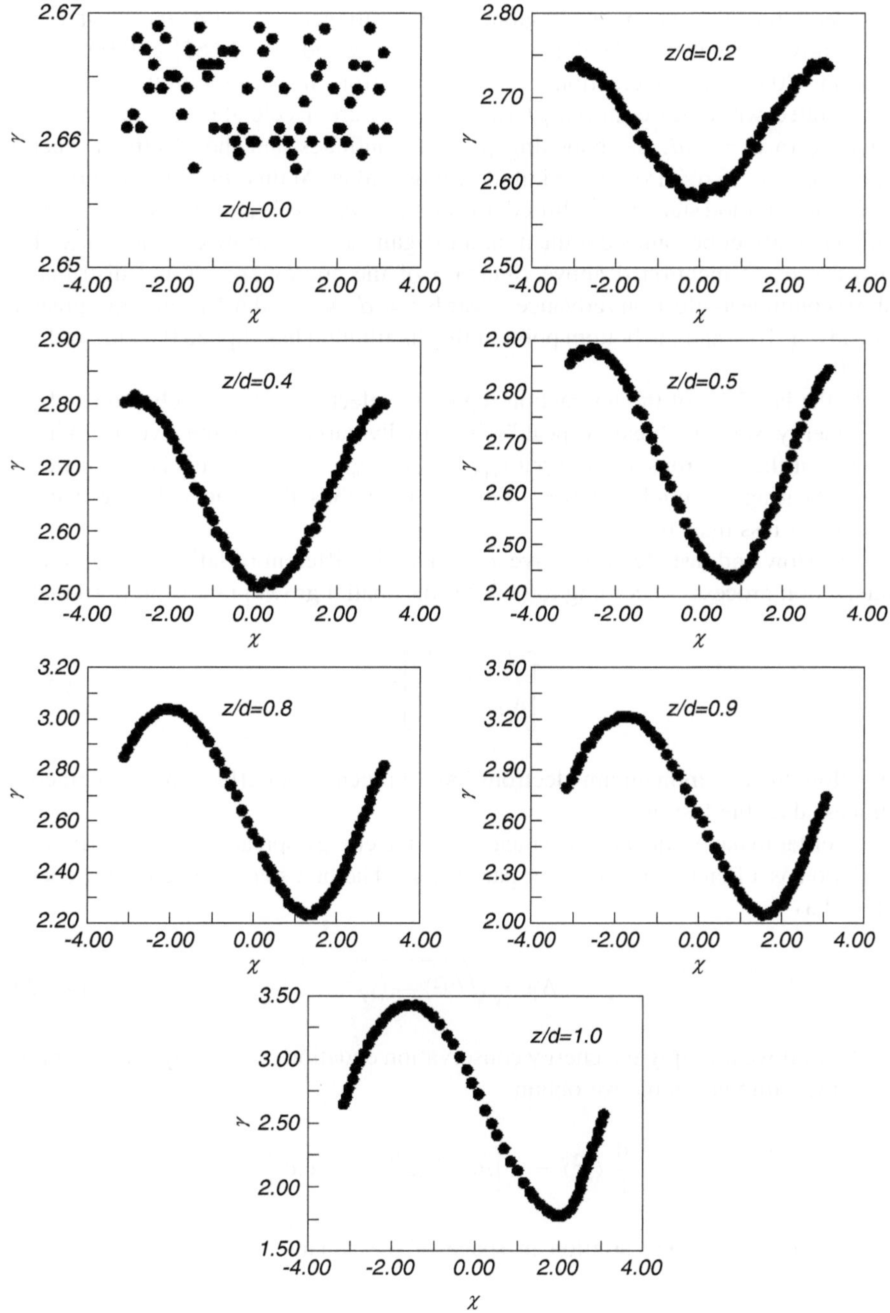

Fig. 4.7 Phase-space distribution at various locations along the interaction region

evolves along the interaction region. At the entrance $z = 0.0d$, the clusters are uniformly distributed in the domain $-\pi < \chi < \pi$ and $|\gamma - 2.665| < 0.005$. After crossing 20% of the interaction region, the electrons *in-phase* with the wave were decelerated while those in *anti-phase* ($\chi = \pm\pi$) are accelerated. As the electrons advance to $z = 0.4d$, the bunching process continues and the electrons' energy spread is now $\pm 6\%$ around the initial average value. At this stage the bottom point of the distribution starts to be shifted towards $\chi = \pi$. This is also the point where the *collective* effect becomes dominant and the gain starts to grow exponentially. The two processes mentioned above (increase of the energy spread and distribution shift) continue as electrons advance towards $z = d$. At $z = 0.6d$ the energy spread is already $\pm 10\%$ and the bottom point of the distribution has slipped 0.8 radians from $\chi = 0$.

In the last 20% of the interaction region the electrons are strongly bunched and the energy spread ("peak-to-peak") is actually larger than the average kinetic energy of the electrons at the input ($\gamma_{\max} = 3.5, \gamma_{\min} = 1.75$). This is a remarkable result bearing in mind that the efficiency (and thus the change in the average energy) is less than 10%.

The slow and fast electrons have a completely different (relative) weight on the interaction process. According to (4.3.25) the spatial growth rate is proportional to

$$\left[\left\langle \frac{1}{(\beta_i \gamma_i)^3} \right\rangle\right]^{1/3}, \tag{4.3.27}$$

therefore the low momentum electrons have a much larger effect on the interaction process than the fast ones.

In order to have a quantitative measure of the energy spread we can determine its variation as a function of the other parameters. The first step is to define the energy spread as

$$\Delta\gamma = \sqrt{\langle \gamma^2 \rangle - \langle \gamma \rangle^2}. \tag{4.3.28}$$

Next, if we multiply the energy conservation equation (4.3.21) by γ_i and average over the entire ensemble we obtain

$$\frac{\mathrm{d}}{\mathrm{d}\zeta} \langle \gamma_i^2 \rangle = -\left[a(\zeta) \langle \gamma_i \mathrm{e}^{j\chi_i(\zeta)} \rangle + \mathrm{c.c.} \right]. \tag{4.3.29}$$

In a similar way, the variation in space of the square of the average energy is given by

$$\frac{\mathrm{d}}{\mathrm{d}\zeta} \langle \gamma_i \rangle^2 = -\left[a(\zeta) \langle \gamma_i \rangle \langle \mathrm{e}^{j\chi_i(\zeta)} \rangle + \mathrm{c.c.} \right]. \tag{4.3.30}$$

If we now subtract (4.3.30) from (4.3.29) and use the definition of the energy spread we find

$$\frac{\mathrm{d}}{\mathrm{d}\zeta}\Delta\gamma^2 = -a(\zeta)\left[\langle\gamma_i \mathrm{e}^{j\chi_i(\zeta)}\rangle - \langle\gamma_i\rangle\langle\mathrm{e}^{j\chi_i(\zeta)}\rangle\right] - a^*(\zeta)\left[\langle\gamma_i \mathrm{e}^{-j\chi_i(\zeta)}\rangle - \langle\gamma_i\rangle\langle\mathrm{e}^{-j\chi_i(\zeta)}\rangle\right]. \tag{4.3.31}$$

This expression indicates that the energy spread is controlled by two principal quantities (1) the amplitude of the radiation field which is obvious since the latter determines the modulation. But this is not sufficient since (2) the phase and energy of all particles have to be correlated i.e.,

$$|\langle\gamma_i \mathrm{e}^{j\chi_i(\zeta)}\rangle| - \left|\langle\gamma_i\rangle\langle\mathrm{e}^{j\chi_i(\zeta)}\rangle\right| > 0, \tag{4.3.32}$$

in order to cause any variation of the energy spread. Otherwise, even for a large amplitude of the radiation field, the change in the energy spread is negligible.

4.3.3 Phase-Space Distribution: Saturation

Saturation process is the next topic to consider. For this purpose we extend the total length by 50% such that $d = 30\,\mathrm{cm}$. Saturation occurs when the electrons start to "absorb" energy from the wave. This will happen when the electrons, which were initially decelerated, reach the point of π phase with the wave and they start to be accelerated. To examine the conditions for saturation we first resort to the hydrodynamic model. Consider the longitudinal component of the electric field which is a self consistent solution of the interaction process near resonance: $E(z) \simeq E_0 \cos[\omega t - z(\omega/\mathrm{v}_0 - K_0/2)]\mathrm{e}^{\sqrt{3}K_0 z/2}$. The amplitude of the oscillation can be estimated by substituting in the equation of motion: $m\mathrm{v}_0(\mathrm{v}_0 K_0)^2\gamma^3\delta z = eE_0\mathrm{e}^{\sqrt{3}z/2}$. Accordingly the saturation length is defined by $\frac{1}{2}K_0\delta z(z = d_{\mathrm{sat}}) = \pi$ hence

$$\frac{d_{\mathrm{sat}}}{d} = \frac{1}{\sqrt{3}K_0 d/2}\ln\left[2\pi\frac{K_0 d(\gamma\beta)^3}{a_0}\right]. \tag{4.3.33}$$

For the present parameters $d_{sat}/d = 0.7$.

We examined the saturation within the framework of the macro-particle approach and the result is illustrated in Figs. 4.8 and 4.9. Figure 4.8 illustrates the phase-space distribution only in the last 33% of the interaction region since in the first 67% it is identical with what we presented in Fig. 4.7. We observe that the bottom of the distribution ($z = 0.7$d) is almost at the π point. Beyond this point slow electrons are accelerated. This is accompanied by a decrease in the gain and efficiency as illustrated in Fig. 4.9.

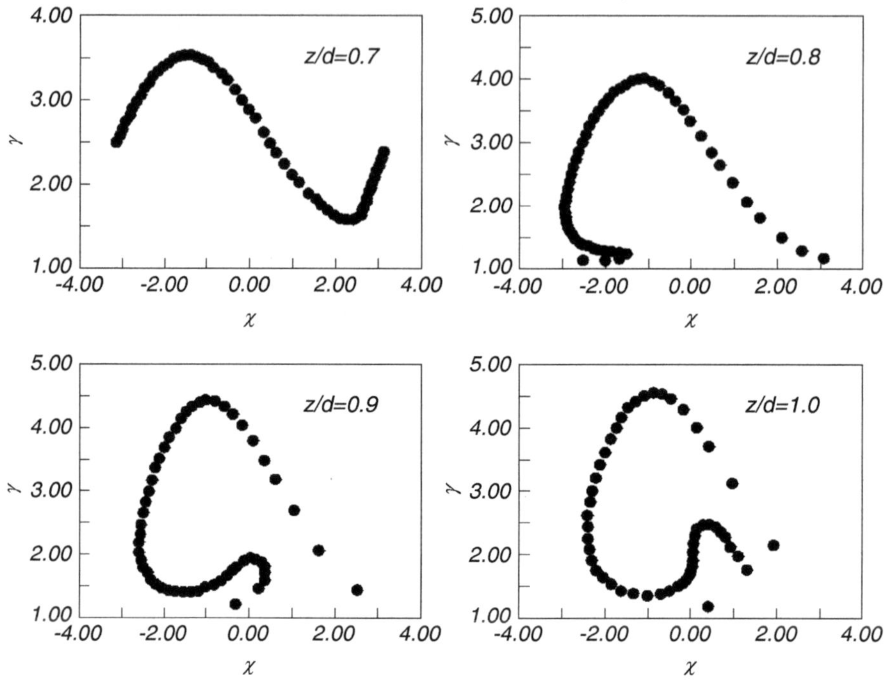

Fig. 4.8 Phase-space distribution at the end of the interaction region

Fig. 4.9 Gain and efficiency in a 50% longer device

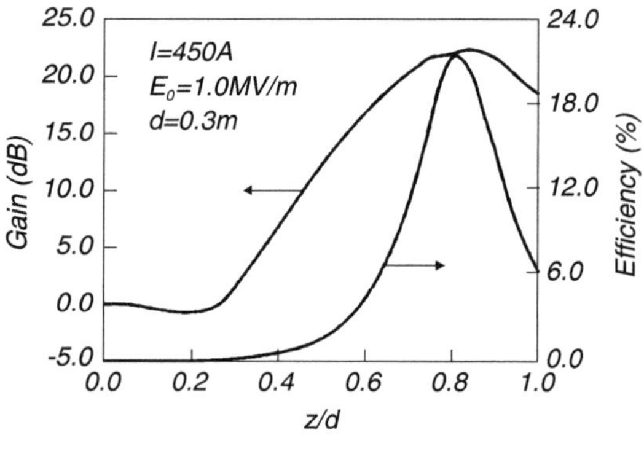

4.3.4 *Interaction in a Slowly Tapered Structure*

In order to increase the efficiency (and consequently the gain) we have to compensate for the energy lost by the electrons. The velocity drop associated with this process and the collective effect itself, cause the phase shift we presented above. In order to adjust the relative phase between the wave and the slow electrons we can *taper* the slow-wave structure. In terms of the dielectric loaded waveguide this can be done by changing the dielectric coefficient along the z axis or changing the radius of the waveguide and/or dielectric slab. Let us now examine the case whereby the dielectric coefficient, $\varepsilon_r(z)$, varies in space. We assume that this variation is weak such that

$$\frac{\omega}{c} \gg \left| \frac{d\varepsilon_r(z)}{dz} \right| \frac{1}{\varepsilon_r(z)}. \qquad (4.3.34)$$

Subject to this condition, the equations that describe the dynamics of the amplitude and the particles' dynamics remain unchanged and only the phase equation becomes

$$\frac{\mathrm{d}}{\mathrm{d}\zeta}\chi_i(\zeta) = \frac{\Omega}{\beta_i(\zeta)} - K(\zeta), \qquad (4.3.35)$$

since the normalized wave-number is $K^2(\zeta) = \varepsilon_r(\zeta)\Omega^2 - p_1^2(d/R)^2$. To simplify the analysis even further we assume a linear variation in space of the dielectric coefficient ε_r namely

$$\varepsilon_r(\zeta) = \varepsilon_r(0) + C_1\zeta. \qquad (4.3.36)$$

We may now ask: what is the optimal value of the slope C_1, given the initial electromagnetic field and the beam characteristics, in order to obtain maximum efficiency and gain. We expect such an optimal value to occur from inspection of Fig. 4.9: when the phase velocity is constant, at $z = 0.8d$ the phase shift is such that they are in anti-phase. Gradually slowing down the phase velocity, by increasing ε_r, we may push the saturation beyond $z = d$. Increasing ε_r too much could cause the wave to be too slow, leading to a weak beam–wave coupling, and consequently, the system does not reach saturation, thus remaining in the linear regime without extracting maximum energy from the beam.

For the parameters mentioned above, we found that the peak occurs at $C_1 = 0.65$ and the efficiency was increased from the 6% in the uniform case to 31% as illustrated in Fig. 4.10. This increase in efficiency is accompanied by 10 dB increase in gain. Figure 4.11 illustrates the gain and the efficiency for a slope

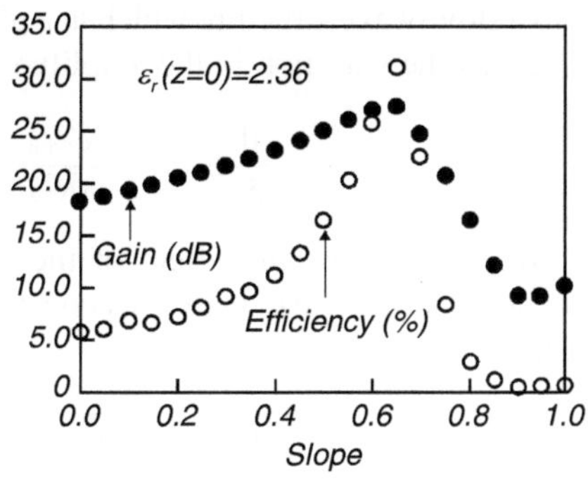

Fig. 4.10 Gain and efficiency as function of the slope of the dielectric coefficient as defined in (4.3.36)

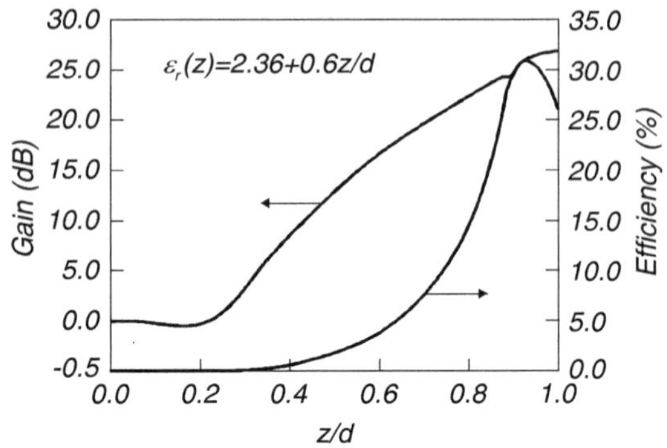

Fig. 4.11 Variation of gain and efficiency in space for linear tapering somewhat below the optical value

which is somewhat below the optimal value $C_1 = 0.6$. Comparing to the uniform case (Fig. 4.10) the saturation point was shifted from $z = 0.8d$ to $z = 0.9d$.

Two comments are in place at this stage:

Comment 4.8. Linear taper is not necessarily the optimal approach and this will be discussed in some detail subsequently in the context of the resonant-particle approximation.

Comment 4.9. Any taper is optimized to a certain set of initial conditions. Deviations from these conditions may lead to performance inferior to that of a uniform structure.

4.3.5 Noise

One of the disadvantages of the hydrodynamic approximation is that the beam is conceived as a fluid and as such, the particle character of the electron is lost. As a result, for evaluation of noise effects one has to postulate velocity or density fluctuations – see discussion by Haus (1959). In the present approach, the individual character of the particles to some extent is preserved. If a single electron is launched into a slow-wave structure which at a given frequency has an interaction impedance Z_{int}, then the variation in the amplitude $E(z)$ is given by

$$\frac{\mathrm{d}}{\mathrm{d}z} E_i(z) = \frac{eZ_{\text{int}}}{\pi R^2 T} \mathrm{e}^{-j[\chi_i(0) + \int_0^z \mathrm{d}z' (\omega/\mathrm{v}_i(z') - k)]}. \tag{4.3.37}$$

We ignore now the effect of the radiation field on the particles such that in the phase term we can take v_i as constant. At the output of the structure the electric field is

$$E_i(d) = \frac{eZ_{int}}{\pi R^2 T} d\mathrm{e}^{-j\chi_i(0)} \mathrm{e}^{-j(\omega/\mathrm{v}_i - k)d/2} \mathrm{sinc}\left[\left(\frac{\omega}{\mathrm{v}_i} - k\right)\frac{d}{2}\right]; \tag{4.3.38}$$

here $\mathrm{sinc}(x) = \sin(x)/x$. For a uniform distribution of electrons the average electric field at the output is zero due to the random phase of the particles relative to the wave: $\langle \mathrm{e}^{-j\chi_i(0)} \rangle = 0$. Nevertheless, each such electron emits spontaneous radiation whose power level at the output is given by

$$P_i = \frac{\pi R^2}{2Z_{\mathrm{int}}} |E_i(d)|^2. \tag{4.3.39}$$

The *total average power* of spontaneous radiation emitted by a uniform beam is

$$P_{\mathrm{sp}} = \left[\frac{1}{2} Z_{\mathrm{int}} I \frac{e}{T}\right] \frac{d^2}{\pi R^2} \mathrm{sinc}^2\left[\left(\frac{\omega}{\mathrm{v}_0} - k\right)\frac{d}{2}\right]. \tag{4.3.40}$$

This power is linearly proportional to the number of electrons (since $I = eN/T$).

4.3.6 Super-Radiant Emission

If for a *uniform* beam the power emitted was proportional to the number of electrons N, in the case of a *pre-bunched* beam, the emitted power is proportional to N^2 this is also known as super-radiant emission. For demonstrating this effect in the framework of the present formulation, we ignore the effect of the radiation field on the electrons. Contrary to the previous case where the low level of emitted power justifies completely this assumption, in this case, it is no longer justified and for an adequate solution one has to take into account the variation in the electrons' phase-space distribution. Nevertheless, in order to have a zero order estimate we do ignore the effect of the radiation on the electrons. In the framework of this approximation the amplitude equation reads

$$\frac{\mathrm{d}}{\mathrm{d}z} E(z) = \frac{I Z_{\mathrm{int}}}{\pi R^2} \langle \mathrm{e}^{-j[\chi_i(0) + z(\omega/\mathrm{v}_i - k)]} \rangle; \tag{4.3.41}$$

the phase ($\chi_i(0)$) and the energy γ_i are correlated and for simplicity we consider a cold bunch (very small energy spread) which has a phase distribution $-\pi < -\chi_0 < \chi_i(0) < \chi_0 < \pi$, hence

$$E(d) = \frac{I Z_{\mathrm{int}}}{\pi R^2} \mathrm{sinc}(\chi_0) \mathrm{e}^{-j(\omega/\mathrm{v}_0 - k)d/2} \mathrm{sinc}\left[\left(\frac{\omega}{\mathrm{v}_0} - k\right)\frac{d}{2}\right]. \tag{4.3.42}$$

Note that at the limit $\chi_0 = \pi$ we obtain the result of a uniformly distributed beam namely, $E(d) = 0$. The power emitted in this case is

$$P_{\mathrm{sr}} \simeq \left[\frac{1}{2} Z_{\mathrm{int}} I^2\right] \frac{d^2}{\pi R^2} \mathrm{sinc}^2\left[\left(\frac{\omega}{\mathrm{v}_0} - k\right)\frac{d}{2}\right] \mathrm{sinc}^2 \chi_0. \tag{4.3.43}$$

This crude estimate of the super-radiant emission reveals its relation to the spontaneous radiation power as

$$P_{sr} \propto N \mathrm{sinc}^2 \chi_0 P_{sp}, \tag{4.3.44}$$

Emphasizing our statement at the beginning of this sub-section regarding the factor N difference between the power emitted in the two processes.

4.3.7 Resonant Particle Model

In Sect. 4.3.2 we found that as electrons lose energy to the wave, their velocity is decreased and therefore they slip from the resonance condition. We have also shown, using a very simple model, that this effect can be corrected if the slow-wave structure is tapered. Let us now examine this process in a more systematic way. Our goal is to determine how the structure should vary in space in order to extract maximum energy from a given distribution of electrons and a given input field. As stated, this requirement (in general) will be very difficult to meet however, we can solve the problem for a limited set of distributions. In particular, we can solve for a very narrow phase-space distribution, which is approximated by a single macro-particle and this solution gives us a crude design as for how the structure should vary in space. With such a design, we can release somewhat the constraint on the initial particle distribution and address distributions that are more practical.

The equations, which describe the dynamics of a system that consists of a single macro particle and an electromagnetic wave, are given by

$$\begin{aligned}
\frac{\mathrm{d}}{\mathrm{d}\zeta} a_r(\zeta) &= \alpha \mathrm{e}^{-j\chi_r(\zeta)}, \\
\frac{\mathrm{d}}{\mathrm{d}\zeta} \gamma_r(\zeta) &= -\frac{1}{2}\left[a_r(\zeta)\mathrm{e}^{j\chi_r(\zeta)} + \text{c.c.}\right], \\
\frac{\mathrm{d}}{\mathrm{d}\zeta} \chi_r(\zeta) &= \frac{\Omega}{\beta_r(\zeta)} - K(\zeta).
\end{aligned} \tag{4.3.45}$$

The coupling coefficient α, is considered to be constant but the wave number in the phase term is allowed to vary (the important variations are assumed to be controlled by the phase term). We further assume that (1) the electrons are ideally bunched such that they form a single macro-particle which (2) remain "glued" together along the entire interaction region. (3) The initial velocity of the macro-particle is equal to the phase velocity of the wave. The problem is to determine the necessary variation in the wave-number of the slow-wave structure in order to keep the macro-particle in resonance along the entire interaction region. This last condition can be mathematically formulated as

$$\frac{\mathrm{d}}{\mathrm{d}\zeta} \chi_r(\zeta) = \frac{\Omega}{\beta_r(\zeta)} - K(\zeta) = 0; \tag{4.3.46}$$

subscript r indicates the resonant particle. Because of this condition, the model is called the *resonant particle model*. From the first term in this equation, we conclude that the phase is constant (χ_r) along the interaction region, therefore the integration of the amplitude equation is straightforward

$$a_r(\zeta) = a(0) + \alpha\zeta e^{-j\chi_r}. \tag{4.3.47}$$

Our natural next step is to substitute this expression into the single particle equation of motion; the result is

$$\gamma_r(\zeta) = \gamma_r(0) - \zeta a(0) \cos\chi_r - \alpha\frac{\zeta^2}{2} \tag{4.3.48}$$

and is illustrated Fig. 4.12. In order to determine the wave-number of the structure, the last result is substituted in (4.3.46) and we obtain

$$K(\zeta) = \Omega\frac{\gamma_r(\zeta)}{\sqrt{\gamma_r^2(\zeta) - 1}}. \tag{4.3.49}$$

According to the initial phase of the resonant particle with the wave the latter will gain ($\chi_r = \pi$) or lose ($\chi_r = 0$) energy and the system operates as an amplifier (left frame) or an accelerator (right frame). In either one of the two cases, the solution above satisfies the global energy conservation i.e.,

$$\gamma_r(\zeta) - 1 + \frac{1}{2\alpha}|a_r(\zeta)|^2 = \gamma_r(0) - 1 + \frac{1}{2\alpha}|a_r(0)|^2. \tag{4.3.50}$$

In an amplifier the normalized kinetic energy ($\gamma_r - 1$) at the input is much larger than the electromagnetic energy per particle ($|a_r(0)|^2/2\alpha$). The situation is reversed in an accelerator. However, in practice the loading effect (change in the electromagnetic energy) in the case of an accelerator is usually small. As an example consider a system which operates as an amplifier at 10 GHz, its length is 2 m and the normalized coupling coefficient is $\alpha = 2.5$. The initial energy of the electrons

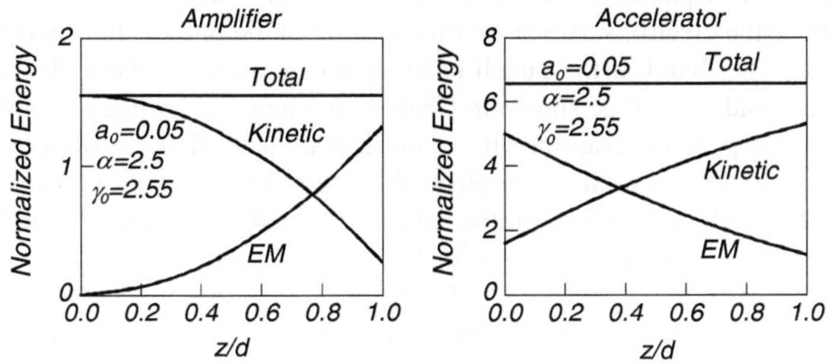

Fig. 4.12 Solution corresponding to the resonant particle model: amplifier left frame and accelerator right frame

corresponds to $\gamma_r = 2.55$ and the input radiation power corresponds to an initial amplitude of $a_r(0) = 0.05$. According to (4.3.47)–(4.3.48) the energy at the output (for $\chi_r = 0$) will correspond to $\gamma_r = 1.25$ thus the efficiency is more than 80%.

The resonant particle model, as presented above, is obviously an idealization of a realistic system which has a finite spread on the initial phase distribution. It is used for the design stage when it is required to calculate the parameters of the structure i.e., the variation in space of the wave number in (4.3.49). In a realistic system, the electrons are not "glued" together therefore they spread. We formulate next the equations for the deviations from the ideal model: $\delta a(\zeta) \equiv a(\zeta) - a_r(\zeta)$ represents the change in the amplitude of the radiation field. In a similar way $\delta\gamma_i(\zeta) \equiv \gamma_i(\zeta) - \gamma_r(\zeta)$ and $\delta\chi_i(\zeta) \equiv \chi_i(\zeta) - \chi_r$. These deviations satisfy the following set of equations:

$$\begin{aligned}
\frac{d}{d\zeta}\delta a(\zeta) &= \alpha e^{-j\chi_r}\left[\langle e^{-j\delta\chi_i(\zeta)}\rangle - 1\right],\\
\frac{d}{d\zeta}\delta\gamma_i(\zeta) &= -\frac{1}{2}\left\{e^{j\chi_r}\left[[a_r(\zeta) + \delta a(\zeta)]e^{j\delta\chi_i(\zeta)} - a_r(\zeta)\right] + \text{c.c.}\right\},.\\
\frac{d}{d\zeta}\delta\chi_i(\zeta) &= \Omega\left[\frac{1}{\beta_i(\zeta)} - \frac{1}{\beta_r(\zeta)}\right]
\end{aligned} \tag{4.3.51}$$

It is instructive to examine this set of equations in a regime where these deviations are small

$$\begin{aligned}
\frac{d}{d\zeta}\delta a(\zeta) &= -j\alpha e^{-j\chi_r}\langle\delta\chi_i(\zeta)\rangle,\\
\frac{d}{d\zeta}\delta\gamma_i(\zeta) &= -\frac{1}{2}[a_r(\zeta)j\delta\chi_i(\zeta) + \delta a(\zeta)e^{j\chi_r} + \text{c.c.}],\\
\frac{d}{d\zeta}\delta\chi_i(\zeta) &= \frac{\Omega}{[\gamma_r(\zeta)\beta_r(\zeta)]^3}\delta\gamma_i(\zeta).
\end{aligned} \tag{4.3.52}$$

It is evident from the first two equations that if the average phase distribution $\langle\delta\chi_i(\zeta)\rangle$ vanishes or is very small, the system behaves as if driven by a single macro particle. The third equation indicates that the phase distribution tends to spread as the momentum of the electrons decreases, diminishing in the process the energy conversion. On the other hand, if the bunch is being accelerated, the phase deviations are much smaller and, as will be shown in Chap. 8, the bunch is actually compressed.

The next step is to examine the operation of a realistic system which has the same parameters as in the example at the beginning ot the previous paragraph. The simulation is based on the solution of (4.3.51) and is performed as follows: we take 10,240 macro-particles uniformly distributed in the range $|\delta\gamma_i(0) = \gamma_i(0) - \gamma_r(0)| < \gamma_r(0)/80$ and $|\delta\chi_i(0)| < \pi/36$. Figure 4.13 illustrates the phase-space plot at three different locations along the interaction region. First frame illustrates the bunch at the input. After 70 cm the bunch lost about 10% of its momentum and it still maintains its shape. After another 70 cm the bunch has lost a total of about 30% of its initial momentum.

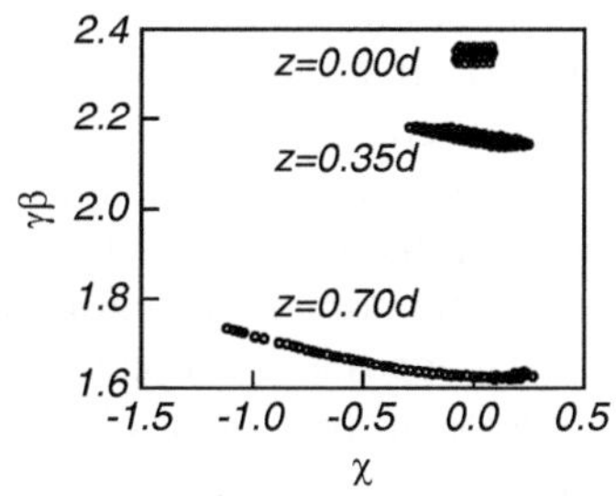

Fig. 4.13 Phase-space distribution in a tapered structure designed based on the resonant particle model

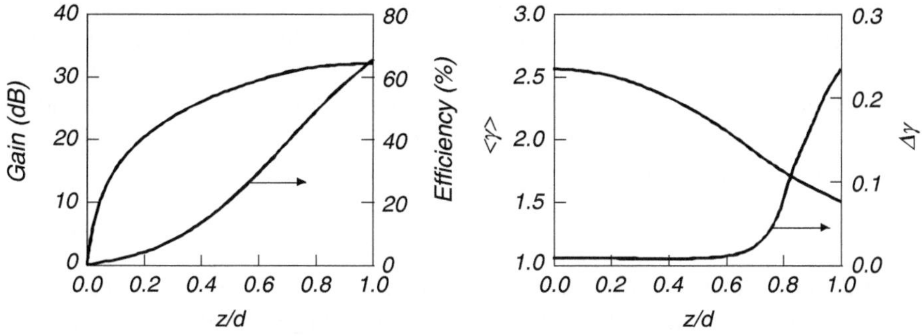

Fig. 4.14 Gain and efficiency variation along the interaction region (*left frame*). Average energy and energy spread in the *right frame*

Finally, we present in Fig. 4.14 the gain and the efficiency (left frame) and in the right the average energy of the electrons as well as their energy spread. Several characteristics are evident (1) The gain starts to grow in space immediately without the *spatial lethargy* required in an amplifier driven by a uniform beam. This result is obvious bearing in mind that the wave at the input does not have to bunch the beam – the latter is already bunched.

The gain is about 33 dB in comparison to 34 dB predicted by the resonant particle model and 16 dB using a uniform structure. (2) The efficiency is more than 65% compared to more than 80% predicted by the resonant particle model or a typical 10% in a uniform structure. (3) From the energy spread we conclude that the bunch maintains its shape for about 70% of the interaction region. Beyond this point there is a significant increase in the energy spread.

4.4 Amplifier and Oscillator: A Unified Approach

In Sect. 4.3 we formulated the interaction in an amplifier based on the single-particle equation of motion and ignoring variation in time i.e., reflections. The next step is to include these effects in the analysis. The motivation for this generalization was introduced already in Sect. 4.2.1 where we discussed the effect of reflections within the framework of the hydrodynamic model and it was shown that one manifestation of their effect is the product *gain* × *bandwidth* which was proved to be constant. Another consequence of a wave being reflected is amplitude variations which occur at the input, according to the time it takes the reflected wave to traverse the distance between the output and input end. Thus, when reflections are not negligible, the assumption of no time variations in a realistic amplifier is not justified.

The opposite situation occurs in oscillators. "Mirrors" at the two ends impose the variation in space of the electromagnetic field. Thus the field amplitude is considered constant in space and the beam–wave interaction determines the temporal growth rate. But the beam which enters the system is presumably unbunched therefore, it will take some space for this beam to become bunched. If so, the modulation amplitude is expected to vary in space and consequently, the amplitude of the radiation field will vary in space. As before, this is in contradiction to the initial assumption.

In order to emphasize even further the difference between an amplifier and an oscillator, we recall that within the framework of the hydrodynamic model, the beam–wave interaction was formulated in terms of a dispersion relation

$$D_{\rm act}(k,\omega)=0. \tag{4.4.1}$$

In an ideal amplifier we assumed that there are *no variations in time of the amplitude* thus the frequency is set for us by an external generator ($\omega=\omega_0$) and we have to determine the variation in space represented by a set of k's which can be complex and they are a solution of:

$$D_{\rm act}(k,\omega=\omega_0)=0. \tag{4.4.2}$$

This is one "extreme" among the regimes of beam–wave interaction. The opposite extreme happens in an ideal oscillator. There it is assumed that there are *no variations in space of the amplitude* since the wave-number k is determined by the separation of the mirrors (d) i.e., $k=\pi n/d$ where n is an integer. Consequently, we have to determine the variation in time represented by a set of frequencies, which can be complex and they are a solution of:

$$D_{\rm act}\left(k=\frac{\pi n}{d},\omega\right)=0. \tag{4.4.3}$$

We now include the role of reflections on the beam–wave interaction and in this process we generalize the formulation which will allow us to derive the operation of an amplifier or an oscillator from one set of equations.

4.4.1 Simplified Set of Equations

As in the previous case, the starting point is Poynting's theorem

$$\nabla\cdot\mathbf{S}+\frac{\partial}{\partial t}W=-\mathbf{J}\cdot\mathbf{E}. \tag{4.4.4}$$

Assuming that the walls of the system are made of an ideal metal, all the power flux flows in the z direction therefore we can integrate over the cross-section of the system:

$$\frac{\partial}{\partial z}\left[2\pi\int_0^R \mathrm{d}rrS_z(r,z,t)\right]+\frac{\partial}{\partial t}\left[2\pi\int_0^R \mathrm{d}rrW(r,z,t)\right]$$
$$=-2\pi\int_0^R \mathrm{d}rrJ_z(r,z,t)E_z(r,z,t). \quad (4.4.5)$$

The first term,

$$P(z,t)=2\pi\int_0^R \mathrm{d}rrS_z(r,z,t), \quad (4.4.6)$$

is the total instantaneous power which flows in the system and

$$W(z,t)=2\pi\int_0^R \mathrm{d}rrW(r,z,t), \quad (4.4.7)$$

represents the total instantaneous energy per unit length stored in the system. As before, we assume that the oscillation is longitudinal and the transverse variations in the electric field are negligible on the scale of the beam thickness. Thus, for a thin pencil beam of radius R_b, the right-hand term of (4.4.5) reads

$$-2\pi E_z(R_\mathrm{b},z,t)\int_0^{R_b} \mathrm{d}rrJ_z(r,z,t). \quad (4.4.8)$$

Now, since the current density is given by

$$J_z(r,z,t)=-e\sum_i \mathrm{v}_i(t)\delta[z-z_i(t)]\frac{1}{2\pi r}\delta[r-r_i(t)], \quad (4.4.9)$$

the integration over the transverse coordinate becomes trivial by virtue of the Dirac delta function. As a result, (4.4.8) reads

$$eE_z(R_\mathrm{b},z,t)\sum_{i=1}^{N} \mathrm{v}_i(t)\delta[z-z_i(t)], \quad (4.4.10)$$

where it was tacitly assumed that there is no transverse motion. According to our assumptions and definitions so far, Poynting's theorem is given by

$$\frac{\partial}{\partial z}P(z,t)+\frac{\partial}{\partial t}W(z,t)=eE_z(R_\mathrm{b},z,t)\sum_{i=1}^{N} \mathrm{v}_i(t)\delta[z-z_i(t)]. \quad (4.4.11)$$

When reflections are non-negligible, the longitudinal electric field, which acts on the electrons has two components: one which is propagating parallel to the electrons and another which is propagating anti-parallel:

$$E_z(R_{\rm b}, z, t) = E_+(z, t)\cos[\omega t - kz - \psi_+(z, t)] + E_-(z, t)\cos[\omega t + kz - \psi_-(z, t)]. \tag{4.4.12}$$

It will be further assumed that the amplitudes and the phases (E_+, E_-, ψ_+ and ψ_-) vary slowly comparing to the trigonometric function i.e.:

$$\begin{aligned} \left|\frac{\partial}{\partial t}E_\pm\right| &\ll \omega|E_\pm|, & \left|\frac{\partial}{\partial z}E_\pm\right| &\ll k|E_\pm|; \\ \left|\frac{\partial}{\partial z}\psi_\pm\right| &\ll k|\psi_\pm|, & \left|\frac{\partial}{\partial t}\psi_\pm\right| &\ll \omega|\psi_\pm|. \end{aligned} \tag{4.4.13}$$

Among the two waves, only the one propagating parallel has an average net effect, therefore the right-hand side of (4.4.11) simplifies to

$$eE_+(z, t)\cos[\omega t - kz - \psi_+(z, t)]\sum_{i=1}^{N} {\rm v}_i(t)\delta[z - z_i(t)]. \tag{4.4.14}$$

Without loss of generality, we can use the trigonometric properties of the cos function to write

$$\begin{aligned} &= eNE_+(z, t)\cos\left[\psi_+(z, t)\right]\langle {\rm v}_i(t)\cos[\chi_i(t)]\delta[z - z_i(t)]\rangle \\ &\quad + eNE_+(z, t)\sin\left[\psi_+(z, t)\right]\langle {\rm v}_i(t)\sin[\chi_i(t)]\delta[z - z_i(t)]\rangle, \end{aligned} \tag{4.4.15}$$

where

$$\chi_i(t) = \omega t - kz_i(t), \tag{4.4.16}$$

is the phase of the ith particle relative to the wave at the time t and $\langle\cdots\rangle = \frac{1}{N}\sum_{i=1}^{N}\cdots$. The notation in (4.4.15) indicates that if the particles move with a velocity which is close to the phase velocity of the wave, all quantities are slow varying functions of z and t. This is in contrast to the left-hand side of the expression for Poynting's theorem, which consists of both slow and fast terms. Naturally, the next step is to eliminate the contribution of the fast oscillation.

The fast variations of the electromagnetic field are determined by the angular frequency $\omega = 2\pi/T$. We can use the definition of the interaction impedance to average out these fast variations in the total power i.e.,

$$\bar{P}(z, t) = \frac{1}{T}\int_{t-T/2}^{t+T/2} {\rm d}t' P(z, t') = \frac{1}{T}\int_{t-T/2}^{t+T/2} {\rm d}t' \frac{\pi R^2}{Z_{\rm int}} E_z^2(R_{\rm b}, z, t'); \tag{4.4.17}$$

the result is

$$\bar{P}(z,t) = \frac{\pi R^2}{2Z_{\rm int}}\left[E_+^2(z,t) - E_-^2(z,t)\right]. \quad (4.4.18)$$

The cross term proportional to (E_+E_-) was neglected since it varies rapidly in space. In addition, we took into account the fact that the power carried by the backward wave is in the opposite direction to that of the forward propagating wave. In a similar way the average energy stored per unit length is given by

$$\bar{W}(z,t) = \frac{1}{2}\pi R^2 \varepsilon_0 \varepsilon_{\rm int}\left[E_+^2(z,t) + E_-^2(z,t)\right]. \quad (4.4.19)$$

Consequently, we can write for the slow varying components of Poynting's theorem

$$\begin{aligned} &\frac{\partial}{\partial z}\left[E_+^2(z,t) - E_-^2(z,t)\right] + \frac{1}{{\rm v}_{\rm en}}\frac{\partial}{\partial t}\left[E_+^2(z,t) + E_-^2(z,t)\right] \\ &\quad = \frac{2eNZ_{\rm int}}{\pi R^2}\left\{E_+(z,t)\cos\left[\psi_+(z,t)\right]\left\langle {\rm v}_i(t)\delta[z - z_i(t)]\cos[\chi_i(t)]\right\rangle\right\} \\ &\quad + \frac{2eNZ_{\rm int}}{\pi R^2}\left\{E_+(z,t)\sin\left[\psi_+(z,t)\right]\left\langle {\rm v}_i(t)\delta[z - z_i(t)]\sin[\chi_i(t)]\right\rangle\right\}. \end{aligned} \quad (4.4.20)$$

Here we used (2.3.33) which relates the interaction impedance with the interaction dielectric coefficient and the energy velocity: $Z_{\rm int}\varepsilon_{\rm int} = \eta_0/\beta_{\rm en}$.

Before we proceed and simplify the amplitude equation, it will be more convenient to use a normalized notation. For this purpose, we examine the single-particle energy conservation ignoring the effect of the backward wave on the motion of the electrons,

$$\frac{\rm d}{{\rm d}t}\gamma_i = -\frac{e}{mc^2}{\rm v}_i(t)E_+[z = z_i(t), t]\cos\left\{\omega t - kz_i(t) - \psi_+[z = z_i(t), t]\right\}; \quad (4.4.21)$$

this justifies the normalization of the electric field according to

$$a_\pm(\zeta,\tau) = \frac{eE_\pm(z,t)d}{mc^2}{\rm e}^{-j\psi_\pm(z,t)}, \quad (4.4.22)$$

of a new spatial variable $\zeta = z/d$ where d is the total length of the interaction region, of a new time variable $\tau = tc/d$ and of the coupling coefficient α as

$$\alpha = \frac{eIZ_{\rm int}}{mc^2}\frac{d^2}{\pi R^2} \quad (4.4.23)$$

where I is the average current in one period of the wave. Subject to this normalized notation (4.4.20)–(4.4.21) read

$$\frac{\partial}{\partial\zeta}\left[|a_+(\zeta,\tau)|^2 - |a_-(\zeta,\tau)|^2\right] + \frac{1}{\beta_{\text{en}}}\frac{\partial}{\partial\tau}\left[|a_+(\zeta,\tau)|^2 + |a_-(\zeta,\tau)|^2\right]$$
$$= \alpha\left[a_+(\zeta,\tau)\left\langle\beta_i(\tau)\delta[\zeta - \bar{z}_i(t)]\mathrm{e}^{j\chi_i(\tau)}\right\rangle_i + \text{c.c.}\right], \tag{4.4.24}$$

and

$$\frac{\mathrm{d}}{\mathrm{d}\tau}\gamma_i = -\beta_i(\tau)\frac{1}{2}\left[a_+[\zeta = \bar{z}_i(\tau),\tau]\mathrm{e}^{j\chi_i(\tau)} + \text{c.c.}\right], \tag{4.4.25}$$

as well as

$$\frac{\mathrm{d}}{\mathrm{d}\tau}\chi_i(\tau) = \Omega - K\beta_i(\tau). \tag{4.4.26}$$

The amplitudes of the forward wave $[a_+(\zeta,\tau)]$ and backward wave $[a_-(\zeta,\tau)]$ are correlated at both ends of the structure by the reflection process. In the interaction region itself the two amplitudes are not coupled since we indicated that the electrons are interacting only with the forward wave. Consequently, the energy conservation associated with this wave reads

$$\left[\frac{\partial}{\partial\zeta} - \frac{1}{\beta_{\text{en}}}\frac{\partial}{\partial\tau}\right]|a_-(\zeta,\tau)|^2 = 0, \tag{4.4.27}$$

that in turn implies that the equation for the amplitude of the forward wave is given by

$$\frac{\partial}{\partial\zeta}a_+(\zeta,\tau) + \frac{1}{\beta_{\text{en}}}\frac{\partial}{\partial\tau}a_+(\zeta,\tau) = \alpha\langle\mathrm{e}^{-j\chi_i(\tau)}\beta_i(\tau)\delta[\zeta - \bar{z}_i(t)]\rangle. \tag{4.4.28}$$

In order to determine the effect of reflections we denote by a_0 the amplitude of the forward wave present in the system in the absence of the beam. The reflection process is represented by a scalar reflection coefficient at the input (ρ_{in}) and output (ρ_{out}) end. At any instant τ, the *change* in the forward wave amplitude is reflected from the output end towards the input according to

$$[a_+(\zeta = 1,\tau) - a_0]\rho_{out}\mathrm{e}^{-jK}. \tag{4.4.29}$$

As already indicated, the backward wave is not directly affected by the beam and therefore it propagates towards the input as if no electrons were present. The time it takes this change in the energy (variation in the amplitude) to reach the input end is determined by the energy velocity of the cold structure. In our normalized notation this delay is $1/\beta_{\rm en}$. Consequently, after this delay, the change mentioned above undergoes an additional reflection – this time from the input end. The contribution of the reflection to the amplitude of the forward wave at the input end ($\zeta = 0$) is given by

$$a_+(\zeta = 0, \tau) - a_0 = \bar{\rho}[a_+(\zeta = 1, \tau - 1/\beta_{en}) - a_0], \tag{4.4.30}$$

where $\bar{\rho} \equiv \rho_{\rm in}\rho_{\rm out}{\rm e}^{-2jK}$ is the feedback term of the passive (no beam) system. Let us now summarize the generalized set of equations, which describe the dynamics of the field and the electrons when reflections are included:

$$\begin{aligned}
&\left[\frac{\partial}{\partial\zeta} + \frac{1}{\beta_{\rm en}}\frac{\partial}{\partial\tau}\right] a_+(\zeta,\tau) = \alpha\langle {\rm e}^{-j\chi_i(\tau)}\beta_i(\tau)\delta[\zeta - \bar{z}_i(t)]\rangle,\\
&\frac{\rm d}{{\rm d}\tau}\gamma_i(\tau) = -\beta_i(\tau)\frac{1}{2}\left[a_+[\zeta = \bar{z}_i(\tau),\tau]{\rm e}^{j\chi_i(\tau)} + {\rm c.c.}\right],\\
&\frac{\rm d}{{\rm d}\tau}\chi_i(\tau) = \Omega - K\beta_i(\tau),\\
&a_+(\zeta = 0,\tau) - a_0 = \bar{\rho}[a_+(\zeta = 1, \tau - 1/\beta_{\rm en}) - a_0].
\end{aligned} \tag{4.4.31}$$

From this generalized formulation, we can obtain the equations which were developed in the previous sections for an ideal amplifier.

4.4.2 *Ideal Amplifier*

In an ideal amplifier, we expect no reflections and thus no time variations, of the amplitude ($\partial/\partial\tau \sim 0$). The amplitude equation is averaged on time and since the integration over the Dirac delta function is straightforward thus

$$\frac{\rm d}{{\rm d}\zeta}a_+(\zeta) = \alpha\langle {\rm e}^{-j\chi_i(\zeta)}\rangle. \tag{4.4.32}$$

Since no time variations are assumed, then according to the definition of ${\rm d}/{\rm d}\tau (\equiv \partial/\partial\tau + \beta_i\partial/\partial\zeta \sim \beta_i\partial/\partial\zeta = \beta_i{\rm d}/{\rm d}\zeta)$ it is sufficient to describe only the space variation and in this framework $\beta_i(\zeta)$ represents the velocity of the ith electron at ζ. Consequently, the equation of motion reads

$$\frac{\rm d}{{\rm d}\zeta}\gamma_i = -\frac{1}{2}\left[a_+(\zeta){\rm e}^{j\chi_i(\zeta)} + {\rm c.c.}\right]. \tag{4.4.33}$$

In a similar way,

$$\frac{\rm d}{{\rm d}\zeta}\chi_i(\zeta) = \frac{\Omega}{\beta_i} - K, \tag{4.4.34}$$

and finally the reflections equation is identically satisfied since there are no reflections thus $\bar{\rho} = 0$ and the amplitude at the input is always the same and it equals a_0.

4.4.3 Ideal Oscillator

For an ideal oscillator we assume that no variations in space occur and therefore the amplitude of the electric field experienced by the electrons does not depend on the location of any individual electron. As a result, we replace $a_+[\bar{z}_i(\tau), \tau]$, in the single particle energy conservation (4.4.31), with its value at the input – $a_+(0, \tau)$ – hence

$$\frac{\mathrm{d}}{\mathrm{d}\tau}\gamma_i = -\beta_i(\tau)\frac{1}{2}\left[a_+(0,\tau)\mathrm{e}^{j\chi_i(\tau)} + \mathrm{c.c.}\right]. \tag{4.4.35}$$

The reflection coefficients from both ends are *unity* and the boundary conditions imply $K = \pi n$. As a result, $\bar{\rho} = 1$ and the reflections equation (4.4.31) reads

$$a_+(0,\tau) = a_+(1, \tau - 1/\beta_{\mathrm{en}}). \tag{4.4.36}$$

In order to determine the dynamics of the amplitude in an oscillator we average the amplitude equation [in (4.4.31)] over the interaction region:

$$\frac{1}{\beta_{\mathrm{en}}}\frac{\mathrm{d}}{\mathrm{d}\tau}a_+(0,\tau) = \alpha\langle\beta_i(\tau)\mathrm{e}^{-j\chi_i(\tau)}\rangle, \tag{4.4.37}$$

and as in the case of an ideal amplifier, we calculate the normalized growth rate. For this purpose we take twice the derivative of (4.4.37), neglect terms which oscillate rapidly and obtain the following expression for $a_+(0, \tau)$

$$\begin{aligned}
&\frac{\mathrm{d}^3 a_+}{\mathrm{d}\tau^3} + \frac{1}{2}\alpha'\left\langle\frac{1}{\gamma_i^3}\right\rangle\frac{\mathrm{d}a_+}{\mathrm{d}\tau} - \frac{1}{2}\alpha'\left\langle\frac{1}{\gamma_i^3}(\Omega - K\beta_i)\right\rangle a_+ \\
&\quad + \frac{1}{2}j\alpha' K\left\langle\frac{\beta_i}{\gamma_i^3}\right\rangle a_+ - \frac{3}{4}\alpha' a_+\left[\left\langle\frac{\beta_i}{\gamma_i^3}e^{j\chi_i}\right\rangle a_+ + \left\langle\frac{\beta_i}{\gamma_i^3}e^{-j\chi_i}\right\rangle a_+^*\right] \\
&\quad = -\alpha'\left\langle\beta_i(\Omega - K\beta_i)^2 e^{-j\chi_i}\right\rangle,
\end{aligned} \tag{4.4.38}$$

where $\alpha' = \alpha\beta_{\mathrm{en}}$. A simple evaluation of the growth rate in the linear regime of operation is possible by ignoring the fifth term (since it is non-linear); near resonance, where we expect the growth rate to be maximum, the third term is much smaller than the fourth. Finally, for relativistic electrons and long interaction length (such that $K \gg 1$) the second term is negligible relative to the fourth, therefore we have

$$\left[\frac{\mathrm{d}^3}{\mathrm{d}\tau^3} + \frac{1}{2} j\alpha' K \left\langle \frac{\beta_i}{\gamma_i^3} \right\rangle \right] a_+ = -\alpha' \left\langle \beta_i (\Omega - K\beta_i)^2 \mathrm{e}^{-j\chi_i} \right\rangle. \tag{4.4.39}$$

Assuming that $\langle \beta_i \gamma_i^{-3} \rangle$does not vary significantly in time, the normalized growth rate is

$$\bar{\omega} = \frac{\sqrt{3}}{2} \left[\frac{1}{2} \alpha \beta_{en} K \left\langle \frac{\beta_i}{\gamma_i^3} \right\rangle \right]^{1/3}. \tag{4.4.40}$$

This expression is identical to that calculated in Sect. 4.2.1 developed using the hydrodynamic approximation.

4.4.4 Global Energy Conservation

Global energy conservation is obtained by multiplying the equation of motion [in (4.4.31)] by $\delta[\zeta - \bar{z}_i(\tau)]$ and averaging over the entire ensemble of particles. In the resulting expression,

$$\left\langle \delta[\zeta - \bar{z}_i(\tau)] \frac{\mathrm{d}}{\mathrm{d}\tau} \gamma_i \right\rangle = -\frac{1}{2} \left[a_+[\zeta = \bar{z}_i(\tau), \tau] \left\langle \beta_i(\tau) \mathrm{e}^{j\chi_i(\tau)} \delta[\zeta - \bar{z}_i(\tau)] \right\rangle + \text{c.c.} \right], \tag{4.4.41}$$

we substitute the equation for the amplitude (4.4.31) and get

$$\left\langle \delta[\zeta - \bar{z}_i(\tau)] \frac{\mathrm{d}}{\mathrm{d}\tau} \gamma_i(\tau) \right\rangle = -\frac{1}{2\alpha} \left[\frac{\partial}{\partial \zeta} |a_+(\zeta, \tau)|^2 + \frac{1}{\beta_{en}} \frac{\partial}{\partial \tau} |a_+(\zeta, \tau)|^2 \right] \tag{4.4.42}$$

In order to bring this last equation to a more familiar form we note that the left-hand side term, after differentiation by parts, reads

$$\frac{\mathrm{d}}{\mathrm{d}\tau} \langle \delta[\zeta - \bar{z}_i(\tau)] \gamma_i \rangle - \left\langle \gamma_i \frac{\mathrm{d}}{\mathrm{d}\tau} \delta[\zeta - \bar{z}_i(\tau)] \right\rangle, \tag{4.4.43}$$

and the last term is zero by virtue of the definition of $\mathrm{d}/\mathrm{d}\tau$ namely

$$\begin{aligned} \frac{\mathrm{d}}{\mathrm{d}\tau} \delta[\zeta - \bar{z}_i(\tau)] &= \frac{\partial}{\partial \tau} \delta[\zeta - \bar{z}_i(\tau)] + \beta_i \frac{\partial}{\partial \zeta} \delta[\zeta - \bar{z}_i(\tau)] \\ &= \frac{\partial}{\partial \tau} \delta[\zeta - \bar{z}_i(\tau)] - \frac{\partial}{\partial \tau} \delta[\zeta - \bar{z}_i(\tau)] \\ &= 0. \end{aligned} \tag{4.4.44}$$

Using the same definition, we can write for (4.4.43)

$$\frac{\partial}{\partial\tau}\langle\gamma_i\delta[\zeta-\bar{z}_i(\tau)]\rangle+\frac{\partial}{\partial\zeta}\langle\beta_i\gamma_i\delta[\zeta-\bar{z}_i(\tau)]\rangle, \tag{4.4.45}$$

which finally allows us to present (4.4.42) in the familiar form of a conservation law i.e.,

$$\frac{\partial}{\partial\tau}\left[\langle\gamma_i\delta[\zeta-\bar{z}_i(\tau)]\rangle+\frac{1}{2\alpha\beta_{\rm en}}|a_+(\zeta,\tau)|^2\right]$$
$$+\frac{\partial}{\partial\zeta}\left[\langle\beta_i\gamma_i\delta[\zeta-\bar{z}_i(\tau)]\rangle+\frac{1}{2\alpha}|a_+(\zeta,\tau)|^2\right]=0\,. \tag{4.4.46}$$

As in the hydrodynamic approximation, we can identify the average energy of an electron and its energy flux ($\langle\gamma_i\delta[\zeta-\bar{z}_i]\rangle$, $\langle\beta_i\gamma_i\delta[\zeta-\bar{z}_i]\rangle$ correspondingly) as well as the normalized electromagnetic power per particle $|a_+|^2/2\alpha$ and the normalized electromagnetic energy per particle $|a_+|^2/2\alpha\beta_{\rm en}$.

4.4.5 Reflections in an Amplifier

There are two processes, which may cause significant time variations in an amplifier: saturation and reflections. Saturation occurs when the initial amplitude of the radiation field is large; we shall not consider here variation in time caused by saturation without reflections involved. An amplifier is designed to operate below the saturation level. However, if in the design process reflections are disregarded, then until the first reflection reaches the input the system will probably operate as designed. But as the first reflection adds to the initial amplitude it may bring the system to saturation.

In this sub-section, we investigate the variation in time caused by reflections in an amplifier. The process is as follows: before the electron beam is injected into the structure, the amplitude of the forward propagating wave is uniform in space and constant in time. Let us denote it by a_0. Ignoring effects associated with the pulse front, we may expect this amplitude to be amplified according to the equations determined previously for an ideal amplifier. The change in this amplitude is propagating with the energy velocity $v_{\rm en}\equiv c\beta_{\rm en}$ so it will take $t_{\rm fb}=d/v_{\rm en}$ to the amplified field to approach the input end. We denote by g_1 the first one-pass gain, which is the ratio of the amplitude at the output of the interaction region to the input point. The contribution of reflections at the input is denoted by b_ν – where ν is the index that numerates the bouncing process therefore, it can be considered as a discrete (normalized) time variable. During the first period, the reflections have no contribution and clearly, $b_0=0$. The amplitude at $\zeta=1$ is $g_1(a_0+b_0)e^{-jkd}$. Without the beam present the amplitude at this point is a_0e^{-jkd}. Therefore, only the difference is reflected backwards. After an additional reflection from the input end, we may write the contribution of the first reflection to the amplitude at the input as $b_1 = \bar{\rho}[g_1(a_0 + b_0) - a_0]$. Before this reflection arrives ($t<2t_{fb}$) the amplitude at the input is constant and its value is a_0. Until the next reflection arrives the amplitude at the input has two contributions which are constant in time, one from the generator and the other from the reflection, i.e., $a_0 + b_1$.

After the ν'th reflection the amplitude at the input is $a_0 + b_\nu$ and at the output end $g_{\nu+1}(a_0 + b_\nu)e^{-jkd}$. As a result, the contribution of the reflections to the input amplitude after $\nu + 1$ steps is

$$b_{\nu+1} = \bar{\rho}[g_{\nu+1}(a_0 + b_\nu) - a_0]. \tag{4.4.47}$$

This expression is a discrete formulation of the reflections equation introduced in (4.4.31); the process is summarized in Fig. 4.15. Note that at the limit of a very long pulse and a linear gain such that $g_\nu = g$ for any ν, we have for $b_\nu = a_0\bar{\rho}(g - 1)/(1 - \bar{\rho}g)$. It implies that the amplitude at the input reads

$$a_0 + b_\nu = a_0 \frac{1 - \bar{\rho}}{1 - \bar{\rho}g}, \tag{4.4.48}$$

exactly as predicted by a linear (steady state) theory.

Let us now consider an example: an amplifier which is 20 cm long, operates at 8.8 GHz, driven by a 800 kV, 1 kA beam. In the absence of reflections the gain of the system is 32 dB for 80 kW at the input. Without loss of generality we chose both the reflection coefficients to be equal $\rho_{in} = \rho_{out} = \rho$. The electrons' pulse is 100 ns long. In Fig. 4.16 we can see how the one-pass gain (squares) and the total gain

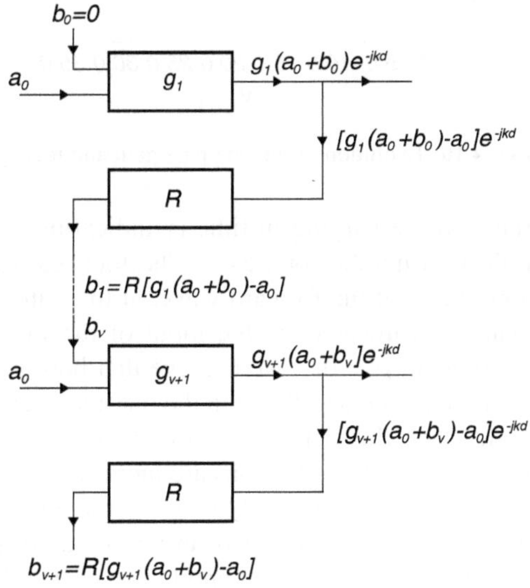

Fig. 4.15 Schematics of the beam–wave interaction in the presence of reflections

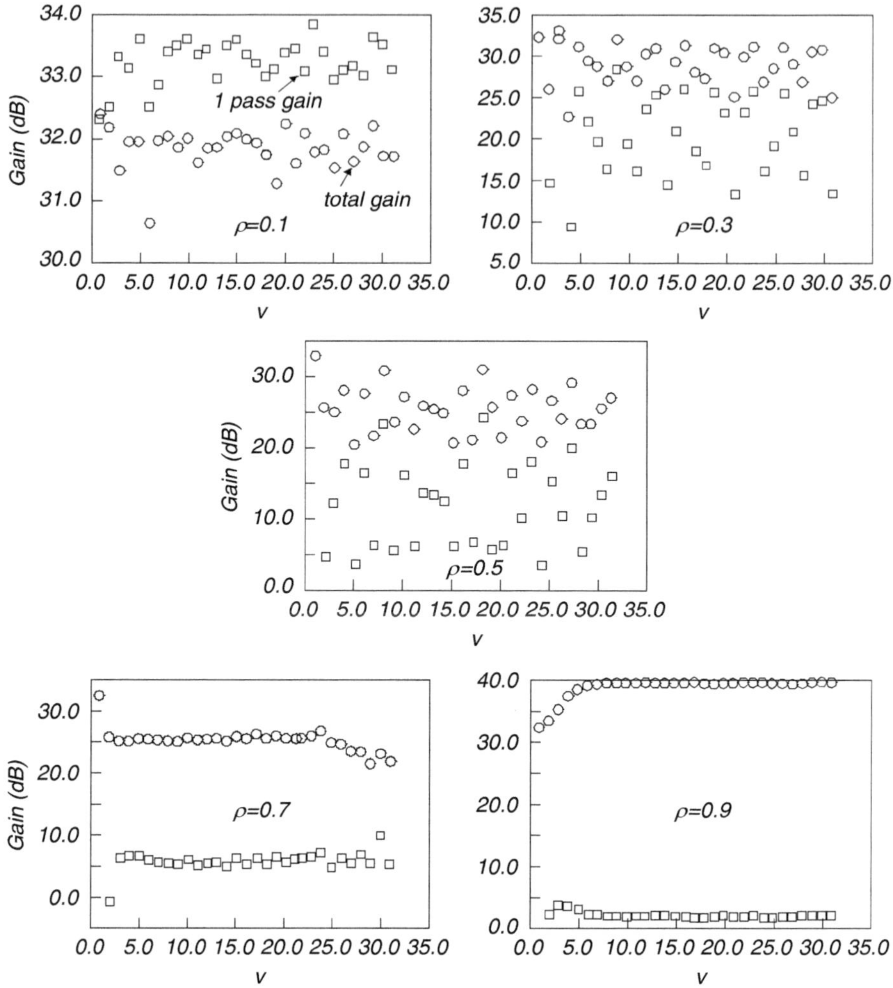

Fig. 4.16 Evolution of the one-pass gain and total gain in the interaction region

(circles) are varying in time (v indicating the index of the reflection i.e., $v = 1$ is reflection number one etc.). The total gain is the ratio between the accumulated amplitude, of the forward wave, at the output and the initial amplitude (before the beam was injected) at the input of the interaction region. For a small reflection coefficient, $\rho = 0.1$, we observe that both gains are relatively stable. The fact that the total gain is smaller than the one-pass gain is not of particular significance at this point since this depends on the phase accumulated by the wave in its round trip. However, as the reflections are increased, the total amplitude at the input increases, saturation is reached and therefore, the one-pass gain is systematically smaller than the total gain. There exists an intermediary point, $\rho = 0.5$, where the system acts very unstable whereas at another, $\rho = 0.7$, the system appears to be very stable in

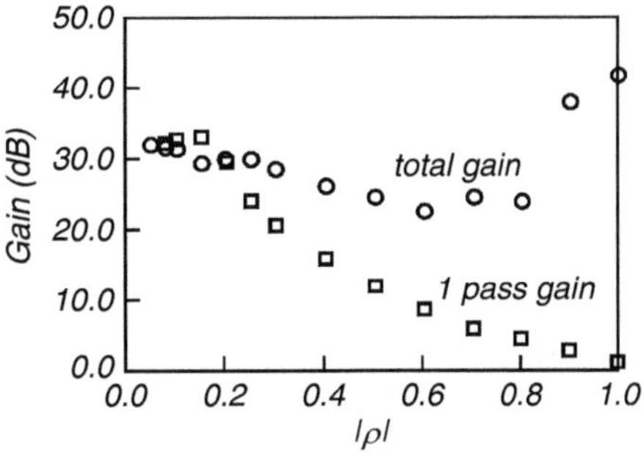

Fig. 4.17 Average one-pass gain and total gain as a function of the reelection coefficient

spite of the fact that the reflection is higher. This is a direct result of the phase dependence of the reflected amplitude.

Ultimately, at high reflection ($\rho = 0.9$) the system reveals an immediate increase of the amplitude in time associated with practically zero one-pass gain, indicating that the system is operating as an oscillator. Note that regardless the actual value of the reflection coefficient, before the first reflection arrives, the one-pass gain and the total gain are equal.

In order to show the general influence of the reflection coefficient on the total gain and the one-pass gain we have averaged out these two quantities over the entire number of reflections for different values of the reflection coefficient. Figure 4.17 illustrates this result. We observe here that the average one-pass gain is monotonically decreasing when increasing the reflection coefficient. The average total gain is stable for small ρ corresponding to a linear regime of operation; it slightly decreases for intermediary reflections – corresponding to saturation and it increases again when the reflection is so high that the system practically operates as an oscillator. Note that in this case the one-pass gain is practically zero.

An additional insight of the physical process can be achieved by examining the spectrum of the signal as illustrated in Figs. 4.18 and 4.19. The power in each frequency component of the signal is normalized to the power in the central frequency (8.8 GHz). When the reflection is low ($\rho < 0.15$) the power in all the other frequencies is 30 dB below the level of the main signal. For $\rho = 0.2$ the eigen-frequencies of the "oscillator" are less than 15 dB below the central frequency. The power in the sidebands is increasing monotonically with the reflection coefficient ρ, and at $\rho = 0.4$ they dominate. It warrant to reiterate that the interaction is in the *linear regime* nevertheless, reflections cause temporal amplitude variations that manifest as *spectrum broadening*. According to the reflection process and the phase accumulation there are longitudinal modes that are "selected" and other are suppressed.

4.4.6 Spatial Variations in an Oscillator

Part of the energy in an oscillator is extracted since intentionally the mirror(s) have a reflection coefficient smaller than unity. As a result, the amount of electromagnetic energy available for interaction with the electrons decreases. Since this

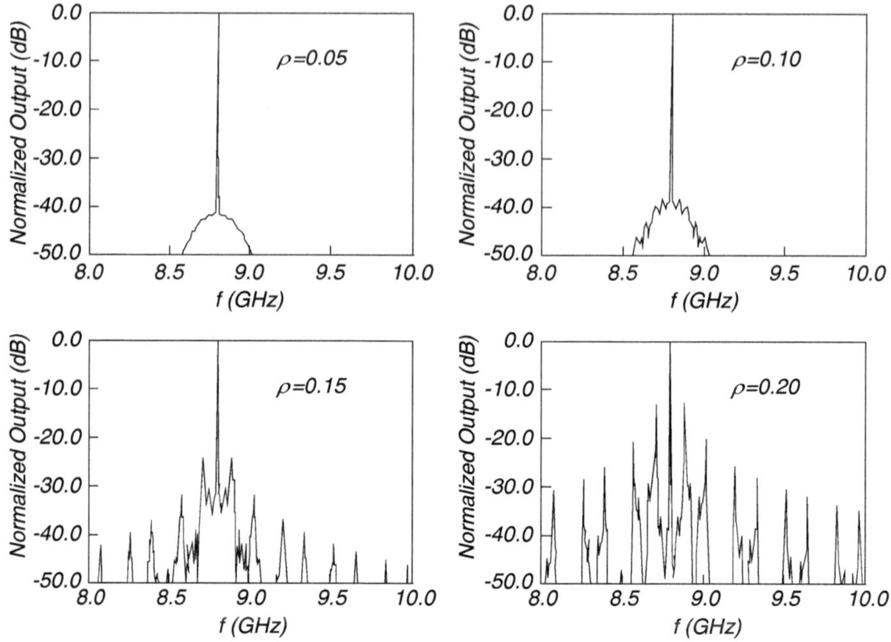

Fig. 4.18 Normalized output power spectrum for various reflection coefficients

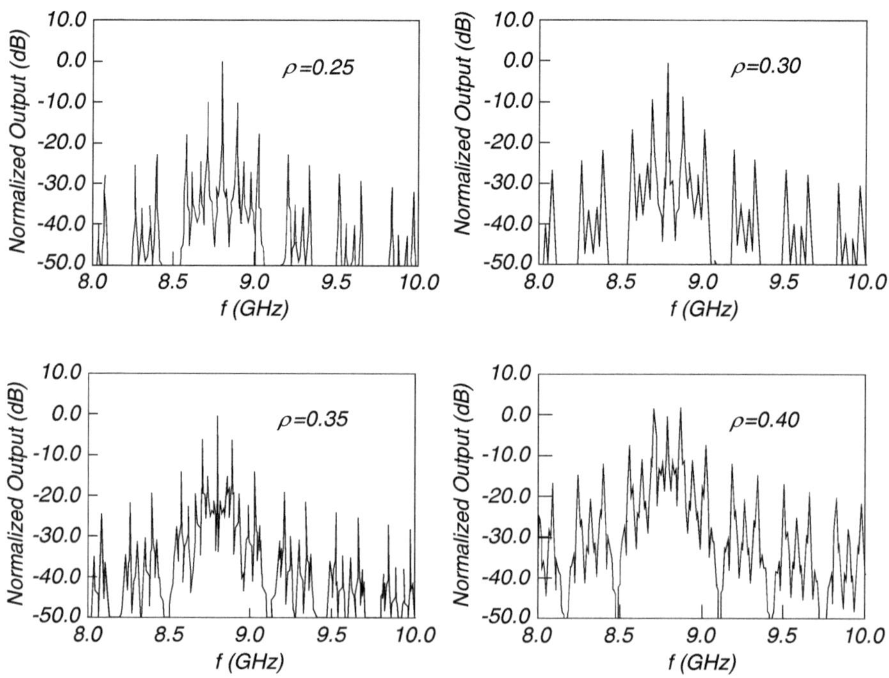

Fig. 4.19 Normalized output power spectrum for various reflection coefficients

power is extracted at the ends, it is revealed as an effective variation of the field amplitude. In order to illustrate the effect of the spatial variation on the operation of an oscillator we start by integrating the equation, which describes the dynamics of the amplitude (4.4.24) over the entire length of the oscillator:

$$\frac{\mathrm{d}}{\mathrm{d}\tau}|a_+(1,\tau)|^2 + \beta_{\mathrm{en}}[|a_+(1,\tau)|^2 - |a_+(0,\tau)|^2] = \alpha\beta_{\mathrm{en}}\langle\beta_i(\tau)\mathrm{e}^{-j\chi_i(\tau)}\rangle. \tag{4.4.49}$$

Next we substitute the reflections equation from (4.4.31). The result is

$$\begin{aligned}&\frac{\mathrm{d}}{\mathrm{d}\tau}|a_+(1,\tau)|^2 + \beta_{\mathrm{en}}[|a_+(1,\tau)|^2 - |a_0(1-\bar{\rho}) + \bar{\rho}a_+(1,\tau - 1/\beta_{\mathrm{en}})|^2]\\ &\quad = \alpha\beta_{\mathrm{en}}\langle\beta_i(\tau)\mathrm{e}^{-j\chi_i(\tau)}\rangle.\end{aligned} \tag{4.4.50}$$

Expanding in Taylor series with respect to $1/\beta_{\mathrm{en}}$ (this normalized characteristic time is assumed to be much shorter than the pulse duration) and assuming that $a_0 = 0$ we finally get

$$\left[\frac{\mathrm{d}}{\mathrm{d}\tau} + \beta_{\mathrm{en}}\frac{1-|\bar{\rho}|^2}{1+|\bar{\rho}|^2}\right]a_+(1,\tau) = \frac{\alpha\beta_{\mathrm{en}}}{1+|\bar{\rho}|^2}\langle\beta_i(\tau)\mathrm{e}^{-j\chi_i(\tau)}\rangle. \tag{4.4.51}$$

This expression replaces (4.4.37) in the description of a non-ideal oscillator. Note that the second term on the left-hand side of (4.4.51) represents the "radiation" loss due to the finite transmission from both ends of the oscillator. This becomes even more evident from the expression for general energy conservation

$$\frac{\mathrm{d}}{\mathrm{d}\tau}\left[\langle\gamma_i\rangle + \frac{1}{2\alpha\beta_{\mathrm{en}}}(1+|\bar{\rho}|^2)|a_+(1,\tau)|^2\right] = -\frac{1}{\alpha}(1-|\bar{\rho}|^2)|a_+(1,\tau)|^2 \tag{4.4.52}$$

as revealed by the right-hand side term.

The only source of energy in the oscillator is the beam and when the mirrors are ideal, all the kinetic energy converted in radiation power is confined to the volume of the oscillator. If part of this energy is allowed to flow out, then self-sustained oscillation is possible only if the current injected is above a threshold value that depends on the reflection coefficients. In order to determine the threshold current we first have to realize that the radiation loss is associated with an exponential decay with a coefficient [see (4.4.51)] $\beta_{\mathrm{en}}(1-|\bar{\rho}|^2)/(1+|\bar{\rho}|^2)$. For self-sustained oscillation this decay has to be compensated by the exponential increase due to the interaction – as determined in (4.4.40), i.e.,

$$\beta_{\mathrm{en}}\frac{1-|\bar{\rho}|^2}{1+|\bar{\rho}|^2} < \bar{\omega} = \frac{\sqrt{3}}{2}\left[\frac{1}{2}\alpha\beta_{\mathrm{en}}K\left\langle\frac{\beta_i}{\gamma_i^3}\right\rangle\right]^{1/3}. \tag{4.4.53}$$

Therefore, the condition for self-sustained oscillation can be formulated as

$$I > I_{\text{th}} \equiv \frac{16}{3^{3/2}} \frac{mc^2}{eZ_{\text{int}}\Omega} \beta_{\text{en}}{}^2 \beta^2 \gamma^3 \frac{\pi R^2}{d^2} \left[\frac{1 - |\bar{\rho}|^2}{1 + |\bar{\rho}|^2} \right]^3 . \tag{4.4.54}$$

Note that in case of "radiation loss" the threshold current is quadratic in the energy velocity, therefore the lower β_{en}, the lower the current required for the system to oscillate.

Energy extracted from both ends is one mechanism responsible for spatial variation but it is not the only one. Another mechanism is associated with the fact that electrons entering the oscillator are un-bunched and their build-up into bunches is not "immediate" in space but it takes some portion of the interaction length. After this transient region, there will be no variations in space, provided that the system does not reach saturation – which will be not considered here. In order to illustrate this effect, we examine the same system as in the case of the amplifier; in this case however the input power P_{in} is zero, the pulse length is 50 ns instead of 100 ns in the amplifier and the "mirrors" at both ends have a reflection coefficient $\rho = 0.9$. The entire pulse was assumed to consist of $35,000$ macro-particles, 512 of those being at any time in the oscillator. In Fig. 4.20 we illustrate the phase space of these electrons which are in the interaction region. In the first 20% of the pulse duration there is not sufficient electromagnetic field built in the oscillator in order to affect significantly the electron's distribution (although there is a small increase in the momentum spread). After 40% of the pulse has passed, we clearly see the spatial transient in the interaction region. At this point in time, the constant amplitude regime is achieved after about 20% of the total interaction length. The normalized momentum spread which at the beginning is less than 0.06 is now larger than 0.35. Later the bunches continue to grow – the momentum spread is further increased approaching 3 at the end of the electrons' pulse.

Before we conclude, we wish to emphasize the difference between the two transients which occur in an oscillator. One is the temporal lethargy which we have discussed already and it is indicative of the time it takes for the exponential growth of the electromagnetic energy to become dominant. However the transient presented in Fig. 4.20 is a *spatial transient in an oscillator*. It is not a result of the three eigen-modes mentioned above since in an (ideal) oscillator these modes have a constant amplitude in space. As we mentioned above, this is a result of the finite length it takes the radiation field to bunch the "fresh" electrons.

The last two sub-sections indicate that the convenient picture of a traveling-wave tube operating either as an amplifier or as an oscillator is too simplistic. In fact we have shown that these two regimes are the extreme cases and any system operates somewhere in between corresponding to the reflection coefficients at both ends, the phase accumulated in one round trip and the gain. Furthermore, in the absence of reflections and saturation in an amplifier it is justified to assume that the amplitude of the electromagnetic wave remains constant in time. However, even a low reflection coefficient may affect the performance of an amplifier if the gain is

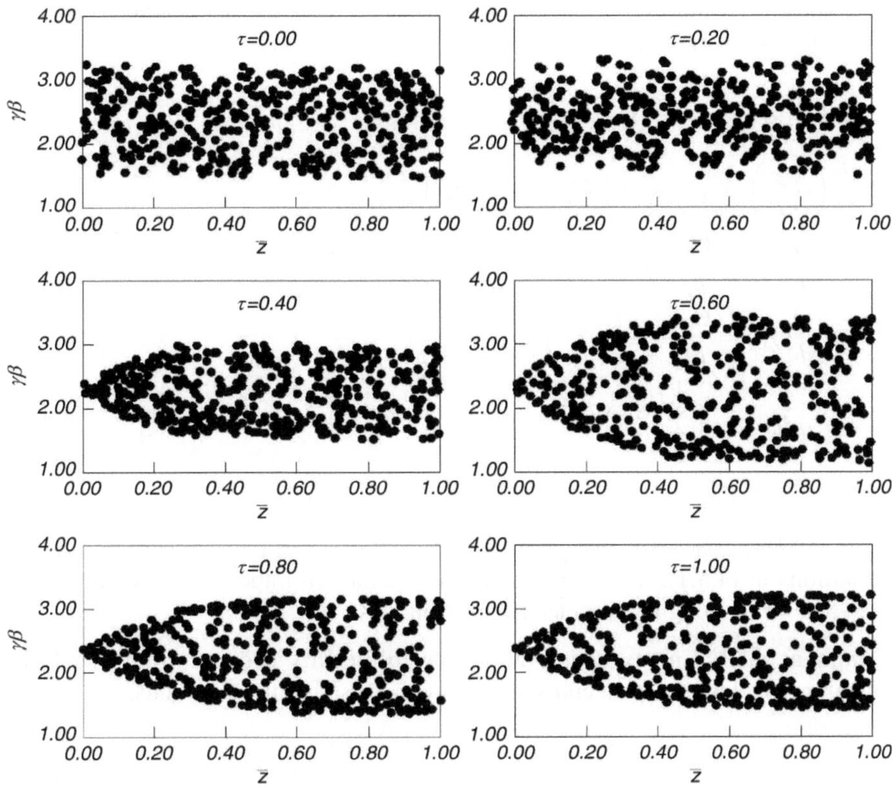

Fig. 4.20 Phase-space distribution in an oscillator at different instants of time

high enough. When reflections were included in the analysis, the amplitude was shown to vary in time. The resulting spectrum revealed peaks at other frequencies. These peaks are symmetric to the initial frequency and their separation is determined by the feedback time, namely the energy velocity.

4.5 Parasitic Hybrid Mode

Similar to the occurrence of hybrid modes in the case of dielectric waveguide due to the asymmetry of the modes, so can hybrids of TM and TE develop in a partially loaded waveguide or in a disk loaded waveguide in spite of the fact, that the structure itself, is azimuthally symmetric. In this section we postulate the existence of such a mode and assume that the beam-line intersects the dispersion curve of both TM_{01} and HEM_{11} modes at (ω_1, k_1) and (ω_2, k_2) respectively. Both modes have amplitudes $E_{z,1}$ and $E_{z,2}$ and correspondingly, the interaction impedance is

$Z_{\text{int},1}$ and $Z_{\text{int},2}$. Subject to these assumptions and ignoring space-charge effects, the dynamics of the system is described by

$$\begin{aligned}
\frac{d}{d\xi}\left(\frac{a_1}{\sqrt{\alpha_1}}\right) &= \sqrt{\alpha_1}\left\langle \exp(-j\chi_{i,1})I_0(\bar{\Gamma}_1\bar{r}_i)\right\rangle_i,\\
\frac{d}{d\xi}\left(\frac{a_2}{\sqrt{\alpha_2}}\right) &= \sqrt{\alpha_2}\left\langle \exp(-j\chi_{i,2}-j\phi_i)I_1(\bar{\Gamma}_2\bar{r}_i)\right\rangle,\\
\frac{d}{d\xi}\chi_{i,1} &= \frac{\Omega_1}{\beta_i} - K_1,\\
\frac{d}{d\xi}\chi_{i,2} &= \frac{\Omega_2}{\beta_i} - K_2,\\
\frac{d}{d\xi}\gamma_i &= -\frac{1}{2}\left[a_1\exp(j\chi_{i,1})I_0(\bar{\Gamma}_1\bar{r}_i) + a_2\exp(j\chi_{i,2}+j\phi_i)I_1(\bar{\Gamma}_2\bar{r}_i) + c.c.\right].
\end{aligned} \tag{4.5.1}$$

Here $\zeta = z/d$, $\bar{\Gamma}_\nu = R_{\text{int}}\sqrt{k_\nu^2 - \omega^2/c^2}$, $a_\nu = eE_{z,\nu}d/mc^2$ the interaction coupling is $\alpha_\nu = \left(IZ_{\text{int},\nu}/mc^2\right)\left(d^2/\pi R_{\text{int}}^2\right)$, $\bar{r}_i = r_i/R_{\text{int}}$ with $\nu = 1,2$. According to the definitions in (4.5.1), it is evident that the interaction impedance is defined with the value of the longitudinal electric field on axis.

Comment 4.10. Energy conservation may be readily deduced from (4.5.1) by averaging the last equation and substituting the first and the third thus,

$$\frac{d}{d\xi}\left[\langle\gamma_i\rangle + \frac{|a_1|^2}{2\alpha_1} + \frac{|a_2|^2}{2\alpha_2}\right] = 0. \tag{4.5.2}$$

In fact, a similar set of equations may also describe the interaction of two TM modes with a beam of electrons.

Comment 4.11. In the framework of this section, we focus only on the longitudinal motion. The effect of both the longitudinal as well as the radial motion was considered in detail by Banna et al. (2000a, b); the details are left as an exercise to the reader (Exercise 4.6). At this point we wish to emphasize that the detrimental aspect of the excitation of HEM_{11}. Specifically, the fact that it has a non-zero transverse magnetic field on axis and this causes deflection of the beam. In the accelerators physics community this is effect is known as beam break-up (BBU) as described by Helm and Loew (1970).

As in Sect. 4.3.1 the *spatial growth* in the system may be readily deduced by evaluating the third derivative of the amplitude. The result being

$$\begin{aligned}
\frac{d^3a_1}{d\xi^3} + \frac{j}{2}\alpha_1\Omega_1(p_1a_1 + Ua_2) &\simeq -\alpha_1\left\langle\left(\frac{\Omega_1}{\beta_i} - K_1\right)^2 e^{-j\chi_{i,1}}\mathrm{I}_0(\bar{\Gamma}_1\bar{r}_i)\right\rangle,\\
\frac{d^3a_2}{d\xi^3} + \frac{j}{2}\alpha_2\Omega_2[p_2a_2 + U^*a_1] &\simeq -\alpha_2\left\langle\left(\frac{\Omega_2}{\beta_i} - K_2\right)^2 e^{-j\chi_{i,2}+j\phi_i}\mathrm{I}_1(\bar{\Gamma}_2\bar{r}_i)\right\rangle.
\end{aligned} \tag{4.5.3}$$

In this expression $U \equiv \left\langle e^{-j(\chi_{i,1}-\chi_{i,2}+\phi_i)}(\gamma_i\beta_i)^{-3}\mathrm{I}_0(\bar{\Gamma}_1\bar{r}_i)\mathrm{I}_1(\bar{\Gamma}_2\bar{r}_i)\right\rangle$, $p_1 \equiv \left\langle \mathrm{I}_0^2(\bar{\Gamma}_1\bar{r}_i)(\gamma_i\beta_i)^{-3}\right\rangle$, $p_2 \equiv \left\langle \mathrm{I}_1^2(\bar{\Gamma}_2\bar{r}_i)(\gamma_i\beta_i)^{-3}\right\rangle$ and the terms where the phase varies rapidly were neglected. Ignoring the two "noise" terms in the right hand side of both equations, we may calculate the eigen-wave number of the coupled system, by assuming solutions of the form $a_\nu = \bar{a}_\nu \exp(-js\zeta)$, hence

$$\begin{pmatrix} s^3 + \frac{1}{2}\alpha_1\Omega_1 p_1 & \frac{1}{2}\alpha_1\Omega_1 U \\ \frac{1}{2}\alpha_2\Omega_2 U^* & s^3 + \frac{1}{2}\alpha_2\Omega_2 p_2 \end{pmatrix} \cdot \begin{pmatrix} \bar{a}_1 \\ \bar{a}_2 \end{pmatrix} = 0. \tag{4.5.4}$$

As clearly revealed by this matrix, the term U represents the coupling between the cold-structure eigen-modes (TM_{01} & HEM_{11}). From its definition it is realized that U is determined by the correlation between the two phases $(\chi_{i,1}, \chi_{i,2})$ and also by the correlation of the azimuthal, radial and momentum distribution of the electrons. When the coupling between the modes is zero, each one of the modes (TM_{01} and HEM_{11}) develops independently according to $s^3 + S_1^3 = 0$ and $s^3 + S_2^3 = 0$ respectively where $S_\mu^3 \equiv \frac{1}{2}\alpha_\mu\Omega_\mu p_\mu$. The coupling between the TM_{01} and HEM_{11} is controlled by a single parameter

$$\bar{u} \equiv \sqrt{\frac{UU^*}{p_1 p_2}} = \frac{\left|\left\langle e^{-j(\chi_{i,1}-\chi_{i,2}+\phi_i)}(\gamma_i\beta_i)^{-3}\mathrm{I}_0(\bar{\Gamma}_1\bar{r}_i)\mathrm{I}_1(\bar{\Gamma}_2\bar{r}_i)\right\rangle\right|}{\sqrt{\left\langle \mathrm{I}_0^2(\bar{\Gamma}_1\bar{r}_i)(\gamma_i\beta_i)^{-3}\right\rangle\left\langle \mathrm{I}_1^2(\bar{\Gamma}_2\bar{r}_i)(\gamma_i\beta_i)^{-3}\right\rangle}}. \tag{4.5.5}$$

Clearly, the solution of the dispersion relation of the *coupled* system can be determined from $s^3 + S_\pm^3 = 0$ where

$$S_\pm^3 = -\frac{1}{2}(S_1^3 + S_2^3) \pm \frac{1}{2}\sqrt{(S_1^3 - S_2^3)^2 + 4S_1^3 S_2^3 \bar{u}^2}\,. \tag{4.5.6}$$

In these expressions S_+ corresponds to the HEM_{11}-like solution since at the limit $\bar{u} = 0$, $S_+ = S_2$ whereas S_- corresponds to the TM_{01}-like solution.

The solid-lines in Figure 4.21 illustrate the variation of the spatial growth per cell, $g_\pm \equiv (L/d)20\log\{\exp[\mathrm{Im}(\sqrt{3}S_\pm/2)]\}$, as a function of the parameter $\bar{u}$. The parameters in this calculation are $I = 300\,\mathrm{A}$, $V = 850\,\mathrm{kV}$, $R_{\mathrm{int}} = 3.5\,\mathrm{mm}$, $R_{\mathrm{ext}} = 5\,\mathrm{mm}$, $R_{\mathrm{beam}} = 2\,\mathrm{mm}$, $L = 1.98\,\mathrm{mm}$, $f_{\mathrm{TM}_{01}} = 35\,\mathrm{GHz}$, $f_{\mathrm{HEM}_{11}} = 38.63\,\mathrm{GHz}$, $Z_{\mathrm{int}}^{\mathrm{TM}_{01}} = 374\Omega$, $Z_{\mathrm{int}}^{\mathrm{HEM}_{11}} = 1.61k\Omega$, and it was assumed that the electrons have a vanishingly small velocity spread. When the modes are completely correlated ($\bar{u} = 1$) the spatial growth of the HEM_{11}-like mode is zero whereas the TM_{01}-like is slightly larger than the case when there is no coupling ($\bar{u} = 0$). Although, the HEM_{11}-like wave becomes unimportant, we have to remember that the TM_{01}-like mode is not a *pure* TM_{01} mode but rather a linear superposition of TM_{01} and HEM_{11} therefore, the impact of the HEM's components are destructive since they have the same spatial growth as the pure TM_{01} as they share the same eigen wave-number.

At the best, the HEM_{11} may prevent the system from reaching high efficiency and in the worst case it deflects the beam to the wall. In order to suppress the asymmetric mode, *selective* damping is required. This is to say that the damping mechanism is transparent to TM_{01} mode but it suppresses the asymmetric mode. In order to envision the impact of such a mechanism on the interaction process we may represent this mechanism by a damping parameter, $\bar{\sigma}$, that in the absence of the beam, causes a decay corresponding to $\exp(-\zeta/\bar{\sigma})$ of the asymmetric mode only. Consequently, in the amplitude equation of the HEM_{11} mode we may replace $\frac{d\bar{a}_2}{d\xi} \rightarrow \frac{d\bar{a}_2}{d\xi} + \frac{1}{\bar{\sigma}}\bar{a}_2$; following the same approach as before we find in stead of (4.5.4)

$$\begin{pmatrix} s^3 + \frac{1}{2}\alpha_1\Omega_1 p_1 & \frac{1}{2}\alpha_1\Omega_1 U \\ \frac{1}{2}\alpha_2\Omega_2 U^* & s^3 + \frac{j}{\bar{\sigma}}s^2 + \frac{1}{2}\alpha_2\Omega_2 p_2 \end{pmatrix} \begin{pmatrix} \bar{a}_1 \\ \bar{a}_2 \end{pmatrix} = 0. \tag{4.5.7}$$

The dashed-lines in Fig. 4.21 illustrate the spatial growth per cell (in dB) in the case of damping the HEM_{11} mode ($\bar{\sigma} \approx 0.05$ corresponding to 1.7 dB per cell in the absence of the beam). Two facts are evident: first, the "HEM_{11}-like" mode is substantially suppressed though not as one would expect from the cold attenuation and second, the "TM_{01}-like" mode is almost independent of the $\bar{u}$ indicating that the "TM_{01}-like" is close to the pure TM_{01} mode.

Figure 4.22 illustrates one possible implementation of such a selective damping that uses a series of choke loaded cavities (Shintake 1992). This structure has a high quality factor Q at the frequency where the TM_{01} operates and low quality factor otherwise. The choke ensures that, in our case, the 35 GHz wave experiences an "ideal" periodic structure whereas any wave of different frequency in its narrow vicinity, is absorbed in the radial transmission line. In the optical regime this concept is implemented in the so called photonic band-gap structures whereby an ensemble of obstacles ensure the propagation of a specific mode and virtually all the other modes are propagate radially.

A detailed discussion on beam breakup (BBU) is presented in Sect. (8.1.5).

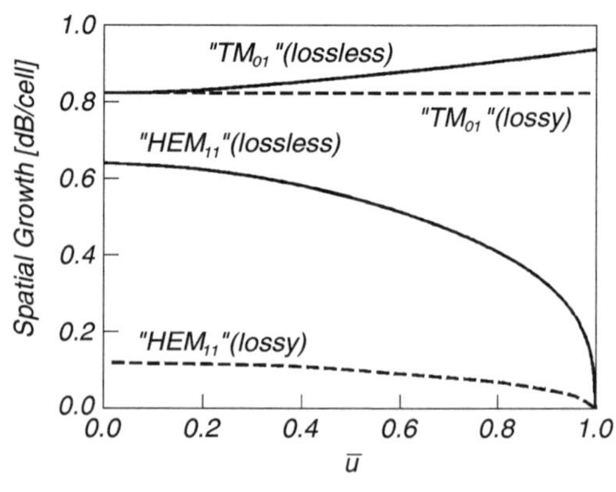

Fig. 4.21 Spatial growth rate per cell oh TM and HEM mode with and without damping of the latter

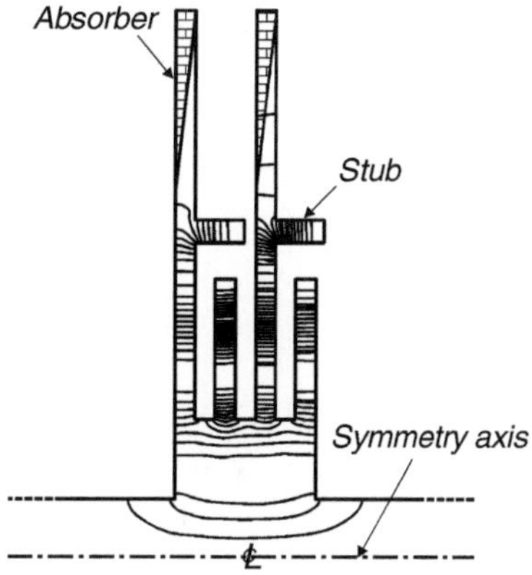

Fig. 4.22 One possible implementation of selective damping of an HEM mode – Shintake cavity

Exercises

4.1 Show that the modes in a partially dielectric loaded waveguide form a (complete) orthogonal set of functions.

4.2 Calculate the interaction impedance, as defined in (2.3.29), in a partially loaded waveguide for two cases (1) pencil and (2) annular beam. Express the coupling coefficient in terms of this impedance.

4.3 Develop the amplitude equation (as in Sect. 4.3) using in one case, the non-homogeneous wave equation for the magnetic vector potential and in the second case, wave equation for E_z.

4.4 Formulate the set of equations, which determine the beam–wave interaction in the framework of the macro-particle approach of an annular beam in a slow-wave structure – Sect. 4.3.1.

4.5 Formulate, based on Sect. 4.4, the set of equations, which describe the interaction in a backward-wave oscillator.

4.6 Formulate based on Sect. 4.5, the set of equations which describe the 3D motion of particles and 1D dynamics of TM_{01} as well as the two adjacent HEM_{11} modes. Hint: use Banna et al. (2000a, b)

Chapter 5
Periodic Structures

One of the conditions for distributed beam-wave interaction to occur is phase velocity smaller than c. There are two relatively simple ways to slow down the phase velocity: (1) load a waveguide with dielectric material or (2) load a waveguide with periodic metallic or dielectric obstacles. The periodic metallic structure is usually the preferred solution in *microwave* devices since it has relatively low loss, it may sustain relatively high gradients and it may drain any stray electrons. Dielectric structures are virtually the only solution in the optical regime since metals have much higher loss. In addition, breakdown is not the major impediment but rather non-linear effects.

A periodic geometry may be conceived as a set of obstacles delaying the propagation of the wave due to the multi-reflection process and as a result, an infinite spectrum of spatial harmonics develops. A few of these harmonics may propagate with a phase velocity larger or equal to c but the absolute majority has a slower phase velocity.

This chapter presents various characteristics of periodic structures with emphasis on these aspects relevant to interaction with electrons. In particular, the interaction impedance, Z_{int}, and the interaction dielectric coefficient, ε_{int}, are calculated and analyzed since in the previous chapter we have shown that they play an important role in the collective beam-wave interaction. We assume that only a single *mode* participates in the interaction and from the infinite spectrum of spatial harmonics of this single mode, only one *harmonic* interacts directly with the electrons. In the accelerators context, we evaluate in the second part of the chapter the wake generated by one bunch or a train of bunches in various periodic structures. Since our treatment of periodic structures is limited to the objectives of the above, we refer the reader to Elachi (1976) for a broader review on periodic structures. Tutorial discussion of this subject can be found in a book by Brillouin (1953) and aspects associated with solid state physics are presented by Kittel (1956) or Ashcroft and Mermin (1976).

In the first section we present the basic theorem of periodic structures namely, Floquet's theorem. This is followed by an investigation of closed periodic

L. Schächter, *Beam-Wave Interaction in Periodic and Quasi-Periodic Structures*, Particle Acceleration and Detection, DOI 10.1007/978-3-642-19848-9_5,

structures in Sect. 5.2 and open structures in the third. Smith-Purcell effect is considered as a particular case of a Green's function calculation for an open structure and a simple scattering problem is also considered. A simple dielectric configuration is analyzed as an introduction to optical accelerators to be discussed in Chap. 8. The chapter concludes with transient phenomena in periodic structure, which is of importance in accelerators where wake fields left behind one bunch, may affect trailing bunches.

5.1 The Floquet Theorem

A periodic function, $f(z)$, is a function whose value at a given point z is equal to its value at a point $z+L$ i.e.,

$$f(z) = f(z+L), \tag{5.1.1}$$

where L is the periodicity of the function. Any periodic function can be represented as a series of trigonometric functions $\exp(-j2\pi nz/L)$ and since this is an orthogonal and complete set of functions, it implies

$$f(z) = \sum_{n=-\infty}^{\infty} f_n \exp\left(-j2\pi n\frac{z}{L}\right). \tag{5.1.2}$$

The amplitudes f_n are determined by the value of the function $f(z)$ in a *single* cell. Specifically, we multiply (5.1.2) by $\exp(+j2\pi nz/L)$ and integrate over one cell i.e.,

$$\int_0^L dz f(z) \exp\left(j2\pi m\frac{z}{L}\right) = \int_0^L dz \exp\left(j2\pi m\frac{z}{L}\right) \sum_{n=-\infty}^{\infty} f_n \exp\left(-j2\pi n\frac{z}{L}\right). \tag{5.1.3}$$

Using the orthogonality of the trigonometric function, we conclude that

$$f_m = \frac{1}{L}\int_0^L dz f(z) \exp\left(j2\pi m\frac{z}{L}\right). \tag{5.1.4}$$

This presentation is called the Fourier series representation and it is valid for a *static* phenomenon in the sense that the value of $f(z)$ at the same relative location in two different cells is identical. For describing a *dynamic* system, the function $f(z)$ has to satisfy

$$f(z) = \xi f(z+L), \tag{5.1.5}$$

which means that the value of the function is proportional to the value of the function in the adjacent cell up to a constant, ξ, whose absolute value has to be unity otherwise at $z \to \pm\infty$ the function diverges or is zero as can be concluded from

$$f(z) = \xi^n f(z + nL), \tag{5.1.6}$$

where n is an arbitrary integer. Consequently, the coefficient ξ can be represented as a phase term of the form $\xi = \exp(j\psi)$ hence

$$f(z) = e^{j\psi} f(z + L); \tag{5.1.7}$$

ψ is also referred to as the *phase advance per cell.* Without loss of generality one can redefine this phase to read $\psi = kL$. Since *a-priori* we do not know ψ, this definition does not change the information available. Nonetheless based on the Fourier series in (5.1.2) we can generalize the representation of a dynamic function in a periodic structure to

$$f(z) = \sum_{n=-\infty}^{\infty} f_n \exp\left(-j2\pi n\frac{z}{L} - jkz\right), \tag{5.1.8}$$

and realize that it satisfies

$$f(z) = e^{jkL} f(z + L), \tag{5.1.9}$$

which is identical with the expression in (5.1.7). The last two expressions are different representations of the so-called *Floquet's Theorem.* Subsequently we shall mainly use the form presented in (5.1.8), however in order to illustrate the use of Floquet's theorem in its latter representation, we investigate next the propagation of a TM wave in a periodically loaded waveguide.

Consider a waveguide of radius R, which is loaded with dielectric layers: a representative cell ($0 \le z \le L$) consists of a region, $0 \le z \le g$, filled with a dielectric, ε_r, and the remainder is vacuum – see Fig. 5.1. Our goal is to determine the dispersion relation of this structure and for this purpose, we write the solution of the magnetic vector potential and electromagnetic field (steady state) in the dielectric ($0 \le z \le g$):

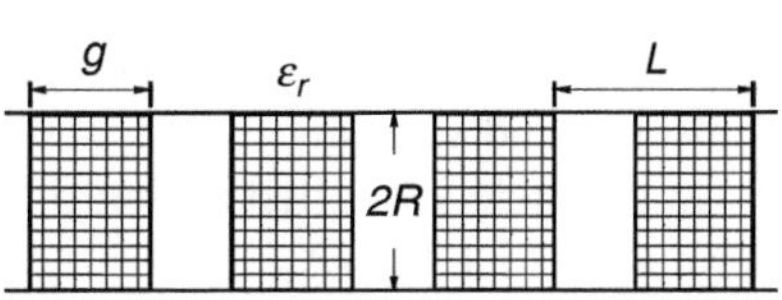

Fig. 5.1 Periodically loaded waveguide

$$
\begin{aligned}
A_z(r,z) &= \sum_{s=1}^{\infty} \mathrm{J}_0\left(p_s \frac{r}{R}\right)\left[A_s \mathrm{e}^{-\Gamma_{d,s}z} + B_s \mathrm{e}^{+\Gamma_{d,s}z}\right],\\
E_r(r,z) &= \frac{c^2}{j\omega\varepsilon_\mathrm{r}} \sum_{s=1}^{\infty} \frac{p_s}{R}\Gamma_{d,s}\mathrm{J}_1\left(p_s \frac{r}{R}\right)\left[A_s \mathrm{e}^{-\Gamma_{d,s}z} - B_s \mathrm{e}^{+\Gamma_{d,s}z}\right],\\
E_z(r,z) &= -\frac{c^2}{j\omega\varepsilon_\mathrm{r}} \sum_{s=1}^{\infty} \Gamma_{d,s}^2\mathrm{J}_0\left(p_s \frac{r}{R}\right)\left[A_s \mathrm{e}^{-\Gamma_{d,s}z} - B_s \mathrm{e}^{+\Gamma_{d,s}z}\right],\\
H_\phi(r,z) &= \frac{1}{\mu_0} \sum_{s=1}^{\infty} \frac{p_s}{R}\mathrm{J}_1\left(p_s \frac{r}{R}\right)\left[A_s \mathrm{e}^{-\Gamma_{d,s}z} + B_s \mathrm{e}^{+\Gamma_{d,s}z}\right],
\end{aligned}
\tag{5.1.10}
$$

where $\Gamma_{d,s}^2 = (p_s/R)^2 - \varepsilon_\mathrm{r}(\omega/c)^2$. In a similar way, we have in the vacuum $(g < z < L)$:

$$
\begin{aligned}
A_z(r,z) &= \sum_{s=1}^{\infty} \mathrm{J}_0\left(p_s \frac{r}{R}\right)\left[C_s \mathrm{e}^{-\Gamma_s(z-g)} + D_s \mathrm{e}^{+\Gamma_s(z-g)}\right],\\
E_r(r,z) &= \frac{c^2}{j\omega} \sum_{s=1}^{\infty} \frac{p_s}{R}\Gamma_s\mathrm{J}_1\left(p_s \frac{r}{R}\right)\left[C_s \mathrm{e}^{-\Gamma_s(z-g)} - D_s \mathrm{e}^{+\Gamma_s(z-g)}\right],\\
E_z(r,z) &= -\frac{c^2}{j\omega} \sum_{s=1}^{\infty} \Gamma_s^2\mathrm{J}_0\left(p_s \frac{r}{R}\right)\left[C_s \mathrm{e}^{-\Gamma_s(z-g)} - D_s \mathrm{e}^{+\Gamma_s(z-g)}\right],\\
H_\phi(r,z) &= \frac{1}{\mu_0} \sum_{s=1}^{\infty} \frac{p_s}{R}\mathrm{J}_1\left(p_s \frac{r}{R}\right)\left[C_s \mathrm{e}^{-\Gamma_s(z-g)} + D_s \mathrm{e}^{+\Gamma_s(z-g)}\right],
\end{aligned}
\tag{5.1.11}
$$

with $\Gamma_s^2 = (p_s/R)^2 - (\omega/c)^2$. At this point, we limit the discussion to the first mode TM_{01}, thus the continuity of the radial electric field at $z = g$ implies

$$
\frac{1}{\varepsilon_\mathrm{r}}\Gamma_{d,1}\left[A_1 \mathrm{e}^{-\Gamma_{d,1}g} - B_1 \mathrm{e}^{+\Gamma_{d,1}g}\right] = \Gamma_1[C_1 - D_1],
\tag{5.1.12}
$$

and in a similar way, the continuity of the azimuthal magnetic field reads

$$
A_1 \mathrm{e}^{-\Gamma_{d,1}g} + B_1 \mathrm{e}^{+\Gamma_{d,1}g} = C_1 + D_1.
\tag{5.1.13}
$$

Last two equations express the relation between the amplitudes of the field in the dielectric and vacuum.

In the dielectric filled region of next cell ($L \le z \le L + g$) the field has a similar form as in (5.1.10) i.e.,

$$
\begin{aligned}
A_z(r,z) &= \sum_{s=1}^{\infty} \mathrm{J}_0\left(p_s \frac{r}{R}\right)\left[A_s' \mathrm{e}^{-\Gamma_{d,s}(z-L)} + B_s' \mathrm{e}^{+\Gamma_{d,s}(z-L)}\right],\\
E_r(r,z) &= \frac{c^2}{j\omega\varepsilon_r} \sum_{s=1}^{\infty} \frac{p_s}{R} \Gamma_{d,s} \mathrm{J}_1\left(p_s \frac{r}{R}\right)\left[A_s' \mathrm{e}^{-\Gamma_{d,s}(z-L)} - B_s' \mathrm{e}^{+\Gamma_{d,s}(z-L)}\right],\\
E_z(r,z) &= -\frac{c^2}{j\omega\varepsilon_r} \sum_{s=1}^{\infty} \Gamma_{d,s}^2 \mathrm{J}_0\left(p_s \frac{r}{R}\right)\left[A_s' \mathrm{e}^{-\Gamma_{d,s}(z-L)} - B_s' \mathrm{e}^{+\Gamma_{d,s}(z-L)}\right],\\
H_\phi(r,z) &= \frac{1}{\mu_0} \sum_{s=1}^{\infty} \frac{p_s}{R} \mathrm{J}_1\left(p_s \frac{r}{R}\right)\left[A_s' \mathrm{e}^{-\Gamma_{d,s}(z-L)} + B_s' \mathrm{e}^{+\Gamma_{d,s}(z-L)}\right].
\end{aligned}
\tag{5.1.14}
$$

The prime indicates that this amplitude represents the solution in the adjacent cell. Accordingly, the boundary conditions at $z = L$ read

$$
\frac{1}{\varepsilon_r}\Gamma_{d,1}\left[A_1' - B_1'\right] = \Gamma_1\left[C_1 \mathrm{e}^{-\Gamma_1(L-g)} - D_1 \mathrm{e}^{\Gamma_1(L-g)}\right], \tag{5.1.15}
$$

and

$$
A_1' + B_1' = C_1 \mathrm{e}^{-\Gamma_1(L-g)} + D_1 \mathrm{e}^{\Gamma_1(L-g)}. \tag{5.1.16}
$$

The relation between the amplitudes of the wave in the second cell ($L < z < 2L$) and the first cell can be represented in a matrix form

$$
\mathbf{a}' = \mathbf{T}\mathbf{a}, \tag{5.1.17}
$$

where the components of $\mathbf{a}'$ are A_1' and B_1' and similarly, the components of $\mathbf{a}$ are A_1 and B_1. According to Floquet's theorem (5.1.9) the two vectors are expected to be related by

$$
\mathbf{a}' = \mathrm{e}^{-jkL}\mathbf{a}, \tag{5.1.18}
$$

thus e^{-jkL} represents the eigen-values of the single cell transmission matrix $\mathbf{T}$:

$$
|\mathbf{T} - \mathrm{e}^{-jkL}\mathbf{I}| = 0. \tag{5.1.19}
$$

Explicitly this reads

$$
\mathrm{e}^{-2jkL} - \mathrm{e}^{-jkL}(T_{11} + T_{22}) + T_{11}T_{22} - T_{12}T_{21} = 0. \tag{5.1.20}
$$

For a *passive system*, the determinant of the matrix $\mathbf{T}$ is unity, thus

$$
\mathrm{e}^{-2jkL} - \mathrm{e}^{-jkL}(T_{11} + T_{22}) + 1 = 0. \tag{5.1.21}
$$

The fact that the last term in this equation is unity indicates that if k is a solution of (5.1.21) $-k$ is also a solution. Consequently, we can write

$$\cos(kL) = \frac{1}{2}(T_{11} + T_{22}). \tag{5.1.22}$$

Note that this is an explicit expression for k as a function of the frequency and the other geometric parameters. In principle, there are ranges of parameters where the right-hand side is larger than unity and there is no real k, which satisfies this relation. As a function of the frequency, the absolute value of the right hand side of (5.1.22) can be either larger or smaller than unity. In the former case there is a real solution for k which entails that in the corresponding frequency range the wave is allowed to propagate – this frequency range is called the *pass band*. In the frequency ranges k is complex therefore the amplitudes are identically zero, otherwise the solution diverges. This frequency range is called the *forbidden band* or *band-gap*. Explicitly, the right-hand side of (5.1.22) reads

$$\frac{1}{2}(T_{11} + T_{22}) = \frac{(Z_1 + Z_2)^2}{4Z_1Z_2}\cosh(\psi + \chi) - \frac{(Z_1 - Z_2)^2}{4Z_1Z_2}\cosh(\psi - \chi), \tag{5.1.23}$$

where $\psi = \Gamma_1(L-g)$, $\chi = \Gamma_{d,1}g$, the characteristic impedances are

$$Z_1 = \eta_0\frac{c\Gamma_{d,1}}{j\omega\varepsilon_r}, \quad Z_2 = \eta_0\frac{c\Gamma_1}{j\omega}, \tag{5.1.24}$$

and $\eta_0 = 377[\text{Ohm}]$ is the impedance of the vacuum. Figure 5.2 illustrates the right-hand side of (5.1.22) as a function of the frequency ($\varepsilon_r = 10$, $R = 2\,\text{cm}$, $L = 1\,\text{cm}$ and $g = L/2$). The blocks at the bottom, illustrate the forbidden frequencies, namely at these frequencies TM waves can not propagate. In Fig. 5.3 the dispersion

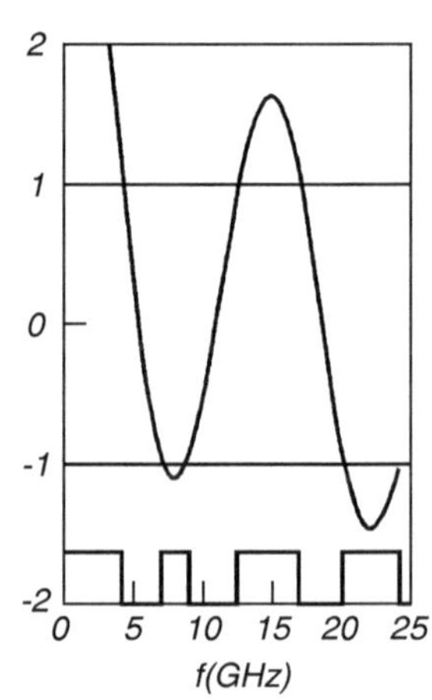

Fig. 5.2 Right-hand side of (5.1.22). For the frequencies marked at the bottom, no electromagnetic wave can propagate in the system

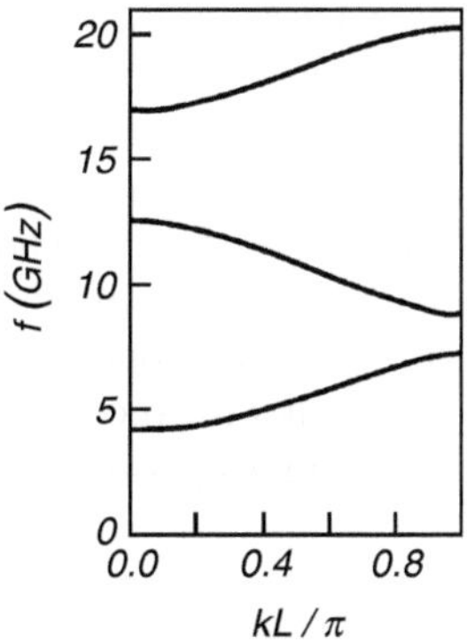

Fig. 5.3 The dispersion relation corresponding to the same parameters as those for which Fig. 5.2 was plotted; the parameters $R = 2\,\mathrm{cm}$, $\varepsilon_r = 10$, $L = 1\,\mathrm{cm}$ and $g = 0.5\,\mathrm{cm}$

relation of the first three passbands are presented; these branches correspond only to the TM_{01} mode. Higher symmetric or asymmetric modes have additional contributions in this frequency range.

Comment 5.1. The expression in (5.1.22) is the dispersion relation of the periodic structure we introduced. From this simple example however we observe that *the dispersion relation of a periodic structure is itself periodic in k with a periodicity* $2\pi/L$. This is a general feature which can be deduced from (5.1.9). If the latter is satisfied for $k = k_0$ then (5.1.9) is satisfied also for $k = k_0 + 2\pi/L$ as shown next

$$\begin{aligned} f(z+L)e^{j(k_0+2\pi/L)L} &= f(z+L)e^{jk_0L}e^{j2\pi}, \\ &= f(z+L)e^{jk_0L} = f(z). \end{aligned} \tag{5.1.25}$$

Consequently, since the dispersion relation is periodic in k, it is sufficient to represent its variation with k in the range $-\pi/L \le k \le \pi/L$; this k domain is also called the *first Brillouin zone*.

Comment 5.2. Bearing in mind the last comment, we can re-examine the expression in (5.1.8) and realize that $f(z)$ is represented by a superposition of *spatial harmonics* $\exp(-jk_nz)$ where

$$k_n = k + \frac{2\pi}{L}n, \tag{5.1.26}$$

which all correspond to the solution of the dispersion relation of the system. According to this definition the phase velocity of each harmonic is

$$\mathrm{v}_{\mathrm{ph},n} = \frac{\omega}{ck_n}, \tag{5.1.27}$$

and for a high harmonic index, n, this velocity decreases as n^{-1}. Furthermore, all harmonics with negative index correspond to waves, which propagate backwards. In addition, note that the zero harmonic ($n = 0$) has a positive group velocity for $\pi/L > k > 0$ and negative in the range $-\pi/L < k < 0$. This is a characteristic of *all* spatial harmonics. Since the group velocity is related to the energy velocity, one can conclude that although the wave number of a particular space harmonic is positive, the power it carries may flow in the negative direction (if the group velocity is negative). This opens a completely new family of devices in which the power flows anti-parallel to the beam – the generic device is called backward-wave oscillator (BWO) and it was discussed in Sect. 4.2.4. Note also that at all π-points, i.e., $kL = \pi n$, the group velocity is zero.

Another instructive example that was considered in what follows corresponds to a continuous periodic dielectric structure

$$\varepsilon(z) = \sum_{n=-\infty}^{\infty} \bar{\varepsilon}_n \exp\left(-j2\pi n \frac{z}{L}\right) \tag{5.1.28}$$

that supports a symmetric transverse magnetic mode confined by an ideal circular waveguide of radius R. In this simplified model, the TM mode is determined by the magnetic vector potential that is a solution of

$$\left[\frac{1}{r}\frac{\partial}{\partial r}r\frac{\partial}{\partial r} + \frac{\partial^2}{\partial z^2} + \varepsilon(z)\frac{\omega^2}{c^2}\right]A_z(r,z) = 0 \tag{5.1.29}$$

For simplicity sake we confine the discussion to the first mode such that $A_z(r,z) = a_z(z)J_0(p_1 r/R)$ thus

$$\left[\frac{d^2}{dz^2} - \frac{p_1^2}{R^2} + \varepsilon(z)\frac{\omega^2}{c^2}\right]a_z(z) = 0. \tag{5.1.30}$$

Employing Floquet theorem, $a_z(z) = \exp(-jkz)\sum_{n=-\infty}^{\infty} \bar{a}_n \exp(-j2\pi nz/L)$ and the orthogonality of the trigonometric functions in one period of the structure we get

$$\left[\left(k + \frac{2\pi n}{L}\right)^2 + \frac{p_1^2}{R^2}\right]\bar{a}_n = \left(\frac{2\pi}{\lambda}\right)^2 \sum_{n'=-\infty}^{\infty} \bar{\varepsilon}_{n-n'}\bar{a}_{n'}. \tag{5.1.31}$$

We may now define the normalized vacuum wavelength $\bar{\lambda} = \lambda/L$ and the characteristic matrix $M_{n,n'} = \bar{\varepsilon}_{n-n'}\left[(kL/2\pi + n)^2 + (p_1 L/2\pi R)^2\right]^{-1}$ implying that for a given k, the normalized vacuum wavelength (representing the frequency) is determined by the eigen-values of the matrix M namely,

$$\left[\mathbf{M} - \bar{\lambda}^2\mathbf{I}\right]\vec{a} = 0. \tag{5.1.32}$$

Contrary to the previous formulation whereby for a given frequency we could establish the wave-number (k), in the framework of this formulation, for a given wave-number$k$, (5.1.32) determines the discrete spectrum of allowed frequencies.

Evidently, this last approach is particularly useful when the periodic structure is represented by a smooth function described by a small number of harmonics.

5.2 Closed Periodic Structure

Based on what was shown in the previous section one can determine the dispersion relation of a TM_{01} mode, which propagates in a more practical periodic structure namely, a corrugated waveguide (Brillouin 1948). Its periodicity is L, the inner radius is denoted by $R_{\rm int}$ and the external by $R_{\rm ext}$; the distance between two cavities (the drift region) is d – see Fig. 3.17. Using Floquet's Theorem (5.1.8) we can write for the magnetic potential in the inner cylinder ($0<r<R_{\rm int}$) the following expression

$$A_z(r,z) = \sum_{n=-\infty}^{\infty} A_n \mathrm{e}^{-jk_n z} \mathrm{I}_0(\Gamma_n r), \tag{5.2.1}$$

and accordingly, the electromagnetic field components read

$$\begin{aligned} E_r(r,z) &= \frac{c^2}{j\omega} \sum_{n=-\infty}^{\infty} (-jk_n\Gamma_n) A_n \mathrm{e}^{-jk_n z} \mathrm{I}_1(\Gamma_n r), \\ E_z(r,z) &= \frac{c^2}{j\omega} \sum_{n=-\infty}^{\infty} (-\Gamma_n^2) A_n \mathrm{e}^{-jk_n z} \mathrm{I}_0(\Gamma_n r), \\ H_\phi(r,z) &= -\frac{1}{\mu_0} \sum_{n=-\infty}^{\infty} \Gamma_n A_n \mathrm{e}^{-jk_n z} \mathrm{I}_1(\Gamma_n r). \end{aligned} \tag{5.2.2}$$

In these expressions,

$$\Gamma_n^2 = k_n^2 - \frac{\omega^2}{c^2}, \tag{5.2.3}$$

and $\mathrm{I}_0(x), \mathrm{I}_1(x)$ are the zero and first order modified Bessel functions of the first kind respectively. This choice of the radial functional variation is dictated by the condition of convergence of the electromagnetic field on axis.

In each individual groove, the electromagnetic field can be derived from the following magnetic vector potential:

$$A_z^{(\sigma)}(r,z) = \sum_{\nu=0}^{\infty} B_\nu^{(\sigma)} \cos[q_\nu(z - z_\sigma - d)] \mathcal{R}_{0,\nu}(r), \tag{5.2.4}$$

where $q_\nu = \pi\nu/(L-d)$,

$$\mathcal{R}_{0,\nu}(r) = \mathrm{I}_0(\Lambda_\nu r)\mathrm{K}_0(\Lambda_\nu R_{\rm ext}) - \mathrm{K}_0(\Lambda_\nu r)\mathrm{I}_0(\Lambda_\nu R_{\rm ext}), \tag{5.2.5}$$

and $\Lambda_\nu^2 = q_\nu^2 - (\omega/c)^2$. The electromagnetic field reads

$$E_r^{(\sigma)}(r,z) = \frac{c^2}{j\omega}\sum_{\nu=0}^{\infty}(-q_\nu)\Lambda_\nu B_\nu^{(\sigma)}\sin[q_\nu(z-z_\sigma-d)]\mathcal{R}_{1,\nu}(r),$$
$$E_z^{(\sigma)}(r,z) = \frac{c^2}{j\omega}\sum_{\nu=0}^{\infty}(-\Lambda_\nu^2)B_\nu^{(\sigma)}\cos[q_\nu(z-z_\sigma-d)]\mathcal{R}_{0,\nu}(r), \quad (5.2.6)$$
$$H_\phi^{(\sigma)}(r,z) = -\frac{1}{\mu_0}\sum_{\nu=0}^{\infty}\Lambda_\nu B_\nu^{(\sigma)}\cos[q_\nu(z-z_\sigma-d)]\mathcal{R}_{1,\nu}(r).$$

In these expressions $\mathcal{R}_{1,\nu}(r)$ is the derivative of $\mathcal{R}_{0,\nu}(r)$ defined by

$$\mathcal{R}_{1,\nu}(r) = \mathrm{I}_1(\Lambda_\nu r)\mathrm{K}_0(\Lambda_\nu R_{\mathrm{ext}}) + \mathrm{K}_1(\Lambda_\nu r)\mathrm{I}_0(\Lambda_\nu R_{\mathrm{ext}}), \quad (5.2.7)$$

and except at $r = R_{\mathrm{int}}$, all the boundary conditions are satisfied; the index σ labels the "cavity".

5.2.1 Dispersion Relation

Our next step is to impose the continuity of the boundary conditions at the interface ($r = R_{\mathrm{int}}$). The continuity of the longitudinal component of the electric field $[E_z(r = R_{\mathrm{int}}, -\infty < z < \infty)]$ reads

$$\frac{c^2}{j\omega}\sum_{n=-\infty}^{\infty}(-\Gamma_n^2)A_n e^{-jk_n z}\mathrm{I}_0(\Gamma_n R_{\mathrm{int}})$$
$$= \begin{cases} 0 & \text{for } z_\sigma < z < z_\sigma + d, \\ -\frac{c^2}{j\omega}\sum_{\nu=0}^{\infty}\Lambda_\nu^2 B_\nu^{(\sigma)}\cos[q_\nu(z-z_\sigma-d)]\mathcal{R}_{0,\nu}(R_{\mathrm{int}}) & \text{for } z_\sigma + d < z < z_\sigma + L, \end{cases} \quad (5.2.8)$$

and the azimuthal magnetic field $[H_\phi(r = R_{\mathrm{int}}, z_\sigma + d < z < z_\sigma + L)]$ reads

$$-\frac{1}{\mu_0}\sum_{n=-\infty}^{\infty}\Gamma_n A_n e^{-jk_n z}\mathrm{I}_1(\Gamma_n R_{\mathrm{int}}) = \frac{1}{\mu_0}\sum_{\nu=0}^{\infty}\Lambda_\nu B_\nu^{(\sigma)}\cos[q_\nu(z-z_\sigma-d)]\mathcal{R}_{1,\nu}(R_{\mathrm{int}}). \quad (5.2.9)$$

From these boundary conditions the dispersion relation of the structure can be developed. For this purpose, we analyze the solution in the grooves having Floquet's theorem in mind. The latter implies that the longitudinal electric field in the σ's groove has to satisfy the following relation:

$$
\begin{aligned}
&-\frac{c^2}{j\omega}\sum_{\nu=0}^{\infty}\Lambda_\nu^2 B_\nu^{(\sigma)}\cos[q_\nu(z-z_\sigma-d)]\mathcal{R}_{0,\nu}(r)\\
&=-\frac{c^2}{j\omega}\sum_{\nu=0}^{\infty}\Lambda_\nu^2 B_\nu^{(\sigma+1)}\mathrm{e}^{jkL}\cos[q_\nu(z+L-z_{\sigma+1}-d)]\mathcal{R}_{0,\nu}(r).
\end{aligned} \tag{5.2.10}
$$

But by definition $z_{\sigma+1}-z_\sigma=L$ therefore, the last expression implies that

$$
B_\nu^{(\sigma)}=B_\nu \mathrm{e}^{-jkz_\sigma}. \tag{5.2.11}
$$

This result permits us to restrict the investigation to a single cell and without loss of generality we chose $z_{\sigma=0}=0$ since if we know B_ν in one cell, the relation in (5.2.11) determines the value of this amplitude in all other cells.

Comment 5.3. In practice we could have harnessed this implication of Floquet theorem already in (5.2.4) but didactically we believe that when used for the first time one should avoid short cuts.

With this result in mind we multiply (5.2.8) by $\exp(jk_m z)$ and integrate over one cell; the result is

$$
\sum_{n=-\infty}^{\infty}\Gamma_n^2 A_n\delta_{n,m}L\mathrm{I}_0(\Gamma_n R_{\mathrm{int}})=\sum_{\nu=0}^{\infty}\Lambda_\nu^2 B_\nu\mathcal{R}_{0,\nu}(R_{\mathrm{int}})\int_d^L dz\mathrm{e}^{jk_m z}\cos[q_\nu(z-d)], \tag{5.2.12}
$$

here $\delta_{n,m}$ is the Kroniker delta function which equals 1 if $n=m$ and zero otherwise; we also used the orthogonality of the Fourier spatial harmonics.

A similar procedure is adopted when imposing the continuity of the magnetic field with one difference, (5.2.9) is defined only in the groove aperture thus we utilize the orthogonality of the trigonometric function $\cos[q_\nu(z-d)]$. Accordingly, (5.2.9) is multiplied by $\cos[q_\mu(z-d)]$ and we integrate over $d<z<L$; the result is

$$
\begin{aligned}
&\sum_{n=-\infty}^{\infty}\Gamma_n A_n\mathrm{I}_1(\Gamma_n R_{\mathrm{int}})\int_d^L dz\cos\left[q_\mu(z-d)\right]\mathrm{e}^{-jk_n z}\\
&=\sum_{\nu=0}^{\infty}\Lambda_\nu B_\nu\mathcal{R}_{1,\nu}(R_{\mathrm{int}})(L-d)g_\mu\delta_{\nu,\mu}
\end{aligned} \tag{5.2.13}
$$

where $g_0=1$ and $g_{\mu\neq 0}=0.5$ otherwise. It is convenient to define the quantity

$$
\mathcal{L}_{n,\nu}(k)=\frac{1}{L-d}\int_d^L dz\cos[q_\nu(z-d)]\mathrm{e}^{jk_n z}, \tag{5.2.14}
$$

which allows us to write (5.2.12) as

$$A_n = \frac{1}{\Gamma_n^2 \mathrm{I}_0(\Gamma_n R_{\mathrm{int}})} \frac{L-d}{L} \sum_{\nu=0}^{\infty} \Lambda_\nu^2 \mathcal{R}_{0,\nu}(R_{\mathrm{int}}) \mathcal{L}_{n,\nu}(k) B_\nu, \tag{5.2.15}$$

and (5.2.13) as

$$B_\nu = \frac{1}{\Lambda_\nu \mathcal{R}_{1,\nu}(R_{\mathrm{int}}) g_\nu} \sum_{n=-\infty}^{\infty} A_n \Gamma_n \mathrm{I}_1(\Gamma_n R_{int}) \mathcal{L}_{n,\nu}^*(k). \tag{5.2.16}$$

These are two equations for two unknown sets of amplitudes (A_n, B_ν) and the dispersion relation can be represented in two equivalent ways: one possibility is to substitute (5.2.16) in (5.2.15) and get

$$\sum_{m=-\infty}^{\infty} \left[\delta_{n,m} - \frac{L-d}{L} \frac{\Gamma_m R_{\mathrm{int}} \mathrm{I}_1(\Gamma_m R_{\mathrm{int}})}{\Gamma_n^2 R_{\mathrm{int}}^2 \mathrm{I}_0(\Gamma_n R_{\mathrm{int}})} \sum_{\nu=0}^{\infty} \frac{\mathcal{R}_{0,\nu}(R_{\mathrm{int}}) \Lambda_\nu R_{\mathrm{int}}}{\mathcal{R}_{1,\nu}(R_{\mathrm{int}}) g_\nu} \mathcal{L}_{n,\nu} \mathcal{L}_{m,\nu}^* \right] A_m = 0, \tag{5.2.17}$$

whereas the other possibility is to substitute (5.2.15) in (5.2.16) and obtain

$$\sum_{\mu=0}^{\infty} \left[\delta_{\nu,\mu} - \frac{L-d}{L} \frac{\Lambda_\mu^2 R_{\mathrm{int}}^2 \mathcal{R}_{0,\mu}(R_{\mathrm{int}})}{\Lambda_\nu R_{\mathrm{int}} \mathcal{R}_{1,\nu}(R_{\mathrm{int}}) g_\nu} \sum_{n=-\infty}^{\infty} \frac{\mathrm{I}_1(\Gamma_n R_{\mathrm{int}})}{(\Gamma_n R_{\mathrm{int}}) \mathrm{I}_0(\Gamma_n R_{\mathrm{int}})} \mathcal{L}_{n,\nu}^* \mathcal{L}_{n,\mu} \right] B_\mu = 0. \tag{5.2.18}$$

In both cases, the dispersion relation is calculated from the requirement that the determinant of the matrix, which multiplies the vector of amplitudes, is zero.

Although the two methods are equivalent, at the practical level, we found that the latter expression to be by far more efficient for practical calculation because of the number of modes required to represent adequately the field in the groove compared to the number of spatial harmonics required to represent the field in the inner section. In the case of single mode operation, we found that 1–3 modes are sufficient for description of the field in the grooves and about 40 spatial harmonics are generally used in the inner section. As indicated by these numbers it will be much easier to calculate the determinant of a 3×3 matrix rather than 40×40 one; we quantify this statement later. At this point, we consider the design of a disk-loaded structure assuming that the number modes in the grooves and harmonics in the inner space are sufficient.

Let assume that we want to determine the geometry of a disk-loaded structure which enables a wave at 10 GHz to be in resonance with electrons of $\beta = 0.9$ and the phase advance per cell is assumed to be $kL = 2\pi/3$. These two conditions determine the period of the structure – in our case $L = 9\,\mathrm{mm}$. There are three additional geometric parameters to be determined: $R_{\mathrm{ext}}, R_{\mathrm{int}}$ and d. The last two have a dominant effect on the width of the passband and for the lowest mode, the

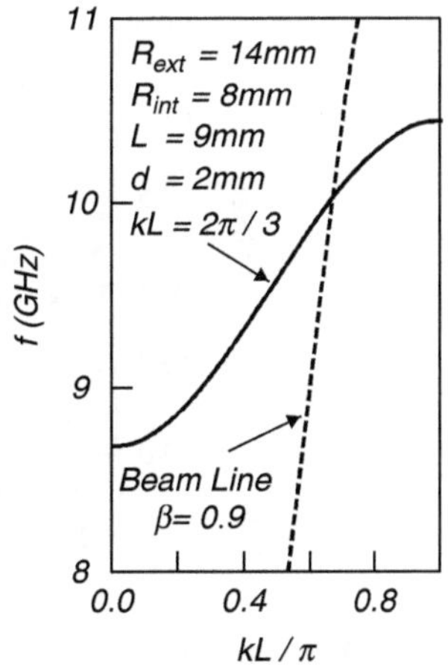

Fig. 5.4 A solution of the dispersion relation in (5.2.18). The geometry chose corresponds to: $R_{\rm int} = 8\,$mm, $L = 9\,$mm, $R_{\rm int} = 13.96\,$mm and $d = 2\,$mm

passband increases with increasing $R_{\rm int}$ and decreases with increasing d. The passband, $\Delta\omega$, of a mode sets a limit on the maximum group velocity as can be seen bearing in mind that the half width of the first Brillouin zone is $\Delta k = \pi/L$. Consequently, $v_{\rm gr} = \Delta\omega/\Delta k < \Delta\omega L/\pi$. A solution of the dispersion relation in (5.2.18) is illustrated in Fig. 5.4, the geometry chosen is: $R_{\rm int} = 8\,$mm and $d = 2\,$mm and from the condition of phase advance per cell of 120° at 10 GHz, we determined, using the dispersion relation, the value of the external radius to be $R_{\rm ext} = 13.96\,$mm.

In the remainder of this section we consider only a single mode in the groove. Therefore, before we conclude this subsection, it remains to quantify the effect of higher modes in the groove. The first mode in the groove ($\nu = 0$) represents a TEM mode which propagates in the radial direction. Other modes ($TM_{0,\nu>0}$) are either propagating or evanescent. The amplitudes of the magnetic and electric field (E_z) of the TEM mode are constant at the groove aperture thus the choice of using a single mode in the groove is equivalent to the averaging the field at the aperture – approach usually adopted in the literature. Figure 5.5 illustrates the dependence of upper and lower cut-off frequency on the number of harmonics used in the calculation; the number of modes in the grooves is a parameter. For the geometry presented above, the number of harmonics required is 20 or larger; typically about 40 harmonics are being used. The effect of the $\nu = 1$ mode is negligible in this case as seen for both upper and lower cut-off frequencies. The effect of the higher mode introduces a correction on the order of 1% which for most practical purposes is sufficient.

5.2.2 *Spatial Harmonics Coupling*

Contrary to uniform dielectric structures, here each mode consists of a superposition of an infinite number of spatial harmonics. These harmonics are all coupled by

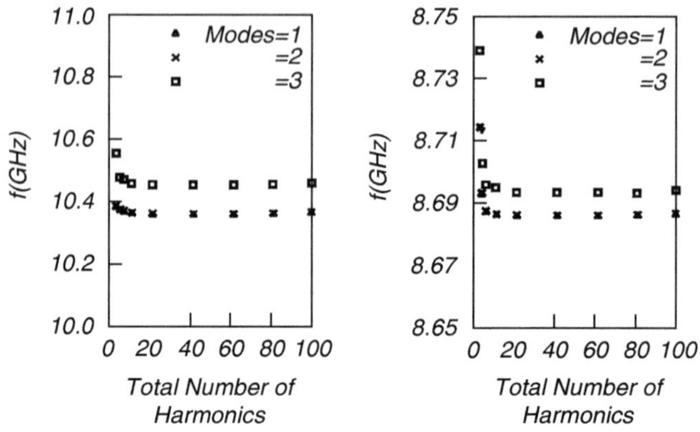

Fig. 5.5 The dependence of the upper and lower cut-off frequency on the number of harmonics used

the conditions imposed on the electromagnetic field by the geometry at $r = R_{\rm int}$. We limit the investigation to the accuracy associated with a single mode taken in the groove, therefore according to (5.2.15), we have

$$A_n = -\frac{1}{\Gamma_n^2 \mathrm{I}_0(\Gamma_n R_{int})} \frac{\omega^2}{c^2} \mathcal{R}_{0,0}(R_{int}) \mathcal{L}_{n,0}(k) B_0, \tag{5.2.19}$$

and in this particular case

$$\mathcal{L}_{n,0}(k) = \frac{L-d}{L} \operatorname{sinc}\left[\frac{1}{2} k_n (L-d)\right] \exp\left[j \frac{1}{2} k_n (L+d)\right]. \tag{5.2.20}$$

Let us compare the first few spatial harmonics relative to the zero harmonic. For this purpose we take $f = 10$ GHz, $\mathrm{v}_0 = 0.9c$, $R_{\rm int} = 8\,\mathrm{mm}$, $L = 9\,\mathrm{mm}$ and $d = 2\,\mathrm{mm}$. The ratio of the first few amplitudes is

$$\left|\frac{A_{-1}}{A_0}\right| = 8 \times 10^{-3}, \left|\frac{A_1}{A_0}\right| = 3 \times 10^{-6}, \left|\frac{A_{-2}}{A_0}\right| = 2 \times 10^{-6}, \left|\frac{A_2}{A_0}\right| = 1 \times 10^{-8}. \tag{5.2.21}$$

This result indicates that *on axis*, the amplitude of the interacting harmonic is dominant. At the interface with the grooves ($r = R_{\rm int}$) the ratio between the contribution of the zero and nth harmonic is much closer to unity and it can be checked that it reads

$$\frac{|E_{z,n}(r = R_{\rm int})|}{|E_{z,0}(r = R_{\rm int})|} = \frac{|\operatorname{sinc}[k_n(L-d)/2]|}{|\operatorname{sinc}[k_0(L-d)/2]|}, \tag{5.2.22}$$

which is virtually unity. It also reveals that there is a significant amount of energy in the high spatial harmonics which may cause breakdown due to the associated gradients on the metallic surface.

A more instructive picture is obtained by examining the average power flowing along one cell:

$$P = 2\pi \int_0^{R_{\rm int}} drr \frac{1}{L} \int_0^L dz \left[\frac{1}{2} E_r(r,z) H_\phi^*(r,z) \right]. \qquad (5.2.23)$$

According to the definition in (5.2.2) we have

$$P = \frac{\pi}{\eta_0} \sum_{n=-\infty}^{\infty} |cA_n|^2 \frac{ck_n}{\omega} \int_0^{\Gamma_n R_{\rm int}} dxx{\rm I}_1^2(x); \qquad (5.2.24)$$

the integral can be calculated analytically (Abramowitz and Stegun 1968, p.484) and it reads

$$U(\xi) \equiv \int_0^{\xi} dxx{\rm I}_1^2(x) = \xi {\rm I}_0(\xi){\rm I}_1(\xi) + \frac{1}{2}\xi^2[{\rm I}_1^2(\xi) - {\rm I}_0^2(\xi)]. \qquad (5.2.25)$$

Based on these definitions we can calculate the average power carried by each harmonic as

$$P_n = \frac{\pi}{\eta_0} |cA_n|^2 \frac{ck_n}{\omega} U(\Gamma_n R_{\rm int}), \qquad (5.2.26)$$

and the result is listed below

$$\frac{P_{-2}}{P_0} = -3 \times 10^{-3}, \ \frac{P_{-1}}{P_0} = -0.16, \ \frac{P_1}{P_0} = 1 \times 10^{-4}, \ \frac{P_2}{P_0} = 3 \times 10^{-3}. \qquad (5.2.27)$$

Although there is a total flow of power along (the positive) direction of the z axis, a substantial amount of power is actually flowing *backwards*. In this numerical example for all practical purposes, we can consider only the lowest two harmonics and write the total power that flows, normalized to the power in the zero harmonic. Thus if the latter is unity, then the power in the forward is $1 - 0.16 = 0.84$. This result indicates that if we have a finite length structure with finite reflections from the input end, then in this periodic structure we have an inherent feedback even if the output end is perfectly matched.

5.2.3 Interaction Parameters

Even if an electron beam interacts only with a single mode the latter consists of an infinite number of harmonics and with this regard, we distinguish between direct and indirect interaction. By *direct interaction,* we refer to the harmonic to which the electron transfers energy directly. For example, for a pencil beam (on axis) it is primarily the $n = 0$ harmonic, which interacts with the beam, namely, it has a phase velocity close to the velocity of the electrons. Because of the interaction we have shown in Chap. 4 that the wave-number of the mode becomes complex, which in a periodic structure, implies that the projection of the wave-number in the first Brillouin zone (k) becomes complex. However, this corresponds to all harmonics, which finally implies that they *all* grow in space by the same relative amount such that locally the boundary conditions are satisfied. With this regard, the beam indirectly interacts with all harmonics and this is referred to as *indirect interaction.* The condition for the beam to interact directly only with a single harmonic can be formulated in terms of the velocity spread of the beam and resonance condition: the latter reads in general $\omega/\mathrm{v}_0 - k_n \simeq 0$ whose variation for a constant frequency reads $\omega|\Delta\mathrm{v}|/\mathrm{v}_0^2 = |\Delta k|$. Since two harmonics are separated by $\Delta k = 2\pi/L$, we conclude that the condition for single harmonic direct operation is that $(\omega L/c)(|\Delta\beta|/\beta^2) \ll 2\pi$. Subject to this condition, the beam-wave interaction is described primarily by a single parameter: the interaction impedance introduced in Sect. 2.3.3 (2.3.29). For a pencil beam of radius R_b the effective field that acts on the electrons in a uniform periodic structure is

$$|E|^2 \equiv \frac{2}{R_b^2}\int_0^{R_\mathrm{b}} drr|E_{z,n=0}(r,z)|^2. \tag{5.2.28}$$

Using the explicit expression for E_z in (5.2.2) we find

$$|E|^2 \equiv \frac{c^4}{\omega^2}|A_0|^2\Gamma_0^4\frac{2}{R_b^2}\int_0^{R_\mathrm{b}} drr\mathrm{I}_0^2(\Gamma_0 r); \tag{5.2.29}$$

the integral can be evaluated exactly (Abramowitz and Stegun 1968, p.484) and it reads

$$W_1(x) \equiv \int_0^x d\xi\xi\mathrm{I}_0^2(\xi) = \frac{1}{2}x^2[\mathrm{I}_0^2(x) - \mathrm{I}_1^2(x)]. \tag{5.2.30}$$

With this expression the effective electric field reads

$$|E|^2 \equiv \frac{2c^4\Gamma_0^2}{\omega^2R_\mathrm{b}^2}|A_0|^2W_1(\Gamma_0R_\mathrm{b}), \tag{5.2.31}$$

and finally we can determine the explicit expression for the interaction impedance in a periodic structure

$$\begin{aligned} Z_\mathrm{int} &\equiv \frac{1}{2}\frac{|E|^2(\pi R_\mathrm{int}^2)}{P}, \\ &= \eta_0\left(\frac{R_\mathrm{int}}{R_b}\right)^2\left(\frac{c\Gamma_0}{\omega}\right)^2 W_1(\Gamma_0R_\mathrm{b})\left[\sum_{n=-\infty}^{\infty}U(\Gamma_nR_\mathrm{int})\left(\frac{ck_n}{\omega}\right)\frac{|A_n|^2}{|A_0|^2}\right]^{-1}. \end{aligned} \tag{5.2.32}$$

The ratio $|A_n/A_0|$ can be deduced from (5.2.19). Furthermore, the last expression reveals the effect of the beam radius on the interaction impedance. The latter can be formulated as

$$Z_{\rm int}(\Gamma_0 R_{\rm b}) = Z_{\rm int}(0)\left[{\rm I}_0^2(\Gamma_0 R_{\rm b}) - {\rm I}_1^2(\Gamma_0 R_{\rm b})\right], \tag{5.2.33}$$

where

$$Z_{\rm int}(0) \equiv \eta_0(\Gamma_0 R_{\rm int})^2\left(\frac{\Gamma_0 c}{\omega}\right)^2\left[\sum_{n=-\infty}^{\infty} U(\Gamma_n R_{\rm int})\left(\frac{ck_n}{\omega}\right)\frac{|A_n|^2}{|A_0|^2}\right]^{-1}. \tag{5.2.34}$$

The interaction impedance increases monotonically with the beam radius; this fact has been discussed also in Chap. 4 in the context of the interaction in a dielectrically loaded waveguide. Another aspect of the same phenomenon is illustrated in Fig. 5.6 where we present the interaction impedance as a function of the internal radius keeping L, d and the frequency ($f = 10$ GHz) constant; the external radius is determined from the resonance condition and the phase advance per cell (which is chosen to be 120° for reasons which will be clarified in Chap. 8). In addition, the beam radius is taken to be $R_{\rm b} = 3\,{\rm mm}$. Note the rapid decrease of the interaction impedance with the increase in the internal radius of the structure. Again, this is a direct result of the exponential decay of the slow (evanescent) wave from the corrugated surface inwards.

The other parameter of interest is the *interaction dielectric coefficient* that is a measure of the total electromagnetic energy stored in one cell of the structure as defined in (2.3.31):

$$\varepsilon_{\rm int} \equiv \frac{W_{\rm em}}{\frac{1}{2}\varepsilon_0|E|^2\pi R_{\rm int}^2}. \tag{5.2.35}$$

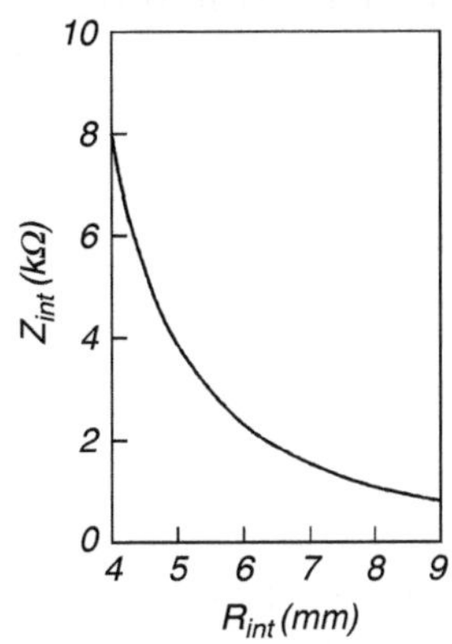

Fig. 5.6 $Z_{\rm int}$ as a function of the internal radius. The period L, the tooth width d and the frequency are kept constant. $R_{\rm ext}$ is determined from the resonance and phase advance conditions

This parameter is important in the description of the operation of an oscillator. However, we have to bear in mind that according to (5.2.33) ε_{int} and Z_{int} are related.

Both the interaction impedance (Z_{int}) and the dielectric coefficient of the interaction (ε_{int}) are illustrated in Fig. 5.7 for these frequencies for which the phase velocity of the wave is smaller than c; the geometric parameters are: $R_{int} = 8\,\text{mm}$, $R_{ext} = 13.96\,\text{mm}$, $L = 9\,\text{mm}$, $d = 2\,\text{mm}$, $R_b = 3\,\text{mm}$ and the number of harmonics used is 13 ($-6 \le n \le 6$). Note that the interaction impedance has a *minimum* at a frequency which is higher than the frequency where the system was designed to operate ($f = 10\,\text{GHz}$). Close to the π-point ($kL = \pi$) the interaction impedance increases since the amount of power which can flow in the system diminishes. At the same time, the dielectric coefficient of the interaction increases, which means that the ratio of energy stored in the system to the electric field acting on the particles, increases. Consequently, in this frequency range the system will tend to oscillate.

ε_{int} has two contributions: the first from the energy stored in the grooves and the second represents the energy stored in the inner cylinder. Figure 5.8 illustrates again ε_{int} and the contribution of each region. We observe that the effect of the groove is dominant at all frequencies of interest, emphasizing its cavity role. We conclude

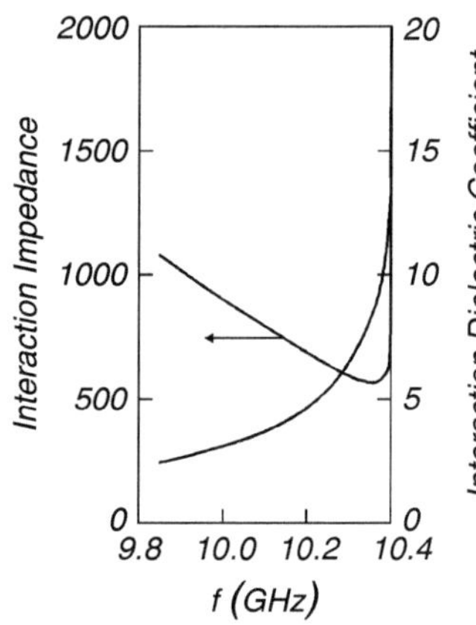

Fig. 5.7 ε_{int} and Z_{int} as a function of the frequency

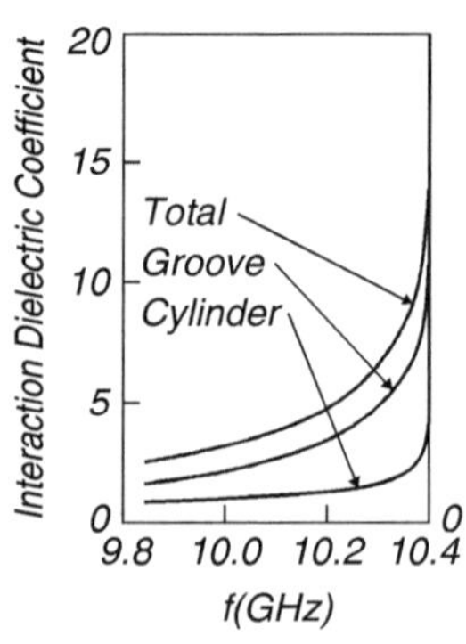

Fig. 5.8 The contribution of the cylinder and the groove to ε_{int}

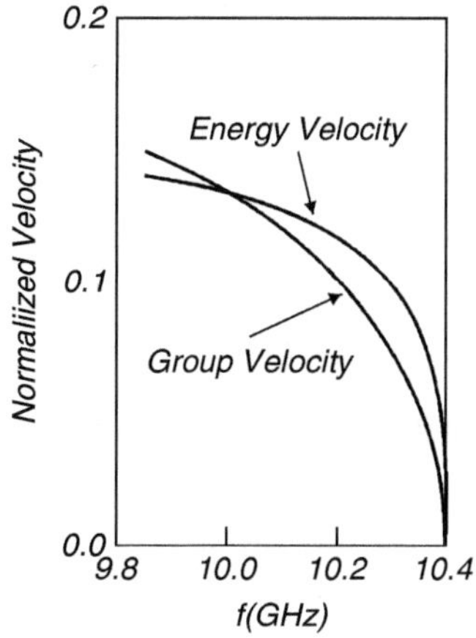

Fig. 5.9 Comparison of energy and group velocity as a function of the frequency

this subsection with a comparison of the group and energy velocity as illustrated in Fig. 5.9. Within the framework of our approximation, the two are close but not identical.

5.3 Open Periodic Structure

In this section, an analysis similar to that in Sect. 5.2 is applied to an open periodic structure. As we shall see the number of modes, which may develop in such a structure is small and therefore, mode competition is minimized. This competition is a byproduct of the necessity to generate high power radiation, which in the case of a single mode operation generates high gradients on the metallic surface. In order to avoid breakdown, it is necessary to increase the volume of the waveguide. Doing so, we allow more than one mode to coexist at the same frequency. Furthermore, the beam line intersects higher modes at frequencies higher than the operating one and these modes may deflect the electrons, as will be briefly discussed in Chap. 8.

We consider a system in which the wave propagates along the periodic structure forming a disk-loaded wire, as illustrated in Fig. 5.10. Its periodicity is L, the inner radius is denoted by $R_{\rm int}$, the external by $R_{\rm ext}$ and the distance between two cavities (the drift region) is d. Floquet's theorem as formulated in (5.1.8) allows us to write for the magnetic potential in the external region ($\infty > r \geq R_{\rm ext}$) the following expression

$$A_z(r,z) = \sum_{n=-\infty}^{\infty} A_n \exp(-jk_n z) \mathrm{K}_0(\Gamma_n r), \tag{5.3.1}$$

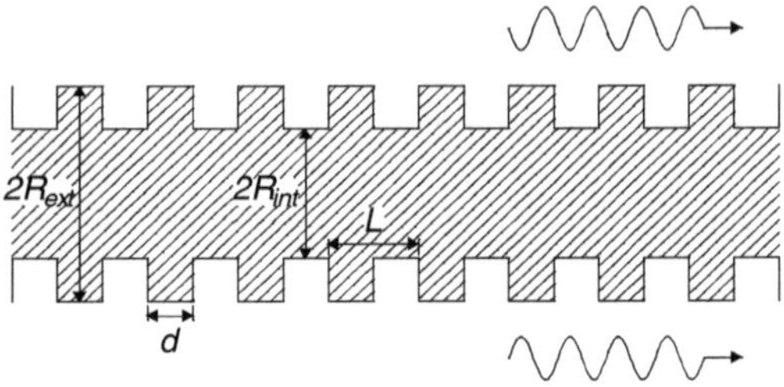

Fig. 5.10 Schematics of a cylindrical open periodic structure; the wave propagates along the outer region

and accordingly, the electromagnetic field components read

$$\begin{aligned}
E_r(r,z) &= \frac{c^2}{j\omega}\sum_{n=-\infty}^{\infty}(jk_n\Gamma_n)A_n\exp(-jk_nz)\mathrm{K}_1(\Gamma_nr),\\
E_z(r,z) &= \frac{c^2}{j\omega}\sum_{n=-\infty}^{\infty}(-\Gamma_n^2)A_n\exp(-jk_nz)\mathrm{K}_0(\Gamma_nr),\\
H_\phi(r,z) &= -\frac{1}{\mu_0}\sum_{n=-\infty}^{\infty}(-\Gamma_n)A_n\exp(-jk_nz)\mathrm{K}_1(\Gamma_nr).
\end{aligned} \tag{5.3.2}$$

In these expressions $\mathrm{K}_0(x), \mathrm{K}_1(x)$ are the zero and first order modified Bessel functions of the second kind respectively and $\Gamma_n^2 = k_n^2 - (\omega/c)^2$. This choice of the radial functional variation is dictated by the condition of convergence of the electromagnetic field far away from the structure.

As pointed out in the previous subsection, based on Floquet theorem, it is sufficient to determine the field in one groove therefore, the magnetic vector potential in a given groove is

$$A_z(r,z) = \sum_{\nu=0}^{\infty}B_\nu\cos[q_\nu(z-d)]\mathcal{R}_{0,\nu}(r), \tag{5.3.3}$$

where $q_\nu = \pi\nu/(L-d)$,

$$\mathcal{R}_{0,\nu}(r) = \mathrm{I}_0(\Lambda_\nu r)\mathrm{K}_0(\Lambda_\nu R_{\mathrm{int}}) - \mathrm{K}_0(\Lambda_\nu r)\mathrm{I}_0(\Lambda_\nu R_{\mathrm{int}}), \tag{5.3.4}$$

and $\Lambda_\nu^2 = q_\nu^2 - (\omega/c)^2$. The electromagnetic field reads

$$\begin{aligned}
E_r(r,z) &= \frac{c^2}{j\omega}\sum_{\nu=0}^{\infty}(-q_\nu)\Lambda_\nu B_\nu\sin[q_\nu(z-d)]\mathcal{R}_{1,\nu}(r),\\
E_z(r,z) &= \frac{c^2}{j\omega}\sum_{\nu=0}^{\infty}(-\Lambda_\nu^2)B_\nu\cos[q_\nu(z-d)]\mathcal{R}_{0,\nu}(r),\\
H_\phi(r,z) &= -\frac{1}{\mu_0}\sum_{\nu=0}^{\infty}\Lambda_\nu B_\nu\cos[q_\nu(z-d)]\mathcal{R}_{1,\nu}(r).
\end{aligned} \tag{5.3.5}$$

In these expressions we used

$$\mathcal{R}_{1,\nu}(r) = \mathrm{I}_1(\Lambda_\nu r)\mathrm{K}_0(\Lambda_\nu R_{\mathrm{int}}) + \mathrm{K}_1(\Lambda_\nu r)\mathrm{I}_0(\Lambda_\nu R_{\mathrm{int}}). \tag{5.3.6}$$

The solution above satisfies all boundary conditions with the exception of $r = R_{\mathrm{ext}}$

5.3.1 Dispersion Relation

Our next step is to impose the continuity of the boundary conditions at the interface ($r = R_{\mathrm{ext}}$). Continuity of the longitudinal component of the electric field implies $E_z(r = R_{\mathrm{ext}}, -\infty < z < \infty)$, reads

$$\frac{c^2}{j\omega}\sum_{n=-\infty}^{\infty}(-\Gamma_n^2)A_n \exp(-jk_n z)\mathrm{K}_0(\Gamma_n R_{\mathrm{ext}}) = \begin{cases} 0 & \text{for} \quad 0 < z < d, \\ -\dfrac{c^2}{j\omega}\sum_{\nu=0}^{\infty}\Lambda_\nu^2 B_\nu \cos[q_\nu(z - z_\sigma - d)]\mathcal{R}_{0,\nu}(R_{\mathrm{ext}}) & \text{for} \quad d < z < L, \end{cases} \tag{5.3.7}$$

and the azimuthal magnetic field, $H_\phi(r = R_{ext}, d < z < L)$, reads

$$\begin{aligned} \frac{1}{\mu_0}\sum_{n=-\infty}^{\infty}\Gamma_n A_n \exp(-jk_n z)\mathrm{K}_1(\Gamma_n R_{\mathrm{ext}}) \\ = -\frac{1}{\mu_0}\sum_{\nu=0}^{\infty}\Lambda_\nu B_\nu \cos[q_\nu(z - z_\sigma - d)]\mathcal{R}_{1,\nu}(R_{\mathrm{ext}}). \end{aligned} \tag{5.3.8}$$

As in the previous section, we multiply (5.3.7) by $e^{jk_m z}$ and integrate over one cell; the result is

$$\sum_{n=-\infty}^{\infty}\Gamma_n^2 A_n \delta_{n,m} L \mathrm{K}_0(\Gamma_n R_{\mathrm{ext}}) = \sum_{\nu=0}^{\infty}\Lambda_\nu^2 B_\nu \mathcal{R}_{0,\nu}(R_{\mathrm{ext}})\int_d^L dz \exp(jk_m z)\cos[q_\nu(z-d)]. \tag{5.3.9}$$

We follow a similar procedure when imposing the continuity of the magnetic field; the difference in this case is that (5.3.8) is defined only in the groove's aperture thus we shall utilize the orthogonality of the trigonometric function $\cos[q_\nu(z-d)]$. Accordingly, (5.3.8) is multiplied by $\cos[q_\mu(z-d)]$ and we integrate over $d < z < L$; the result is

$$\sum_{n=-\infty}^{\infty} \Gamma_n A_n \mathrm{K}_1(\Gamma_n R_{\mathrm{ext}}) \int_d^L dz \cos[q_\mu(z-d)] \exp(-jk_n z) \\ = -\Lambda_\mu B_\mu \mathcal{R}_{1,\mu}(R_{\mathrm{ext}})(L-d) g_\mu \tag{5.3.10}$$

In this expression $g_0 = 1$ and $g_{n\neq 0} = 0.5$. It is convenient to define the quantity

$$\mathcal{L}_{n,\nu}(k) = \frac{1}{L-d} \int_d^L dz \cos[q_\nu(z-d)] \exp(jk_n z), \tag{5.3.11}$$

by whose means, (5.3.9) reads

$$A_n = \frac{1}{\Gamma_n^2 \mathrm{K}_0(\Gamma_n R_{\mathrm{ext}})} \frac{L-d}{L} \sum_{\nu=0}^{\infty} \Lambda_\nu^2 \mathcal{R}_{0,\nu}(R_{\mathrm{ext}}) \mathcal{L}_{n,\nu}(k) B_\nu, \tag{5.3.12}$$

whereas (5.3.10) reads

$$B_\nu = -\frac{1}{\Lambda_\nu \mathcal{R}_{1,\nu}(R_{ext}) g_\nu} \sum_{n=-\infty}^{\infty} A_n \Gamma_n \mathrm{K}_1(\Gamma_n R_{\mathrm{ext}}) \mathcal{L}_{n,\nu}^*(k). \tag{5.3.13}$$

These are two equations for two unknown sets of amplitudes (A_n, B_ν). As before, the dispersion relation can be represented in two equivalent ways: One possibility is to substitute (5.3.13) in (5.3.12) and obtain one equation for the amplitudes of the various harmonics

$$\sum_{m=-\infty}^{\infty} \left[\delta_{n,m} + \frac{L-d}{L} \frac{(\Gamma_m R_{\mathrm{ext}}) \mathrm{K}_1(\Gamma_m R_{\mathrm{ext}})}{(\Gamma_n R_{\mathrm{ext}})^2 \mathrm{K}_0(\Gamma_n R_{\mathrm{ext}})} \sum_{\nu=0}^{\infty} \frac{\mathcal{R}_{0,\nu}(R_{\mathrm{ext}}) \Lambda_\nu R_{\mathrm{ext}}}{\mathcal{R}_{1,\nu}(R_{\mathrm{ext}}) g_\nu} \mathcal{L}_{n,\nu} \mathcal{L}_{m,\nu}^* \right] A_m = 0. \tag{5.3.14}$$

The other possibility is to substitute (5.3.12) in (5.3.13) and obtain one equation for the amplitudes of the various modes in the groove

$$\sum_{\mu=0}^{\infty} \left[\delta_{\nu,\mu} + \frac{L-d}{L} \frac{(\Lambda_\mu R_{\mathrm{ext}})^2 \mathcal{R}_{0,\mu}(R_{\mathrm{ext}})}{(\Lambda_\nu R_{\mathrm{ext}}) \mathcal{R}_{1,\nu}(R_{\mathrm{ext}}) g_\nu} \sum_{n=-\infty}^{\infty} \frac{\mathrm{K}_1(\Gamma_n R_{\mathrm{ext}}) \mathcal{L}_{n,\nu}^* \mathcal{L}_{n,\nu}}{(\Gamma_n R_{\mathrm{ext}}) \mathrm{K}_0(\Gamma_n R_{\mathrm{ext}})} \right] B_\mu = 0. \tag{5.3.15}$$

In both cases, the dispersion relation is calculated from the requirement that the determinant of the matrix which multiplies the vector of amplitudes, is zero. As in the closed structure, the two methods are equivalent, but the last expression is by far more efficient for practical calculation.

There is one substantial difference between open and closed periodic structures. In the latter case, the radiation is guided by the waveguide and there is an infinite discrete spectrum of frequencies that can propagate along the system. In open structures, modes can propagate provided that the projection of the wave-numbers

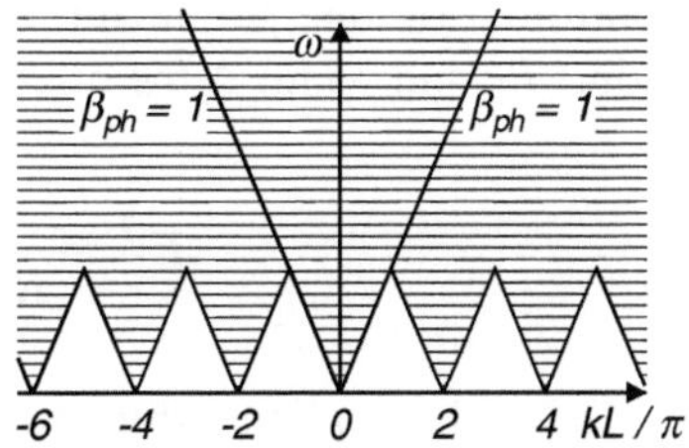

Fig. 5.11 Only the wave-numbers which are in the white triangles correspond to waves supported by an open structure

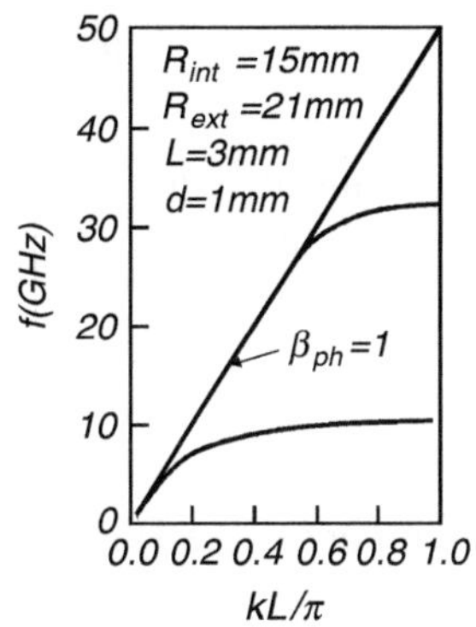

Fig. 5.12 Dispersion relation of the open periodic structure

of *all harmonics* in the first Brillouin zone corresponds to waves whose phase velocity is smaller than c; in other words, no radiation propagates outwards (radially). Figure 5.11 illustrates the two regions of interest: in the shadowed region no solutions are permissible and in the remainder the solution is possible with an adequate choice of the geometric parameters. It is evident from this picture that waves at frequencies higher than

$$f \geq \frac{1}{2}\frac{c}{L}, \tag{5.3.16}$$

cannot be supported by a disk-loaded wire, regardless of the geometrical details of the cavity. With this regard, an open structure forms a low pass filter. Figure 5.12 illustrates the dispersion relation of such a system for $L = 3$ mm, $d = 1$ mm, $R_{int} = 15$ mm and $R_{ext} = 21$ mm. For comparison, in the same frequency range (0–50 GHz) there are 6 symmetric TM modes which can propagate in a closed system of the same geometry; obviously there are many others at higher frequencies.

5.3.2 Interaction Parameters

Provided that the electrons are interacting primarily with one harmonic (say $n = 0$) then we assume that the spatial component of the interaction in an amplifier is controlled by one parameter: the interaction impedance. For an annular beam of radius R_b and width Δ the effective field that acts on the electrons is

$$|E|^2 \equiv \frac{1}{R_\mathrm{b}\Delta}\int_{R_\mathrm{b}-\Delta/2}^{R_\mathrm{b}+\Delta/2} drr|E_{z,n=0}(r,z)|^2. \tag{5.3.17}$$

Using the explicit expression for E_z in (5.3.2) we find

$$|E|^2 \equiv \frac{c^4}{\omega^2}|A_0|^2\Gamma_0^2\frac{1}{R_\mathrm{b}\Delta}\int_{\Gamma_0(R_\mathrm{b}-\Delta/2)}^{\Gamma_0(R_\mathrm{b}+\Delta/2)} d\xi\xi\mathrm{K}_0^2(\xi); \tag{5.3.18}$$

assuming that the variations of the wave across the beam section are negligible the integral reads

$$\int_{\Gamma_0(R_\mathrm{b}-\Delta/2)}^{\Gamma_0(R_\mathrm{b}+\Delta/2)} d\xi\xi\mathrm{K}_0^2(\xi) \simeq (\Gamma_0^2R_\mathrm{b}\Delta)\mathrm{K}_0^2(\Gamma_0R_\mathrm{b}); \tag{5.3.19}$$

With this expression the effective electric field reads

$$|E|^2 \equiv \frac{c^4\Gamma_0^4}{\omega^2}|A_0|^2\mathrm{K}_0^2(\Gamma_0R_\mathrm{b}). \tag{5.3.20}$$

In order to determine the explicit expression for the interaction impedance the total power, which flows along the structure, has to be determined. According to (5.3.2) it is given by

$$P = \frac{1}{2}(2\pi)\frac{1}{\eta_0}\sum_{n=-\infty}^{\infty}|cA_n|^2\left(\frac{ck_n}{\omega}\right)\int_{\Gamma_nR_\mathrm{ext}}^{\infty} d\xi\xi\mathrm{K}_1^2(\xi). \tag{5.3.21}$$

The last integral can be evaluated analytically (Abramowitz and Stegun 1968, p.484) and it reads

$$W_3(x) \equiv \int_x^{\infty} d\xi\xi\mathrm{K}_1^2(\xi) = x\mathrm{K}_0(x)\mathrm{K}_1(x) + \frac{x^2}{2}\left[\mathrm{K}_0^2(x) - \mathrm{K}_1^2(x)\right] \tag{5.3.22}$$

hence

$$\begin{aligned} Z_\mathrm{int} &\equiv \frac{1}{2}\frac{|E|^2(\pi R_\mathrm{ext}^2)}{P} = \frac{1}{2}\eta_0\left[\frac{\mathrm{K}_0(\Gamma_0R_\mathrm{b})}{\mathrm{K}_0(\Gamma_0R_\mathrm{ext})}\right]^2 \\ &\times\left[\sum_{n=-\infty}^{\infty}\frac{(\frac{\omega}{c}R_\mathrm{ext})(k_nR_\mathrm{ext})W_3(\Gamma_nR_\mathrm{ext})}{(\Gamma_nR_\mathrm{ext})^4\mathrm{K}_0^2(\Gamma_nR_\mathrm{ext})}\frac{\mathrm{sinc}^2\left[\frac{1}{2}k_n(L-d)\right]}{\mathrm{sinc}^2\left[\frac{1}{2}k_n(L-d)\right]}\right]^{-1} \end{aligned} \tag{5.3.23}$$

The other parameter of interest in an oscillator is the interaction dielectric coefficient, which is a measure of the total electromagnetic energy, stored in the open structure and is defined by

$$\varepsilon_{\text{int}} \equiv \frac{W_{\text{em}}}{\frac{1}{2}\varepsilon_0 |E|^2 (\pi R_{\text{ext}}^2)}. \tag{5.3.24}$$

Note that in the open system the effect of the beam distance from the structure is represented by

$$Z_{\text{int}} \propto \left[\frac{\text{K}_0(\Gamma_0 R_{\text{b}})}{\text{K}_0(\Gamma_0 R_{\text{ext}})}\right]^2, \tag{5.3.25}$$

which for large arguments of the modified Bessel function implies $Z_{int} \propto \exp[-2\Gamma_0(R_{\text{b}} - R_{ext})]$ whereas the dielectric coefficient of the interaction,

$$\varepsilon_{\text{int}} \propto \left[\frac{\text{K}_0(\Gamma_0 R_{\text{ext}})}{\text{K}_0(\Gamma_0 R_{\text{b}})}\right]^2, \tag{5.3.26}$$

is proportional to $\varepsilon_{\text{int}} \propto \exp[2\Gamma_0(R_{\text{b}} - R_{\text{ext}})]$ for large arguments.

The two frames in Fig. 5.13 illustrate the interaction impedance of the two modes, which are supported by the structure introduced above. In contrast to closed structure where the impedance has a minimum, in the open structure presented here, the impedance has a maximum as a function of the frequency. The peak of the lower (frequency) branch occurs at 9.575 GHz, the phase velocity is 0.58 c, the interaction impedance, for $R_{\text{b}} = 25\,\text{mm}$, is $Z_{\text{int}} = 196$ and the coupling coefficient $K_0 = 29.2\,\text{m}^{-1}$ see (4.1.18) when the total current is 500 A. The peak at the upper branch occurs at 31.75 GHz and $Z_{\text{int}} = 31.9$ corresponding to $K_0 = 9.3\,\text{m}^{-1}$; the phase velocity in this case is 0.877 c.

We indicated previously that the advantage of an open periodic structure is that it supports the propagation of a small number of modes. Clearly, it would be

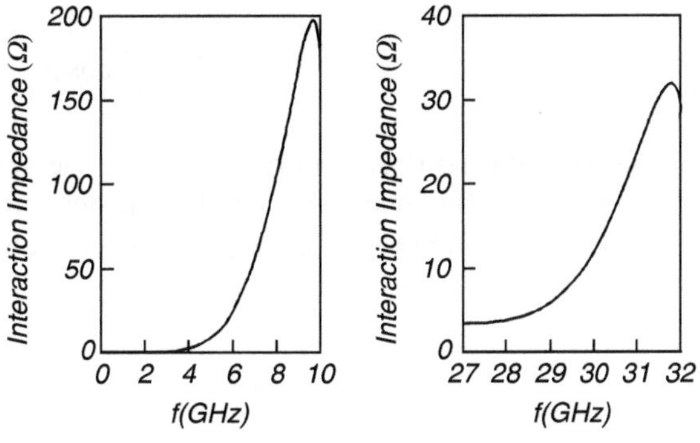

Fig. 5.13 The interaction impedance of the two modes presented in Fig. 5.12

convenient to utilize this fact to generate radiation at high frequency. From the previous example, we observe that we could generate radiation at 32 GHz if we can suppress the lower frequency and no additional frequencies (of TM like modes) can develop in the system. In order to suppress the lower frequency we can take advantage of the fact that the interacting waves decay exponentially in the radial direction and a high frequency wave decays radially more rapidly than a low frequency one. It implies that in principle, we can put an absorbing wall (say at $R = 30\,\mathrm{mm}$) which will virtually absorb all the energy from the low frequency wave but practically it will not affect the higher frequency mode since its amplitude is virtually zero at its location. Obviously, the coupling to the beam is weaker because of the exponential radial decay associated to the evanescent wave.

5.3.3 Green's Function: The Smith-Purcell Effect

When we investigated the electromagnetic field generated by a charged particle in its motion near a dielectric material it was shown that radiation could be generated if the velocity of the particle exceeds the phase velocity of the plane wave in the medium. A similar process may occur in a metallic periodic structure. Qualitatively the process is as follows: it was indicated in Sect. 2.2.4 that a point charge moving at a velocity v_0 generates a continuous spectrum of evanescent (non-radiating) waves; these waves impinge upon the grating whose periodicity is L. The incident wave-number in the direction parallel to the motion of the particle is given by

$$k_z^{\mathrm{inc}} = \frac{\omega}{\mathrm{v}_0}. \tag{5.3.27}$$

Although the great majority of the reflected waves are evanescent, under certain circumstances, there might be a few, which can propagate. An observer, located far away from the grating at an angle θ relative to the motion of the particle (z direction), measures the outgoing radiation. The projection on the z direction of the wave-number as measured by this observer is

$$k_z^{\mathrm{obs}} = \frac{\omega}{c}\cos\theta. \tag{5.3.28}$$

The periodic structure couples between the wave-numbers in the z-direction therefore the difference between the incident and observed (scattered) wave-numbers is attributed to the grating and it is an integer number ν of grating wave-numbers $2\pi\nu/L$,

$$k_z^{\mathrm{inc}} - k_z^{\mathrm{obs}} = \frac{2\pi\nu}{L}. \tag{5.3.29}$$

Substituting the previous two equations into the latter, we obtain

$$\frac{\omega}{c} = \frac{2\pi}{L} \frac{v}{\beta^{-1} - \cos\theta}. \tag{5.3.30}$$

It indicates that for a given velocity and a given periodicity, different frequencies are emitted in different directions. Smith and Purcell (1953) first reported the effect. Toraldo di Francia (1960) has formulated the problem in terms of the coupling between evanescent and propagating waves and Van den Berg (1973) has calculated the effect numerically. Salisbury (1970) observed a similar spectrum of radiation but in his experiment, he found that there is a correlation between the radiation intensity and the current associated with electrons scattered by the grating. His interpretation is based on the oscillation of the electrons in the periodic potential induced by the scattered electrons. However, estimates of the acceleration associated with this process indicate that it cannot account for the intensity of the observed radiation – Chang and McDaniel (1989). Later, the Smith-Purcell effect was re-examined at much higher energies (3.6 MeV) and in the angle range $56^\circ - 150^\circ$; the agreement between the dispersion relation and the experiment was excellent – see Doucas et al. (1992). A renew interest in the Smith-Purcell effect was motivated by the increased interest in terahertz radiation (Andrews et al. 2005; Korbly et al. 2005; Kim 2007; Shin et al. 2007). In this subsection we discuss in detail the dynamics of the Smith-Purcell effect as a particular case of Green's function formulation of the electromagnetic problem in an open periodic structure.

Consider a *train of charged rings* of radius $R_b > R_{\rm ext}$ moving at a constant velocity v_0 along a periodic structure identical to the one illustrated in Fig. 5.10. Consequently, the current density is periodic in space and time therefore

$$J_z(r,z,t) = \frac{-q\mathrm{v}_0}{2\pi r}\delta(r - R_{\rm b})\frac{1}{L}\sum_{n=-\infty}^{\infty} \exp\left[-j\frac{2\pi n}{L}(z - \mathrm{v}_0 t)\right]. \tag{5.3.31}$$

This current density excites the longitudinal component of the magnetic vector potential that has a homogeneous and a non-homogeneous component. For evaluating the latter, we may ignore the structure thus

$$A_z^{(nh)}(r,z,t) = \sum_{n=-\infty}^{\infty} A_n(r)\exp\left[-j\frac{2\pi n}{L}(z - \mathrm{v}_0 t)\right] \tag{5.3.32}$$

where $a_n(r)$ is a solution of

$$\left[\frac{1}{r}\frac{\mathrm{d}}{\mathrm{d}r}r\frac{\mathrm{d}}{\mathrm{d}r} - \left(\frac{2\pi n}{\gamma L}\right)^2\right]A_n(r) = \frac{q\mathrm{v}_0\mu_0}{2\pi L}\frac{1}{r}\delta(r - R_{\rm b}) \tag{5.3.33}$$

and defining $\chi_n = 2\pi|n|/\gamma L$, it reads

$$A_n(r) = -\frac{q v_0 \mu_0}{2\pi L} \mathrm{I}_0(\chi_n R_b)\mathrm{K}_0(\chi_n R_b)\begin{cases} \mathrm{K}_0(\chi_n r)/\mathrm{K}_0(\chi_n R_b) & r > R_b, \\ \mathrm{I}_0(\chi_n r)/\mathrm{I}_0(\chi_n R_b) & r < R_b, \end{cases} \tag{5.3.34}$$
$$a_n \equiv \mathrm{I}_0(\chi_n R_{\mathrm{ext}})\mathrm{K}_0(\chi_n R_b).$$

The homogeneous solution has to have the same temporal dependence since together they must satisfy the boundary conditions at any instant thus

$$A_z^{(h)}(r,z,t) = -\frac{q v_0 \mu_0}{2\pi L} \sum_{n=-\infty}^{\infty} \exp\left(j\frac{2\pi n}{L} v_0 t\right) \sum_{m=-\infty}^{\infty} \exp\left(-j\frac{2\pi m}{L} z\right)$$
$$\times\, b_m(n)\mathrm{K}_0\left[\frac{2\pi r}{L}\sqrt{m^2 - n^2\beta^2}\right]. \tag{5.3.35}$$

Imposing the boundary conditions for E_z and H_ϕ in a similar way as done in this section for establishing the dispersion relation, we may formulate the two non-homogeneous equations in terms of a *reflection matrix*

$$b_m = \sum_{m'} \overline{\mathcal{R}}_{m,m'} a_{m'} \tag{5.3.36}$$

a_m is the normalized amplitude of the wave impinging upon the periodic structure. We are not aware of an analytic expression for the reflection matrix and in general, numerical methods are necessary for its evaluation. However, for the geometry employed here, quasi-analytic formulation is possible adopting an approach identical to that used to determine the dispersion relation. An explicit formulation of this matrix is left as an exercise. In the remainder of this subsection we determine the expression for: (1) the power emitted radially and (2) the decelerating force acting on one ring assuming that the reflection matrix is known.

For the power radiated radially the relevant field components are

$$E_z(r,z,t) = \frac{q v_0 \mu_0}{2\pi L} \sum_{n=-\infty}^{\infty} \exp\left(j\frac{2\pi n}{L} v_0 t\right) \sum_{m=-\infty}^{\infty} \exp\left(-j\frac{2\pi m}{L} z\right) \mathrm{b}_m(n)$$
$$\times\, \mathrm{K}_0\left[2\pi\frac{r}{L}\sqrt{m^2 - n^2\beta^2}\right]\left\{\frac{c^2}{j\frac{2\pi n}{L} v_0}\left[\left(\frac{2\pi m}{L}\right)^2 - \left(\frac{2\pi n}{L}\beta\right)^2\right]\right\}$$
$$H_\phi(r,z,t) = -\frac{q v_0 \mu_0}{2\pi L}\frac{1}{\mu_0} \sum_{n=-\infty}^{\infty} \exp\left(j\frac{2\pi n}{L} v_0 t\right) \sum_{m=-\infty}^{\infty} \exp\left(-j\frac{2\pi m}{L} z\right) \mathrm{b}_m(n)$$
$$\times\, \mathrm{K}_1\left[2\pi\frac{r}{L}\sqrt{m^2 - n^2\beta^2}\right]\left(\frac{2\pi}{L}\sqrt{m^2 - n^2\beta^2}\right) \tag{5.3.37}$$

thus the energy radiated ($r \to \infty$) is

$$W_{\text{rad}} \equiv \int_0^{L/v_0} dt 2\pi r \int_0^L dz S_r(r,z,t) = -2\pi r \int_0^{L/v_0} dt \int_0^L dz E_z(r,z,t) H_\phi(r,z,t)$$
$$= \frac{q^2}{4\pi\varepsilon_0 L} 8\pi^2 \sum_{n=1,m=0}^{\infty} \frac{n^2\beta^2 - m^2}{n} |\mathrm{b}_m(n)^2| h(n^2\beta^2 - m^2) \tag{5.3.38}$$

whereas the decelerating force on one charged ring is $F_{\text{dec}} = W_{\text{rad}}/L$ or explicitly

$$F_{\text{dec}} = \frac{q^2}{4\pi\varepsilon_0 (L/2\pi)^2} 2 \sum_{n=1,m=0}^{\infty} \frac{n^2\beta^2 - m^2}{n} |\mathrm{b}_m(n)^2| h(n^2\beta^2 - m^2). \tag{5.3.39}$$

As clearly reflected from the step function, in both cases only harmonics corresponding to propagating waves contribute. The radiation energy as determined in (5.3.38) is indicative of the *coherent Smith-Purcell radiation* as generated by the train of charged rings moving in the close vicinity of the disk-loaded wire. Our next step is to generalize this approach in order to determine the electromagnetic field generated by a non-periodic source in the presence of a periodic structure

5.3.4 Periodic Structure and Non-periodic Source

The reflection matrix formulation introduced in the previous subsection although developed with a specific geometry in mind is general and it may be employed for an *arbitrary geometry*. In this subsection we aim to determine the maximum decelerating and transverse force on a charged line (Q/Δ_y) moving at a constant velocity v_0 at a height h from the top of a grating of arbitrary geometry but of periodicity L – see Fig. 5.14.

As above, the non-homogeneous component of the magnetic vector potential is

$$A_z^{(nh)}(x,z,t) = \int_{-\infty}^{\infty} \mathrm{d}\omega \int_{-\infty}^{\infty} \mathrm{d}k A(\omega,k) \frac{1}{4\pi\Gamma} \exp(j\omega t - jkz - \Gamma|x-h|) \tag{5.3.40}$$

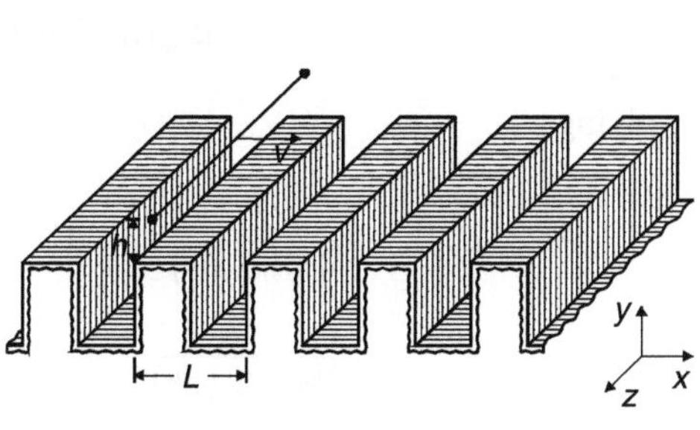

Fig. 5.14 Rectangular grating of periodicity L

where $\Gamma = \sqrt{k^2 - (\omega/c)^2}$, $A(\omega, k) = -(Q/\Delta_y)\mu_0 \mathrm{v}_0 \delta(\omega - \mathrm{v}_0 k)$ and the homogeneous counterpart that, also represents the scattered waves, is given by

$$A_z^{(h)}(x,z,t) = \int_{-\infty}^{\infty} \mathrm{d}\omega \int_{-\infty}^{\infty} \mathrm{d}k\, B(\omega,k) \exp(j\omega t - jkz - \Gamma x)\,. \quad (5.3.41)$$

Without loss of generality, the amplitude of the scattered field, $B(\omega, k)$, may be expressed in terms of Floquet harmonics $\overline{B}_n(\omega, k') \equiv B(\omega, k' + 2\pi n/L)$ where $k'$is the projection of the wave number on the first Brillouin zone. It is related to the incident field in terms of a reflection matrix

$$\bar{B}_n(\omega,k') = \sum_m \bar{\mathcal{R}}_{nm}(\omega,k')\bar{A}_m(\omega,k') \quad (5.3.42)$$

where

$$\begin{aligned}\overline{A}_m(\omega,k') &= A\left(\omega, k' + \frac{2\pi m}{L}\right)\frac{\exp(-\Gamma_m h)}{4\pi\Gamma_m},\\ \Gamma_m &= \sqrt{k_m^2 - \frac{\omega^2}{c^2}},\\ k_m &= k' + \frac{2\pi m}{L}.\end{aligned} \quad (5.3.43)$$

As a second step we determine the average longitudinal gradient (per unit length),

$$E_\parallel \equiv \frac{\mathrm{v}_0}{L}\int_{-\mathrm{v}_0/2L}^{\mathrm{v}_0/2L} \mathrm{d}t\, E_z^{(h)}(x = h, z = \mathrm{v}_0 t,\ t). \quad (5.3.44)$$

which explicitly reads

$$E_\parallel = \frac{Q}{4\pi\varepsilon_0\Delta_y}\frac{1}{\gamma}\int_{-\pi/L}^{\pi/L} \mathrm{d}k' \sum_{n=-\infty}^{\infty} \bar{\mathcal{R}}_{nn}(\omega = \mathrm{v}_0 k_n, k')\frac{|k_n|}{jk_n}\exp(-2|k_n|h/\gamma)\,. \quad (5.3.45)$$

At this point we assume that the absolute value of each diagonal term of the reflection matrix is smaller than unity,

$$|\mathcal{R}_{nn}(\omega = \mathrm{v}_0 k_n, k')| \leq 1 \quad (5.3.46)$$

implying

$$\begin{aligned}E_\parallel \leq E_\parallel^{(\max)} &\equiv \frac{Q}{4\pi\varepsilon_0\Delta_y h}\int_{-\pi/L}^{\pi/L} \mathrm{d}k' h/\gamma \sum_{n=-\infty}^{\infty}\exp(-2|k_n|h/\gamma)\\ &= \frac{Q}{2\pi\varepsilon_0\Delta_y(2h)}\,.\end{aligned} \quad (5.3.47)$$

In a similar way we may calculate the average transverse field

$$E_{\perp} \equiv \frac{\mathrm{v}_0}{L} \int_{-L/2\mathrm{v}_0}^{L/2\mathrm{v}_0} \mathrm{d}t \left[E_x - \mathrm{v}_0 \mu_0 H_y\right]_{z=vt,x=h,t}. \quad (5.3.48)$$

As in the longitudinal case

$$E_{\perp} = \frac{Q}{2\pi\varepsilon_0 \Delta_y (2h)} \frac{1}{\gamma} \int_{-\pi/L}^{\pi/L} \mathrm{d}k'(h/\gamma) \sum_{n=-\infty}^{\infty} \bar{\mathcal{R}}_{nn}(\omega = \mathrm{v}_0 k_n, k') \exp(-2|k_n|h/\gamma)\,, \quad (5.3.49)$$

and subject to the condition in (5.3.46), we get

$$E_{\perp} \leq E_{\perp}^{(\max)} = \frac{Q}{2\pi\varepsilon_0 \Delta_y (2h)} \frac{1}{\gamma} \quad (5.3.50)$$

The last two results (5.3.47) and (5.3.50) provide us with extremum values for the transverse and longitudinal gradients as a charged-line traverses an open periodic structure of *arbitrary* geometry. Several aspects are evident subject to our assumption in (5.3.46): (1) both gradients are independent of the period or any other geometric parameter of the structure. (2) Both gradients are inversely proportional to the image-charge gradient (corresponding to a flat surface) i.e. are inversely proportional to the height (h). (3) The extremum *longitudinal* force is independent of the kinetic energy of the particle $(\gamma \gg 1)$. (4) This is in particular important for ultra-relativistic particles since the reflection matrix required for the calculation becomes very large. (5) For a motionless particle the longitudinal gradient vanishes. (6) The extremum *transverse* gradient corresponds to an image force that is inversely proportional to γ and for a motionless charge-line this extremum equals the exact gradient in the case of a flat plane.

5.4 Bragg Waveguides

Bragg reflection waveguides are one-dimensional periodic structures, designed to guide light in a low refractive index surrounded by alternating layers of high refractive index. Two possible configurations are depicted in Fig. 5.15, one is the planar Bragg reflection waveguide, and the other is the cylindrical, which is also known as the *Bragg Fiber*. The theory of Bragg reflection waveguides was developed by Yeh and Yariv (1976) and recently there has been a growing interest in using such hollow cylindrical structures as low-loss optical fibers in long distance communications as well as for optical structures for particles acceleration.

Most of the studies carried out on Bragg reflection waveguides dealt with configurations where all the dielectric layers are transverse quarter-wave thick.

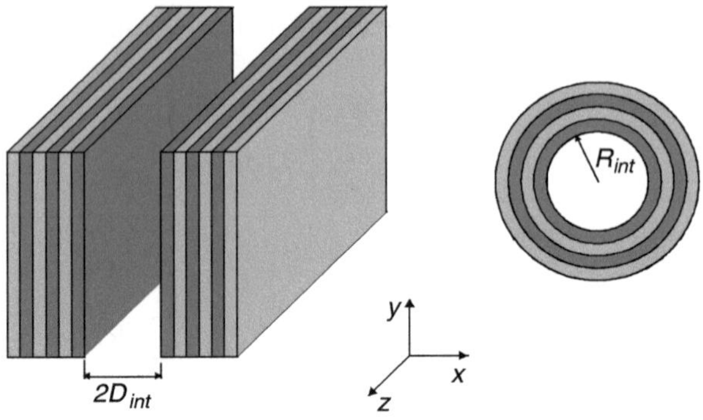

Fig. 5.15 Planar and cylindrical Bragg reflection waveguides

However, controlling the dispersion properties in Bragg fibers was demonstrated by creating a defect in the form of changing one or more of the layer widths. In this section, a systematic adjustment of the layer adjacent to the core, will be shown to change the waveguide properties to fulfill specified requirements.

A metallic waveguide supports modes with phase velocity which is always greater than c, and therefore for a metallic waveguide to serve as a particle accelerator, the electromagnetic wave must be "slowed down". Such an effect can be achieved by either designing an appropriate metallic periodic structure, or by partially filling the waveguide with a dielectric material. If at the vicinity of the operation wavelength in a Bragg reflection waveguide, the reflector acts similarly to a metallic wall, as was shown in previous studies, the phase velocity is expected to be greater than c, and therefore some adaptation is required for this to change. Motivated by the requirement $v_{ph} = c$, we develop in this section a general method for designing the Bragg waveguide for a given phase velocity, given the core dimension and a set of dielectric materials. The core's dimension itself may be dictated by other considerations, such as the maximum field, allowed to develop within the core to prevent material breakdown, and the interaction efficiency. The modes of interest are the symmetric modes, TM and TE, the hollow core field distributions of which are summarized in Table 5.1. As a special case, which received modest attention in waveguide literature, the field components corresponding to $v_{ph} = c$, are written explicitly. All the demonstration presented in this section will be on waveguides made from dielectric materials with refractive indices 1.6 and 4.6. In addition, we neglect the losses due to the finite cladding.

5.4.1 Matching Layer

We consider the *planar* Bragg reflection waveguide ($\partial/\partial y = 0$), with core half-width $D_{\rm int}$, as depicted in Fig. 5.15. Let us assume that at some specific wavelength λ_0 with a corresponding angular frequency ω_0, this waveguide is required to support a symmetric TM mode with a specific phase velocity $v_{ph} = \omega_0/k_z$,

Table 5.1 Hollow core symmetric modes. The transverse wave-numbers are $k_\perp = \sqrt{\omega^2/c^2 - k_z^2}$. In the planar case $k_x = k_\perp$ whereas in the cylindrical configuration $k_r = k_\perp$

	General symmetric mode $\times \exp(-jk_z z)$	Special case $v_{ph} = c \times \exp(-j\omega z/c)$
Planar TM	$E_z = E_0 \cos(k_\perp x)$	$E_z = E_0$
	$E_x = j\left(\frac{k_z}{k_\perp}\right) E_0 \sin(k_\perp x)$	$E_x = j\left(\frac{\omega}{c}x\right) E_0$
	$H_y = j\left(\frac{\omega}{ck_\perp}\right)\frac{1}{\eta_0} E_0 \sin(k_\perp x)$	$H_y = \frac{j}{\eta_0}\left(\frac{\omega}{c}x\right) E_0$
Planar TE	$H_z = H_0 \cos(k_\perp x)$	$H_z = H_0$
	$H_x = j\left(\frac{k_z}{k_\perp}\right) H_0 \sin(k_\perp x)$	$H_x = j\left(\frac{\omega}{c}x\right) H_0$
	$E_y = -j\left(\frac{\omega}{ck_\perp}\right)\eta_0 H_0 \sin(k_\perp x)$	$E_y = -j\left(\frac{\omega}{c}x\right)\eta_0 H_0$
Cyl. TM	$E_z = E_0 J_0(k_\perp r)$	$E_z = E_0$
	$E_r = j\left(\frac{k_z}{k_\perp}\right) E_0 J_1(k_\perp r)$	$E_r = j\left(\frac{1}{2}\frac{\omega}{c}r\right) E_0$
	$H_\phi = j\left(\frac{\omega}{ck_\perp}\right)\frac{1}{\eta_0} E_0 J_1(k_\perp r)$	$H_\phi = j\left(\frac{1}{2}\frac{\omega}{c}r\right)\frac{1}{\eta_0} E_0$
Cyl. TE	$H_z = H_0 J_0(k_\perp r)$	$H_z = H_0$
	$H_r = j\left(\frac{k_z}{k_\perp}\right) H_0 J_1(k_\perp r)$	$H_r = j\left(\frac{1}{2}\frac{\omega}{c}r\right) H_0$
	$E_\phi = -j\left(\frac{\omega}{ck_\perp}\right)\eta_0 H_0 J_1(k_\perp r)$	$E_\phi = -j\left(\frac{1}{2}\frac{\omega}{c}r\right)\eta_0 H_0$

k_z being the longitudinal wave-number. Equivalently, a specific field distribution in the hollow core may be required, determined by the transverse wave-number $k_x = \sqrt{\omega^2/c^2 - k_z^2}$, as shown in Table 5.1. Expressing the phase velocity in terms of the transverse wave-number in the core, we obtain

$$v_{ph} = \frac{c}{\sqrt{1 - \left(\frac{k_x \lambda_0}{2\pi}\right)^2}}. \tag{5.4.1}$$

The electromagnetic field components in the layer adjacent to the core, which has a dielectric coefficient ε_1, are given by

$$\begin{aligned}
E_z &= [A_1 \exp(-jk_1 x) + B_1 \exp(+jk_1 x)] \exp(-jk_z z),\\
E_x &= -\frac{k_z}{k_1}[A_1 \exp(-jk_1 x) - B_1 \exp(+jk_1 x)] \exp(-jk_z z),\\
H_y &= -\frac{1}{Z_1}[A_1 \exp(-jk_1 x) - B_1 \exp(+jk_1 x)] \exp(-jk_z z),
\end{aligned} \tag{5.4.2}$$

wherein the transverse wave-number is $k_1 = \sqrt{\varepsilon_1 \omega^2/c^2 - k_z^2}$ and the transverse impedance is $Z_1 = \eta_0 k_1 c/\omega\varepsilon_1$. The required electromagnetic field in the *vacuum* core dictates by virtue of the boundary conditions on E_z and H_y at the interface between the core and the adjacent dielectric layer, the amplitudes in the *first layer*. Imposing the boundary conditions on E_z and H_y at $x = D_{\rm int}$, the amplitudes are found to be given by

$$\begin{aligned} A_1/E_0 &= (B_1/E_0)^* \\ &= \frac{1}{2}\exp(jk_1 D_{\rm int})\cos(k_x D_{\rm int}) - j\frac{k_1}{2\varepsilon_1 k_x}\exp(jk_1 D_{\rm int})\sin(k_x D_{\rm int}). \end{aligned} \tag{5.4.3}$$

The electromagnetic field required in the core entails the amplitudes of the outgoing and incoming transverse waves, as calculated above. It is now our goal to ensure that the complete structure, including the Bragg reflector, indeed supports the required field at the given wavelength as an eigen-mode.

The Bragg reflector can be analyzed from the perspective of a pure periodic structure according to the Floquet theorem (Yeh and Yariv 1976). This analysis gives the eigen-vectors and the eigen-values of the periodic structure, and determines the band-gaps of the system, where the waves are evanescent. The strongest exponential decay is found to be when each material of the two is chosen to be a quarter of wavelength thick. In case of a Bragg reflection waveguide it is a quarter of the transverse wavelength, meaning that this thickness of layer v with dielectric coefficient ε_v is given by

$$\Delta_v = \frac{\pi}{2\sqrt{\frac{\omega_0^2}{c^2}\varepsilon_v - k_z^2}}. \tag{5.4.4}$$

The eigen-values in this optimal decay case are given by the ratios of the two transverse impedances in the two materials. In the optimal confinement case, each one of the electromagnetic field components either peaks or vanishes at the interface between *any* two dielectrics. Going back to the Bragg reflection waveguide, since the amplitudes in the first dielectric layer are already known, the interface between the first and the second dielectric layers may be considered as an *entrance* to a periodic structure, to which the wave must enter in one of the eigen-vectors for the mode to be supported. Explicitly, this condition is given by

$$\left(\begin{array}{ll} E_z(x = D_{\rm int} + \Delta_1) = 0 & Z_1 > Z_2 \\ \dfrac{\partial E_z}{\partial x}(x = D_{\rm int} + \Delta_1) = 0 & Z_1 < Z_2, \end{array} \right. \tag{5.4.5}$$

where Δ_1 is the first layer width, and Z_1, Z_2 are the transverse impedances of the first and second layers respectively. This condition was pointed out by Mizrahi and Schächter (2004a). Setting the first layer width according to the above condition will ensure that the required mode at the given wavelength will indeed be supported by the waveguide. The first layer may therfore be conceived as a *matching layer* between the *vacuum region* and the subsequent *periodic structure*, as it rotates the amplitude vector dictated by the vacuum mode, to overlap the eigen-vector of the periodic structure.

Given the amplitudes, as required by (5.4.3), it is now straightforward to determine the points where E_z peaks or vanishes. The resulting expression for the first layer width reads

$$\Delta_1^{(\mathrm{TM})} = \begin{cases} \dfrac{1}{k_1} \arctan\left(\dfrac{\varepsilon_1 k_x}{k_1} \cot(k_x D_{\mathrm{int}})\right) & Z_1 > Z_2 \\ \dfrac{1}{k_1} \arctan\left(-\dfrac{k_1}{\varepsilon_1 k_x} \tan(k_x D_{\mathrm{int}})\right) & Z_1 < Z_2. \end{cases} \tag{5.4.6}$$

In the above expression, the smallest positive value of the arctan function is chosen. It should be noted that k_x may be purely imaginary, meaning that the transverse waves in the core are evanescent, and the expression still holds, as long as the transverse wave-numbers in the dielectric layers are real. A special case of this expression is when $k_x D_{\mathrm{int}} = \pi, \pi/2$, and then the matching layer is transverse quarter-wavelength thick similarly to the outer layers. For the special case where the phase velocity equals the speed of light ($k_z = \omega_0/c$), the expression for the first layer width reads

$$\Delta_1^{(\mathrm{TM})} = \begin{cases} \dfrac{1}{k_1} \arctan\left[\left(\dfrac{Z_1}{\eta_0}\dfrac{\omega_0}{c} D_{\mathrm{int}}\right)^{-1}\right] & Z_1 > Z_2 \\ \dfrac{1}{k_1} \arctan\left(-\dfrac{Z_1}{\eta_0}\dfrac{\omega_0}{c} D_{\mathrm{int}}\right) & Z_1 < Z_2. \end{cases} \tag{5.4.7}$$

Figure 5.16 illustrates at the bottom curve the planar TM first layer width as a function of the core half-width D_{int}, for the requirement that $v_{\mathrm{ph}} = c$. The first layer was set to have a refractive index of $n_1 = \sqrt{\varepsilon_1} = 1.6$, and the other material was taken to be of refractive index $n_2 = \sqrt{\varepsilon_2} = 4.6$. The first layer width is normalized by $\Delta_{\mathrm{q}} \equiv \lambda_0/4\sqrt{\varepsilon_1 - 1}$, which is the transverse quarter-wavelength width in the

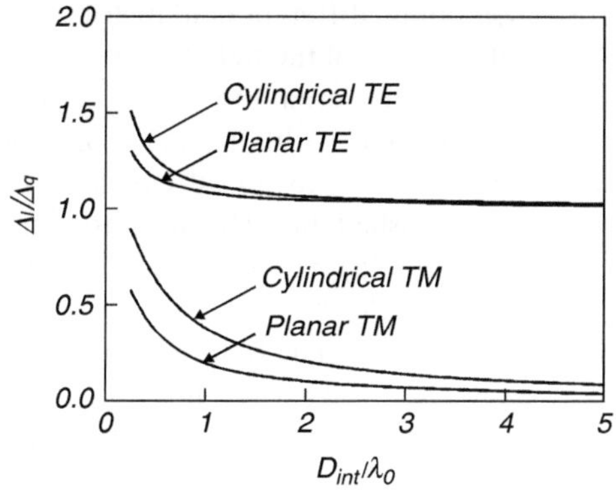

Fig. 5.16 First layer width for $v_{\mathrm{ph}} = c$, normalized by $\Delta_{\mathrm{q}} \equiv \lambda_0/4\sqrt{\varepsilon_1 - 1}$. The layer adjacent to the core has a refractive index of $n_1 = 1.6$ and the other material has $n_2 = 4.6$

$v_{ph} = c$ case. The choice of placing the lower refractive index first entails that first case of (5.4.7) is used in the calculation of the planar TM first layer width.

The same principles can easily be implemented for TE modes, planar and cylindrical. From the requirement that H_z either peaks or vanishes the planar TE matching layer width is found to be

$$\Delta_1^{(\mathrm{TE})} = \begin{cases} \dfrac{1}{k_1} \arctan\left(\dfrac{k_x}{k_1} \cot(k_x D_{\mathrm{int}})\right) & Y_1 > Y_2 \\ \dfrac{1}{k_1} \arctan\left(-\dfrac{k_1}{k_x} \tan(k_x D_{\mathrm{int}})\right) & Y_1 < Y_2, \end{cases} \tag{5.4.8}$$

wherein $Y_{1,2} = ck_{1,2}/\omega_0\eta_0$ are the transverse admittances; $k_{1,2}$ are the transverse wave-numbers in the first and second layers. The difference between the above expression and the TM expression is in a ε_1 factor in the arctan argument. The second curve from the top of Fig. 5.16 illustrates the planar TE case. We observe that the TM curves approach zero, whereas the TE curve is above the $\Delta_1/\Delta_{\mathrm{q}} = 1$ line, and approaches it for larger core widths. This situation is reversed according to the given analytical expressions, should the material of the layer adjacent to the core is chosen to be of the higher refractive index of the two mentioned.

To summarize, the design procedure is as follows: According to the required k_z, a Bragg mirror is designed so that all layers are transverse quarter-wavelength thick. In order to match the mirror to the desired core field, the layer adjacent to the core is adjusted to the width given above by analytic expressions. A similar approach may be employed for cylindrical structure and the reader can refer Mizrahi and Schächter (2004a) for details.

5.4.2 *Field Distribution*

With the same given set of two dielectric materials and a given core dimension, the above design procedure makes it possible to achieve different phase velocities, and correspondingly, different field distributions across the core. As a demonstration of the ability to control the field behavior in the core, the symmetric planar TM mode will next be considered. Figure 5.17 presents different configurations, where in all cases the core half-width is $D_{\mathrm{int}} = 1\lambda_0$, and the two materials used have refractive indices of 1.6 and 4.6. In all cases, E_z is marked by a solid line whereas H_y is marked by a dashed line. The dielectric layers are depicted in gray, where the higher refractive index layers are indicated by the darker gray.

As an example, we consider a structure which supports a mode with a phase velocity equal to c. The transverse impedance for the first layer, which is given by $Z_1 = \eta_0 ck_1/\omega\varepsilon_1$, takes the form $Z_1 = \eta_0\sqrt{\varepsilon_1 - 1}/\varepsilon_1$, when $k_z = \omega_0/c$. For the materials chosen here, this entails that the higher refractive index material has the

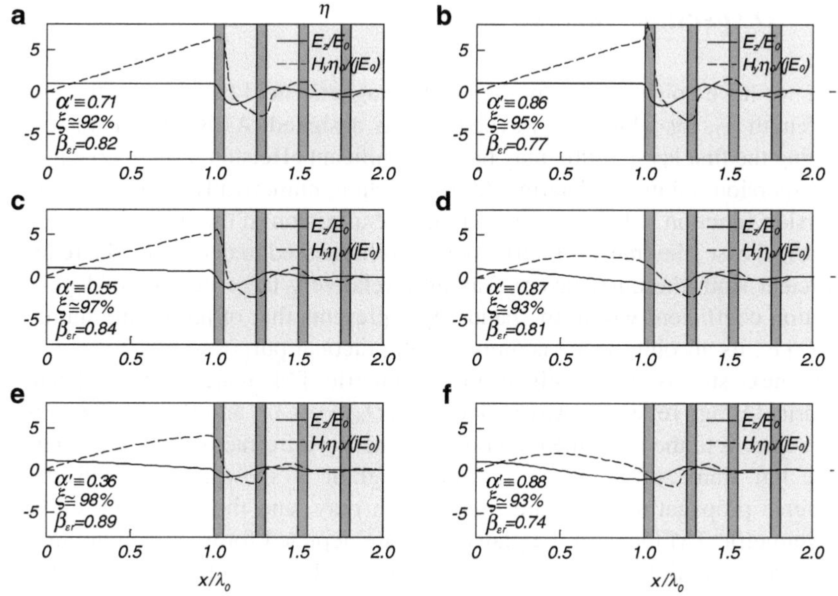

Fig. 5.17 Planar TM profiles. (**a**) $k_xD_{\text{int}} = 0$ low refractive index first (**b**) $k_xD_{\text{int}} = 0$ high refractive index first (**c**) $k_xD_{\text{int}} = \pi/3$ (**d**) $k_xD_{\text{int}} = 3\pi/4$ (**e**) $k_xD_{\text{int}} = \pi/2$ (metallic-like walls) (**f**) $k_xD_{\text{int}} = \pi$ (magnetic-like walls). The normalized decay parameter $\alpha' = \alpha\lambda_0/\tan\delta$ and ξ is the ratio of the power flowing in the core to the total power

lower transverse impedance and vice versa. We locate the material with the lower refractive index adjacent to the hollow core. Moreover, setting the matching layer width according to (5.4.7) for the $v_{\text{ph}} = c$ case, the field profile depicted in Fig. 5.17a, in which E_z is uniform across the core, is obtained. As indicated by (5.4.5) the longitudinal electric field vanishes and the transverse magnetic field peaks at the interface between the first and the second dielectric layers, identically to the case of a metallic wall located at that interface. Maintaining the same field distribution in the core itself, Fig. 5.17b illustrates the case where for $v_{\text{ph}} = c$, the material with the higher refractive index borders the core. The second case of (5.4.7) is used, and the picture obtained is as if a perfect magnetic wall is placed at the interface between the first and the second layers. As examples of arbitrary field profiles that can be achieved setting the matching layer width according to (5.4.6), Fig. 5.17c, d, where the transverse wave-numbers were chosen to be $k_xD_{\text{int}} = \pi/3$ and $k_xD_{\text{int}} = 3\pi/4$ respectively, are shown. Finally, as a special case of (5.4.6), the field distributions when all the layers are transverse quarter-wave thick, are shown. Figure 5.17e illustrates the case $k_xD_{\text{int}} = \pi/2$, in which the field in the core behaves as if the core boundary is a metallic wall. Figure 5.17f illustrates the case $k_xD_{\text{int}} = \pi$, which has an identical field to a perfect magnetic wall at the core boundary. Common to all cases presented is that at every interface between any two dielectrics, each one of the fields either peaks or vanishes.

5.4.3 Dispersion Curves

So far we have considered only the electromagnetic field behavior at the specific wavelength λ_0, for which the waveguide was designed. As is demonstrated next, adjusting the first layer width, may have a significant effect on the dispersion curve. The dispersion points are determined by searching numerically for the zeros of the dispersion function, which has an analytical expression in the planar case. For the cylindrical case, the transfer matrix method is harnessed to determine the reflection coefficient from the outer layers. Taking a relatively large number of layers, the reflection coefficient within the band-gap represents that of an infinite number of layers, i.e., its absolute value is unity for all practical purposes.

Our next step is to investigate the symmetric TM mode of both planar and cylindrical Bragg reflection waveguides with $D_{\rm int} = 0.3\lambda_0$ and $R_{\rm int} = 0.3\lambda_0$. For the layer adjacent to the core, the material with the lower refractive index was chosen. In the left frame of Fig. 5.18, a band diagram is shown, where the allowed transverse propagation areas are indicated in gray, and the dispersion curves of the symmetric TM mode in the planar case are depicted for two configurations. In the first configuration, all the layers are $\lambda_0/(4\sqrt{\varepsilon - 1})$ thick, meaning that the Bragg mirror is designed for $v_{\rm ph} = c$, but without using a matching layer to match between the mirror and the core field. The result is that the red solid line does not intersect the point $(\omega/\omega_0, ck_z/\omega_0) = (1, 1)$, as is required. Nevertheless, this dispersion curve intersects the light-line at a lower frequency. Operating the waveguide at that frequency is not desirable since the mirror is not optimal, i.e., the transverse exponential decay is weaker than could be achieved. When the first layer is adjusted

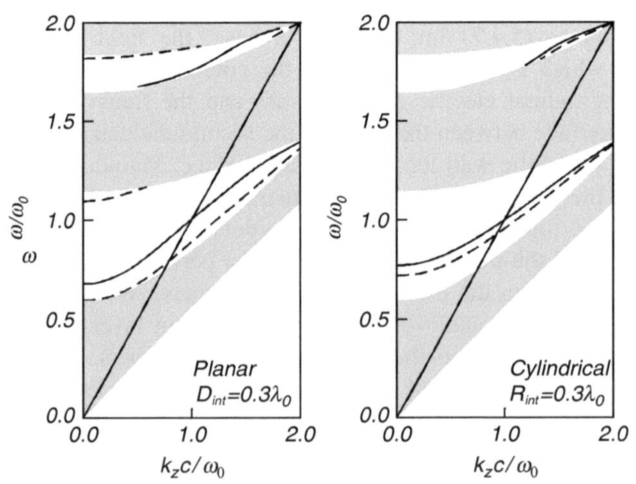

Fig. 5.18 Symmetric TM mode dispersion diagram for: *Left frame*: Planar waveguide with $D_{\rm int} = 0.3\lambda_0$. *Right frame*: cylindrical waveguide with $R_{\rm int} = 0.3\lambda_0$. In both cases the dashed curves are obtained with no design procedure (layer identical to the structure's), and the solid curves correspond to a $v_{\rm ph} = c$ design procedure

according to the design procedure describe above, the blue curve is obtained. It is seen that changing the first layer thickness shifted the dispersion curve so that there is now an intersection with the point $(\omega/\omega_0, ck_z/\omega_0) = (1, 1)$. A similar picture is obtained for the cylindrical case shown in the right frame of Fig. 5.18.

5.4.4 Quasi-TEM Mode

In a planar waveguide, it is possible to create a power distribution in the core with either monotonically increasing or monotonically decreasing profile, and non-zero on axis. A special case is when the power profile is completely uniform within the core, implying that inside the core the field is TEM, having both $E_z \equiv 0$ and $H_z \equiv 0$. In the dielectric layers, the modes are either TM or TE, and the waveguide may be matched to either. By computing the amplitudes in the layer adjacent to the core, we find for the TEM-TM that the first layer should either be transverse quarter-wavelength thick for $Z_1 < Z_2$ (higher refractive index first for the materials chosen here), meaning that actually no matching to the Bragg mirror is needed, or transverse half-wavelength thick for $Z_1 > Z_2$. A TEM-TE mode would have H_z, H_x, and E_y, and no matching to the mirror is needed if the higher TE admittance material is first, and a transverse half-wavelength is required if lower admittance is first – see the field profiles in Fig. 5.19.

5.4.5 Forces on the Layers and Discontinuities

Electromagnetic power injected into a Bragg waveguide exerts forces on the dielectric layers and on the discontinuities. In this subsection, we evaluate these

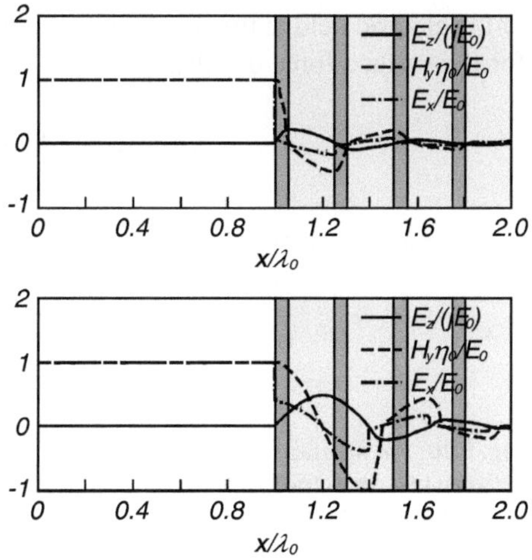

Fig. 5.19 Planar TEM-TM profiles: higher refractive index first (*top*) and lower refractive index first (*bottom*)

forces for a general regime but the examples are motivated by an acceleration structure. For simplicity sake, the discussion will be limited to a planar structure.

Within the dielectric layers, the Lorentz *volume force density* is given by

$$\mathbf{f} = \rho\mathbf{E} + \mathbf{J} \times \mathbf{B}, \tag{5.4.9}$$

where ρ is the instantaneous electric charge density, $\mathbf{J}$ is the instantaneous electric current density, $\mathbf{E}$ is the instantaneous electric field, and $\mathbf{B}$ is the instantaneous magnetic induction. In a polarizable material with instantaneous polarization density $\mathbf{P}$, the macroscopic effective charge density is $\rho = -\nabla \cdot \mathbf{P}$, and the effective current density is $\mathbf{J} = \partial\mathbf{P}/\partial t$. Since in a dielectric material $\varepsilon_0\varepsilon_r\mathbf{E} = \varepsilon_0\mathbf{E} + \mathbf{P}$ and $\nabla \cdot \mathbf{E} = 0$ as there is no free charge, it follows that $\rho = 0$. Therefore, only the second term in (5.4.9) is nonzero, and the volume force density, using $\mathbf{B} = \mu_0\mathbf{H}$, reads

$$\mathbf{f} = \varepsilon_0(\varepsilon_r - 1)\frac{\partial\mathbf{E}}{\partial t} \times \mu_0\mathbf{H}. \tag{5.4.10}$$

Specifically, for time harmonic fields of time-dependence $\exp(j\omega t)$, the time-average force density is given by $\langle\mathbf{f}\rangle = \frac{1}{2}\mathrm{Re}[-j\omega\varepsilon_0(\varepsilon_r - 1)\mathbf{E}^* \times \mu_0\mathbf{H}]$, where we kept the same notations for the phasors of the two field components.

At the interface between any two dielectric layers, a polarization surface charge is created, giving rise to a *surface force density*. The instantaneous polarization surface charge between layer v and layer $v+1$, as shown in Fig. 5.20, is given by

$$\rho_{s,v} = -\mathbf{1}_x \cdot \left(\mathbf{P}^{(-)}_{v+1} - \mathbf{P}^{(+)}_{v}\right), \tag{5.4.11}$$

where $\mathbf{P}^{(-)}_{v+1}$ and $\mathbf{P}^{(+)}_{v+1}$ are the polarization densities at the interface in layer $v+1$ and layer v respectively. The Lorentz force per unit area is obtained by multiplying the polarization surface charge density by the *average* of the perpendicular electric fields from both sides of the discontinuity. Defining $E^{(+)}_{x,v}$ and $E^{(-)}_{x,v}$ as the x components of the electric field at the boundary in layer v and layer $v+1$ respectively, we obtain for the surface force density

$$F_{x,v} = \rho_{s,v}\frac{1}{2}\left(E^{(+)}_{x,v} + E^{(-)}_{x,v+1}\right). \tag{5.4.12}$$

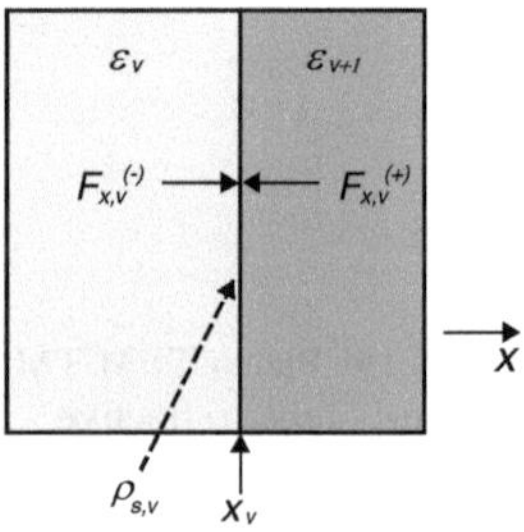

Fig. 5.20 An interface between two dielectric layers, where a polarization surface charge density is formed

Using the boundary condition $\varepsilon_\nu E^{(+)}_{x,\nu} = \varepsilon_{\nu+1} E^{(-)}_{x,\nu+1}$ together with (5.4.12), the surface force density at the boundary reads

$$F_{x,\nu} = \frac{1}{2}\varepsilon_0 E^{(+)2}_{x,\nu}\left(\frac{\varepsilon^2_\nu}{\varepsilon^2_{\nu+1}} - 1\right). \tag{5.4.13}$$

This total surface force density may be conveniently divided into two contributions from the two polarization densities of each layer at the interface. For this purpose, we may postulate the existence of an infinitesimal vacuum gap between the two layers, and then the force density is calculated on each of the two surface polarization charges. The force density on the surface charge of layer ν is given by

$$F^{(-)}_{x,\nu} = \frac{1}{2}\varepsilon_0 E^{(+)2}_{x,\nu}\left(\varepsilon^2_\nu - 1\right) > 0, \tag{5.4.14}$$

and the force density on the polarization surface charge of layer $\nu + 1$ is

$$F^{(+)}_{x,\nu} = -\frac{1}{2}\varepsilon_0 E^{(+)2}_{x,\nu+1}\left(\varepsilon^2_{\nu+1} - 1\right) < 0, \tag{5.4.15}$$

and $F_{x,\nu} = F^{(-)}_{x,\nu} + F^{(+)}_{x,\nu}$. In the two above inequalities, we have assumed that $\varepsilon_\nu > 1$, leading to the conclusion that the effect of these forces is to pull each of the two layers at the interface towards the other, as illustrated in Fig. 5.20.

Finally, the *total* transverse pressure exerted on all the layers may be found by integrating the Maxwell stress-tensor (Stratton 1941) over a closed surface. Within the vacuum core, the relevant Maxwell stress-tensor component T_{xx} for a TM_{0n} mode reads

$$T_{xx} = \frac{1}{2}\varepsilon_0 E^2_x - \frac{1}{2}\varepsilon_0 E^2_z - \frac{1}{2}\mu_0 H^2_y, \tag{5.4.16}$$

In the remainder of this sub-section, we consider a planar acceleration structure as an example. The explicit expressions for the field components of the TM more were presented in the right column of Table 5.1. For this field distribution the time-averaged Maxwell stress-tensor is $\langle T_{xx}\rangle = -(\varepsilon_0/4)|E_0|^2$. Enclosing one side of the waveguide by a rectangular surface, only the T_{xx} component contributes to the integral, as the T_{xz} component has zero contribution due to symmetry, and the T_{xy} component is identically zero. Assuming that the laser field decays to zero at $x = \pm\infty$, the time-averaged transverse pressure exerted by the guided mode on the Bragg structure located at $x = D_{\rm int}$ is

$$\langle F_{x,\rm T}\rangle = \frac{1}{4}\varepsilon_0 |E_0|^2; \tag{5.4.17}$$

the subscript T indicates that this is the *total* pressure. Hence, for a given accelerating gradient E_0, the total transverse pressure is repelling and is independent of the details of the structure. For comparison purposes, we note that this pressure is 1/4 of the pressure exerted by a plane wave of amplitude E_0 incident *perpendicularly* upon a perfect metallic plate. Assuming that the gradient of interest is $E_0 = 1$ GV/m, the total pressure is $\langle F_{x,\mathrm{T}} \rangle \simeq 2.2 \times 10^{-6}$ N/μm^2.

Based on (5.4.10) the time-averaged volume force densities associated with the TM acceleration mode are

$$\langle f_x \rangle = \frac{1}{2} \mathrm{Re}\left[j\omega_0 \varepsilon_0 (\varepsilon_r - 1) E_z^* \mu_0 H_y \right] \tag{5.4.18}$$

whereas the longitudinal component of the volume force density is zero

$$\langle f_z \rangle = \frac{1}{2} \mathrm{Re}\left[-j\omega_0 \varepsilon_0 (\varepsilon_r - 1) E_x^* \mu_0 H_y \right] = 0, \tag{5.4.19}$$

and so is the horizontal component $\langle f_y \rangle \equiv 0$

The *surface force densities* are computed using the time-average of (5.4.14) and it warrants to point out that the total pressure is the sum of all transverse force densities, and explicitly,

$$\langle F_{x,\mathrm{T}} \rangle = \sum_{v=0}^{\infty} \left[\langle F_{x,v} \rangle + \int_{x_v}^{x_{v+1}} dx \langle f_x(x) \rangle \right]; \tag{5.4.20}$$

x_v denotes the boundary between layer v and layer $v+1$ and $v = 0$ denotes the core.

The above expressions for the force densities may now be utilized for Bragg acceleration structures, and particularly, the analysis of two structures made of SiO_2 ($\sqrt{\varepsilon_r} = 1.45$) and Si ($\sqrt{\varepsilon_r} = 3.45$) with core half-width $D_{\mathrm{int}} = 0.3\lambda_0$ is given in Fig. 5.21. In each figure, the frames in the left column correspond to a structure having the SiO_2 as the matching layer adjacent to the core, and the right column corresponds to a structure with Si as the layer adjacent to the core. When the lower refractive index is used for the matching layer, the maximum of the volume force density is obtained inside the second layer (Fig. 5.21c), whereas in the second case, the maximum is obtained inside the matching layer (Fig. 5.21d).

The total internal pressure at some point within the layers against an external mechanical enforcement is given by the *cumulative* sum of both surface and volume force densities starting from the vacuum core up to the point of interest, similarly to the sum of (5.4.20), which is up to $x = \infty$. This pressure, which we denote by $\langle F_{\mathrm{pr}} \rangle$, is depicted in Fig. 5.21e–f. It is seen that in the layers that are close to the vacuum core, the pressure is negative, pulling these layers towards the core. Farther away from the core, the pressure becomes positive and approaches $\langle F_{x,\mathrm{T}} \rangle = \frac{1}{4}\varepsilon_0 |E_0|^2$ (not seen clearly in the figure due to the scale).

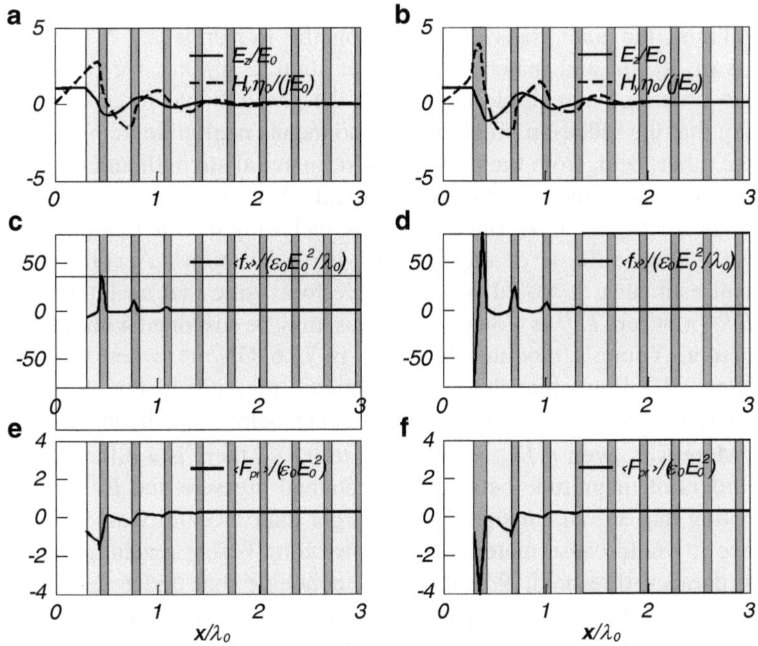

Fig. 5.21 Volume force densities in two Bragg structures made of SiO_2 (indicated by lighter *gray*) and Si. (**a–b**) Two of the electromagnetic field components, (**c–d**) volume force densities, and (**e–f**) total internal pressure. The left column frames correspond to a structure with SiO_2 as the matching layer, and the right column corresponds to Si as the matching layer

By calculating the fields inside the dielectric layers given the vacuum field, it is possible to show that the behavior of the maximal volume force density takes the form

$$|\langle f_x\rangle|_{\max} = \kappa\frac{\pi}{2}\left[1+\left(\frac{\omega_0 D_{\text{int}}}{c}\right)^2\frac{\varepsilon_1-1}{\varepsilon_1^2}\right]\frac{\varepsilon_0|E_0|^2}{\lambda_0}, \tag{5.4.21}$$

where κ is a constant independent of D_{int}. When the maximum is obtained in the matching layer, $\kappa = \varepsilon_1\sqrt{\varepsilon_1-1}$, and when the maximum is in the second layer, $\kappa = (\varepsilon_2-1)\bar{Z}_2/\bar{Z}_1^2$, where $\bar{Z}_\nu = \sqrt{\varepsilon_\nu - 1}/\varepsilon_\nu$, $\nu = 1, 2$ are the normalized transverse impedances. Since the maximal surface force density occurs at the vacuum-dielectric interface, we may use this quantity as reference

$$|\langle F_x\rangle|_{\max} = \frac{1}{4}\left(\frac{\omega_0 D_{\text{int}}}{c}\right)^2\frac{\varepsilon_1^2-1}{\varepsilon_1^2}\varepsilon_0|E_0|^2. \tag{5.4.22}$$

We have already stated that the total transverse pressure on one of the Bragg mirrors is $\langle F_{x,\text{T}}\rangle \simeq 2.2\times 10^6$ N/m^2 if we consider an accelerating field of

$E_0 = 1\,\mathrm{GV/m}$. Assuming that one Bragg mirror in the structure has an area of $1\,\mathrm{mm} \times 1\,\mathrm{mm}$, the total transverse force on the mirror is 2.2 N. For a mirror thickness of 50 μm and material density of about 2 gr/cm^3, we obtain that this force is 6 orders of magnitude larger than the gravitational force on the mirror, indicating that the radiation pressure is by no means negligible.

On the other hand, from the perspective of material strength and the possibility of crack formation, the situation is different. The order of magnitude of the pressures in the $D_{\mathrm{int}} = 0.3\lambda_0$ Bragg structure including the internal pressure $\langle F_{\mathrm{pr}} \rangle$, as shown in Fig. 5.21, is of the order of $10^6 - 10^7$ N/m^2. According to rough theoretical estimates, it would be reasonable to assume that an internal pressure below $E^{(Y)}/\pi$, where $E^{(Y)}$ is Young's modulus, may be sustained without damage to the structure. Young's modulus for SiO_2 is 72.6 GN/m^2, whereas for Si it is 162 GN/m^2. It follows that the electromagnetic pressure in the structure under consideration is at least 3 orders of magnitude below the theoretical threshold $E^{(Y)}/\pi$. Moreover, even if D_{int} is increased to $0.8\lambda_0$, there is a difference of more than 2 orders of magnitude between the obtained pressure and $E^{(Y)}/\pi$. It is also worth noting that a gradient significantly larger than 1 GV/m would be unacceptable since it would cause material breakdown long before reaching the radiation pressure damage threshold. We, therefore, conclude that under the assumptions considered here, the electromagnetic forces on the planar Bragg acceleration structure do not pose a significant threat to the operation of an optical Bragg accelerator. A detailed analysis was published by Mizrahi and Schächter (2006).

5.5 Transients and Wakes

When several bunches of electrons are injected in a structure, as is the case in an accelerator they not only interact with the electromagnetic field, which was prepared for their acceleration, but they also generate a whole spectrum of waves at different frequencies. These form a so-called *wake* field, which in turn decelerates the bunch. In order to visualize the process, imagine a pulse consisting of two bunches. When the first enters the periodic structure it generates a wake-field and if this is not "drained" fast enough then it may affect the interaction of the trailing bunch according to the distance between the two.

Propagation of a pulse in a disk-loaded waveguide should, in principle, account for all the modes and all the reflections from the disks. The difficulties in the analysis of transients generated by charged particles in periodic closed structures arise from the fact that (1) the frequency spectrum of a moving point-charge is infinite and (2) although the spectrum of frequencies in a closed periodic structure is discrete, it spans to infinity. The analysis is somewhat simplified by the fact that in the transverse direction the (evanescent) wave decays exponentially $\exp(-\omega r/c\gamma\beta)$ therefore, the contribution of the high frequencies might be small – at least at low energies. The situation is different in open periodic structures where, as we already indicated, the spectrum is discrete and finite. Therefore, potentially less energy is induced in the system. In this context Smith-Purcell effect can be regarded as a transient generated by a moving particle. We also consider the wakes generated in a Bragg-reflection structure. What we do not consider here is azimuthally asymmetric modes that may cause deflection forces on the electrons. A qualitative discussion on this subject is presented in Sects. 4.5 and 8.1.5.

5.5.1 Propagation of a Wave-Packet in a Periodic Structure

In order to illustrate the effect of the periodicity on the propagation of a *wave* packet we consider at $t = 0$ the same wave-packet $a(z)$, in free space and in a periodic structure. The propagation in *free space* is represented by a dispersion relation $k^2 = \omega^2/c^2$, therefore a scalar wave function $\Psi(z,t)$ is given by

$$\Psi(z,t) = \int_{-\infty}^{\infty} dk\,\psi(k)\exp(-jkz)\frac{1}{2}[\exp(jkct) + \exp(-jkct)]. \tag{5.5.1}$$

Since at $t = 0$ this function equals $a(z)$, the amplitudes $\psi(k)$ can be readily determined using the inverse Fourier transform hence

$$\psi(k) = \frac{1}{2\pi}\int_{-\infty}^{\infty} dz a(z)\exp(jkz). \tag{5.5.2}$$

Substituting back into (5.5.1) we find that

$$\Psi(z,t) = \frac{1}{2}[a(z - ct) + a(z + ct)], \tag{5.5.3}$$

which indicates that the pulse moves at the speed of light in both directions and asymptotically, it preserves its shape.

In a *periodic structure* the description of the wave packet is complicated by the dispersion relation which in its lowest order approximation (e.g., first TM symmetric mode in a waveguide) can be approximated by

$$\omega(k) = \bar{\omega} - \delta\omega\cos(kL), \tag{5.5.4}$$

where $\bar{\omega} = (\omega_0 + \omega_\pi)/2$ is the average frequency between the low ($kL = 0$) cut-off denoted by ω_0 and the high cut-off ($kL = \pi$) denoted by ω_π. The quantity $\delta\omega = (\omega_\pi - \omega_0)/2$ is half the pass-band width and L is the period of the structure. Contrary to the previous case k here denotes the wave-number in the first Brillouin zone. In the framework of this approximation, we can use Floquet's representation to write

$$\Psi(z,t) = \mathrm{Re}\left\{\sum_{n=-\infty}^{\infty}\int_{-\pi/L}^{\pi/L} dk\psi_n(k)\exp[j\omega(k)t - jk_n z]\right\}, \tag{5.5.5}$$

where $k_n = k + 2\pi n/L$. The amplitudes $\psi_n(k)$ are determined by the value of the function at $t = 0$ hence

$$\psi_n(k) = \frac{1}{2\pi}\int_{-\infty}^{\infty} dza(z)\exp(jk_n z). \tag{5.5.6}$$

Substituting back into (5.5.5) we have

$$\Psi(z,t) = \mathrm{Re}\left[\begin{array}{l}\frac{1}{2\pi}\int_{-\infty}^{\infty} d\zeta a(\zeta) \\ \times \sum_{n=-\infty}^{\infty}\int_{-\pi/L}^{\pi/L} dk\exp\{jt[\bar{\omega} - \delta\omega\cos(kL)] - jk_n(z-\zeta)\}\end{array}\right]. \tag{5.5.7}$$

At this point, we can take advantage of the generating Bessel function and simplify the last equation

$$\begin{aligned}\Psi(z,t) = \mathrm{Re}\Bigg[\frac{1}{2\pi}\int_{-\infty}^{\infty} d\zeta a(\zeta)\sum_{n=-\infty}^{\infty}\int_{-\pi/L}^{\pi/L} dk\exp(j\bar{\omega}t) \\ \times \sum_{v=-\infty}^{\infty} \mathrm{J}_v(\delta\omega t)\exp[j(kL-\pi/2)v]\exp(-jk_n(z-\zeta))\Bigg],\end{aligned} \tag{5.5.8}$$

which after the evaluation of the integrals and summation (over n) reads

$$\Psi(z,t) = \sum_{v=-\infty}^{\infty} a(z - vL)\mathrm{J}_v(\delta\omega t)\cos(\bar{\omega}t - \pi v/2). \tag{5.5.9}$$

Figure 5.22 illustrates the propagation of two wave-packets in vacuum (dashed line) and in a periodic structure. The latter is characterized by $\bar{\omega} = 2\pi \times 10[\mathrm{GHz}]$, $\delta\omega = \bar{\omega}/30$ and a spatial periodicity of $L = 1\,\mathrm{cm}$. At $t = 0$ the distribution is a Gaussian, $a(z) = \exp[-(z/L)^2]$. In each one of the frames $\Psi(z,t)$ was plotted at a different time as a function of z. Characteristic to all the frames is the relatively large peak following the front of the pulse.

It is evident that although the front of the pulse propagates at the speed of light (as in vacuum) the main pulse propagates slower. In fact, a substantial fraction of the energy remains at the origin even a long time after $t = 0$. For the parameters used, the amplitude of the signal at the origin $(z = 0)$ is dominated by the zero order Bessel function i.e., $\mathrm{J}_0(\delta\omega t)$ therefore the energy is drained on a time scale which is determined by the asymptotic behavior of the Bessel function namely $\propto 1/\sqrt{\delta\omega t}$. Clearly the wider the pass-band the faster the energy is drained from the origin.

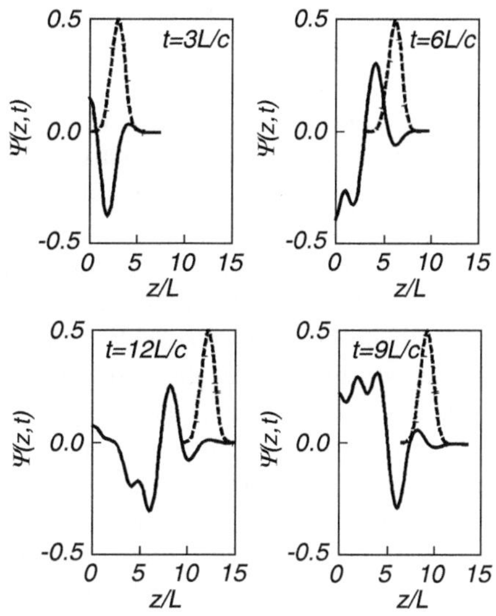

Fig. 5.22 Propagation of the wave-packet in vacuum (*dashed line*) and in the periodic structure at four instants

5.5.2 Wake in a Closed Periodic Structure

A point-charge moving in free space generates a broad spectrum of evanescent waves. This spectrum is attached to the charge and as long as near the particle there is no obstacle, this spectrum is not altered and the charge may move at a constant velocity. In the presence of obstacles, this spectrum is partially scattered and these scattered waves affect the motion of the particle. It is therefore natural to characterize the wake generated by the particle in a periodic structure in terms of the external power (or force) necessary to apply in order to preserve constant motion.

The electromagnetic wake generated by a relativistic bunch of particles in a periodic structure or single cavity, that RF accelerators consist of, was the subject of many studies. However due to the complexity of the problem there are only a few analytic or quasi-analytic solutions: for example Bane et al. (1981), has developed a simple model that describes the energy loss of a bunched beam traversing a cavity attached to a cylindrical waveguide. The calculation was subject to the assumption of an azimuthally symmetric structure. Later Dome (1985) has extended the approach to asymmetric modes.

Another quasi-analytic approach was developed by Dome et al. (1991) for the evaluation of the diffraction of the electromagnetic field created by a charge traveling on the axis of circular apertures in a set of perfectly conducting infinite planes. The total field was assumed to be a superposition of the evanescent waves traveling with the charge itself and the radiation from the plates. Or in other words, the total field is a superposition of two components: a part generated by the charge in free space and a part reflected from the plates; together they satisfy the boundary conditions.

In recent years wakes have attracted the attention of many researchers. Several reviews summarize in a systematic way many of the phenomena and the methods involved: Heifets and Kheifets (1991), Chao (1993) and Zotter and Kheifets (1998). For a detailed discussion the reader should consult these publications. In the present sub-section we focus on an analytic and intuitive approach yet we try to keep the discussion as general as possible. Explicitly, we determine the *maximum* field acting on a point-charge moving in a cylindrical waveguide with periodic wall of *arbitrary* but azimuthally symmetric geometry. The gradient is a result of the electromagnetic field scattered by the periodic wall where the latter is electromagnetically described by a *reflection matrix*. The extremum determined here is subject to the assumption that the absolute value of each diagonal term of this reflection matrix is smaller than unity.

Consider a point-charge (Q) moving at a constant velocity v_0 along the axis of an azimuthally symmetric structure of periodicity L and arbitrary geometry otherwise. The nearest point of the waveguide's wall to the axis is denoted by R_{int} – see Fig. 5.23. The current density associated with this charge is given by $J_z(r,z,t) = -Q\mathrm{v}_0\delta(z-\mathrm{v}_0t)\delta(r)/2\pi r$ and it excites the longitudinal component of the magnetic vector potential. The latter has two components: a non-homogeneous part (superscript nh) that based on the free-space Green's function may be written as

$$A_z^{(nh)}(r,z,t) = \int_{-\infty}^{\infty} \mathrm{d}\omega\, \exp(j\omega t) \int_{-\infty}^{\infty} \mathrm{d}k\, A(\omega,k)\exp(-jkz)\mathrm{K}_0(\Gamma r) \qquad (5.5.10)$$

where $\Gamma = \sqrt{k^2 - (\omega/c)^2}$, $A(\omega,k) = Q\mu_0\mathrm{v}_0\delta(\omega - \mathrm{v}_0k)/(2\pi)^2$ and the homogeneous counterpart (superscript h) that reads

$$A_z^{(h)}(r,z,t) = \int_{-\infty}^{\infty} \mathrm{d}\omega\, \exp(j\omega t) \int_{-\infty}^{\infty} \mathrm{d}k\, B(\omega,k)\exp(-jkz)\mathrm{I}_0(\Gamma r). \qquad (5.5.11)$$

The boundary conditions at $r = R_{\mathrm{int}}$ impose the relation between $A(\omega,k)$ and $B(\omega,k)$. In general, this relation may be expressed in terms of the *reflection operator* ($\mathcal{R}$)

$$B(\omega,k) = \int_{-\infty}^{\infty} \mathrm{d}k'\, \mathcal{R}(\omega;k,k')A(\omega,k'). \qquad (5.5.12)$$

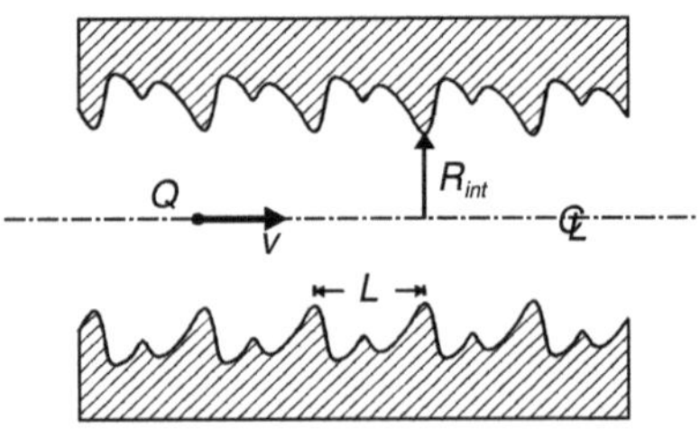

Fig. 5.23 Schematics of the structure under consideration

Since by virtue of Poynting's theorem we know that the amount of power in the wake-field is entirely determined by the reaction-field on the particle, we consider the longitudinal electric field linked to the homogeneous solution:

$$E_z^{(h)}(r,z,t) = \int_{-\infty}^{\infty} \mathrm{d}\omega \, \exp(j\omega t) \int_{-\infty}^{\infty} \mathrm{d}k \left(\frac{-c^2\Gamma^2}{j\omega}\right) \exp(-jkz) \mathrm{I}_0(\Gamma r)$$
$$\times \int_{-\infty}^{\infty} \mathrm{d}k' \, \mathcal{R}(\omega;k,k') A(\omega,k'). \tag{5.5.13}$$

At this point we may take advantage of the explicit expresion for $A(\omega,k)$ that includes a Dirac delta function, therefore at $r=0$

$$E_z^{(h)}(z,t) = \frac{Q\mu_0 \mathrm{v}_0}{(2\pi)^2} \int_{-\infty}^{\infty} \mathrm{d}k \mathrm{d}k' e^{j(k'\mathrm{v}_0 t - kz)} \frac{-(\Gamma')^2 c^2}{jk'\mathrm{v}_0} \mathcal{R}(\omega = k'\mathrm{v}_0;k,k') \tag{5.5.14}$$

wherein $\Gamma' \equiv \sqrt{k^2 - \beta^2 (k')^2}$. This expression is general (subject to the previously mentioned assumptions) and at this point it will be applied to a specific class of problems i.e., periodic geometry. In this case, we limit the discussion to the decelerating field averaged over one period (L) of the structure

$$E_{\|} \equiv \frac{\mathrm{v}_0}{L} \int_{-L/2\mathrm{v}_0}^{L/2\mathrm{v}_0} \mathrm{d}t E_z^{(h)}(r=0, z=\mathrm{v}_0 t, t) \tag{5.5.15}$$

and take advantage of the the Floquet decomposition of the continuous integral $\int_{-\infty}^{\infty} \mathrm{d}k \ldots = \int_{-\pi/L}^{\pi/L} \mathrm{d}q \sum_{n=-\infty}^{\infty} \ldots..$ as well as the simplification of the reflection operator

$$\mathcal{R}(\omega;k,k') \equiv 2\pi\delta(q-q')\rho_{n,m}(\omega;q,q'). \tag{5.5.16}$$

The latter reflects the fact that a wave scattered by a periodic structure preserves its frequency and wave-length projection in the the first Brillouin zone but the harmonics are coupled by the *reflection matrix* – $\rho_{n,m}(\omega,q)$.

Averaging over the period of time it takes the particle to traverse one period of the structure, resorting to the Floquet decomposition and using the explicit expression for $\mathcal{R}$ in (5.5.16) the expression for the decelerating field is given by

$$E_{\|} = \frac{Q}{2\pi\varepsilon_0\gamma^2} \int_{-\pi/L}^{\pi/L} \mathrm{d}q \sum_n jk_n \rho_{nn}(\omega = \mathrm{v}_0 k_n, q) \,. \tag{5.5.17}$$

It should be emphasized that this is an *exact* expression for the average decelerating field. Practical evaluation of (5.5.16) is relatively easy at non-relativistic velocities since if the period (L) and the internal radius (R_{int}) are of the same order of magnitude then the argument of the modified Bessel function becomes significant for $|n| > 2$ and as a result, only the contribution of the first few harmonics is not negligible. As the particle becomes relativistic, the number of terms (harmonics) that have a non-negligible contribution becomes larger and accurate evaluation of a larger reflection matrix is required. For example if $\gamma \sim 10^5$, $R_{int}/L \sim 0.5$ then ρ is a $10^6 \times 10^6$ matrix and obviously the calculation is fairly time-consuming particularly bearing in mind that for each q and each n the entire matrix has to be re-calculated. It is therefore natural to search for an upper value of this integral. With this purpose in mind, we assume that in the ultra relativistic regime, the absolute value of each *diagonal* term of the reflection matrix satisfies,

$$|\rho_{nn}(\omega = \mathrm{v}_0 k_n, q)| \le \mathrm{K}_0^2(|k_n| R_{\mathrm{int}}/\gamma), \tag{5.5.18}$$

and consequently we found that

$$E_{\|} \le E_{\|}^{(max)} \equiv \frac{Q}{4\pi\varepsilon_0 R_{int}^2} \times \left[4\int_0^\infty \mathrm{d}\xi \xi K_0^2(\xi)\right] = \frac{Q}{4\pi\varepsilon_0 R_{int}^2} \times 2. \tag{5.5.19}$$

The last result provides us with extremum value for the longitudinal decelerating field as a point-charge traverses a periodic structure of *arbitrary* geometry. Several aspects are evident: the field is independent of the period or any other geometric parameter of the structure with exception of R_{int}. The result in (5.5.19) may be generalized to include the effect of bunch of finite sizes. For a finite length (Δ_z) the average decelerating field may be generalized and it is given by

$$\begin{aligned} E_{\|}^{(max)} &= \frac{Q}{4\pi\varepsilon_0 R_{\mathrm{int}}^2} \times \left[4\int_0^\infty \mathrm{d}\xi \xi K_0^2(\xi) \mathrm{sinc}^2\left(\frac{1}{2}\gamma \frac{\Delta_z}{R_{\mathrm{int}}}\xi\right)\right] \\ &\simeq \frac{Q}{4\pi\varepsilon_0 R_{\mathrm{int}}^2} \times \frac{2}{1 + \left(\frac{1}{2\pi}\gamma \frac{\Delta_z}{R_{\mathrm{int}}}\right)^{3/2}} \end{aligned} \tag{5.5.20}$$

Note that if $\Delta_z \sim 0.4 R_{\mathrm{int}}$ and $\gamma \sim 100$, the decelerating field is less than 6% of the original value ($\Delta_z = 0$).

In a similar way we may determine the maximum power generated by M micro-bunches each one carrying a charge q and the spacing between two adjacent micro-bunches is L_b namely

$$\begin{aligned} P^{(\max)} &= \frac{-q^2 \mathrm{v}_0}{4\pi\varepsilon_0 R_{\mathrm{int}}^2} \times \left[4\int_0^\infty \mathrm{d}\xi \xi K_0^2(\xi) \left|\sum_{m=1}^{M} \exp\left(-j\xi\gamma \frac{L_b}{R_{\mathrm{int}}} m\right)\right|^2\right] \\ &= \frac{-q^2 \mathrm{M}^2 \mathrm{v}_0}{4\pi\varepsilon_0 R_{\mathrm{int}}^2} \times \left[4\int_0^\infty \mathrm{d}\xi \xi K_0^2(\xi) \frac{\mathrm{sinc}^2\left(\frac{1}{2} M\xi\gamma \frac{L_b}{R_{\mathrm{int}}}\right)}{\mathrm{sinc}^2\left(\frac{1}{2}\xi\gamma \frac{L_b}{R_{\mathrm{int}}}\right)}\right] \end{aligned} \tag{5.5.21}$$

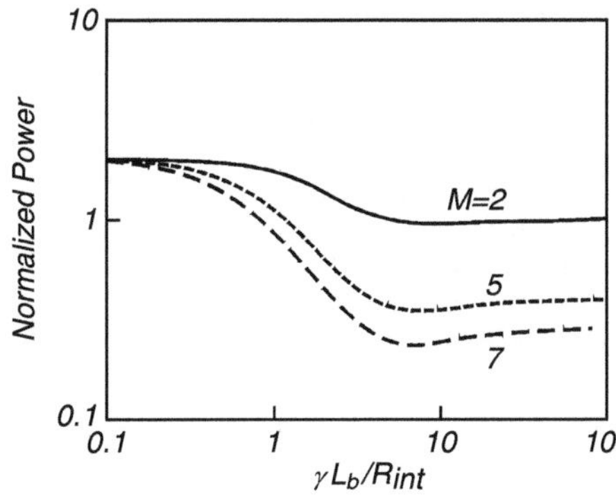

Fig. 5.24 Maximum normalized power as a function of the bunch spacing parameter $\gamma L_b/R_{\rm int}$; the number of bunches is a parameter $M = 2, 5, 7$

At the limit $\gamma L_b/R_{\rm int} \gg 1$ the integral reaches an asymptotic value $(2/M)$ or explicitly

$$P^{(\max)} = \frac{-q^2 \mathrm{M}^2 \mathrm{v}_0}{4\pi\varepsilon_0 R_{\rm int}^2} \times \frac{2}{M} \tag{5.5.22}$$

Figure 5.24 reveals the dependence of the normalized power as a function of the bunch spacing $\gamma L_b/R_{\rm int}$. We clearly observe that for large normalized spacing the term in the square brackets reaches its asymptotic value.

5.5.3 *Wake Effect in a Bragg Waveguide*

Bragg reflection waveguides rely on perfect reflection in predesigned frequency range(s). Outside this range the electromagnetic power may escape the structure therefore the effect of a single bunch on the trailing bunches may diminish. This is one of the main differences between a Bragg reflection waveguide and closed structures. In this subsection the attention will be directed towards the wake generated by a train of relativistic bunches with special emphasis on the effect of the Bragg reflections properties. Figure 5.25 reveals schematically the configuration under investigation: A train of M bunches each one carrying a charge q moves on the axis of a vacuum tunnel of radius $R_{\rm int}$ surrounded by a dielectric medium $\varepsilon_{\rm r}$; the spacing between adjacent bunches is denoted by L_b. Beyond the matching layer, there is a Bragg structure represented here by a reflection coefficient $\mathcal{R}$ thus the magnetic vector potential in the matching layer is given by

$$\begin{aligned} A_z(r,z,t) = & \int_{-\infty}^{\infty} d\omega A(\omega) \exp\left[j\omega\left(t - \frac{z}{\mathrm{v}_0}\right)\right] \\ & \times \left[H_0^{(2)}\left(r\frac{\omega}{c}\sqrt{\varepsilon_{\rm r}-1}\right) + \mathcal{R}(\omega) H_0^{(1)}\left(r\frac{\omega}{c}\sqrt{\varepsilon_{\rm r}-1}\right)\right]. \end{aligned} \tag{5.5.23}$$

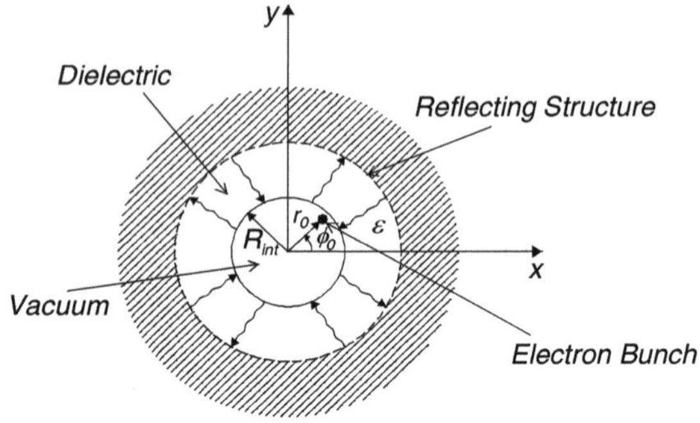

Fig. 5.25 A train of M bunches each one carrying a charge q moving on the axis ($r_0 = 0$) of a vacuum tunnel of radius $R_{\rm int}$ surrounded by a dielectric medium ε_r; the spacing between adjacent bunches is denoted by L_b. Beyond the matching layer, there is a Bragg structure

After defining $\bar{\omega} \equiv (\omega R_{\rm int}/c)\sqrt{\varepsilon_{\rm r} - 1}/2\varepsilon_{\rm r}$ the power generated by the wake is given by

$$\begin{aligned} P &= \frac{-q^2 c}{4\pi\varepsilon_0 R_{\rm int}^2}\frac{2}{\pi}\int_{-\infty}^{\infty} d\bar{\omega}\frac{1+\mathcal{R}(\bar{\omega})}{1+j\bar{\omega}-(1-j\bar{\omega})\mathcal{R}(\bar{\omega})}\left|\sum_{n=1}^{M}\exp(-j\bar{\omega}\bar{L}_b n)\right|^2 \\ &= \frac{-q^2 M^2 c}{4\pi\varepsilon_0 R_{\rm int}^2}\frac{2}{\pi}\int_{-\infty}^{\infty} d\bar{\omega}\frac{1+\mathcal{R}(\bar{\omega})}{1+j\bar{\omega}-(1-j\bar{\omega})\mathcal{R}(\bar{\omega})}\frac{\operatorname{sinc}^2\left(\frac{1}{2}\bar{\omega}\bar{L}_b M\right)}{\operatorname{sinc}^2\left(\frac{1}{2}\bar{\omega}\bar{L}_b\right)} \end{aligned} \tag{5.5.24}$$

where $\bar{L}_b \equiv 2\varepsilon_{\rm r}(L_b/R_{\rm int})/\sqrt{\varepsilon_{\rm r} - 1}$ denotes the normalized bunch spacing.

Clearly, in the absence of reflections i.e., $\mathcal{R} = 0$,

$$\begin{aligned} P_0 &= \frac{-q^2 c}{4\pi\varepsilon_0 R_{\rm int}^2}\frac{2}{\pi}\sum_{n,m=1}^{M}\int_{-\infty}^{\infty} d\bar{\omega}\frac{\exp[j\bar{\omega}\bar{L}_b(n-m)]}{1+j\bar{\omega}} \\ &= \frac{-q^2 c}{4\pi\varepsilon_0 R_{\rm int}^2} \times 2\left\{2\sum_{n,m=1}^{M}\exp[-\bar{L}_b(n-m)]h(n-m)\right\} \\ &= \frac{-q^2 c}{4\pi\varepsilon_0 R_{\rm int}^2} \times 2f(M,\bar{L}_b) \end{aligned} \tag{5.5.25}$$

where

$$h(x) = \begin{cases} 0 & x<0 \\ 0.5 & x=0 \\ 1 & x>0 \end{cases} \tag{5.5.26}$$

is the Heaviside step function and $f(M,\bar{L}_b)$ is the train form factor that can be evaluated analytically

$$\begin{aligned} f(M,\bar{L}_b) &\equiv 2\sum_{n,m=1}^{M} \exp[-\bar{L}_b(n-m)]h(n-m) \\ &= M + 2(M-1)\exp(-\bar{L}_b) + \ldots\ldots + 2\exp[-\bar{L}_b(M-1)] \\ &= \frac{M[1-\exp(-2\bar{L}_b)] - 2\exp(-\bar{L}_b)[1-\exp(-M\bar{L}_b)]}{[1-\exp(-\bar{L}_b)]^2} \end{aligned} \tag{5.5.27}$$

Four regimes are evident: (1) in case of a single bunch the form factor is unity, (2) if the length of the train is much smaller than the effective radius of the tunnel, the form factor is proportional to M^2 this is to say that the wake consists of coherent radiation. (3) In between the form factor is determined by the expression in the third line

$$f(M,\bar{L}_b) = \begin{cases} 1 & M = 1 \\ M^2 & M\bar{L}_b \ll 1 \\ [M\bar{L}_b - 1 + \exp(-M\bar{L}_b)]/\bar{L}_b^2 & \bar{L}_b \ll 1 \\ M & \bar{L}_b \gg 1 \end{cases} \tag{5.5.28}$$

whereas (4) if the train length is large the power is proportional to the number of bunches in the train. Note that each bunch is assumed to be a point-charge.

Let us now repeat the analysis from the above but accounting for reflections from the Bragg structure, namely

$$P = \frac{-q^2 c}{4\pi\varepsilon_0 R_{\text{int}}^2}\frac{2}{\pi}\int_{-\infty}^{\infty} d\bar{\omega}\frac{1}{\frac{1-\mathcal{R}(\bar{\omega})}{1+\mathcal{R}(\bar{\omega})} + j\bar{\omega}}\left|\sum_{n=1}^{M}\exp(-j\bar{\omega}\bar{L}_b n)\right|^2 \tag{5.5.29}$$

In order to account for the special character of the reflection matrix describing a Bragg structure let us assume that there is a set of normalized frequencies and their vicinity $|\bar{\omega} - \bar{\omega}_i| \le \delta\bar{\omega}_i/2$ wherein $\mathcal{R} = 1$ and zero otherwise. For an analytic estimate we push this assumption to the extreme and consider the *limiting case* when the reflection coefficient is unity over the entire spectrum

$$P_1 = \frac{-q^2 c}{4\pi\varepsilon_0 R_{\text{int}}^2}\frac{2}{\pi}\int_{-\infty}^{\infty} d\bar{\omega}\frac{1}{j\bar{\omega}}\left|\sum_{n=1}^{M}\exp(-j\bar{\omega}\bar{L}_b n)\right|^2 \tag{5.5.30}$$

implying

$$P_1 = \frac{-q^2 c}{4\pi\varepsilon_0 R_{\text{int}}^2} \times 2\left[2\sum_{n,m=1}^{M} h(n-m)\right] = \frac{-q^2 c}{4\pi\varepsilon_0 R_{\text{int}}^2} \times 2M^2. \tag{5.5.31}$$

Obviously, for a train of M point-charges when the radiation is reflected from the structure back to axis where the electrons are, the emitted power reaches the maximum value. In case of an open structure ($\mathcal{R} = 0$), the power generated by the electrons reaches the limit of the closed structure only if the train length is much shorter that the tunnel radius.

For a more realistic assessment of the reflections effect, one may consider a perfect reflector ($E_z = 0$) located at $r = R_{\text{ext}} > R_{\text{int}}$ for which the reflection coefficient is

$$\mathcal{R} = -\frac{H_0^{(2)}(\Lambda R_{\text{ext}})\, H_0^{(1)}(\Lambda R_{\text{int}})}{H_0^{(1)}(\Lambda R_{\text{ext}})\, H_0^{(2)}(\Lambda R_{\text{int}})} \simeq -e^{-2j\bar{\omega}\bar{r}} \tag{5.5.32}$$

wherein it is tacitly assumed that $\Lambda R_{\text{int}} \gg 1$ and $\bar{r} \equiv 2\varepsilon_{\text{r}}(R_{\text{ext}} - R_{\text{int}})/R_{\text{int}}$; this last parameter represents the normalized distance where the reflection occurs. Explicitly, the power is given by

$$P = \frac{-q^2 c}{4\pi\varepsilon_0 R_{\text{int}}^2} \frac{2}{\pi} \int_{-\infty}^{\infty} d\bar{\omega} \frac{1}{\frac{1}{j\tan(\bar{\omega}\bar{r})} + j\bar{\omega}} \left| \sum_{n=1}^{M} \exp(-j\bar{\omega}\bar{L}_b n) \right|^2 \tag{5.5.33}$$

and its detailed analysis is formulated as an exercise (5.10) at the end of this chapter.

Exercises

5.1 Based on the solution for $A_z(r,z)$ in Sect. 5.1 determine the Floquet representation of the magnetic vector potential (TM_{01}). In other words write

$$A_z(r,z) = \mathrm{J}_0\left(p_1 \frac{r}{R}\right) \sum_n a_n(k) \exp(-jk_n z)$$

and determine $a_n(k)$.

5.2 Find all the waves which can propagate between $f = 0$ and 20 GHz, including asymmetric modes for the system described in Sect. 5.1. Repeat this exercise for the branches of the TE modes.

5.3 For the parameters in Figs. 5.2 and 5.3 consider N cells and calculate (using the transmission matrix formulation harnessed in Sect. 5.1) the transmission coefficient and analyze it in the frequency range where if $N \to \infty$ the system were in the forbidden gap. Hint: show that the transmission coefficient decays exponentially with N.

5.4 Consider a 3D dielectric structure $\varepsilon(x,y,z)$ periodic in all three dimensions (L_x, L_y, L_z). Based on Maxwell's equations, show that the dispersion relation has the same form as the one in (5.1.32).

(continued)

5.5 Analyze the coupling of spatial harmonics for the system in Sect. 5.3 in a similar way as in Sect. 5.2.2.

5.6 Determine the expression for the reflection matrix defined in (5.3.36) Use it to analyze the decelerating force as prescribe in (5.3.39).

5.7 In the context of Sect. 5.3.3, calculate the force which acts on the moving charge. Based on this expression, determine the total power emitted.

5.8 Repeat the calculation of the propagation of a transient in a periodic structure (Sect. 5.4) but this time for a TEM-like mode. [Hint: consider $\omega = \omega_\pi \sin(kL/2)$.]

5.9 Extend the expression for the maximum power emitted by a point charge in a periodic structure as expressed in (5.5.19) to include the effect of a finite radius (R_b).

5.10 For the evaluation of the power emitted by a train of electrons in a closed structure (5.5.33), define the dispersion relation

$$D(\bar{\omega}) = \bar{\omega} - \cot(\bar{\omega}\bar{r})$$

which determines the poles of the integral, $D(\bar{\omega}_i) = 0$; note that if $\bar{\omega}_i$ is a solution, $-\bar{\omega}_i$ is a solution too. By expanding $D(\bar{\omega}) \simeq (\bar{\omega} - \bar{\omega}_i)\left(\frac{dD}{d\bar{\omega}}\right)_{\bar{\omega}=\bar{\omega}_i}$ in the vicinity of the poles and employing the Cauchy residue theorem show that

$$P = \frac{-q^2 c}{4\pi\varepsilon_0 R_{\rm int}^2} \times 2 \sum_{n,m=1}^{M} \sum_{i=1}^{\infty} \alpha_i \cos[\bar{\omega}_i \bar{L}_b(n-m)] h(n-m)$$

What is α_i? Find an analytic expression for $\sum_{n,m=1}^{M} \cos[\bar{\omega}_i \bar{L}_b(n-m)]$ $h(n-m)$ and compare the train form factor with the two extremes presented in the text.

Chapter 6
Quasi-Periodic Structures

Periodic structures play an important role in the interaction of electrons with waves since they support harmonics of phase-velocity smaller than c and with an adequate design, this velocity can be set equal to the average velocity of the electrons. In particular, in *extraction structures*, as the electrons interact with the wave and lose energy, they slip out of phase and consequently, the interaction is degraded. In order to avoid this situation the phase velocity of the wave has to be adjusted and the geometry change associated with this process should be designed for minimum reflections, otherwise the system oscillates. In a similar way, in *photo-injectors*, electron bunches are accelerated from zero velocity to virtually the speed of light in a relatively short distance (typically 1.5 period) and therefore, the design needs to account for the accelerating bunch such that the latter experiences maximum (rf) electric field.

In a periodic structure, at a given frequency and single mode operation, the electromagnetic wave is characterized by a single wave-number k and quantities like phase velocity, group velocity and interaction impedance are well defined. In principle, if the structure is no longer periodic the field cannot be represented by a single wave-number except if the variations are adiabatic in which case these characteristics are assumed to be determined by the geometry of the *local* cell. Adiabatic perturbations in the geometry may improve the efficiency from a few percent level in uniform structures to the 30% level. But one cannot expect to achieve 60–80% efficiency by moderate variation of the structure, bearing in mind that in contrast to accelerators where these changes occur over many wavelengths, in traveling-wave extraction structures these changes should occur in one or, at the most, two wavelengths.

Non-adiabatic change of geometry dictates a wide spatial spectrum in which case the formulation of the interaction in terms of a single wave with a varying amplitude and phase is inadequate. In fact, the electromagnetic field cannot be expressed in a simple (analytic) form if substantial geometric variations occur from one cell to another. To be more specific: in a uniform or weakly tapered disk-loaded waveguide, the beam-wave interaction is analyzed assuming that the general functional form of the electromagnetic wave is known i.e., $A(z)\cos[\omega t - kz - \phi(z)]$ and

L. Schächter, *Beam-Wave Interaction in Periodic and Quasi-Periodic Structures*,
Particle Acceleration and Detection, DOI 10.1007/978-3-642-19848-9_6,

as indicated in Chap. 4 the beam affects the amplitude $A(z)$ and the phase $\phi(z)$. Furthermore, it is assumed that the variation due to the interaction is small on the scale of one wavelength of the radiation. Both assumptions are not acceptable in the case of a structure designed for high efficiency interaction. In order to emphasize even further this difficulty, we recall that a non-adiabatic local perturbation of geometry affects *global* electromagnetic characteristics, this is to say that a change in a given cell affects the interaction impedance or the group velocity several cells before and after the point where the geometry was altered.

In order to overcome these difficulties, we present in the first part of this chapter, an analytical technique that was developed in order to design and analyze quasi-periodic metallic structures of the type discussed in Chap. 5. The method relies on a model consisting of a cylindrical waveguide to which a number of pill-box cavities and radial arms are attached. In principle, the number of cavities and arms is arbitrary. We formulate the boundary condition problem in terms of the amplitudes of the electromagnetic field in the cavities and arms. The elements of the matrix, which relates these amplitudes with the source term, are analytic functions and no a-priori knowledge of the functional behavior of the electromagnetic field is necessary. In Sect. 6.1 we examine the homogeneous electromagnetic characteristic of quasi-periodic structures. We further develop this technique to include Green's function formulation in Sect. 6.2 followed by the investigation of space-charge waves (Sect. 6.3) within the framework of the linear hydrodynamic approximation for the beam dynamics. In Sect. 6.4 the method is further generalized to include effects of large deviations from the initial average velocity of the electrons by formulating the beam-wave interaction in the framework of the macro-particle dynamics. In Sect. 6.5 we employ the quasi-periodic structure for investigating the electromagnetic wake generated by a bunch moving in the vicinity of a finite roughness surface. In the last section, we present a simple analytic model for another system wherein the quasi-periodic structure plays a crucial role namely, the photo-injector. However, since the beam loading in the framework of the model presented is not dominant, we consider an analytic formulation. Additional aspects of beam-wave interaction in quasi-periodic structures will be discussed in Chap. 7 in the context of a free-electron laser with a tapered wiggler.

The study regarding output structures presented in this chapter was triggered by research conducted at Cornell University. In the introduction to Chap. 4 we indicated that power levels in excess of 200 MW were generated in a 50 MHz bandwidth. The 200 MW generated with this structure were accompanied by gradients larger than 200 MV/m and no rf breakdown was observed experimentally. However, for any further increase in the power levels, it is necessary to increase the volume of the last two or three cells in order to minimize the electric field on the metallic surface. The system becomes then quasi-periodic. In order to envision the process in a clearer way let us assume that 80% efficiency is required from our source. If the initial beam is not highly relativistic, which is the case in most systems, such efficiency implies a dramatic change in the geometry of the structure over a short distance. Specifically, for a 500 keV beam, the initial velocity is $v_0 \sim 0.86c$ and 80% efficiency would imply a phase velocity of $0.55c$ at the output. This corresponds to a 36% change in the phase velocity and a similar change will be required in the geometry, which is by no means an adiabatic change when it occurs in one period of the wave.

Based on our experience there are three main difficulties associated with an extraction section based on a quasi-periodic traveling-wave structure. (1) Reduction of the reflections primarily at the output end of the structure in order to maintain a clean spectrum and to avoid oscillations, (2) taper the output section to avoid breakdown and (3) compensate for the decrease in the velocity of the electrons. The technique presented in the next 4 sections enables us to optimize these conflicting requirements.

6.1 Homogeneous Solution

The model used to analyze a quasi-periodic structure consists of a set of radial arms and pill-box cavities attached to a cylindrical waveguide. Their number and order is arbitrary. However for this presentation we consider a situation in which the input arm is the first cell (subscript 1) and the output arm is the last (subscript N) – as illustrated in Fig. 6.1. Each aperture, whether it corresponds to a cavity or an arm, has a width denoted by d_n where n is the index ascribed to each unit ($n = 1, 2 \cdots N$); N is the total number of cells and arms. The height, width and separation of each cavity can be arbitrary. Only the internal radius ($R_{\rm int}$) has to be the same throughout the device. The height of each cavity is determined by its external radius denoted by $R_{{\rm ext},n}$. A cylindrical coordinates system is used: its origin is chosen in the center of the first aperture. Furthermore, the system is azimuthally symmetric and so is the electromagnetic excitation. Consequently, throughout this chapter we consider only symmetric transverse magnetic (TM) modes. Specifically, in this section we examine the transmission and reflection characteristics.

One way to analyze the electromagnetic characteristics of such a structure is by mode decomposition and formulating the boundary condition problem in terms of a transmission matrix from each discontinuity – see Sect. 2.5.2. This method is addressed in literature (Mittra and Lee 1971 or Lewin 1975), but its performance is poor whenever more than one discontinuity is involved. This is due to the large and small numbers evolving from the evanescent modes associated with each discontinuity and their advance from one discontinuity to another.

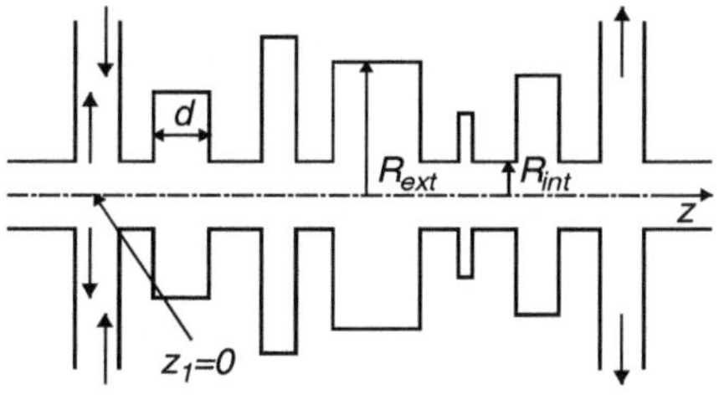

Fig. 6.1 Schematic of the model used for investigation of quasi-periodic structures

6.1.1 Definition of the Model

Contrary to a periodic structure, where the field in the inner cylinder ($0<r<R_{\rm int}$) can be represented by Floquet series, in this system we have to consider the entire spatial spectrum of waves, therefore the magnetic vector potential reads

$$A_z(r,z;\omega)=\int_{-\infty}^{\infty}\mathrm{d}kA(k)\mathrm{I}_0(\Gamma r)\exp(-jkz), \qquad (6.1.1)$$

where $\Gamma^2=k^2-\omega^2/c^2$ and $\mathrm{I}_0(x)$ is the zero order modified Bessel function of the first kind. All the transients are assumed to be zero or in other words, the system has reached a steady state regime thus a phasor notation $\exp(j\omega t)$, is adopted. In the arms or grooves the electromagnetic field should be represented by a superposition of modes which satisfy the boundary conditions on the metallic walls. In principle an infinite number of such modes are required. However, as long as the vacuum wavelength is about 5 times larger than the groove or arm width, the first mode [transverse electric and magnetic (TEM)] is sufficient for most practical purposes. This assumption is by no means critical for the present analysis and the calculation is similar when a larger number of modes are required, however we use it since it makes the presentation simpler. In order to quantify this statement let us give a simple example of a periodic disk-loaded structure: consider the case whereby $R_{\rm ext}=15.9$ mm, $R_{\rm int}=9.0$ mm, the period of the system is 10.0 mm and the disk is 5 mm wide. It is required that the phase advance per cell will be 120° at 9 GHz. With 39 spatial Floquet harmonics, the lower cutoff frequency ($kL=0$) was calculated to be 8.206 GHz using three modes (TEM, TM_{01} and TM_{02}) in the grooves, with two modes (TEM and TM_{01}) the cutoff was 8.192 GHz and 8.192 GHz when only the TEM mode was used. For the higher cutoff ($kL=\pi$) the calculated frequencies were 9.270 GHz, 9.229 GHz and 9.229 GHz correspondingly. Thus in the regime of interest the typical error associated with the higher modes omission in the grooves is expected to be of the order of 1% or less.

Within the framework of this approximation we can write for the magnetic vector potential in the input arm,

$$A_z(r,z;\omega)=A_{\rm in}\mathrm{H}_0^{(1)}\left(\frac{\omega}{c}r\right)+D_1\mathrm{H}_0^{(2)}\left(\frac{\omega}{c}r\right), \qquad (6.1.2)$$

where $\mathrm{H}_0^{(1)}(x)$ and $\mathrm{H}_0^{(2)}(x)$ is the zero order Hankel function of the first and second kind respectively; $A_{\rm in}$ represents the amplitude of the incoming wave and D_1 is the amplitude of the reflected wave which is yet to be determined. In the nth $(1<n<N)$ groove we have

$$A_z^n(r,z;\omega)=D_nT_{0,n}\left(\frac{\omega}{c}r\right), \qquad (6.1.3)$$

where D_n is the amplitude of the magnetic vector potential,

$$T_{0,n}\left(\frac{\omega}{c}r\right) = \mathrm{J}_0\left(\frac{\omega}{c}r\right)\mathrm{Y}_0\left(\frac{\omega}{c}R_{\mathrm{ext,n}}\right) - \mathrm{Y}_0\left(\frac{\omega}{c}r\right)\mathrm{J}_0\left(\frac{\omega}{c}R_{\mathrm{ext,n}}\right) \tag{6.1.4}$$

subsequently, we also use the function

$$T_{1,n}\left(\frac{\omega}{c}r\right) = \mathrm{J}_1\left(\frac{\omega}{c}r\right)\mathrm{Y}_0\left(\frac{\omega}{c}R_{\mathrm{ext,n}}\right) - \mathrm{Y}_1\left(\frac{\omega}{c}r\right)\mathrm{J}_0\left(\frac{\omega}{c}R_{\mathrm{ext,n}}\right) \tag{6.1.5}$$

Finally, in the output arm,

$$A_z(r,z;\omega) = D_N\mathrm{H}_0^{(2)}\left(\frac{\omega}{c}r\right), \tag{6.1.6}$$

represents a cylindrical outgoing wave.

In order to determine the various amplitudes we next impose the boundary conditions in a way that is similar to the case of a periodic structure but we no longer consider a single cell to characterize the entire system, instead we examine each individual region. From the condition of continuity of the *longitudinal electric field* we can conclude that

$$A(k) = -\frac{1}{2\pi}\frac{\alpha^2}{\Delta^2\mathrm{I}_0(\Delta)}\left[A_{\mathrm{in}}\mathrm{H}_0^{(1)}(\alpha)d_1\mathcal{L}_1(k) + \sum_{n=1}^{N} D_n\psi_{0,n}d_n\mathcal{L}_n(k)\right], \tag{6.1.7}$$

where $\alpha = \omega R_{\mathrm{int}}/c$ is the normalized angular frequency, $\Delta = \Gamma R_{\mathrm{int}}$ is the normalized wave-number in the radial direction and

$$\mathcal{L}_n(k) = \frac{1}{d_n}\int_{z_n-d_n/2}^{z_n+d_n/2} \mathrm{d}z\exp(jkz); \tag{6.1.8}$$

z_n is the location of the center of the nth groove or arm and in the first cell its value is zero ($z_1 = 0$). The function

$$\psi_{v,n} = \begin{cases} H_v^{(2)}(\alpha) & n=1 \quad \text{or} \quad n=N, \\ T_{v,n}(\alpha) & n\neq 1 \quad \text{or} \quad n\neq N, \end{cases} \tag{6.1.9}$$

is a generalized function defined in the aperture of either the grooves or the arms and $v = 0, 1$.

Imposing the continuity of the *tangential magnetic field* on each aperture we find

$$A_{\mathrm{in}}\mathrm{H}_1^{(1)}(\alpha)\delta_{n,1} + D_n\psi_{1,n} = -\frac{1}{\alpha}\int_{-\infty}^{\infty} \mathrm{d}kA(k)\Delta\mathrm{I}_1(\Delta)L_n^*(k). \tag{6.1.10}$$

It is now convenient to substitute (6.1.7) in (6.1.10) in order to represent the entire electromagnetic problem in terms of the amplitudes of the mode in the grooves and arms i.e.

$$\sum_{m=1}^{N} M_{n,m} D_m = S_n, \tag{6.1.11}$$

where

$$\begin{aligned} M_{n,m} &= \psi_{1,n}\delta_{n,m} - \psi_{0,m}\chi_{n,m}, \\ S_n &= -\mathrm{H}_1^{(1)}(\alpha)\delta_{n,1}A_{\text{in}} + \mathrm{H}_0^{(1)}(\alpha)\chi_{n,1}A_{\text{in}}, \end{aligned} \tag{6.1.12}$$

and

$$\chi_{n,m} = \frac{d_m \alpha}{2\pi} \int_{-\infty}^{\infty} \mathrm{d}k \frac{\mathrm{I}_1(\Delta)}{\Delta \mathrm{I}_0(\Delta)} \mathcal{L}_n^*(k)\mathcal{L}_m(k). \tag{6.1.13}$$

In principle, with the matrix M established, the electromagnetic problem is solved.

6.1.2 Evaluation of Green's Function

Our next step is to simplify the expression for the matrix M and for this purpose, we evaluate the integral that defines the matrix χ in terms of analytic functions using Cauchy's residue theorem. First we substitute the explicit expressions for $\mathcal{L}_n(k)$ from (6.1.8); the result is

$$\chi_{n,m} = \frac{d_m \alpha}{2\pi} \frac{1}{d_m} \int_{z_m-d_m/2}^{z_m+d_m/2} \mathrm{d}x_1 \frac{1}{d_n} \int_{z_n-d_n/2}^{z_n+d_n/2} \mathrm{d}x_2 \times \int_{-\infty}^{\infty} \mathrm{d}k \frac{\mathrm{I}_1(\Delta)}{\Delta \mathrm{I}_0(\Delta)} \exp[jk(x_1 - x_2)]. \tag{6.1.14}$$

If we now examine the integrand we observe that there are an infinite set of poles which correspond to $\mathrm{I}_0(\Delta) = 0$ since the modified Bessel function and the regular one $[\mathrm{J}_0(x)]$ are related thus we realize that the condition above is satisfied for $k^2 = (\omega/c)^2 - (p_s/R_{\text{int}})^2$; here p_s are all the zeros of the zero order Bessel function of the first kind i.e., $\mathrm{J}_0(p_s) \equiv 0$. According to Cauchy's theorem the contribution to the integral will come from the poles of the integrand thus the last integral in (6.1.14) reads

$$\frac{1}{2\pi} \int_{-\infty}^{\infty} \mathrm{d}k \frac{\mathrm{I}_1(\Delta)}{\Delta \mathrm{I}_0(\Delta)} \exp[jk(x_1 - x_2)] = \frac{1}{\pi R_{\text{int}}^2} \sum_{s=1}^{\infty} \int_{-\infty}^{\infty} \mathrm{d}k \frac{\exp[jk(x_1 - x_2)]}{k^2 + \Gamma_s^2}, \tag{6.1.15}$$

wherein $\Gamma_s^2 = (p_s/R_{int})^2 - (\omega/c)^2$. The last integral corresponds to Green's function for a uniform waveguide and it is easily evaluated as $G(x_1|x_2) = (\pi/\Gamma_s)_s \exp(-\Gamma_s|x_1 - x_2|)$. This result enables us to express the matrix χ in terms of analytic functions. Moreover, the integration over x_1 and x_2 in (6.1.15) can be performed explicitly

$$\chi_{n,m} = \frac{\alpha}{R_{\text{int}}^2} \sum_{s=1}^{\infty} \begin{cases} \frac{2}{\Gamma_s^2}[1 - \exp(-\Gamma_s d_n/2)\text{sinhc}(\Gamma_s d_n/2)] & n = m, \\ (d_m/\Gamma_s)\exp[-\Gamma_s|z_n - z_m|]\text{sinhc}(\Gamma_s d_n/2) \\ \quad \times\text{sinhc}(\Gamma_s d_m/2) & n \neq m; \end{cases} \tag{6.1.16}$$

in this expression $\text{sinhc}(x) = \sinh(x)/x$. The electromagnetic problem was now simplified to the inversion of a matrix whose components are analytic functions.

6.1.3 Transmission and Reflection

In order to test the method we used a set of identical cells. We were able to calculate the pass-band in the transmission coefficient and it fits very well that calculated using the dispersion relation of an infinite periodic structure. The following example illustrates the potential of this method: our first goal is to determine what should be the location of the arms to feed power adequately into a 9 cell structure ($R_{\text{ext}} = 14.2$ mm, $R_{\text{int}} = 6.2$ mm, $L = 12$ mm and $d = 6$ mm). Figure 6.2 illustrates the geometry of the narrow band structure with 9 cavities and two arms. In the first case, the arms are 6 mm from the adjacent cells (see lower system) and we observe that the *average* transmission coefficient, as illustrated in the lower curves, is about -20 dB. Thus, the bandwidth is much narrower than that of a practical source and to this extent, the fact that the peaks reach the 0 dB level becomes irrelevant to any experimental consideration. For this reason, we prefer to consider the average transmission coefficient in a range of frequencies. As the length of the waveguide between the arm and adjacent cell was shortened to 1 mm (both at the output and input), the transmission coefficient increases dramatically to an average value of -3 dB.

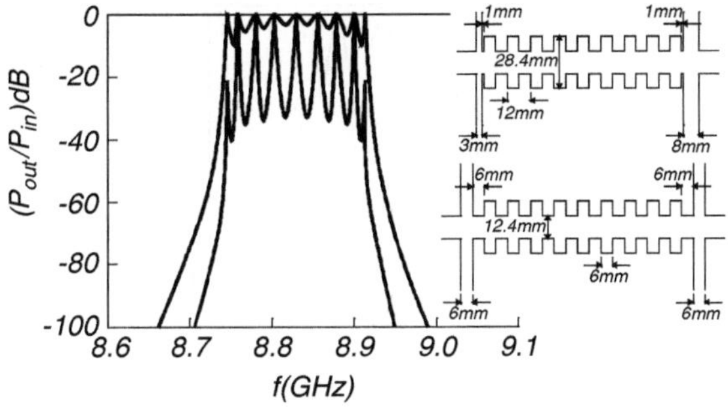

Fig. 6.2 Power transmitted for two geometries. The upper geometry corresponds to the upper curve and the distance between the groove and the arm is 1 mm. In the lower geometry and corresponding curve this distance is 6 mm.

Let us now assume that we have matched the system for a given frequency, i.e., the transmission coefficient in dB, defined by $10\log(|D_N|^2 d_N/|A_{\rm in}|^2 d_1)$, is zero. It is known that in a narrow pass-band structure high gradients may develop in the (high power) interaction process – in particular in the last couple of cells. In order to avoid rf breakdown the volume in which the electromagnetic energy is stored has to be increased, thus reducing in the process the energy density, and consequently reducing the field. As first attempt, we consider a *linear* tapering of the external radius of the last three cells. In the process, the width of these cells and their separation was varied in a wide range of parameters to bring the transmission coefficient to 0 dB at given frequency and the best we could achieve was -3 dB which is not acceptable (see Fig. 6.3). At this stage, we returned to the initial geometry but doubled the external radius of the last two cells. These cavities have two (rather than one) resonant frequencies, one of which is close to that of a cavity in the uniform structure. After some fine-tuning, we obtained the transmission

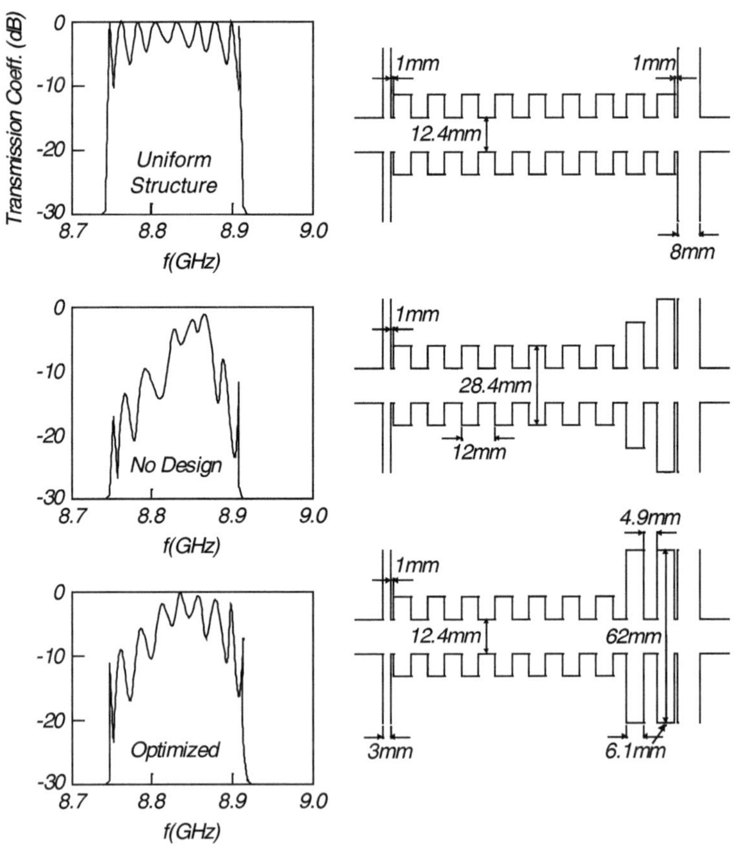

Fig. 6.3 Transmission coefficient for three different geometries

which is optimized to the required frequency – as indicated in the lower frame of Fig. 6.3.

This example emphasizes the dual way we can examine a quasi-periodic structure: as a traveling-wave structure or as a set of coupled cavities. It is the latter which is of great importance in the design of extraction regions since, as we indicated above, quantities like phase or group velocity have practically no meaning when the geometry of the structure *varies rapidly* in space. Such variation is a direct result of the broad spectrum of wave-numbers compared to a single wave-number in a regular periodic structure.

6.2 Non-homogeneous Solution

The homogeneous solution presented above assumes that the source of the electromagnetic field is far away from the structure and the arm guides the electromagnetic energy into the system. In this section, we consider the case when the source is in the structure. By virtue of linearity of Maxwell equations, we may assume that the remote sources are zero and we calculate only the contribution of the inner source. A general solution is obviously a superposition of the two solutions.

6.2.1 *Green's Function*

When a current distribution is present in the structure, we have to solve the non-homogeneous wave equation

$$\left[\nabla^2 + \frac{\omega^2}{c^2}\right] A_z(r, z, \omega) = -\mu_0 J_z(r, z, \omega), \tag{6.2.1}$$

and we proceed by calculating Green's function of the system. For this purpose, consider, instead of the general source of the above, a simple one, namely a narrow ring located at $z = z'$ and $r = r'$ which is a source to a field a_z that satisfies

$$\left[\nabla^2 + \frac{\omega^2}{c^2}\right] a_z(r, z|r', z') = -\frac{1}{2\pi r}\delta(r - r')\delta(z - z'), \tag{6.2.2}$$

and when subject to the same boundary conditions as A_z it is exactly Green's function of the system. In the absence of any boundaries this function is given by (see Sect. 2.4.1)

$$a_z(r, z|r', z') = \int_{-\infty}^{\infty} \mathrm{d}k g_k(r|r') \exp[-jk(z - z')], \tag{6.2.3}$$

where

$$g_k(r|r') = \frac{1}{(2\pi)^2}\begin{cases} \mathrm{I}_0(\Gamma r)\mathrm{K}_0(\Gamma r') & \text{for} \quad 0 \le r \le r', \\ \mathrm{K}_0(\Gamma r)\mathrm{I}_0(\Gamma r') & \text{for} \quad r' \le r < \infty. \end{cases} \tag{6.2.4}$$

Accordingly, the solution for the vector magnetic potential reads

$$\begin{aligned} A_z(r,z,\omega) = 2\pi\mu_0 \int_0^{R_b} \mathrm{d}r' r' \int_{-\infty}^{\infty} \mathrm{d}z' a_z(r,z|r',z') J_z(r',z',\omega) \\ + \int_{-\infty}^{\infty} \mathrm{d}k A(k) \exp(-jkz) \mathrm{I}_0(\Gamma r). \end{aligned} \tag{6.2.5}$$

The second term is the solution of the homogeneous equation, which does not diverge on axis and is a direct result of the presence of the metallic surface; R_b is the radius of the source. In the region outside the source ($r \ge R_b$), this expression can also be written as

$$A_z(r > R_b, z, \omega) = \int_{-\infty}^{\infty} \mathrm{d}k [B(k)\mathrm{K}_0(\Gamma r) + A(k)\mathrm{I}_0(\Gamma r)] \exp(-jkz), \tag{6.2.6}$$

where

$$B(k) = \frac{\mu_0}{2\pi} \int_0^{R_b} \mathrm{d}r' r' \mathrm{I}_0(\Gamma r') \int_{-\infty}^{\infty} \mathrm{d}z' \exp(jkz') J_z(r',z',\omega), \tag{6.2.7}$$

is the spatial Fourier transform of the current density. For the boundary condition problem the relevant components of the electromagnetic field are

$$\begin{aligned} E_z(r > R_b, z, \omega) &= \frac{c^2}{j\omega} \int_{-\infty}^{\infty} \mathrm{d}k (-\Gamma^2) [B(k)\mathrm{K}_0(\Gamma r) + A(k)\mathrm{I}_0(\Gamma r)] \exp(-jkz), \\ H_\phi(r > R_b, z, \omega) &= -\frac{1}{\mu_0} \int_{-\infty}^{\infty} \mathrm{d}k (\Gamma) [-B(k)\mathrm{K}_1(\Gamma r) + A(k)\mathrm{I}_1(\Gamma r)] \exp(-jkz). \end{aligned} \tag{6.2.8}$$

In the grooves and arms the solution is identical with (6.1.2)–(6.1.6) except that $A_{\mathrm{in}} \equiv 0$. When imposing the continuity of the longitudinal component of the electric field, we obtain

$$\begin{aligned} -2\pi \frac{\Delta^2}{\alpha^2} [B(k)\mathrm{K}_0(\Delta) + A(k)\mathrm{I}_0(\Delta)] = D_1 \mathrm{H}_0^{(2)}(\alpha) d_1 \mathcal{L}_1(k) \\ + \sum_{n=2}^{N-1} D_n T_{0,n}(\alpha) d_n \mathcal{L}_n(k) \\ + D_N \mathrm{H}_0^{(2)}(\alpha) d_N \mathcal{L}_N(k), \end{aligned} \tag{6.2.9}$$

and the continuity of the azimuthal magnetic field provides us with an additional set of equations similar to (6.1.10):

$$\alpha D_n \psi_{1,n} = \int_{-\infty}^{\infty} \mathrm{d}k \Delta [B(k)\mathrm{K}_1(\Delta) - A(k)\mathrm{I}_1(\Delta)] \mathcal{L}_n^*(k). \tag{6.2.10}$$

In these two equations $\mathcal{L}_n(k)$ was defined in (6.1.8). Based on these two equations we can determine the amplitudes D_n in the arms and grooves by substituting (6.2.9) in (6.2.10). The result is similar to the homogeneous case:

$$\sum_{m=1}^{N} M_{n,m} D_m = S_n, \tag{6.2.11}$$

except that the source term is now given by

$$S_n = \frac{1}{\alpha} \int_{-\infty}^{\infty} \mathrm{d}k \frac{1}{\mathrm{I}_0(\Delta)} B(k) \mathcal{L}_n^*(k). \tag{6.2.12}$$

In this expression we used the property of the modified Bessel functions: $\mathrm{I}_0(x)\mathrm{K}_1(x) + \mathrm{I}_1(x)\mathrm{K}_0(x) = 1/x$. Expression (6.2.11) indicates that if we know the source term S_n we can determine all the amplitudes D_n using the inverse of exactly the same matrix $\mathbf{M}$ we defined in the previous section. Therefore, we next direct our efforts to simplify the expression for the source term S_n.

Based on the definition of $B(k)$ in (6.2.7) we can write

$$S_n = \frac{\mu_0}{\alpha} \int_0^{R_b} \mathrm{d}r' r' \int_{-\infty}^{\infty} \mathrm{d}z' J_z(r', z', \omega) \sigma_n(r', z'), \tag{6.2.13}$$

where

$$\sigma_n(r', z') = \frac{1}{2\pi} \int_{-\infty}^{\infty} \mathrm{d}k \frac{\mathrm{I}_0(\Gamma r')}{\mathrm{I}_0(\Delta)} \mathcal{L}_n^*(k) \exp(jkz'). \tag{6.2.14}$$

Thus in order to simplify the source term S_n one has first to simplify the function $\sigma_n(r', z')$. We substitute the explicit expression for $\mathcal{L}_n(k)$ and then evaluate the integral on k based on Cauchy's residue theorem; the result is

$$\sigma(r', z') = \frac{1}{2\pi} \frac{1}{d_n} \int_{z_n - d_n/2}^{z_n + d_n/2} d\zeta \sum_{s=1}^{\infty} \frac{2 p_s \mathrm{J}_0(p_s r'/R_{\mathrm{int}})}{R_{\mathrm{int}}^2 \mathrm{J}_1(p_s)} \frac{\pi}{\Gamma_s} \exp(-\Gamma_s |\zeta - z'|). \tag{6.2.15}$$

Our next step is to perform the integration over ζ thus

$$S_n = \frac{\mu_0}{\alpha}\int_0^{R_b} dr'r' \int_{-\infty}^{\infty} dz' J_z(r', z', \omega) \sum_{s=1}^{\infty} \frac{p_s J_0(p_s r'/R_{int})}{\Gamma_s R_{int}^2 J_1(p_s)} \sigma_{s,n}(z'), \tag{6.2.16}$$

where

$$\sigma_{s,n}(z) \equiv \frac{1}{d_n}\int_{z_n-d_n/2}^{z_n+d_n/2} d\xi e^{-\Gamma_s|\xi-z|}$$
$$= \begin{cases} \exp(-\Gamma_s|z-z_n|)\mathrm{sinhc}\left(\frac{1}{2}\Gamma_s d_n\right) & \text{for} |z-z_n| > \frac{d_n}{2}, \\ \frac{2}{\Gamma_s d_n}\left\{1-\exp\left(-\frac{1}{2}\Gamma_s d_n\right)\cosh[\Gamma_s(z-z_n)]\right\} & \text{for} |z-z_n| < \frac{d_n}{2}. \end{cases} \tag{6.2.17}$$

Formally, this concludes the formulation of the boundary condition problem in (6.2.11) and in the remainder we present two examples.

6.2.2 Stationary Dipole

Further simplification of the analysis is possible by making the following assumptions: (1) the current density varies very slowly in the transverse direction such that it can be considered constant. (2) We examine a Dirac delta function current distribution in the longitudinal direction, such that the field due to any other current distribution can be represented as a superposition of such point sources i.e.

$$J_z(r, z) = \frac{I}{\pi R_b^2}\Delta_z \delta(z - z_c) h(R_b - r); \tag{6.2.18}$$

$h(x)$ is the regular step function. In this expression I is the dipole's current, Δ_z is its characteristic length and z_c is its longitudinal location. This is a stationary (motionless) dipole, which oscillates at an angular frequency ω. With these assumptions and bearing in mind that

$$\int_0^{R_b} dr r J_0(p_s r/R_{int}) = R_b R_{int} \frac{1}{p_s} J_1\left(p_s \frac{R_b}{R_{int}}\right), \tag{6.2.19}$$

the source term in (6.2.11) is given by

$$S_n = \frac{\mu_0 I \Delta_z}{\pi \alpha R_b} \sum_{s=1}^{\infty} \frac{J_1(p_s R_b/R_{int})}{J_1(p_s)} \frac{1}{\sqrt{p_s^2 - \alpha^2}} \sigma_{s,n}(z_c). \tag{6.2.20}$$

In order to present the radiation emitted by such a stationary dipole it is convenient to normalize both the source term and the amplitude with the term $a \equiv \mu_0 I\Delta_z/(\pi R_b)$ hence $\bar{S}_n = S_n/a$ and $\bar{D}_n = D_n/a$. The average power which flows through the vth arm is $P_v = 2\omega d_v |D_v|^2/\mu_0$. Accordingly, the average normalized power flowing through each one of the arms in the structure is

$$\bar{P}_v \equiv P_v \left[\frac{1}{2}\eta_0 I^2 \frac{\Delta_z^2}{\pi R_b^2}\right]^{-2} = \frac{4}{\pi}\frac{d_v}{R_{\text{int}}}\alpha|\bar{D}_v(\alpha)|^2; \qquad (6.2.21)$$

here, the index v indicates the input or output arms only i.e., $v = 1$ represents the input arm and $v = N$, the output. First to be examined was the effect of the arm location on the radiation emitted by a single dipole and for this purpose two quantities are defined: the total emitted power, $(P_{\text{tot}})_{dB} \equiv 10\log(\bar{P}_1 + \bar{P}_N)$, and the ratio between the power emitted in the output arm and input arm i.e., $(P_N/P_1)_{dB} \equiv 10\log(\bar{P}_N/\bar{P}_1)$. The geometry considered next is somewhat different than in the previous section: $R_{\text{ext}} = 17.3$ mm, $R_{\text{int}} = 9$ mm, $L = 10.4$ mm and $d = 1.4$ mm. This choice of parameters was determined by the need to increase the internal radius of the structure while at the same time maintain the group velocity relatively low. The phase advance per cell was chosen to be 120° at 9 GHz which is the resonant frequency with a 1 MeV electron.

Figure 6.4 illustrates the power emitted by a dipole oscillating at 9 GHz as its location varies along the structure, for two different geometries. The upper frame represents the case we showed previously to be the optimal from the point of view of feeding the system; namely, minimum distance between the arms and adjacent cavities ($g_{\text{in}} = g_{\text{out}} = 1$ mm). In the lower frame, the separation of the input arm is $g_{\text{in}} = g_1 = 5$ mm. There are several features, which should be emphasized. First, for the upper frame there is a clear pattern of larger emission when the dipole is in the cavity region compared to the case when it oscillates in the space between two cavities. Second, comparing the power in the output arm with that in the input arm for the upper case we observe that both are of the same order of magnitude. Thirdly, breaking the symmetry of the system ($g_{\text{in}} = 5$ mm), causes a preferred direction of emission towards the output (since the input is "blocked") as indicated in the lower frame. Note that although the dipole current is the same, the peak power is larger. In addition, the clear pattern of maximum power obtained when the dipole is in the cavity (see Fig. 6.4, upper frame) is not as clear in the case shown in the lower frame of Fig. 6.4.

Another case of interest is to examine the effect of the length of the drift regions between two adjacent cavities. We increased the distance between the third and the fourth cavity from 9 mm to $g_4 = 20$ mm. The effect is illustrated in the upper frame of Fig. 6.5, and the lower frame shows the case when $g_3 = 20$ mm. After examining the previous case, the results are intuitive: in the first part of the structure, the emission is primarily towards the input arm whereas in the second part, practically all the radiation is emitted through the output arm. It should be mentioned that since the current density is imposed, the emitted power is a direct measure of the

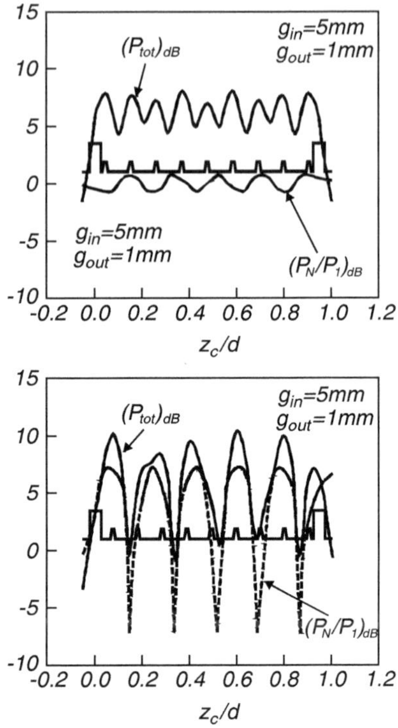

Fig. 6.4 Power emitted by the dipole as a function of its location in two different geometries which differ only by the distance of the first cavity from the input arm

longitudinal component of the electric field in the structure. As such, we observe that the main difference between the upper and lower frame is the field pattern – directly associated with the change in the geometry. We subsequently return to this geometry since it can simulate the operation of a two-stage traveling-wave structure or a klystron with a traveling-wave output.

6.2.3 Distributed Current Density

In a uniform section of a traveling-wave amplifier the modulation amplitude grows exponentially. In this subsection we calculate the electromagnetic field generated when imposing a current density similar to that developing in the interaction process in a traveling-wave amplifier. The current density is given by

$$J_z(r, z; \omega) = J_0 \exp(-jKz) h(R_b - r); \tag{6.2.22}$$

K is a complex wave-number which represents the phase advance and the amplitude variation. According to (6.2.13) the source term is

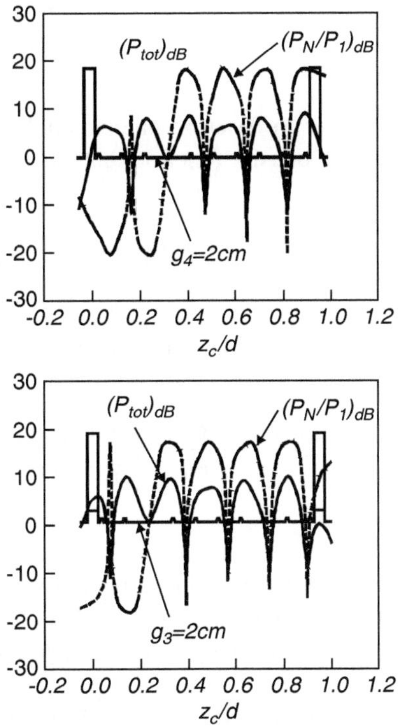

Fig. 6.5 Power emitted by the dipole as a function of its location in two different geometries which differ in the distance between the third and fourth groove (*top*) and second and third groove (*bottom*)

$$S_n = \frac{J_0\mu_0}{\alpha}\frac{R_\mathrm{b}}{R_\mathrm{int}}\sum_{s=1}^{\infty}\frac{R_\mathrm{int}^2}{\sqrt{p_s^2-\alpha^2}}\frac{\mathrm{J}_1(p_sR_\mathrm{b}/R_\mathrm{int})}{\mathrm{J}_1(p_s)} \times \frac{1}{R_\mathrm{int}}\int_{-\infty}^{\infty}\mathrm{d}z\mathrm{e}^{-jKz}\int_{z_n-d_n/2}^{z_n+d_n/2}\frac{\mathrm{d}z'}{d_n}\exp(-\Gamma_s|z-z'|). \tag{6.2.23}$$

Changing the order of integration, we have

$$\frac{1}{R_\mathrm{int}}\int_{-\infty}^{\infty}\mathrm{d}z\mathrm{e}^{-jKz}\cdots = \frac{1}{d_n}\int_{z_n-d_n/2}^{z_n+d_n/2}\mathrm{d}z'\frac{1}{R_\mathrm{int}}\int_{-\infty}^{\infty}\mathrm{d}z\exp(-jKz-\Gamma_s|z-z'|), \tag{6.2.24}$$

thus

$$S_n = 2\frac{J_0R_\mathrm{b}^2\mu_0}{\alpha}\frac{R_\mathrm{int}}{R_b}\sum_{s=1}^{\infty}\frac{\mathrm{J}_1(p_sR_\mathrm{b}/R_\mathrm{int})}{\mathrm{J}_1(p_s)}\frac{\mathrm{sinc}(Kd_n/2)}{p_s^2-\alpha^2+(KR_\mathrm{int})^2}\mathrm{e}^{-jKz_n}. \tag{6.2.25}$$

According to the coefficient in this expression we define the normalization factor $a \equiv 2J_0\mu_0R_\mathrm{b}R_{int}$ which entails that $\bar{D}_n = D_n/a$ and $\bar{S}_n = S_n/a$. With these quantities the normalized average emitted power reads

$$\bar{P}_v = \frac{P_v}{\frac{1}{2}\eta_0 (J_0 \pi R_b^2)^2} = \frac{16}{\pi^2} \frac{d_v}{R_b} \frac{R_{\rm int}}{R_b} \alpha |\bar{D}_v(\alpha)|^2. \tag{6.2.26}$$

In order to describe phenomenologically the saturation effect we can consider a current density function that has the form

$$J_z(r, z; \omega) = J_0 \left(1 - \frac{z}{d_{\rm sat}}\right) e^{-jKz} h(R_b - r); \tag{6.2.27}$$

in which case the source term is

$$S_n = 2\frac{J_0 R_b^2 \mu_0}{\alpha} \frac{R_{\rm int}}{R_b} \times \sum_{s=1}^{\infty} \frac{J_1(p_s R_b / R_{\rm int})}{J_1(p_s)} \left[1 + j\frac{1}{d_{sat}} \frac{d}{dK}\right] \frac{\mathrm{sinc}(K d_n/2)}{p_s^2 - \alpha^2 + (K R_{\rm int})^2} e^{-jKz_n}. \tag{6.2.28}$$

Next we examine quantitatively the radiation emitted at 9 GHz by the current distribution in (6.2.27) for the following parameters: $d_{\rm sat} = 1.3 d_{\rm tot}$, $K = 0.5K_0(1 + j\sqrt{3}) + \omega/c\beta$, $\beta = 0.94$, $K_0 = 80\,\mathrm{m}^{-1}$ and $d_{\rm tot}$ is the total length of the system. The total power emitted by this current distribution when in a uniform structure is $P_{\rm tot} = 54.9\,\mathrm{dB}$ (see definition in previous section), and most of this power is emitted forward due to the spatial phase correlation and the varying amplitude. The asymmetry associated with the current distribution is $P_N/P_1 = 8.6\,\mathrm{dB}$. As in the first section, we now increase the volume of the last cavity by increasing the width of the cell $d_{10} = 5\,\mathrm{mm}$. Its separation from the previous cavity remains the same ($g_9 = 9\,\mathrm{mm}$). Figure 6.6 illustrates the total power and the arms power ratio as the external radius of the 10th cavity is varied. It is evident from this figure the resonant character of the structure.

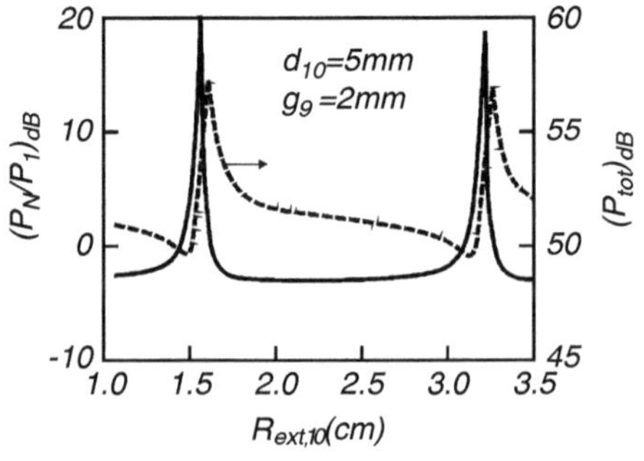

Fig. 6.6 Power emitted by the dipole as a function of the external radius of the last cavity

6.3 Beam-Wave Interaction: Hydrodynamic Approximation

In the previous subsection the current density was *imposed* and the effect of saturation was included phenomenologically. At this point, we extend our investigation to a self-consistent solution of the current density and the electromagnetic field in the framework of the (linear) hydrodynamic approximation. This facilitates to examine the propagation of space-charge waves in quasi-periodic structures.

6.3.1 Definition of the Model

In the framework of this model, the beam is considered to be an active linear medium which satisfies

$$J_z(r,k;\omega) = -j\omega\varepsilon_0 \frac{\omega_{\rm p}^2}{(\omega - k{\rm v}_0)^2} E_z(r,k;\omega), \tag{6.3.1}$$

namely, it is considered to be a fluid. The relativistic plasma frequency is defined as

$$\omega_{\rm p}^2 = \frac{\eta_0 eI}{mc^2} \frac{c^2}{\pi R_{\rm b}^2} \frac{1}{\beta\gamma^3}. \tag{6.3.2}$$

With (6.3.1) and the definition of the longitudinal electric field in terms of A_z [i.e., $E_z = -j\omega A_z + j(ck)^2 A_z/\omega$], the non-homogeneous wave equation for the magnetic vector potential

$$\left[\nabla^2 + \frac{\omega^2}{c^2}\right] A_z(r,z,\omega) = -\mu_0 J_z(r,z,\omega), \tag{6.3.3}$$

becomes homogeneous and its solution (for a pencil beam) reads

$$A_z(r,z;\omega) = \int_{-\infty}^{\infty} {\rm d}k A(k) \exp(-jkz) {\rm I}_0(\Lambda r), \tag{6.3.4}$$

with

$$\Lambda^2 = \Gamma^2 \left[1 - \frac{\omega_{\rm p}^2}{(\omega - k{\rm v}_0)^2}\right]. \tag{6.3.5}$$

The longitudinal electric field and the azimuthal magnetic field read

$$E_z(r,z;\omega) = -\frac{c^2}{j\omega}\int_{-\infty}^{\infty} \mathrm{d}k\Gamma^2 A(k)\exp(-jkz)\mathrm{I}_0(\Lambda r),$$
$$H_\phi(r,z;\omega) = -\frac{1}{\mu_0}\int_{-\infty}^{\infty} \mathrm{d}k\Lambda A(k)\exp(-jkz)\mathrm{I}_1(\Lambda r). \tag{6.3.6}$$

In the gap between the beam and the metallic surface ($R_\mathrm{b} < r < R_\mathrm{int}$) the solution of the magnetic vector potential reads

$$A_z(r,z,\omega) = \int_{-\infty}^{\infty} \mathrm{d}k[B(k)\mathrm{I}_0(\Gamma r) + C(k)\mathrm{K}_0(\Gamma r)]\exp(-jkz), \tag{6.3.7}$$

and correspondingly, the field components relevant for the boundary condition problem are

$$E_z(r,z,\omega) = -\frac{c^2}{j\omega}\int_{-\infty}^{\infty} \mathrm{d}k\Gamma^2[B(k)\mathrm{I}_0(\Gamma r) + C(k)\mathrm{K}_0(\Gamma r)]\exp(-jkz),$$
$$H_\phi(r,z,\omega) = -\frac{1}{\mu_0}\int_{-\infty}^{\infty} \mathrm{d}k\Gamma[B(k)\mathrm{I}_1(\Gamma r) - C(k)\mathrm{K}_1(\Gamma r)]\exp(-jkz). \tag{6.3.8}$$

Continuity of these two components at $r = R_\mathrm{b}$ implies

$$\frac{1}{\Gamma}\frac{B(k)\mathrm{I}_0(b_\mathrm{v}) + C(k)\mathrm{K}_0(b_\mathrm{v})}{B(k)\mathrm{I}_1(b_\mathrm{v}) - C(k)\mathrm{K}_1(b_\mathrm{v})} = \frac{1}{\Lambda}\frac{\mathrm{I}_0(b_\mathrm{b})}{\mathrm{I}_1(b_\mathrm{b})}, \tag{6.3.9}$$

where $b_\mathrm{b} = \Lambda R_\mathrm{b}$ and $b_\mathrm{v} = \Gamma R_\mathrm{b}$ (subscript v stands for vacuum and subscript b for beam). This expression determines the relation between $B(k)$ and $C(k)$:

$$\rho(k) \equiv \frac{C(k)}{B(k)} = \frac{b_\mathrm{v}\mathrm{I}_0(b_\mathrm{b})\mathrm{I}_1(b_\mathrm{v}) - b_\mathrm{b}\mathrm{I}_1(b_\mathrm{b})\mathrm{I}_0(b_\mathrm{v})}{b_\mathrm{v}\mathrm{I}_0(b_\mathrm{b})_1(b_\mathrm{v}) + b_\mathrm{b}\mathrm{I}_1(b_\mathrm{b})_0(b_\mathrm{v})}. \tag{6.3.10}$$

It is now convenient to extract $B(k)$ from the brackets of (6.3.7)–(6.3.8) and define the radial functions:

$$\bar{I}_0(k,r) \equiv \mathrm{I}_0(\Gamma r) + \rho(k)\mathrm{K}_0(\Gamma r),$$
$$\bar{I}_1(k,r) \equiv \mathrm{I}_1(\Gamma r) - \rho(k)\mathrm{K}_1(\Gamma r). \tag{6.3.11}$$

These can be considered generalizations of the modified Bessel functions we used in the homogeneous case therefore the magnetic vector potential and the field components relevant to the boundary condition problem are given by

$$\begin{aligned} A_z(r,z,\omega) &= \int_{-\infty}^{\infty} \mathrm{d}k B(k)\bar{I}_0(k,r)\exp(-jkz), \\ E_z(r,z,\omega) &= -\frac{c^2}{j\omega}\int_{-\infty}^{\infty} \mathrm{d}k\Gamma^2 B(k)\bar{I}_0(k,r)\exp(-jkz), \\ H_\phi(r,z,\omega) &= -\frac{1}{\mu_0}\int_{-\infty}^{\infty} \mathrm{d}k\Gamma B(k)\bar{I}_1(k,r)\exp(-jkz). \end{aligned} \tag{6.3.12}$$

In the grooves and arms, the functional form of the solution is identical with that established in Sect. 6.1. Therefore, the formulation now is similar to the case when no beam is present and we can use the formal result we presented in Sect. 6.1 namely

$$\sum_{m=1}^{N} M_{n,m} D_m = S_n, \tag{6.3.13}$$

where

$$M_{n,m} = \psi_{1,n}\delta_{n,m} - \psi_{0,m}\chi_{n,m}, \tag{6.3.14}$$

$$S_n = -\mathrm{H}_1^{(1)}(\alpha)\delta_{n,1}A_{\mathrm{in}} + \mathrm{H}_0^{(1)}(\alpha)\chi_{n,1}A_{\mathrm{in}}, \tag{6.3.15}$$

and

$$\chi_{n,m} = \frac{d_m\alpha}{2\pi}\int_{-\infty}^{\infty} \mathrm{d}k \frac{\bar{I}_1(k,R_{\mathrm{int}})}{\Delta\bar{I}_0(k,R_{\mathrm{int}})}\mathcal{L}_n^*(k)\mathcal{L}_m(k). \tag{6.3.16}$$

The only difference is that the modified Bessel functions (I_0 and I_1) were replaced by the generalized counterparts $\bar{I}_0$ and $\bar{I}_1$ defined in (6.3.11).

6.3.2 Evaluation of Green's Function

As in the first section, we express the elements of the matrix χ in terms of analytic functions. Our first step is to substitute the explicit expressions for $\mathcal{L}_n(k)$; the result is

$$\chi_{n,m} = \frac{d_m\alpha}{2\pi}\frac{1}{d_m}\int_{z_m-d_m/2}^{z_m+d_m/2} \mathrm{d}x_1 \frac{1}{d_n}\int_{z_n-d_n/2}^{z_n+d_n/2} \mathrm{d}x_2 \times \int_{-\infty}^{\infty} \mathrm{d}k \frac{\bar{I}_1(k,R_{\mathrm{int}})}{\Delta\bar{I}_0(k,R_{\mathrm{int}})}\exp[jk(x_1-x_2)]. \tag{6.3.17}$$

It is convenient to define the following Green's function

$$G(x_1|x_2) = \frac{d_m}{2\pi}\int_{-\infty}^{\infty} \mathrm{d}k \frac{\alpha}{\Delta}\frac{\bar{I}_1(k,R_{\mathrm{int}})}{\bar{I}_0(k,R_{\mathrm{int}})}\exp[jk(x_1-x_2)], \tag{6.3.18}$$

hence,

$$\chi_{n,m} = \frac{1}{d_m}\int_{z_m-d_m/2}^{z_m+d_m/2} dx_1 \frac{1}{d_n}\int_{z_n-d_n/2}^{z_n+d_n/2} dx_2 G(x_1|x_2). \tag{6.3.19}$$

If in the previous sections we evaluated Green's function G using a "simple" set of poles which were the zeros of $\mathrm{I}_0(\Delta)$, in this case we have to examine the poles of

$$\tilde{G} = \frac{\alpha}{\Delta}\frac{\bar{I}_1(k, R_{\mathrm{int}})}{\bar{I}_0(k, R_{\mathrm{int}})}. \tag{6.3.20}$$

For the sake of simplicity we consider the case when the beam fills the entire waveguide i.e., $R_{\mathrm{b}} = R_{\mathrm{int}}$, in which case

$$\tilde{G} = \frac{\alpha}{\Delta}\frac{b}{\Delta}\frac{\mathrm{I}_1(b)}{\mathrm{I}_0(b)}, \tag{6.3.21}$$

where $b = \Lambda R_{\mathrm{int}}$. The poles of this expression correspond to the zeros of the dispersion relation of a waveguide filled with a beam. Since it was shown in Chap. 3 that in a cylindrical waveguide the electromagnetic modes and the space-charge modes are essentially "decoupled", we now determine the poles accordingly. In other words, the expression for $\tilde{G}$ is a superposition of the electromagnetic and space-charge modes

$$\tilde{G} = \tilde{G}_{\mathrm{EM}} + \tilde{G}_{\mathrm{SC}}. \tag{6.3.22}$$

The contribution of the electromagnetic modes is determined by ignoring the presence of the beam ($\omega_{\mathrm{p}} = 0$) and it is identical with what was found in Sects. 6.1–6.2. Using (6.1.16) we have

$$\chi_{n,m}^{(\mathrm{EM})} = \frac{\alpha}{R_{\mathrm{int}}^2}\sum_{s=1}^{\infty}\begin{cases}\dfrac{2}{\Gamma_s^2}\left[1 - \mathrm{e}^{-\Gamma_s d_n/2}\mathrm{sinhc}(\Gamma_s d_n/2)\right] & n = m,\\ \dfrac{d_m}{\Gamma_s}\mathrm{e}^{-\Gamma_s|z_n - z_m|}\mathrm{sinhc}(\Gamma_s d_n/2)\mathrm{sinhc}(\Gamma_s d_m/2) & \text{otherwise}.\end{cases} \tag{6.3.23}$$

Next the contribution of the space-charge waves is evaluated. As in the empty case, we consider the pole around $b = jp_s$, namely

$$b_s \simeq \sqrt{\left(\frac{\omega^2}{c^2}R_{\mathrm{int}}^2\frac{1}{\beta^2} - \frac{\omega^2}{c^2}R_{\mathrm{int}}^2\right)\left[1 - \frac{\omega_{\mathrm{p}}^2}{(\omega - k_s v_0)^2}\right]} \simeq jp_s; \tag{6.3.24}$$

accordingly, Green's function for the space-charge waves reads

$$\tilde{G}_{SC} = -\sum_{s=1}^{\infty} \frac{K_{b,s}^2}{(k-\omega/v_0)^2 - K_{p,s}^2}, \tag{6.3.25}$$

where

$$K_{b,s}^2 = \frac{2\xi_s^2\gamma^2\beta^2}{\alpha(1+\xi_s^2)} K_{p,s}^2, \quad K_{p,s}^2 = \frac{\omega_p^2}{v_0^2} \frac{1}{1+\xi_s^2}, \tag{6.3.26}$$

and $\xi_s = p_s\gamma\beta/\alpha$. The next step is to evaluate Green's function

$$G_{SC}(x_1|x_2) = -\sum_{s=1}^{\infty} \frac{K_{b,s}^2 d_m}{2\pi} \int_{-\infty}^{\infty} dk \frac{\exp[jk(x_1-x_2)]}{(k-\omega/v_0)^2 - K_{p,s}^2}. \tag{6.3.27}$$

After adequate change of variables the latter reads

$$G_{SC}(x_1|x_2) = -\sum_{s=1}^{\infty} \frac{K_{b,s}^2 d_m}{2\pi} \exp\left[j\frac{\omega}{v_0}(x_1-x_2)\right] \int_{-\infty}^{\infty} dk \frac{\exp[jk(x_1-x_2)]}{k^2 - K_{p,s}^2}. \tag{6.3.28}$$

The last integral is identical (except for the fact that the poles are real quantities) to that in (6.1.15); however, for its evaluation we have to be more careful since contrary to the electromagnetic waves, the space-charge modes propagate only along the beam. Therefore, we may expect the integral in (6.3.28) to be identically zero for $x_1 > x_2$ otherwise the solution would indicate a wave propagating against the beam. In order to solve the integral it is convenient to follow the same approach as in Sect. 6.1. The function

$$g_s(x_1|x_2) \equiv \int_{-\infty}^{\infty} dk \frac{\exp[jk(x_1-x_2)]}{k^2 - K_{p,s}^2}, \tag{6.3.29}$$

is defined and it can be shown to satisfy

$$\left[\frac{d^2}{dx_1^2} + K_{p,s}^2\right] g(_s x_1|x_2) = -2\pi\delta(x_1-x_2). \tag{6.3.30}$$

In the case of the space-charge waves we know that there are two waves which form continuous solution at $x_1 = x_2$ whereas their derivatives as determined by integrating (6.3.30) are discontinuous at the same location. This ultimately implies that

$$g_s(x_1|x_2) = -\frac{2\pi}{K_{p,s}} \sin\left[K_{p,s}(x_1-x_2)\right] h(x_1-x_2). \tag{6.3.31}$$

The contribution of the space-charge waves to the χ matrix can now be formulated as

$$\chi_{n,m}^{(\mathrm{SC})} = \sum_{s=1}^{\infty} \int_{z_m-d_m/2}^{z_m+d_m/2} \frac{\mathrm{d}x_1}{d_m} \int_{z_n-d_n/2}^{z_n+d_n/2} \frac{\mathrm{d}x_2}{d_n} \frac{K_b^2 d_m}{K_{p,s}} \exp\left[j\frac{\omega}{\mathrm{v}_0}(x_1 - x_2)\right] \times \sin\left[K_{p,s}(x_1 - x_2)\right] h(x_2 - x_1). \tag{6.3.32}$$

In the evaluation of these integrals we take advantage of the fact that for $x_1 > x_2$ the integrand is zero which means that if $m > n$ then

$$\chi_{n,m}^{(\mathrm{SC})} \equiv 0. \tag{6.3.33}$$

Diagonal terms ($n = m$) of the matrix are given by

$$\chi_{n,n}^{(\mathrm{SC})} = -\sum_{s=1}^{\infty} \frac{K_{b,s}^2 d_n}{4K_{p,s}} \left[\frac{1}{\xi_{s,n,+}}\left(1 - \mathrm{e}^{-j\xi_{s,n,+}} \operatorname{sinc}(\xi_{s,n,+})\right) - \frac{1}{\xi_{s,n,-}}\left(1 - \mathrm{e}^{-j\xi_{s,n,-}} \operatorname{sinc}(\xi_{s,n,-})\right)\right], \tag{6.3.34}$$

where $\xi_{s,n,\pm} = d_n(\omega/\mathrm{v}_0 \pm K_{p,s})/2$ and the off-diagonal non-zero terms ($n > m$) are

$$\chi_{n,m}^{(\mathrm{SC})} = -\sum_{s=1}^{\infty} \frac{jK_{b,s}^2 d_m}{2K_{p,s}} \left[\mathrm{e}^{-j(z_n - z_m)(\omega/\mathrm{v}_0 + K_{p,s})} \operatorname{sinc}(\xi_{s,n,+}) \operatorname{sinc}(\xi_{s,m,+})\right] + \sum_{s=1}^{\infty} \frac{jK_{b,s}^2 d_m}{2K_{p,s}} \left[\mathrm{e}^{-j(z_n - z_m)(\omega/\mathrm{v}_0 - K_{p,s})} \operatorname{sinc}(\xi_{s,n,-}) \operatorname{sinc}(\xi_{s,m,-})\right]. \tag{6.3.35}$$

Finally, the χ matrix is the superposition of the electromagnetic term $\chi^{(\mathrm{EM})}$ defined in (6.3.23) and the space-charge term $\chi^{(\mathrm{SC})}$ defined in (6.3.33)–(6.3.35), i.e.,

$$\chi_{n,m} = \chi_{n,m}^{(\mathrm{EM})} + \chi_{n,m}^{(\mathrm{SC})}. \tag{6.3.36}$$

The *sinc* function in the χ matrix implies that this method is a hybrid of local beam-gap interaction as in a klystron where the cavities are electromagnetically decoupled and distributed beam-wave interaction as in a traveling-wave structure where the cavities are electromagnetically coupled. In the examples presented below, the finite size of the beam is accounted for by introducing the filling factor

$$F_\mathrm{f} = \left(\frac{R_\mathrm{int}}{R_\mathrm{b}}\right)^2 \frac{\mathrm{I}_1^2(\alpha R_\mathrm{b}/R_\mathrm{int}\gamma\beta)}{\mathrm{I}_1^2(\alpha/\gamma\beta)}, \tag{6.3.37}$$

which represents the actual overlap of the beam with the wave and it multiplies the plasma frequency term ($\omega_\mathrm{p}^2 \to \omega_\mathrm{p}^2 F_\mathrm{f}$).

6.3.3 Transmission and Reflection

In the remainder of this section we present a few results from simulations which use this method. The system used is identical with the one in Sect. 6.2.3. A 1 MV, 1 kA beam is propagating through the structure and the wave injected in the system is assumed to be of sufficiently low power such that the system operates in the linear regime. A frequency scan of a symmetric ($g_{in} = g_{out} = 1$ mm) system is illustrated in Fig. 6.7. The gain, defined as the power at the output divided by the power injected in the input arm, has a maximum at 9.0 GHz as designed. Another peak is close to the π-point and it occurs at 9.06 GHz. Note that the gain is relatively low – about 16 dB – and this is also the ratio (in dB) between the power in the output arm compared to the input arm. For comparison, a Pierce-like analysis predicts a growth rate of the order of 2.5 dB/cm which in a 10 cm long structure corresponds to a net gain of approximately 15 dB.

The next stage is to break the symmetry of the system by increasing the distance between the input arm and the first cavity. By doing so, both the gain and the arms power ratio jumped to 24 dB (see Fig. 6.8).

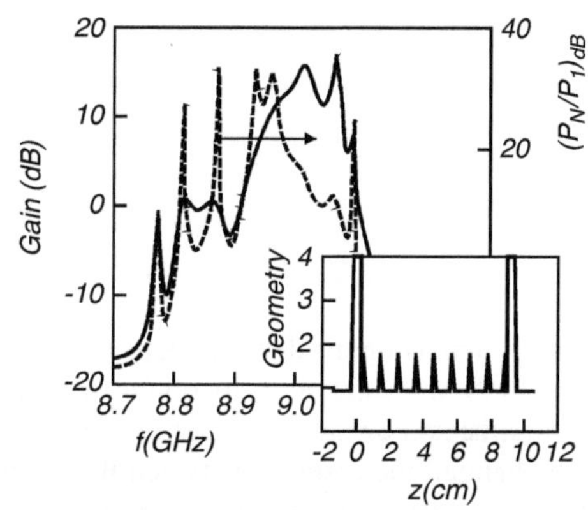

Fig. 6.7 Gain and power in a uniform structure with 1 mm separation between each arm and their adjacent cavities

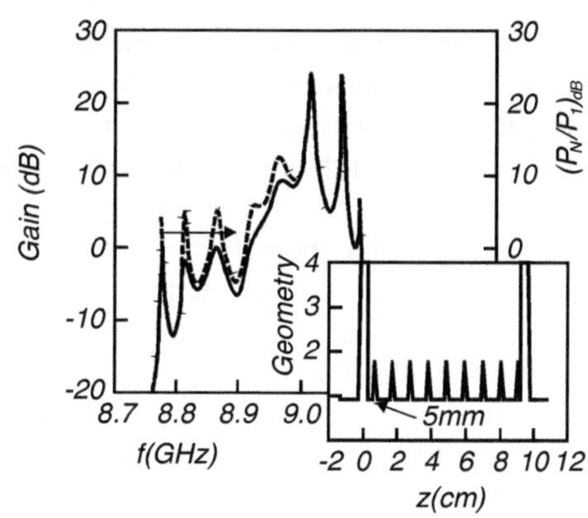

Fig. 6.8 Gain and power in a uniform structure with 5 mm separation between the input arm and its adjacent cavity

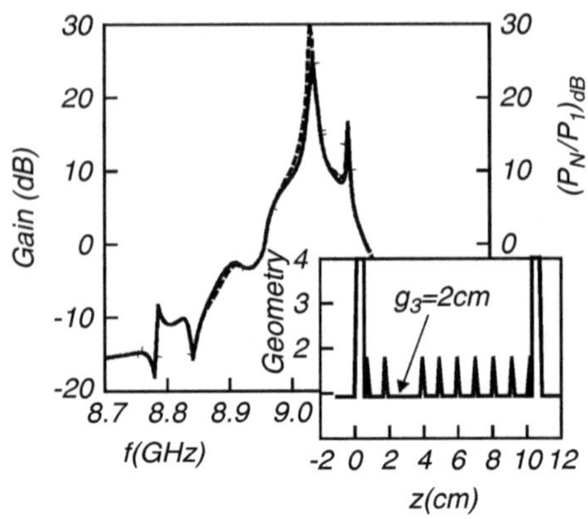

Fig. 6.9 Gain and power in a structure with 2 cm drift region between the second and third groove

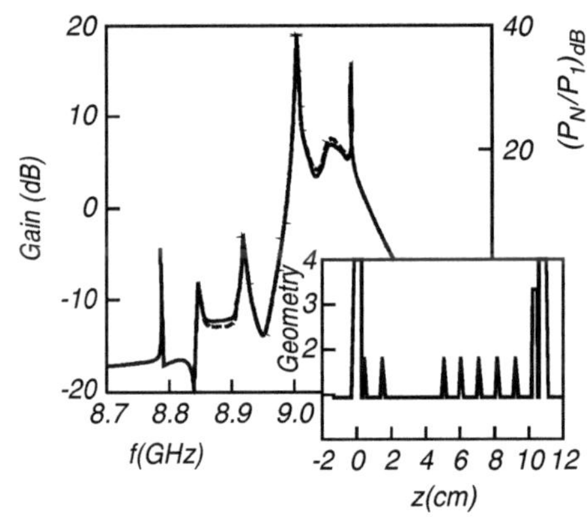

Fig. 6.10 Gain and power in a structure with $g_3 = 50\,\text{mm}$, $g_9 = 11\,\text{mm}$, $g_{10} = 5\,\text{mm}$ and $R_{\text{ext},10} = 32\,\text{mm}$

By increasing the distance between the second and the third cavity to 20 mm (see Fig. 6.9) we were able to obtain a similar gain at the required frequency and minimize somewhat the effect of the higher frequency peak (9.06 GHz). If instead we changed the distance between the third and the fourth cavity, the gain dropped below 20 dB. With the former result in mind, we increased the radius of the last cavity to $R_{\text{ext},10} = 32\,\text{mm}$ and its width to $d_{10} = 5\,\text{mm}$. The gain as a function of the separation from the 9th cavity was found to have an optimum for $g_9 = 11\,\text{mm}$.

Next, we consider the separation between the second and the third cavity. We vary it in order to obtain maximum gain. According to the classical klystron theory, we would expect the maximum to occur around $\lambda_p/4$, which in our case is roughly 4.6 cm. Figure 6.10 illustrates the gain and the arms power ratio for $g_3 = 50\,\text{mm}$. In this case the gain approaches 40 dB. Finally, for the same geometry we calculated the gain as a function of the (normalized) average velocity of the electrons at the input. Figure 6.11 presents this gain and we observe that the gain may actually exceed the 40 dB level.

Before we conclude this section, we wish to emphasize the important steps and the main differences of the method presented here. When no beam is injected, the poles (which determine Green's function) correspond to the electromagnetic modes

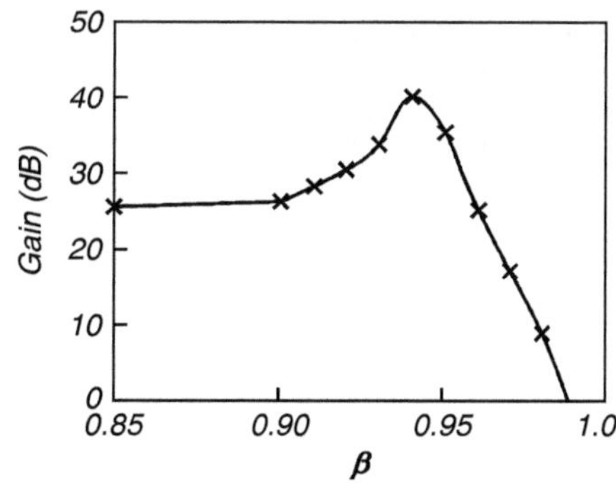

Fig. 6.11 Gain as a function of the beam velocity. The geometric parameters are identical with these in Fig. 6.10

in a cylindrical waveguide. In the presence of the beam, there is an additional set of poles that correspond to the space-charge waves "family". Contrary to the electromagnetic modes, which can propagate in both directions, the space-charge waves propagate only along the beam (forwards). This fact has been addressed in the evaluation of the integrals. It is also important to point out that all poles are *real* (both electromagnetic and space-charge) as they all correspond to the eigen-modes in a cylindrical waveguide. Consequently, the gain in the system is a result of the coupling between all these modes introduced by the cavities and arms – as in a klystron. This is different from the regular approach of beam-wave interaction in traveling-wave structures where the analysis relies strongly on the periodicity of the structure and the poles (eigen-wave-numbers) are *complex* – see Chap. 4.

6.4 Macro-Particle Approach

The investigation of beam-wave interaction in quasi-periodic structures was motivated by the large geometry variations required to obtain high efficiency. The latter in turn implies substantial variation in the kinetic energy of individual electrons from the ensemble average value. Consequently, the beam-wave interaction is investigated here in the framework of the macro-particle approach. Another issue addressed in this section is how one can design and analyze quasi-periodic structures when quantities like phase velocity, group velocity and interaction impedance are not well defined since there is an entire (spatial) spectrum of waves that the electrons interact with. To be more specific, in a periodic structure, for a given frequency, there is a single interacting wave (harmonic), and the interaction impedance is well defined. The question addressed here regards the analog in the case when substantial geometry variations occur.

It is shown that the interaction is controlled by a *matrix interaction impedance*, which is a generalization of the scalar interaction impedance concept, introduced for periodic structures. Its definition is possible after defining a set of functions that are characteristic to each aperture. The number of functions is determined by the number of apertures and number of modes that represent the electromagnetic field

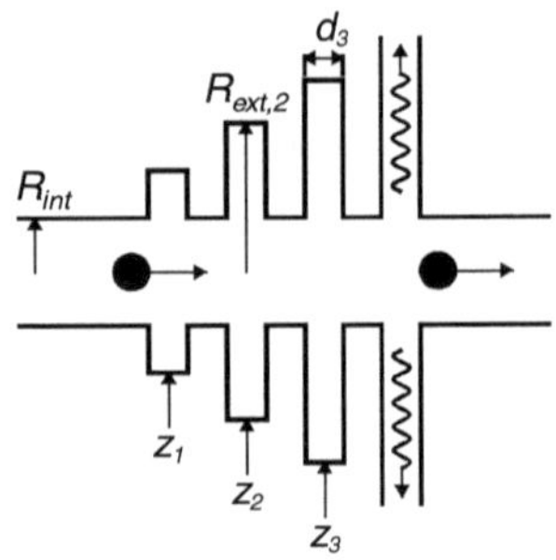

Fig. 6.12 A set of coupled pill-box cavities and an output arm model a quasi-periodic output structure

in the grooves/arms. Each function has its peak at a different aperture but they are not necessarily orthogonal. The matrix interaction impedance is closely related to Green's function of the system in the representation of this set of functions. After we establish the basic formalism we illustrate the design and analysis of a high efficiency (70%) traveling-wave section, including space-charge effects.

6.4.1 *Definition of the Model*

A schematic of the system is presented in Fig. 6.12. It consists of a cylindrical waveguide of radius $R_{\rm int}$ to which an arbitrary number of pill-box cavities and one output arm are attached; all the geometric definitions are like those in Sect. 6.1 with only one difference, there is only a single (output) arm. The system is driven by a modulated beam which in turn is guided by a very strong ("infinite") magnetic field confining the electrons' motion to the z-direction. Consequently, in the inner cylinder ($0 < r < R_{\rm int}$) the only non-zero component of the current density is in this direction i.e., $\mathbf{J}(r,z;t) = J_z(r,z;t)\mathbf{1}_z$ and it is given by

$$J_z(r,z;t) = -e\sum_i \mathrm{v}_i(t)\delta[z - z_i(t)]\frac{1}{2\pi r}\delta[r - r_i(t)]. \qquad (6.4.1)$$

In this expression $r_i(t)$ and $z_i(t)$ is the location of the $i^{\rm th}$ particle at time t and subject to the assumptions above $r_i(t) = r_i(0)$.

The operation of the system as an amplifier, dictates a single frequency operation, thus the time dependence of all electromagnetic field components is assumed to be sinusoidal, $\exp(j\omega t)$; this tacitly implies that all the transients associated with the front of the beam have decayed and for a particular phase-space distribution of electrons, the system has reached steady state. According to the assumptions above, the time Fourier transform of the current density is

$$J_z(r,z;\omega) = \frac{1}{T}\int_0^T \mathrm{d}t \exp(-j\omega t)J_z(r,z;t); \qquad (6.4.2)$$

$T = 2\pi/\omega$ is the period of the wave. The last expression can be further simplified. For this purpose we denote by $\tau_i(z)$ the time it takes the ith particle to reach the point z in the interaction region and by $\mathrm{v}_i(z)$ the velocity of the ith particle at z; the two are related through

$$\tau_i(z) = \tau_i(0) + \int_0^z \mathrm{d}\zeta \frac{1}{\mathrm{v}_i(\zeta)}; \tag{6.4.3}$$

$\tau_i(0)$ is the time the ith particle reaches the $z = 0$ point, which is chosen to be in the center of the first aperture.

Comment 6.1. In (6.4.3) it has been tacitly assumed that no electrons are reflected backwards.

Using these definitions the integral in (6.4.2) can be evaluated analytically and the result is

$$J_z(r, z; \omega) = \frac{-e}{2\pi r T} \sum_i^{N_{\text{el}}} \exp[-j\omega\tau_i(z)]\delta[r - r_i(0)]. \tag{6.4.4}$$

The summation is over all electrons (N_{el}) present in one time period of the wave and $I = eN_{\text{el}}/T$ is the average current. It is convenient to average over the transverse direction thus by denoting the beam radius by R_{b} and assuming that the electrons are uniformly distributed on the beam cross-section we find that

$$J(z) \equiv \frac{2}{R_{\text{b}}^2} \int_0^{R_{\text{b}}} \mathrm{d}r r J_z(r, z; \omega) = -\frac{I}{\pi R_{\text{b}}^2} \langle \exp[-j\omega\tau_i(z)] \rangle_i, \tag{6.4.5}$$

where $\langle \cdots \rangle \equiv N_{\text{el}}^{-1} \sum_{i=1}^{N_{\text{el}}} \cdots$. Finally, subject to the previous assumptions, the current density distribution reads

$$J_z(r, z; \omega) = -\frac{I}{\pi R_{\text{b}}^2} \langle \exp[-j\omega\tau_i(z)] \rangle_i h(R_{\text{b}} - r), \tag{6.4.6}$$

wherein $h(x)$ is the Heaviside step function and in what follows, the expression $\langle \exp[-j\omega\tau_i(z)] \rangle_i$, is referred to as the *normalized current density*.

The longitudinal electric field averaged over the beam cross-section i.e.,

$$E(z) = \frac{2}{R_{\text{b}}^2} \int_0^{R_{\text{b}}} \mathrm{d}r r E_z(r, z; \omega), \tag{6.4.7}$$

determines the dynamics of the particles via the single particle equation of motion which in our case coincides with the single particle energy conservation

$$\frac{\mathrm{d}}{\mathrm{d}z}\gamma_i(z) = -\frac{1}{2}\frac{e}{mc^2}\{E(z)\exp[j\omega\tau_i(z)] + \text{c.c.}\}; \tag{6.4.8}$$

m is the rest mass of the electron. In the next subsection we determine the relation between the longitudinal electric field [averaged over the beam cross-section, (6.4.7)] and the current density (6.4.6).

6.4.2 Evaluation of Green's Function

A magnetic vector potential excited by the current distribution introduced above, satisfies

$$\left[\nabla^2 + \frac{\omega^2}{c^2}\right] A_z(r, z; \omega) = -\mu_0 J_z(r, z; \omega), \tag{6.4.9}$$

in the cylindrical waveguide and

$$\left[\nabla^2 + \frac{\omega^2}{c^2}\right] A_z(r, z; \omega) = 0, \tag{6.4.10}$$

in the grooves or output arm. The solution of the magnetic vector potential in the first region ($0 < r < R_{\mathrm{int}}$) reads

$$\begin{aligned} A_z(r, z; \omega) = 2\pi\mu_0 \int_0^{R_\mathrm{b}} \mathrm{d}r' r' \int_{-\infty}^{\infty} \mathrm{d}z' G_\omega(r, z|r', z') J_z(r, z; \omega) \\ + \int_{-\infty}^{\infty} \mathrm{d}k A(k) \exp(-jkz) \mathrm{I}_0(\Gamma r). \end{aligned} \tag{6.4.11}$$

where $\Gamma^2 = k^2 - (\omega/c)^2$, $G_\omega(r, z|r', z')$ is the vacuum Green's function:

$$G_\omega(r, z|r', z') = \int_{-\infty}^{\infty} \mathrm{d}k \exp[-jk(z - z')] g_{\omega,k}(r|r'), \tag{6.4.12}$$

and

$$g_{\omega,k}(r|r') = \frac{1}{(2\pi)^2} \begin{cases} \mathrm{I}_0(\Gamma r)\mathrm{K}_0(\Gamma r') & \text{for} \quad 0 \leqslant r < r', \\ \mathrm{K}_0(\Gamma r)\mathrm{I}_0(\Gamma r') & \text{for} \quad r' \leqslant r < \infty. \end{cases} \tag{6.4.13}$$

$\mathrm{I}_0(x)$ and $\mathrm{K}_0(x)$ are the zero order modified Bessel function of the first and second kind correspondingly. Due to the azimuthal symmetry of the current distribution and the metallic structure, only symmetric transverse magnetic (TM) modes have been considered.

In the grooves, the electromagnetic field should be represented by a superposition of modes, which satisfy the boundary conditions on the metallic walls.

However, for the same reasons presented in Sect. 6.1, it is sufficient to consider only the first mode in the grooves. Within the framework of this approximation we can write

$$A_z^n(r,z;\omega) = D_n T_{0,n}\left(\frac{\omega}{c}r\right), \tag{6.4.14}$$

for the magnetic vector potential in the grooves, where D_n is the amplitude of the magnetic vector potential and $T_{0,n}(\omega r/c)$ was defined in the context of (6.1.3). In the output arm, the magnetic vector potential reads

$$A_z(r,z;\omega) = D_N \mathrm{H}_0^{(2)}\left(\frac{\omega}{c}r\right), \tag{6.4.15}$$

and $\mathrm{H}_0^{(2)}(x)$ is the zero order Hankel function of the second kind. This functional form is dictated by the boundary conditions, which in this case assume no reflected wave along the output arm.

In order to determine the various amplitudes we next impose the boundary conditions following the same procedure as in the previous sections. From the condition of continuity of the longitudinal electric field, we conclude that

$$A(k)\mathrm{I}_0(\Delta) + B(k)\mathrm{K}_0(\Delta) = -\frac{1}{2\pi}\frac{\alpha^2}{\Delta^2}\sum_{n=1}^{N} D_n \psi_{0,n} d_n \mathcal{L}_n(k), \tag{6.4.16}$$

where $\alpha = \omega R_{int}/c$ is the normalized angular frequency, $\Delta = \Gamma R_{\mathrm{int}}$ is the normalized wave number in the radial direction, $\mathcal{L}_n(k)$ was defined in (6.1.8) and is the normalized spatial Fourier transform of the first mode amplitude (whose amplitude is constant) in the domain of the nth aperture. The function

$$\psi_{\nu,n} = \begin{cases} H_\nu^{(2)}(\alpha) & n = N, \\ T_{\nu,n}(\alpha) & n \neq N, \end{cases} \tag{6.4.17}$$

determined at the internal radius and nth aperture; $\nu(=0,1)$ is the order of the function. In addition, z_n is the location of the center of the nth groove or arm and d_n is the corresponding width.

Imposing the continuity of the tangential magnetic field at each aperture (grooves and arm) we find

$$D_n \psi_{1,n} = -\frac{1}{\alpha}\int_{-\infty}^{\infty} \mathrm{d}k [A(k)\mathrm{I}_1(\Delta) - B(k)\mathrm{K}_1(\Delta)]\Delta \mathcal{L}_n^*(k). \tag{6.4.18}$$

In these expressions

$$B(k) = \frac{\mu_0}{2\pi}\int_0^{R_b} \mathrm{d}r' r' \mathrm{I}_0(\Gamma r') \int_{-\infty}^{\infty} \mathrm{d}z' \exp(jkz') J_z(r', z', \omega), \tag{6.4.19}$$

is the spatial Fourier transform of the current density averaged over the transverse direction with a weighting function which is proportional to the longitudinal electric field.

It is now convenient to substitute (6.4.16) in (6.4.18) in order to represent the entire electromagnetic problem in terms of the amplitudes of the mode in the grooves and output arm i.e.,

$$\sum_{m=1}^{N} M_{n,m} D_m = S_n. \tag{6.4.20}$$

The source term

$$S_n = -\frac{\mu_0 I}{2\pi\alpha} a_n, \tag{6.4.21}$$

is proportional to the average current and the Fourier transform of the normalized current density:

$$a_n = \frac{1}{R_{\text{int}}} \int_{-\infty}^{\infty} dz f_n(z) \langle \exp[-j\omega\tau_i(z)] \rangle_i. \tag{6.4.22}$$

The Fourier transform is with respect to a function

$$f_n(z) = \sum_{s=1}^{\infty} \frac{p_s F_s}{\Delta_s J_1(p_s)} \sigma_{s,n}(z), \tag{6.4.23}$$

which is associated with the nth aperture. In particular, if all the modes in the inner cylinder (index s) are below cutoff, this function peaks in the center of the aperture; p_s are the zeros of the zero order Bessel function of the first kind i.e., $J_0(p_s) = 0$. The function $f_n(z)$ is the product of two components,

$$\begin{aligned}\sigma_{s,n}(z) &\equiv \frac{1}{d_n} \int_{z_n-d_n/2}^{z_n+d_n/2} d\xi e^{-\Gamma_s|\xi - z|} \\ &= \begin{cases} e^{-\Gamma_s|z-z_n|} \operatorname{sinhc}(\Gamma_s d_n/2) & \text{for } |z - z_n| > d_n/2 \\ 2[1 - e^{-\Gamma_s d_n/2} \cosh(\Gamma_s(z - z_n))]/\Gamma_s d_n & \text{for } |z - z_n| < d_n/2 \end{cases}\end{aligned} \tag{6.4.24}$$

is the projection of Green's function (s-mode) on the nth aperture; $\Delta_s^2 = p_s^2 - \alpha^2$. The other component is the filling factor, $F_s \equiv 2J_1(p_s R_b/R_{\text{int}})/(p_s R_b/R_{\text{int}})$. To determine the amplitudes in (6.4.20) one has to multiply the source term by the inverse of the matrix M defined by

$$M_{n,m} = \psi_{1,n} \delta_{n,m} - \psi_{0,m} \chi_{n,m}. \tag{6.4.25}$$

In this expression $\chi_{n,m}$ was defined in (6.1.13), simplified in Sect. 6.1.2 and expressed in terms of analytic functions in (6.1.16). The electromagnetic problem has now been simplified to the inversion of a matrix whose components are analytic functions without a-priori assumption on the form of the electromagnetic field.

6.4.3 The Governing Equations

The motion of the electrons is determined by the longitudinal electric field averaged over the beam cross-section $[E(z)]$ as defined in (6.4.7). In this subsection we use (6.4.20) to simplify the relation between the normalized current density and $E(z)$. The longitudinal component of the electric field is related to the magnetic vector potential by

$$E_z(r,z;\omega) = \frac{c^2}{j\omega}\left[\frac{\omega^2}{c^2} + \frac{\partial^2}{\partial z^2}\right] A_z(r,z;\omega), \tag{6.4.26}$$

which after substituting (6.4.9) reads

$$E_z(r,z;\omega) = \frac{c^2}{j\omega}\left[-\mu_0 J_z(r,z;\omega) - \frac{1}{r}\frac{\partial}{\partial r} r \frac{\partial}{\partial r} A_z(r,z;\omega)\right]. \tag{6.4.27}$$

Thus according to the definition of the effective electric field in (6.4.7), we have

$$E(z) = \frac{c^2}{j\omega}\left\{-\mu_0 J(z) - \frac{2}{R_\mathrm{b}}\left[\frac{\partial}{\partial r} A_z(r,z;\omega)\right]_{r=R_\mathrm{b}}\right\}. \tag{6.4.28}$$

At this stage, we substitute the explicit expression for the magnetic vector potential in (6.4.11) and the result has two contributions: the space-charge term

$$\begin{aligned} E_\mathrm{SC}(z) = &-\frac{1}{j\omega\varepsilon_0}\int_{-\infty}^{\infty} \mathrm{d}z' J(z') \frac{1}{2\pi}\int_{-\infty}^{\infty} \mathrm{d}k \exp[-jk(z-z')] \\ &\times \left\{1 - 2\frac{\mathrm{I}_1(\Delta_b)}{\mathrm{I}_0(\Delta)}[\mathrm{I}_0(\Delta)\mathrm{K}_1(\Delta_b) + \mathrm{K}_0(\Delta)\mathrm{I}_1(\Delta_b)]\right\}, \end{aligned} \tag{6.4.29}$$

and the "pure" electromagnetic term

$$E_\mathrm{EM}(z) = \frac{-j\omega}{2\pi}\int_{-\infty}^{\infty} \mathrm{d}k \exp(-jkz) \frac{1}{\mathrm{I}_0(\Delta)} F(k) \sum_{n=1}^{N} D_n d_n \mathcal{L}_n(k) \psi_{0,n}, \tag{6.4.30}$$

where $F(k) \equiv 2\mathrm{I}_1(\Delta_b)/\Delta_b$ is the filling factor and $\Delta_b = \Gamma R_\mathrm{b}$.

In the framework of the current approximation, we observe that the grooves have no explicit effect on the space-charge term. Taking advantage of this fact we start from Green's function associated with TM_{0s} modes in a cylindrical waveguide (2.4.32) and using the same method as in (6.4.26)–(6.4.28) we obtain

$$E(z) = \frac{-1}{j\omega\varepsilon_0}\left[J(z) - \frac{1}{R_{\rm int}}\int_{-\infty}^{\infty} dz' J(z') \sum_{s=1}^{\infty} \exp(-\Gamma_s|z-z'|)\frac{1}{2\Delta_s}\left(\frac{2J_1(p_s R_b/R_{\rm int})}{J_1(p_s)}\right)^2\right], \tag{6.4.31}$$

this can be simplified if all electromagnetic modes are below cutoff and in particular, for the case when the current density $|J(z)|$ varies slowly in comparison with $\exp(-\Gamma_s|z-z'|)$. Subject to these assumptions, we can assume that the main contribution to the integral is from the region $z = z'$ and therefore $J(z)$ can be extracted from the integral. The result in this case reads

$$E(z) = \frac{-1}{j\omega\varepsilon_0} J(z)\left[1 - \sum_{s=1}^{\infty}\left(\frac{J_1(p_s R_b/R_{\rm int})}{\Delta_s J_1(p_s)}\right)^2\right], \tag{6.4.32}$$

or

$$E_{\rm SC}(z;\omega) = -\frac{1}{j\omega\varepsilon_0}\xi_{\rm SC} J(z), \tag{6.4.33}$$

where the space-charge coefficient $\xi_{\rm SC}$ is given by

$$\xi_{\rm SC} = 1 - \sum_{s=1}^{\infty}\left[\frac{J_1(p_s R_b/R_{\rm int})}{\Delta_s J_1(p_s)}\right]^2, \tag{6.4.34}$$

and is an approximation to the space-charge *reduction factor*.

It is possible to simplify the electromagnetic term by substituting the explicit expression for $\mathcal{L}_n(k)$ and using Cauchy's residue theorem. The result reads

$$E_{\rm EM}(z) = \frac{\eta_0 I}{R_{\rm int}} \sum_{n=1}^{N} f_n(z)\left[\sum_{m=1}^{N} \mathcal{T}_{n,m} a_m\right], \tag{6.4.35}$$

where

$$\mathcal{T}_{n,m} = \frac{j}{2\pi}\frac{d_n}{R_{\rm int}}\psi_{0,n}\left[M^{-1}\right]_{n,m}, \tag{6.4.36}$$

and it can be considered as a "discrete" Green function of the system since a_m is the Fourier transform of the normalized current density with respect to the function $f_m(\zeta)$ – as defined in (6.4.22).

Now that the relation between the effective electric field acting on the particles and the current density has been established,

$$E(z) = E_{\mathrm{SC}}(z) + E_{\mathrm{EM}}(z), \tag{6.4.37}$$

we proceed to analysis of the beam-wave interaction. Substituting this effective field in the single particle energy conservation, defining $\bar{\xi}_{\mathrm{SC}} = \xi_{\mathrm{SC}}(R_{\mathrm{int}}/R_{\mathrm{b}})^2/\alpha\pi$ and $\bar{I} = \eta_0 Ie/mc^2$, we obtain

$$\frac{\mathrm{d}}{\mathrm{d}z}\gamma_i = \frac{-\bar{I}}{2R_{\mathrm{int}}}\left\{\begin{array}{l}\exp[j\omega\tau_i(z)]\left[\sum\limits_{n,m=1}^{N}\mathcal{T}_{n,m}a_m f_n(z) - j\bar{\xi}_{\mathrm{SC}}\langle\exp[-j\omega\tau_\nu(z)]\rangle_\nu\right]\\ +\ \mathrm{c.c.}\end{array}\right\}. \tag{6.4.38}$$

This is an integro-differential equation which describes the dynamics of the electrons. In order to determine γ_i at any given location it is necessary to know the Fourier transform of the normalized current density, a_n, which in turn requires to know the trajectories of all particles over the entire interaction region, as indicated in (6.4.22).

Before we proceed to actually presenting a solution of this set of equations it is important to make two comments which are evident from (6.4.38) and our prior definitions:

Comment 6.2. Global energy conservation implies

$$\langle\gamma_i(\infty)\rangle - \langle\gamma_i(-\infty)\rangle = -\frac{1}{2}\bar{I}\sum_{n,m=1}^{N} a_n^* Z_{n,m} a_m, \tag{6.4.39}$$

where

$$Z_{n,m} \equiv \frac{1}{2}\left[\mathcal{T}_{n,m} + \mathcal{T}_{m,n}^*\right], \tag{6.4.40}$$

is the *interaction impedance matrix*. This expression implies that in case of non-adiabatic changes from periodicity, as is the case in quasi-periodic structures, we can no longer refer to the interaction impedance as a scalar (and local) quantity but rather as a matrix and the interaction at a given location is affected by the geometry elsewhere. Furthermore, since the left-hand side of the global energy conservation [(6.4.39)] is proportional to the overall efficiency, it is evident that the latter is controlled by the interaction impedance matrix. In the example presented next, it will be shown that it is the largest eigen-value of this matrix that determines the efficiency of the interaction.

Comment 6.3. The space-charge term has no *explicit* effect on the global energy conservation.

In order to solve the integro-differential equation in (6.4.38) for a large number of macro-particles (more than 30,000 were used), an iterative way was chosen. Typically a simple distribution is assumed, enabling the calculation of the zero iteration $a_n^{(0)}$. With this quantity, the trajectories of all particles are calculated and in parallel, the "new" $a_n^{(1)}$ is evaluated; at the end of the iteration the two a_n's are compared. If the relative error is less than 1% the simulation is terminated. Otherwise we calculate the equations of motion again but this time using $a_n^{(1)}$ to determine the dynamics of the particles and calculate in parallel $a_n^{(2)}$. If the energy spread of the electrons at the input is not too large, then 3–4 iterations are sufficient for convergence.

Consider now a modulated beam, which drives an extraction structure. The initial energy of the electrons is 850 keV and the structure should extract 70% of their kinetic energy; for the zero order design let us assume that in the interaction region there is only a single macro-particle at a time. Furthermore, the disk thickness is taken to be 1 mm in order to ensure maximum group velocity. For the same reason the phase advance per cell is taken to be 90°. For the preliminary design, a single macro-particle in one period of the wave and its velocity in the interaction region is assumed to satisfy

$$\mathrm{v}(z) = \frac{\mathrm{v}(0)}{1 + qz}; \tag{6.4.41}$$

q and the total interaction length d_{tot} are determined from the required efficiency and the condition that no two bunches will be present in the interaction region at a time. For simplicity, we also assume that the internal and external radius are the same in all cells. Their value is determined by *maximizing the largest eigen-value of the interaction impedance matrix* at 9 GHz – as illustrated in Fig. 6.13 where $R_{int} = 9\,\mathrm{mm}$ and $R_{ext} = 16.47\,\mathrm{mm}$; the other geometrical parameters are $d_1 = 6.5\,\mathrm{mm}$, $d_2 = 6.0\,\mathrm{mm}$, $d_3 = 5.7\,\mathrm{mm}$ and $d_4 = 5.4\,\mathrm{mm}$. Overlaid is also the efficiency assuming a single macro-particle injected into the system in one period of the wave. The dynamics of the particle is calculated numerically (6.4.38).

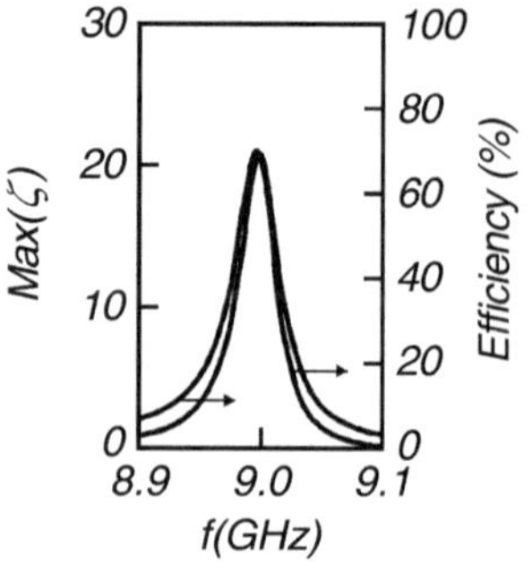

Fig. 6.13 The largest eigen-value of the interaction impedance and the efficiency as a function of the frequency using the resonant particle model

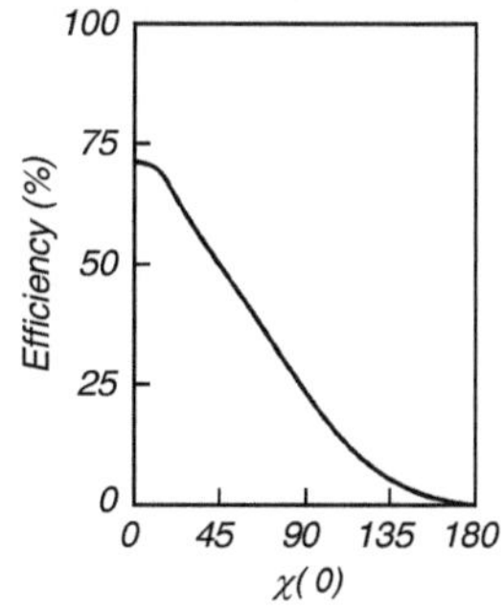

Fig. 6.14 The efficiency as a function of the modulation angle i.e., the fraction of wavelength in degrees occupied by electrons. The case $\chi(0) = 180°$ corresponds to a uniform beam

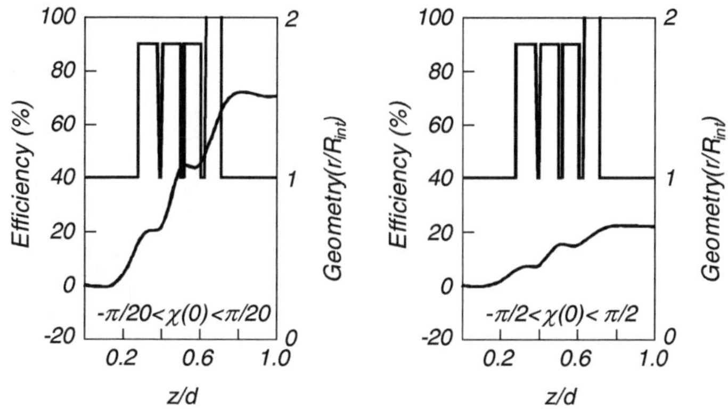

Fig. 6.15 The efficiency as a function of the location along the interaction region for two different initially bunched beams

Figure 6.14 indicates that the efficiency of the electromagnetic energy conversion is strongly dependent on the phase-space distribution at the input; the phase here is defined as $\chi_i(z) \equiv \omega\tau_i(z)$. For a perfectly bunched beam the efficiency is as designed (for $I = 300$ A). However, as the initial phase distribution increases to $-45° < \chi(0) < 45°$ the efficiency drops to 45% and to 25% for $-90° < \chi(0) < 90°$. It drops to virtually zero for a uniform distribution.

An interesting feature is revealed in Fig. 6.15 where we present the variation in space of the efficiency for two initial distributions: $-9° < \chi(0) < 9°$ $-90° < \chi(0) < 90°$. We observe that the general pattern is virtually identical in both cases and only the spatial growth rate is smaller. The reduced efficiency is a result of energy transferred back to electrons, which are actually accelerated as illustrated in Fig. 6.16; clearly, in the narrower initial phase-space distribution all the electrons are decelerated at the output, whereas in the case of broader phase-space distribution a substantial fraction of electrons is accelerated.

Finally, the efficiency is illustrated in Fig. 6.17 as a function of the frequency for $-15° < \chi(0) < 15°$. The curve is virtually identical to that of the single macro-particle case (Fig. 6.14). Overlaid, we present the energy spread

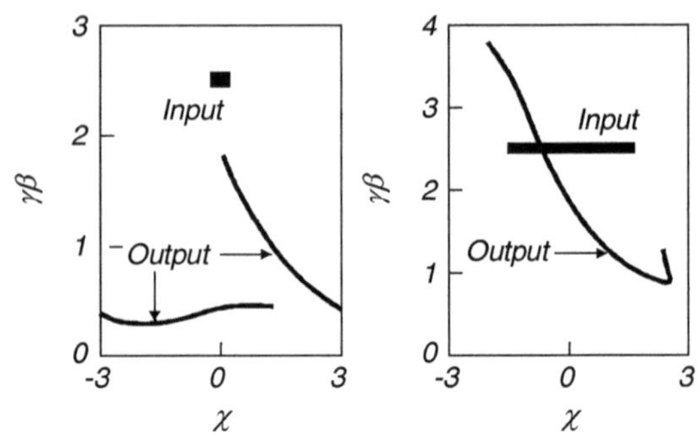

Fig. 6.16 The phase-space plots for the two cases illustrated in Fig. 6.15

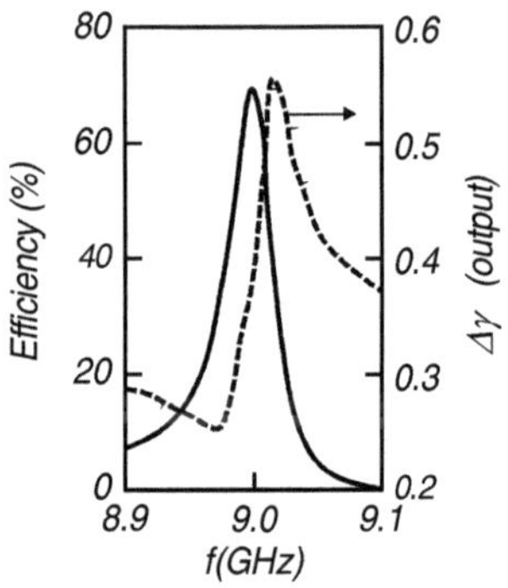

Fig. 6.17 The efficiency (*solid-line*) and the energy spread at the output (*dashed-line*) as a function of the frequency

$\left(\Delta\gamma \equiv \sqrt{\langle\gamma^2\rangle - \langle\gamma\rangle^2}\right)$ at the output and we observe that up to a constant value, this quantity varies as the derivative of the efficiency with respect to the frequency.

6.4.4 *Qualitative Approach*

The approach presented above provides us with a convenient 1D tool for calculating the dynamics of electrons in a quasi-periodic structure. Although this can be used as a design tool, it is usually convenient to apply more qualitative arguments for a zero order design that can be later improved with our model. Let us now follow such a qualitative argument: consider an ideal bunch of electrons which generate a current I. The electrons are mono-energetic and they are initially accelerated by an initial voltage denoted by $\mathcal{E}$. If we require an extraction efficiency η in an interaction region of a length D then the average electric field experienced by the bunch is

$$E = \eta\frac{\mathcal{E}}{D}. \tag{6.4.42}$$

Based on the definition of the interaction impedance in (2.3.29) we conclude that the rf power in the system is

$$P_{\mathrm{rf}} = \frac{\pi R_{\mathrm{int}}^2 E^2}{2Z_{\mathrm{int}}}, \tag{6.4.43}$$

and the power carried by the beam is

$$P_{\mathrm{beam}} = \mathcal{E}I. \tag{6.4.44}$$

These two are related since we assumed an efficiency η and energy conservation implies

$$P_{\mathrm{rf}} = \eta P_{\mathrm{beam}}. \tag{6.4.45}$$

From the expressions above we can determine the interaction impedance of the structure i.e.,

$$Z_{\mathrm{int}} = \frac{\pi}{2}\eta\left(\frac{R_{\mathrm{int}}}{D}\right)^2\frac{\mathcal{E}}{I}, \tag{6.4.46}$$

and we observe that we should design the interaction impedance of the structure in conjunction with the effective impedance of the beam ($\mathcal{E}/I$). In order to have a feeling as for the values of the impedance consider $\mathcal{E} = 0.85\,\mathrm{MV}$, $I = 0.5$ kA, $R_{\mathrm{int}} = 12$ mm, total interaction length $D = 2.75$ cm and efficiency of $\eta = 70\%$; for these parameters $Z_{\mathrm{int}} = 213\Omega$. It should be pointed out that here we tacitly assumed that E is constant thus the dynamics of the particles in space is different than the one prescribed in (6.4.41) and is given by

$$\gamma(z) = \gamma(0) - \frac{eEz}{mc^2} = \gamma(0)(1 - qz). \tag{6.4.47}$$

The length of the structure (D) and q can be determined exactly in the same way as prescribed at the end of the last subsection. Once D is determined and assuming that $\mathcal{E}$ and I are known, then (6.4.46) provides us with a simple relation between the internal radius and the interaction impedance. If we have in mind the disk-loaded structure then this relation in conjunction with the expression for the interaction impedance in (5.2.32) determine one constraint on the geometry of the structure. Thus out of the four geometric parameters ($R_{\mathrm{int}}, R_{\mathrm{ext}}, L$ and d in a periodic structure) we are left with three degrees of freedom. The resonance condition, the phase advance per cell and the group velocity (maximum gradient allowed) at resonance determine three additional constraints that in turn set the "local values" of these parameters. In other words, they roughly determine the geometry of the single cell, which in turn is part of a quasi-periodic structure. Fine-tuning of the design should be made following the approach in Sect. 6.4.3.

6.5 Surface Roughness

Exactly as a bunch of electrons moving in a slow-wave structure generates a wake-field, motion of a similar bunch in a metallic pipe with a rough surface excites an electromagnetic signal. This field may become detrimental to the generating bunch or trailing ones either by affecting their longitudinal or transverse dynamics. Specifically, it may increase the transverse emittance or even enlarge the offset from axis and in certain conditions, it may even affect the energy spread. Shorter bunch may have a more significant impact in particular if its length is comparable to the irregularities of the surface.

In order to envision the obstacles that we may encounter in future optical accelerators, we need to bear in mind that, in the past, the acceleration structure of the so-called next linear collider (NLC) designed to operate at X-band was machined with an accuracy of about 1 μm. Therefore, there are four orders of magnitude between the operating wavelength and the typical surface roughness. This difference will be difficult to maintain in the case of a vacuum optical accelerator operating at 1 μm since it entails engineering the structure at the atomic level (1 Å). Other applications also require high degree of surface "machining": advanced light sources are expected to be sensitive to surface roughness and throughout the years a significant amount of research reports have been published. A brief review of publications on the wake-field generated by surface-roughness can be found in the introduction of the article this section relies upon – compiled by Banna et al. (2004).

In this section, we employ the quasi-analytic approach developed earlier in this chapter facilitating a relatively simple evaluation of the wake-field due to surface roughness of arbitrary size. The model relies on a cylindrical waveguide of radius $R_{\rm int}$ to which a series of grooves are attached; their geometric parameters i.e. width, height and location, are assumed to be *random*. In principle, they can be large on the scale of the typical wavelength of the radiation that drives the system. Simulation results focus on the *average* and *standard-deviation* of the electro-magnetic energy emitted by a relativistic bunch in terms of the roughness parameter.

6.5.1 *Definition of the Model*

In order to analyze the wake-field generated by the surface roughness consider a metallic structure consisting of a random number of pill-box cavities attached to a cylindrical waveguide (internal radius $R_{\rm int}$) as illustrated in Fig. 6.18.

The center of the $n^{\rm th}$ groove is denoted by z_n, its width by d_n and its external radius by $R_{ext,n}$. An electron bunch of radius R_b, length Δ_z and a total charge Q, is moving along the symmetry (z) axis of the structure at a constant velocity v_0.

This bunch generates a current density

$$J_z(r,z;t) = -Qv_0\frac{1}{2\pi}\left[\frac{2}{R_b^2}h(R_b - r)\right]\left[\frac{1}{\Delta_z}h\left(\frac{\Delta_z}{2} - |z - v_0 t|\right)\right], \qquad (6.5.1)$$

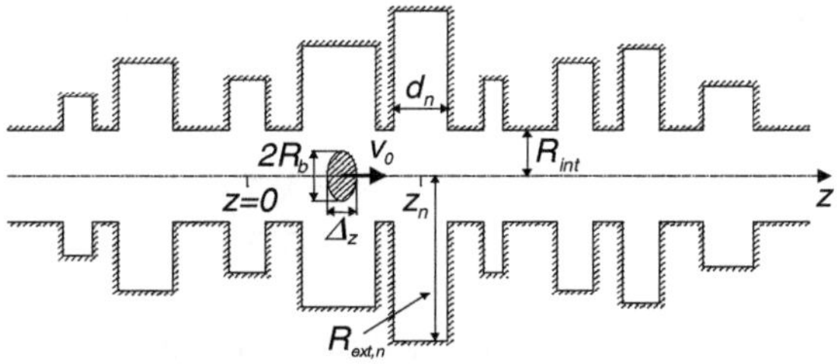

Fig. 6.18 A finite-size bunch moving in vacuum at a constant velocity v_0 along the axis of a cylindrical structure with grooves of random size

where the function $h(\xi)$ is the well-known Heaviside step function. In what follows it will be more convenient to use the Fourier transform of this quantity i.e.

$$J_z(r, z; \omega) = \frac{-Q}{(2\pi)^2} \frac{2}{R_b^2} h(R_b - r) \mathrm{sinc}\left(\frac{1}{2} \frac{\omega}{v_0} \Delta_z\right) \exp\left(-j\omega \frac{z}{v_0}\right), \tag{6.5.2}$$

wherein $\mathrm{sinc}(\xi) \equiv \sin(\xi)/\xi$.

As the only component of the current density is in the z-direction, it is sufficient to consider only the longitudinal magnetic vector potential A_z that satisfies the non-homogeneous wave equation – (6.2.1). Its formal solution was introduced in (6.2.3), (6.2.4) and (6.2.5). Contrary to Sect. 6.2 in this case we have no input or output arms therefore, the boundary condition associated with the longitudinal electric field reads

$$-2\pi \frac{\Delta^2}{\alpha^2} [B(k) \mathrm{K}_0(\Delta) + A(k) \mathrm{I}_0(\Delta)] = \sum_{n=1}^{N} D_n T_{0,n}(\alpha) d_n \mathcal{L}_n(k) \tag{6.5.3}$$

which replaces (6.2.9). Here

$$B(k) = \frac{\mu_0}{2\pi} \int_0^{R_b} dr' r' \mathrm{I}_0(\Gamma r') \int_{-\infty}^{\infty} dz' \exp(jkz') J_z(r', z'; \omega), \tag{6.5.4}$$

is the spatial Fourier transform of the current density, and is given explicitly by

$$B(k) = -\frac{Q\mu_0}{(2\pi)^2} \mathrm{sinc}\left(\frac{1}{2} \frac{\omega}{v_0} \Delta_z\right) \mathrm{I}_c\left(\frac{\omega}{v_0 \gamma} R_b\right) \delta\left(k - \frac{\omega}{v_0}\right), \tag{6.5.5}$$

$\delta(\xi)$ being the Dirac delta function, $\mathrm{I}_c(x) = 2\mathrm{I}_1(x)/x$, $\beta \equiv v_0/c$ and $\gamma = 1/\sqrt{1-\beta^2}$. In each one of the grooves the continuity of the azimuthal magnetic field provides us with an additional set of equations namely,

$$\alpha D_n \mathrm{T}_{1,n}(\alpha) = \int_{-\infty}^{\infty} dk \Delta [B(k) \mathrm{K}_1(\Delta) - A(k) \mathrm{I}_1(\Delta)] \mathcal{L}_n^*(k) \tag{6.5.6}$$

Based on these two equations (6.5.3) and (6.5.6) the amplitudes in the grooves (D_n) may be determined from the following algebraic relation

$$\sum_{m=1}^{N} M_{n,m} D_m = S_n. \tag{6.5.7}$$

Here the source term S_n and the matrix $M_{n,m}$ are given by

$$\begin{aligned} S_n &\equiv \frac{1}{\alpha}\int_{-\infty}^{\infty} dk \frac{1}{\mathrm{I}_0(\Delta)} B(k)\mathcal{L}_n^*(k), \\ M_{n,m} &\equiv \mathrm{T}_{1,n}(\alpha)\delta_{n,m} - \mathrm{T}_{0,m}(\alpha)\chi_{n,m}, \\ \chi_{n,m} &\equiv \frac{d_m\alpha}{2\pi}\int_{-\infty}^{\infty} dk \frac{\mathrm{I}_1(\Delta)}{\Delta \mathrm{I}_0(\Delta)}\mathcal{L}_n^*(k)\mathcal{L}_m(k) \end{aligned} \tag{6.5.8}$$

and the explicit expression for χ was evaluated in (6.1.16).

6.5.2 Emitted Energy

In order to quantify the effect of roughness on the electron bunch it is convenient to focus the discussion on the emitted energy. For this purpose we bear in mind that the emitted power is

$$P(t) = 2\pi \int_0^{R_b} r dr \int_{-\infty}^{\infty} dz J_z(r,z;t) E_z^{(s)}(r,z;t), \tag{6.5.9}$$

where $E_z^{(s)}(r,z;t)$ is the secondary longitudinal electric field generated due to the presence of the metallic surface. With the power, the emitted energy is given by

$$W \equiv \int_{-\infty}^{\infty} dt P(t) = -\frac{Q^2}{4\pi\varepsilon_0 R_{\mathrm{int}}} \mathrm{Re}\left[\int_0^{\infty} d\alpha S(\alpha)\right] \equiv -\frac{Q^2}{4\pi\varepsilon_0 R_{\mathrm{int}}}\bar{W}; \tag{6.5.10}$$

the integrand $[S(\alpha)]$ represents the normalized spectrum of the emitted energy. The latter is directly related to the so-called longitudinal impedance

$$Z_{\parallel}(\omega) \equiv -\frac{1}{Q}\int_{-\infty}^{\infty} dz E_z^{(s)}(z;\omega) \exp\left(j\frac{\omega}{\mathrm{v}_0}z\right) = \frac{\eta_0}{8\pi} S(\alpha). \tag{6.5.11}$$

Evidently, for quantifying the total energy emitted it is necessary to determine the longitudinal electric field $E_z^{(s)}$

$$E_z^{(s)}(r,z;\omega) = \int_{-\infty}^{\infty} dk \exp(-jkz) \mathrm{I}_0(\Gamma r) \frac{-c^2\Gamma^2}{j\omega} A(\omega,k), \\ = \int_{-\infty}^{\infty} dk \exp(-jkz) \mathrm{I}_0(\Gamma r) \mathcal{E}(\omega,k), \tag{6.5.12}$$

wherein $\mathcal{E}(\omega,k)$ is given explicitly by

$$\mathcal{E}(\omega,k) \equiv \frac{-j\omega}{2\pi} \frac{1}{\mathrm{I}_0(\Delta)} \sum_{n=1}^{N} d_n D_n(\omega) \mathrm{T}_{0,n}(\alpha) \mathcal{L}_n(k) \tag{6.5.13}$$

consequently, the normalized spectrum of the emitted energy $S(\alpha)$ reads

$$S(\alpha) = j\frac{8\pi\alpha}{Q\mu_0} \operatorname{sinc}\left(\frac{\alpha}{2\beta}\bar{\Delta}_z\right) \frac{\mathrm{I}_c(\alpha\bar{R}_b/\gamma\beta)}{\mathrm{I}_0(\alpha/\gamma\beta)} \\ \times \sum_{n=1}^{N} \bar{d}_n \mathrm{T}_{0,n}(\alpha) \operatorname{sinc}\left(\frac{\alpha}{2\beta}\bar{d}_n\right) \exp\left(j\frac{\alpha}{\beta}\bar{z}_n\right) D_n(\alpha), \tag{6.5.14}$$

where $\bar{\Delta}_z \equiv \Delta_z/R_{\mathrm{int}}$, $\bar{d}_n \equiv d_n/R_{\mathrm{int}}$, $\bar{R}_b = R_b/R_{\mathrm{int}}$ and $\bar{z}_n \equiv z_n/R_{\mathrm{int}}$. In the analysis that follows it is assumed that the geometrical parameters of each groove are random and are of the same order of magnitude. Explicitly, they are given by

$$\bar{R}_{\mathrm{ext},n} = 1 + \overline{\delta}_n, \qquad \bar{d}_n = \overline{\delta}_n \tag{6.5.15}$$

where $\overline{\delta}_n$ is a random variable that characterized by the average roughness $\langle\overline{\delta}\rangle$ and its standard-deviation $\overline{\Delta\delta}$. The center of the first groove ($\bar{z}_1 \equiv z_1/R_{\mathrm{int}}$) is being chosen as point of reference and accordingly, the location of the $n^{th}(n = 2,3,...,N)$ groove is given by $\bar{z}_{n+1} = \bar{z}_n + \bar{d}_n/2 + \bar{d}_{n+1}/2 + \overline{\delta}_n$. Clearly, if R_{int} is of the order of 0.5 μm and the typical roughness is not expected to be significantly larger than 0.1 μm, then the normalized roughness parameter is expected to be of the order of 0.2 i.e., $0 \le \langle\overline{\delta}\rangle \le 0.2$. Moreover, the accelerated bunch is expected to be of the order of $30^\circ \div 45^\circ$ (namely about 0.1 μm), therefore $0.15 \le \bar{\Delta}_z \le 0.30$. Typically the bunch's normalized radius is chosen to be $\bar{R}_b = 0.5$. In the results presented next, each data point is a result of average over 80 *different distributions*. Increasing the number of distribution did not affect the results of the simulation. In practice, the electron bunch propagates along an extended length traversing a large number of grooves. For the parameters of interest the radiation emitted per obstacle is the same, as was demonstrated by Banna et al. (2004).

When examining the *total energy* emitted by the bunch per obstacle we focus on the dependence of the emitted energy on the roughness characteristics $\left(\langle\overline{\delta}\rangle, \overline{\Delta\delta}\right)$

with $\bar{\Delta}_z$ and γ as parameters. For $\gamma \geq 50$ the emitted energy per groove is virtually independent of γ and it can be approximated by

$$\frac{<W>}{\frac{Q^2}{4\pi\varepsilon_0 R_{\rm int}} \times N} \simeq 0.57 \tanh\left(\frac{45\frac{\langle\delta\rangle}{R_{\rm int}}}{1+20.7\frac{\Delta_z}{R_{\rm int}}}\right) + \frac{1}{0.7+14.5\frac{\Delta_z}{R_{\rm int}}} \tag{6.5.16}$$

and

$$\frac{\sqrt{<W^2>-<W>^2}}{\frac{Q^2}{4\pi\varepsilon_0 R_{\rm int}} \times N} \simeq 0.15\left(\frac{\Delta\delta}{R_{\rm int}}\right)^{1/4} \tanh\left(121.2\frac{\Delta\delta}{R_{\rm int}}\right) \times \left[0.57 \tanh\left(\frac{45\frac{\langle\delta\rangle}{R_{\rm int}}}{1+20.7\frac{\Delta_z}{R_{\rm int}}}\right) + \frac{1}{0.7+14.5\frac{\Delta_z}{R_{\rm int}}}\right] \tag{6.5.17}$$

Note that for a point-charge ($\Delta_z = 0$) and $\langle\delta\rangle, \Delta\delta \geq 0.2R_{\rm int}$ the expression for the average energy per groove reads

$$\frac{<W>}{N} \simeq \frac{Q^2}{4\pi\varepsilon_0 R_{\rm int}} \times 2, \tag{6.5.18}$$

whereas the standard-deviation per groove is now given by

$$\frac{\sqrt{<W^2>-<W>^2}}{N} \simeq \frac{Q^2}{4\pi\varepsilon_0 R_{\rm int}} \times 0.3\left(\frac{\langle\delta\rangle}{R_{\rm int}}\right)^{1/4}. \tag{6.5.19}$$

It should be pointed out that the expression in (6.5.18) is virtually identical to the decelerating field (2.4.78) on a relativistic point-charge propagating in a vacuum tunnel (of radius $R_{\rm int}$) in an otherwise uniform dielectric material – Cerenkov force.

Comment 6.4. For more details regarding electromagnetic wakes associated with surface-rougness, the reader may want to consult Kurennoy (1997), Stupakov (1998), Mostacci et al. (2002) and Bane and Stupakov (2003).

Comment 6.5. The model presented in this chapter was extended to include an arbitrary number of modes in each groove in order to justify the single mode approximation used so far. While there is less than 10% difference between using one mode comparing to two modes but there is negligible difference between 2, 3 and 4 are modes in the grooves.

6.6 Photo-Injector

Laser-driven rf electron guns are operational for a quite long period of time and they are the main source of high-current, low-emittance, short bunch-length electron beams, required in virtually all electrons accelerators. In essence, a high-power laser beam illuminates a photocathode surface placed on an end wall of an rf cavity. The emitted electrons are accelerated in a short distance from rest to a relativistic energy by the strong rf field stored in the cavity. For a simple mathematical description of the physical processes described above, we chose to present in this sub-section an analytic approach inspired by that developed by Kim (1989).

Before presenting the essentials of a mathematical model, let us review some of the milestones in the development of photo-injectors and its typical parameters: following the successful demonstration of the free electron laser (FEL) the need for better electron source has become a necessity and a photo-cathode based gun was suggested. The first experimental demonstration of normal-conductivity photo-injector was performed at Los Alamos National Laboratory as part of the FEL program in the mid eighties. Later, O'Shea et al. (1993) has reported the generation of ultraviolet radiation employing an FEL driven by a photo-injector. At about the same time, the first operation of a photo-cathode was demonstrated in a super-conducting cavity – Michalke et al. (1992). Aiming for higher average power and lower emittance, more recently, electrons emitted from photo-cathodes were accelerated by a dc field, as reported by Siggins et al. (2001). As of today, this is the most advanced electron injector concept harnessed for the new generation of light-sources based on energy recovery linacs (ERL). For a more detailed overview of photo-injectors the reader is referred to Russell (2003). Probably the most widely used photo-injector among many laboratories throughout the world is the so-called "ATF-BNL/SLAC/UCLA – injector". Its typical parameters are as follows: a 3 ps (rms) laser pulse generates up to 1 nC of charge, the peak electric field at the cathode is 120 MV/m and the energy of the electrons at the output is of the order of 6 MeV; the emittance (rms) $\tilde{\varepsilon}_{\perp} \sim 0.8[\text{mm} - \text{mrad}]$.

While in the context of photo-injectors the electron dynamics in a dc field is conceptually different than in an rf field, energy-wise the outcome is virtually identical. Moreover, in the near future it is most probably that rf injectors will keep playing an important role in accelerator systems therefore we now pursue the essentials of bunch generation in an rf photo-injector. Its main advantages are that the time structure of the electron beam is controlled by the laser, facilitating an enhanced rf field, so that the degrading effects due to space-charge repulsion can be minimized. In this section, we limit the discussion to a simple but vigorous model leaving the fine details to numerical codes.

Consider a sequence of rf cells operating in the π-mode. Since the electrons start from rest, the first cavity is actually a half-cell such that by the time the bunch enters the second cell the field reverses sign. Schematic of the structure is illustrated in Fig. 6.19.

On axis, the longitudinal electric field may be approximated by

$$E_z = E_0 \cos\left(\frac{\omega}{c} z\right) \sin(\omega t + \phi_0). \tag{6.6.1}$$

Here it has been tacitly assumed that: (1) this is an accelerating structure such that the phase-velocity is c, (2) the plane where the photo-cathode is located is at

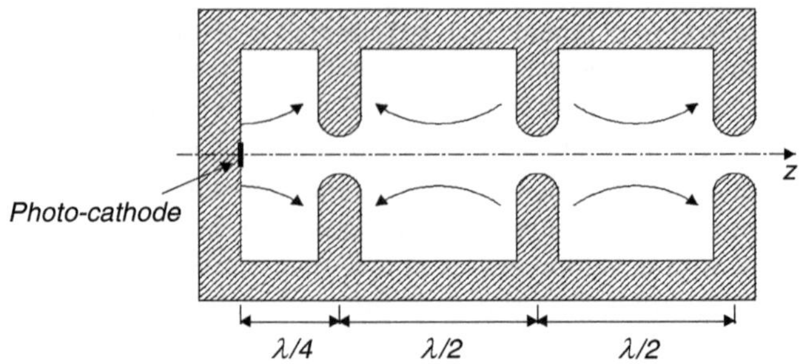

Fig. 6.19 Schematics of an rf gun. The photo-cathode is on the left wall of the half-cell. At the operating frequency the cell operates at the π-mode. The entrance or exit coordinates of the $n+1/2$ cell are $z=(n-1/2)\lambda/2$ and $z=(n+1/2)\lambda/2$ correspondingly

$z=0$ and (3) a particle emitted at $t=0$ experiences a phase ϕ_0. Moreover, contrary to the previous sections since efficiency is not a major concern at this point, the effect of the particles on the rf field, is ignored. Next we define $\tau(z)=\int_0^z \mathrm{v}^{-1}(z')dz'$ which represents the time it takes a particle to reach a point z, tacitly assuming that electrons do not bounce back. We further define the phase of the electron reaching this point as

$$\begin{aligned}\phi(z) &\equiv \phi_0+\omega\tau(z)-\frac{\omega}{c}z\\ &=\phi_0+\frac{\omega}{c}\int_0^z dz'\left(\frac{\gamma(z')}{\sqrt{\gamma^2(z')-1}}-1\right).\end{aligned} \tag{6.6.2}$$

Consequently, the equation of motion reads

$$\frac{d}{dz}\gamma=\frac{eE_0}{2mc^2}\left[\sin\phi+\sin\left(\phi+2\frac{\omega}{c}z\right)\right] \tag{6.6.3}$$

Formally, the phase variation is

$$\begin{aligned}&\frac{d\phi}{dz}=\frac{\omega}{c}\left(\frac{\gamma}{\sqrt{\gamma^2-1}}-1\right)\Rightarrow\gamma^2=\frac{\left(\frac{d\phi}{dz}\frac{c}{\omega}+1\right)^2}{\left(\frac{d\phi}{dz}\frac{c}{\omega}+1\right)^2-1}\\ &\frac{d^2\phi}{dz^2}=-\frac{\omega}{c}\frac{1}{(\gamma^2-1)^{3/2}}\frac{d\gamma}{dz}\Rightarrow\frac{d\gamma}{dz}=\frac{-\frac{d^2\phi}{dz^2}\frac{c}{\omega}}{\left[\left(\frac{d\phi}{dz}\frac{c}{\omega}+1\right)^2-1\right]^{3/2}}\end{aligned} \tag{6.6.4}$$

leading to the following non-linear differential equation

$$\frac{d^2\phi}{dz^2}\frac{c^2}{\omega^2} = -\frac{eE_0}{2mc^2}\frac{c}{\omega}\left[\sin\phi + \sin\left(\phi + 2\frac{\omega}{c}z\right)\right]\left[\left(\frac{d\phi}{dz}\frac{c}{\omega} + 1\right)^2 - 1\right]^{3/2}$$
$$\phi(0) = \phi_0$$
$$\gamma(0) = 1 \tag{6.6.5}$$

Rather than solving this non-linear equation in a self-consistent way, we proceed by adopting an iterative approach.

The second term in (6.6.2) reflects the fact that when the bunch becomes relativistic ($\gamma \gg 1$)the contribution of the integral to the phase of the particle is negligible (except if $|\phi_0| \ll \pi$) therefore, in zero order we may assume that the main contribution to the phase is from the first cell where the major change in velocity occurs. Denoting by $\gamma_{\rm app}$ the approximate solution for γ and $\alpha = eE_0/2mc\omega$ as the dimensionless field parameter representing the strength of the accelerating field acting on an electron on the scale of one wavelength, we get

$$\gamma \simeq \gamma_{\rm app} = 1 + 2\alpha\sin(\phi_0)\left(\frac{\omega}{c}z\right) \tag{6.6.6}$$

Substituting in the phase equation (6.6.2) the phase is given by

$$\phi = \phi_0 + \frac{\gamma_{\rm app} - 1}{2\alpha\sin\phi_0}\left[\sqrt{\frac{\gamma_{\rm app}+1}{\gamma_{\rm app}-1}} - 1\right] \tag{6.6.7}$$

enabling to determine a better approximation for γ namely,

$$\gamma \simeq 1 + \alpha\left\{\frac{\omega}{c}z\sin\phi + \frac{1}{2}\left[\cos\phi - \cos\left(\phi + 2\frac{\omega}{c}z\right)\right]\right\} \tag{6.6.8}$$

note that according to (6.6.7) at the limit $\gamma \gg 1$ the phase is

$$\phi_\infty = \phi_0 + \frac{1}{2\alpha\sin\phi_0} \tag{6.6.9}$$

Comment 6.6. In case of a finite spread $\Delta\phi_0$ around ϕ_0 at the photo-cathode, the spread at the output is

$$\frac{\Delta\phi_\infty}{\Delta\phi_0} = 1 - \frac{\cos\phi_0}{2\alpha\sin^2\phi_0} \tag{6.6.10}$$

For assessment of the transverse motion several assumptions are in place: (1) the initial transverse motion at the cathode is zero ($p_\perp = 0$), (2) the acceleration mode is azimuthally symmetric and (3) at the output end the bunch is relativistic ($\beta \simeq 1$). Subject to these assumptions the radial momentum at the output ($z = z_{\text{out}}$) is

$$p_{r,\text{out}} = \alpha\left(\frac{\omega}{c}r\right)\sin(\phi_{\text{out}}) \tag{6.6.11}$$

In what follows we omit the subscript "out" with the understanding that all quantities are at the output end of the structure. In Chap. 3 we have shown that the transverse motion may be characterized by the emittance and without any loss of generality, we limit the discussion to the x-direction namely, $p_x \equiv \gamma\beta(dx/dz) = (\alpha\omega\sin\phi/c)x$ thus $\varepsilon_x = \sqrt{\langle p_x^2\rangle\langle x^2\rangle - \langle xp_x\rangle^2}$. Further assuming that the transverse and longitudinal dynamics are independent, we conclude that the contribution of the rf field to the emittance is

$$\varepsilon_x = \alpha\frac{\omega}{c}\langle x^2\rangle\sqrt{\langle\sin^2\phi_i\rangle_i - \langle\sin\phi_i\rangle_i^2} \tag{6.6.12}$$

For a small bunch whose average phase is $\langle\phi\rangle$ the phase is $\phi_i = \langle\phi\rangle + \delta\phi_i$ and we will assume that $|\delta\phi_i| \ll \pi$ hence

$$\varepsilon_x = \alpha\frac{\omega}{c}\langle x^2\rangle\sqrt{\sin^2\langle\phi\rangle\left[\langle\cos^2\delta\phi_i\rangle - \langle\cos\delta\phi_i\rangle^2\right] + \cos^2\langle\phi\rangle\left[\langle\sin^2\delta\phi_i\rangle - \langle\sin\delta\phi_i\rangle^2\right]} \tag{6.6.13}$$

It can be readily checked that this expression has a minimum if

$$\langle\phi\rangle = \frac{\pi}{2} \tag{6.6.14}$$

implying that

$$\begin{aligned}\varepsilon_{x,\min} &= \alpha\frac{\omega}{c}\langle x^2\rangle\sqrt{\langle\cos^2\delta\phi_i\rangle - \langle\cos\delta\phi_i\rangle^2},\\ &\simeq \frac{1}{2}\alpha\frac{\omega}{c}\langle x^2\rangle\sqrt{\langle\delta\phi_i^4\rangle - \langle\delta\phi_i^2\rangle^2}.\end{aligned} \tag{6.6.15}$$

From this result, we conclude that minimum emittance is achieved when the transverse momentum is maximum and consequently, it will be necessary to focus the beam immediately after exiting the injector.

According to (6.6.9), the condition for minimal emittance at the output (6.6.14) further implies that the initial phase should match the field strength parameter namely

$$\left(\frac{\pi}{2} - \phi_0\right)\sin\phi_0 = \frac{1}{2\alpha} \tag{6.6.16}$$

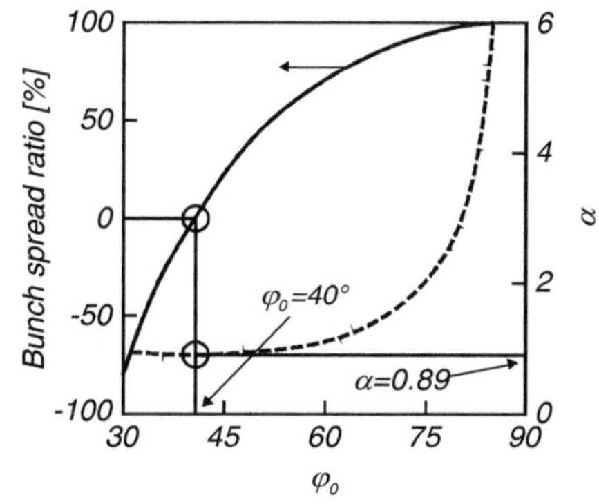

Fig. 6.20 Bunch spread ratio (in [%]) and the field strength parameter (α) as a function of the initial phase of the bunch at the cathode (ϕ_0). The upper circle reveals that choosing the latter to be $\phi_0 \sim 40^o$ leads to zero spread at the output; for this to happen, $\alpha \sim 0.89$

Figure 6.20 illustrates both the bunch spread as defined in (6.6.10) (expressed in percentage) and the field strength parameter (α) as a function of the initial phase of the bunch at the photo-cathode (ϕ_0). It is evident that there is an entire range of angles (ϕ_0) which actually facilitates compression of the bunch $|\Delta\phi_\infty| \leq |\Delta\phi_0|$. In fact, if $\phi_0 \sim 40^{\rm o}$ the spread is virtually zero, $\Delta\phi_\infty \sim 0$; for this to happen the field strength parameter should be $\alpha \simeq 0.89$. In the framework of this estimate, the space-charge effect was ignored and therefore, these results should be considered only as a rough estimate.

Exercises

6.1. Analyze the electromagnetic problem as in Sect. 6.1 but with three modes in each groove and arm.

6.2. Analyze the electromagnetic problem as in Sect. 6.1 but for three arms that are not necessarily located at the ends of the structure.

6.3. Analyze the electromagnetic problem as in Sect. 6.1 but for a rectangular waveguide.

6.4. Analyze the electromagnetic problem as in Sect. 6.1 but for symmetric TE modes in a cylindrical waveguide.

6.5. Analyze the beam-wave interaction problem as in Sect. 6.4 but assume that the metal at $r = R_{\rm ext}$ is of finite conductivity. Determine the effect of resistive wall instability on the energy exchange.

Chapter 7
Free-Electron Laser

In Chap. 1 we have shown that the interaction of electrons with an electromagnetic wave is possible even when the phase velocity of the latter is larger than c, provided that there is a way to conserve simultaneously both energy and momentum. In a free-electron laser (FEL) this is facilitated by the presence of a periodic magnetic field. In most cases, the components of this field are transverse to the initial velocity of the electron. An electron injected in a periodic magnetic field (wiggler) oscillates and, as a result, it emits radiation. The highest frequency is emitted in the forward direction and in zero order it is determined by the periodicity of the wiggler, L, and the electron energy, γ. In Sect. 3.2.3 it was shown that for relativistic electrons ($\beta \sim 1$) this frequency is given by $\omega \simeq 2\gamma^2(2\pi c/L)$.

To the best of our knowledge, the first analysis of the motion of an electron in a wiggler of this kind was performed in the early 1930s by Kapitza and Dirac (1933). The question raised was whether it would be possible to observe stimulated scattered radiation from electrons moving in an electromagnetic wave. For this purpose, the authors considered a low energy beam of electrons injected in a standing wave region and they estimated the number of scattered electrons due to the stimulated radiation. In the early 1950s Motz (1951) investigated the radiation emitted by electrons as they move in a wiggler and later Phillips (1960) built the first coherent radiation source with a wiggler as its central component; it was called the Ubitron. In the late 1960s Pantell et al. (1968) suggested the same concept at much shorter wavelengths and Madey (1971) has proven that laser light can be amplified using this scheme but it was only later at Stanford that Elias et al. (1976) demonstrated experimentally the amplification of a 10.6 μm laser beam and since then the name – free electron laser.

There are numerous textbooks, review articles, proceedings and articles on free-electron lasers a small fraction of which will be mentioned in Sect. 7.5. However, for an introductory guide to the free-electron laser we find the article of Hasegawa (1978) as a good starting point. An excellent tutorial work on the theory of the free-electron laser is the article by Kroll et al. (1981) which in fact inspired many of the topics presented in Sect. 7.4. An overview of the field is presented in an article by

L. Schächter, *Beam-Wave Interaction in Periodic and Quasi-Periodic Structures*, Particle Acceleration and Detection, DOI 10.1007/978-3-642-19848-9_7,

Roberson and Sprangle (1989) and among the books dedicated to FEL's, Marshall (1985) covers the basic theory and the early work done and more recently the book by Freund and Antonsen (1992) also covers advanced topics on free-electron lasers in addition to the basic theory.

In this chapter, we present what we conceive as the basics of free-electron lasers. Alternative schemes of energy conversion from free electrons and some advanced applications as advanced light source are briefly described in the last part of this chapter. Specifically, in the first section we consider the spontaneous emission as an electron traverses an ideal wiggler. This is followed by the investigation of coherent interaction in the low-gain Compton regime. Section 7.3 deals with the high-gain Compton regime, which includes cold and warm beam operation. The macro-particle approach is presented in Sect. 7.4 and we review various alternative schemes of free-electron lasers (Sect. 7.5). A brief description of FEL's as an advanced light sources concludes this chapter.

7.1 Spontaneous Radiation

As an electron is injected into a periodic magnetic field, it oscillates and emits spontaneous radiation. In this section, we examine this process. For this purpose, we consider a transverse periodic magnetic field that is uniform in the transverse direction – at least on the scale of the beam cross-section. A helical undulator field can be derived from the following magnetic vector potential

$$\mathbf{A}_{\mathrm{w}} = -A_{\mathrm{w}}[\mathbf{1}_x \cos(k_{\mathrm{w}} z) + \mathbf{1}_y \sin(k_{\mathrm{w}} z)], \tag{7.1.1}$$

or explicitly,

$$\mathbf{B}_{\mathrm{w}} = B_{\mathrm{w}}[\mathbf{1}_x \cos(k_{\mathrm{w}} z) + \mathbf{1}_y \sin(k_{\mathrm{w}} z)]; \tag{7.1.2}$$

the two amplitudes A_{w} and B_{w} are related via $A_{\mathrm{w}} = B_{\mathrm{w}}/k_{\mathrm{w}}$ where $k_{\mathrm{w}} = 2\pi/L$ is the wiggler's wave-number; $\mathbf{1}_x, \mathbf{1}_y$ are the unit vectors in the x and y directions correspondingly. An electron is injected along the z axis and we examine its motion in the absence of any radiation field. The relativistic Hamiltonian that describes the motion of an electron in the presence of an electromagnetic field was developed in Sect. 3.1 [(3.1.15)] and it is given by

$$H = \sqrt{(\mathbf{p} + e\mathbf{A}_{\mathrm{w}})^2 c^2 + (mc^2)^2} = mc^2\gamma \tag{7.1.3}$$

where collective effects are ignored and since no external voltage is applied and we ignore space-charge effects, the electrostatic potential is taken as zero. In addition, no boundaries are involved.

The canonical momentum **p** has two components: one which is parallel to the major velocity component of the electron and is denoted by $p_{\parallel}$ and the transverse one $p_{\perp}$. As indicated in Sect. 3.1, if the Hamiltonian is not explicitly dependent on the transverse coordinates then the transverse canonical momentum is conserved ($p_{\perp} = \text{const.}$). This canonical momentum has also two contributions, the kinetic and the electromagnetic i.e., $p_{\perp} = m\gamma \mathrm{v}_{\perp} - eA_{\perp}$. Assuming that the electron is born outside the magnetic field and its initial transverse motion is zero, we immediately conclude that $p_{\perp} = 0$, which implies

$$\mathrm{v}_{\perp} = \frac{eA_{\perp}}{m\gamma}. \tag{7.1.4}$$

We also observe that this Hamiltonian does not explicitly depend on time therefore, energy is conserved i.e.,

$$\gamma = \text{const.}. \tag{7.1.5}$$

From the last two relations we can deduce the expressions that describe the motion of an electron in space, they read

$$\begin{aligned} \mathrm{v}_x(z) &= -\frac{eA_{\mathrm{w}}}{\gamma m}\cos(k_{\mathrm{w}}z), \\ \mathrm{v}_y(z) &= -\frac{eA_{\mathrm{w}}}{\gamma m}\sin(k_{\mathrm{w}}z), \\ \mathrm{v}_z(z) &= \mathrm{v}_0. \end{aligned} \tag{7.1.6}$$

It is evident from these expressions that the particle undergoes a helical motion whose amplitude is determined by the amplitude of the wiggler B_{w}, its wave-number k_{w} and the initial energy of the particle. This fact becomes clearer when realizing that in the x–y plane the particles undergo a circular motion as revealed by the first two equations of (7.1.6) which can be rewritten as

$$\mathrm{v}_x^2 + \mathrm{v}_y^2 = \left(\frac{eB_{\mathrm{w}}}{k_{\mathrm{w}}\gamma m}\right)^2. \tag{7.1.7}$$

If we assume that the transverse motion is much slower than the longitudinal component we can assume that $z \simeq \mathrm{v}_{\parallel}t$ and therefore,

$$\begin{aligned} X(t) &= X(0) - \frac{eA_{\mathrm{w}}}{\gamma m}\frac{1}{k_{\mathrm{w}}\mathrm{v}_{\parallel}}\sin(k_{\mathrm{w}}\mathrm{v}_{\parallel}t), \\ Y(t) &= Y(0) + \frac{eA_{\mathrm{w}}}{\gamma m}\frac{1}{k_{\mathrm{w}}\mathrm{v}_{\parallel}}\left[\cos(k_{\mathrm{w}}\mathrm{v}_{\parallel}t) - 1\right]. \end{aligned} \tag{7.1.8}$$

This trajectory of the particle implies that it will interact naturally with a circularly polarized plane wave. However, before we consider the radiation emitted, it is instructive to make one more observation. The energy factor, γ, is defined by

$$\gamma = \frac{1}{\sqrt{1-\beta_{\parallel}^2-\beta_{\perp}^2}}, \tag{7.1.9}$$

and we can also define a similar factor associated only with the longitudinal motion i.e.,

$$\gamma_{\parallel} \equiv \frac{1}{\sqrt{1-\beta_{\parallel}^2}}. \tag{7.1.10}$$

Since the transverse velocity, as determined in (7.1.4), is γ dependent, we find that

$$\gamma_{\parallel} = \frac{\gamma}{\sqrt{1+(eB_{\mathrm{w}}/mck_{\mathrm{w}})^2}}, \tag{7.1.11}$$

which indicates that the effective energy factor ($\gamma_{\parallel}$) can be substantially smaller than γ. For example, if $B_{\mathrm{w}} = 0.5\,\mathrm{T}$ and $L = 2\,\mathrm{cm}$ the longitudinal energy factor is about 70% of the original γ.

An electron that follows the trajectory described by (7.1.8) radiates. In order to calculate the emitted radiation we assume that the current density is given by

$$\begin{bmatrix} J_x(\mathbf{r},t) \\ J_y(\mathbf{r},t) \\ J_z(\mathbf{r},t) \end{bmatrix} = -e \begin{bmatrix} \mathrm{v}_x \\ \mathrm{v}_y \\ \mathrm{v}_{\parallel} \end{bmatrix} \delta(x)\delta(y)\delta(z-\mathrm{v}_{\parallel}t), \tag{7.1.12}$$

where the transverse displacement of the electron was neglected. In free-space the radiation generated by this current density is given by

$$\mathbf{A}(\mathbf{r},t) = \mu_0 \int \mathrm{d}\omega \int \mathrm{d}\mathbf{r}' \frac{\exp\left[j\omega\left(t-\frac{|\mathbf{r}-\mathbf{r}'|}{c}\right)\right]}{4\pi|\mathbf{r}-\mathbf{r}'|}\mathbf{J}(\mathbf{r}',\omega) \tag{7.1.13}$$

The time Fourier transform of the current density in (7.1.12), denoted above by $\mathbf{J}(\mathbf{r},\omega)$, is given by

$$\begin{aligned} J_x(\mathbf{r},\omega) &= -\frac{e}{2\pi}\frac{\mathrm{v}_x(t=z/\mathrm{v}_{\parallel})}{\mathrm{v}_{\parallel}}\exp\left(-j\frac{\omega}{\mathrm{v}_{\parallel}}z\right)\delta(x)\delta(y), \\ J_y(\mathbf{r},\omega) &= -\frac{e}{2\pi}\frac{\mathrm{v}_y(t=z/\mathrm{v}_{\parallel})}{\mathrm{v}_{\parallel}}\exp\left(-j\frac{\omega}{\mathrm{v}_{\parallel}}z\right)\delta(x)\delta(y), \\ J_z(\mathbf{r},\omega) &= -\frac{e}{2\pi}\exp\left(-j\frac{\omega}{\mathrm{v}_{\parallel}}z\right)\delta(x)\delta(y). \end{aligned} \tag{7.1.14}$$

The integrals in (7.1.13) can be simplified for the case when the observer is far away from the wiggler $[(\omega/c)r \gg 1\ ]$ in which case we have

$$\mathbf{A}(\mathbf{r},t) = \frac{\mu_0}{4\pi r}\int \mathrm{d}\omega \exp\left[j\omega\left(t-\frac{r}{c}\right)\right]\int \mathrm{d}\mathbf{r}'\exp\left(j\frac{\omega}{c}z'\cos\theta\right)\mathbf{J}(\mathbf{r}',\omega) \qquad (7.1.15)$$

θ in these expressions is the angle between the vector which connects the center of the wiggler and the observer with the z axis. Substituting the explicit expression for the current densities allows us to evaluate the integral analytically. Neglecting rapidly oscillating terms $\left[(\omega/c)\cos\theta - k_{\mathrm{w}} - \omega/\mathrm{v}_{\|}\right]$ we have

$$\begin{aligned}
A_x(\mathbf{r},t) &= \frac{e\mu_0}{(4\pi)^2 r}\frac{eB_{\mathrm{w}}D}{mk_{\mathrm{w}}\gamma\mathrm{v}_{\|}}\mathrm{Re}\left\{\int \mathrm{d}\omega\exp\left[j\omega\left(t-\frac{r}{c}\right)\right]\mathrm{sinc}\left[\left(\frac{\omega}{c}\cos\theta-\frac{\omega}{\mathrm{v}_{\|}}+k_{\mathrm{w}}\right)\frac{D}{2}\right]\right\},\\
A_y(\mathbf{r},t) &= \frac{e\mu_0}{(4\pi)^2 r}\frac{eB_{\mathrm{w}}D}{mk_{\mathrm{w}}\gamma\mathrm{v}_{\|}}\mathrm{Re}\left\{\int \mathrm{d}\omega\frac{1}{j}\exp\left[j\omega\left(t-\frac{r}{c}\right)\right]\mathrm{sinc}\left[\left(\frac{\omega}{c}\cos\theta-\frac{\omega}{\mathrm{v}_{\|}}+k_{\mathrm{w}}\right)\frac{D}{2}\right]\right\},\\
A_z(\mathbf{r},t) &\simeq 0;
\end{aligned} \qquad (7.1.16)$$

where $\mathrm{sinc}(x) = \sin(x)/x$, the total length of the wiggler is denoted by D and it spans from $-D/2<z<D/2$. The longitudinal component of the magnetic vector potential is negligible since it is proportional to $\mathrm{sinc}[(\cos\theta - 1/\beta_{\|})(\omega D/2c)]$ and this function varies rapidly for a wiggler length D much larger than the wavelength of interest.

The magnetic vector potential determines the electromagnetic field, which in turn enables us to evaluate the power and energy emitted. The Poynting flux is given by

$$S_z(\mathbf{r},t) = E_x(\mathbf{r},t)H_y(\mathbf{r},t) - E_y(\mathbf{r},t)H_x(\mathbf{r},t), \qquad (7.1.17)$$

and the energy emitted per unit area in this process is given by

$$\mathrm{w}(\mathbf{r}) = \int \mathrm{d}t S_r(\mathbf{r},t), \qquad (7.1.18)$$

where $S_z = S_r\cos\theta$. Substituting the explicit expressions for the components of the electromagnetic field, followed by the evaluation of the integral over t simplifies substantially the calculation since the resulting Dirac delta function can be utilized to evaluate the double integration

$$\mathrm{w}(\mathbf{r}) = \frac{4\pi}{\eta_0}\left[\frac{e\mu_0}{(4\pi)^2}\left(\frac{eB_{\mathrm{w}}}{mk_{\mathrm{w}}\gamma\mathrm{v}_{\|}}\right)\frac{D}{r}\right]^2 \times \int \mathrm{d}\omega\omega^2\mathrm{sinc}^2\left[\left(\frac{\omega}{c}\cos\theta-\frac{\omega}{\mathrm{v}_{\|}}+k_{\mathrm{w}}\right)\frac{D}{2}\right]. \qquad (7.1.19)$$

This energy is emitted by a single electron. An ensemble of N electrons in the interaction region carry a current $I = eN\mathrm{v}_{\parallel}/D$ and the energy in an angular interval $\theta \to \theta + d\theta$ is

$$W = Nr^2 \mathrm{w}$$
$$= \frac{eI\eta_0}{(4\pi)^3}\left(\frac{eB_\mathrm{w}}{mck_\mathrm{w}\gamma}\right)^2 \left(\frac{D}{\beta_{\parallel}}\right)^3 \times \int \mathrm{d}\left(\frac{\omega}{c}\right)\frac{\omega^2}{c^2}\mathrm{sinc}^2\left[\left(\frac{\omega}{c}\cos\theta - \frac{\omega}{\mathrm{v}_{\parallel}} + k_\mathrm{w}\right)\frac{D}{2}\right]. \quad (7.1.20)$$

The term

$$g(\xi) \equiv \mathrm{sinc}^2(\xi), \quad (7.1.21)$$

represents the spontaneous emission line shape and we shall encounter it again when considering the power in the low-gain Compton regime. The argument of the sinc function is directly associated with the resonance condition

$$\xi \equiv \frac{D}{2}\left(\frac{\omega}{c}\cos\theta - \frac{\omega}{\mathrm{v}_{\parallel}} + k_\mathrm{w}\right) \quad (7.1.22)$$

and maximum power is emitted when this condition is satisfied i.e., $\xi = 0$. It implies that the frequency emitted in the forward direction ($\theta = 0$) is given by

$$\omega = \omega_\mathrm{res} \equiv ck_\mathrm{w}\frac{\beta_{\parallel}(1+\beta_{\parallel})\gamma^2}{1+\left(\frac{eB_\mathrm{w}}{mk_\mathrm{w}c}\right)^2}, \quad (7.1.23)$$

which clearly depends on the strength of the wiggler. This can be considered the exact resonance condition while the expressions presented in the first and third chapters [(1.1.19) and (3.2.17)] are approximations, which are valid in case of a weak wiggler field i.e. $eB_\mathrm{w}c/k_\mathrm{w} \ll mc^2$.

Rather than considering the whole spectrum of waves emitted in a given direction in space, it is instructive to present the energy emitted in a frequency interval $\omega \to \omega + d\omega$ in one period of the wave i.e.,

$$\frac{\omega}{2\pi}\frac{dW}{d\omega} = \frac{eI\eta_0}{(4\pi)^4}\left(\frac{eB_\mathrm{w}}{mck_\mathrm{w}\gamma}\right)^2 \left(\frac{\omega}{\mathrm{v}_{\parallel}}D\right)^3 \mathrm{sinc}^2\left[\left(\frac{\omega}{c}\cos\theta - \frac{\omega}{\mathrm{v}_{\parallel}} + k_\mathrm{w}\right)\frac{D}{2}\right]. \quad (7.1.24)$$

Assuming operation at resonance we can substitute the explicit expression for the resonant frequency and obtain

$$\overline{W} \equiv \left[\frac{\omega}{2\pi}\frac{\mathrm{d}W}{\mathrm{d}\omega}\right]_{\omega=\omega_\mathrm{res}\theta=0} = \frac{eI\eta_0}{32\pi}\left[\frac{D}{L}(1+\beta_{\parallel})\right]^3 \gamma^4 \frac{\Omega_\mathrm{w}^2}{\left(1+\Omega_\mathrm{w}^2\right)^3}, \quad (7.1.25)$$

where $\Omega_{\mathrm{w}} = eB_{\mathrm{w}}/mck_{\mathrm{w}}$. Note that as a function of Ω_{w}, the emitted power has a maximum at $\Omega_{\mathrm{w}} = 1/\sqrt{2}$ therefore

$$\overline{W}_{\max} = \frac{eI\eta_0}{216\pi}\left[\frac{D}{L}(1+\beta_{\|})\right]^3 \gamma^4. \tag{7.1.26}$$

According to this result, a 10 MeV, 1 kA beam generates $\overline{W}_{\max} = 0.1\ \mu$J of energy in 10 periods of the structure. If the electron pulse duration corresponds to the radiation period 1/35 GHz, then the total energy carried by the beam is 0.28 J which is six orders of magnitude larger than the radiated power.

The energy lost by the electron as it traverses the periodic magnetic field, can be interpreted in terms of an effective gradient that decelerates the moving electron. In order to evaluate this gradient, we integrate (7.1.19) over the spherical envelope

$$\begin{aligned}\tilde{W} &\equiv 2\pi r^2 \int_0^{\pi} \mathrm{d}\theta \sin\theta W(r,\theta) \\ &= \frac{2(4\pi)^2}{\eta_0}\frac{(e\mu_0)^2}{(4\pi)^4}\left(\frac{eB_{\mathrm{w}}}{mck_{\mathrm{w}}\gamma\beta}\right)^2 D^2 \int_0^{\pi} \mathrm{d}\theta \sin\theta \int_{-\infty}^{\infty} \mathrm{d}\omega \\ &\quad \times \omega^2 \mathrm{sinc}^2\left[\frac{D}{2}\left(\frac{\omega}{c}\cos\theta - \frac{\omega}{c}\frac{1}{\beta} + k_{\mathrm{w}}\right)\right].\end{aligned} \tag{7.1.27}$$

For a long interaction region we use

$$\lim_{D\to\infty}\left\{\frac{D}{2}\mathrm{sinc}\left[\frac{D}{2}(k_1 - k_2)\right]\right\} = 2\pi\delta(k_1 - k_2), \tag{7.1.28}$$

thus

$$\tilde{W} = \left(\frac{eB_{\mathrm{w}}}{2mck_{\mathrm{w}}\gamma\beta}\right)^2 D \frac{\mathrm{e}^2}{4\pi\varepsilon_0/k_{\mathrm{w}}^2}\int_0^{\pi} \mathrm{d}\theta \frac{\sin\theta}{(1/\beta - \cos\theta)^3}. \tag{7.1.29}$$

The integral can be calculated analytically and, as we indicated, it is convenient to determine the effective gradient as $E_{\mathrm{eff}} \equiv \tilde{W}/eD$ which reads

$$E_{\mathrm{eff}} = \frac{1}{2}\beta\gamma^2 \frac{ek_{\mathrm{w}}^2}{4\pi\varepsilon_0}\left(\frac{eB_{\mathrm{w}}}{mck_{\mathrm{w}}}\right)^2. \tag{7.1.30}$$

This is the decelerating gradient which acts on the particle. Note that for a relativistic particle it is quadratic in γ, the energy of the particle, and it will become an important factor when discussing acceleration of electrons using the FEL scheme in Chap. 8.

7.2 Low-Gain Compton Regime

If an electromagnetic wave is injected parallel to the beam and its frequency matches the resonance condition, then stimulated radiation may occur. As a first stage, we examine the lowest order effect of the beam on the radiation field. To be more specific we look for the contribution to the radiated power of the first order in ω_{p}^2 term.

The wiggler is the same as in (7.1.1) and the injected wave is circularly polarized:

$$\mathbf{A}_{\mathrm{rf}}(\mathbf{r},t) = A_0[\mathbf{1}_x \cos(\omega t - kz) + \mathbf{1}_y \sin(\omega t - kz)]. \tag{7.2.1}$$

Ignoring space-charge effects and in the absence of boundaries it is justified to omit the electrostatic potential from the expression for the relativistic Hamiltonian (3.1.15) hence

$$H = \sqrt{(\mathbf{p} + e\mathbf{A}_{\mathrm{w}} + e\mathbf{A}_{\mathrm{rf}})^2 c^2 + (mc^2)^2}. \tag{7.2.2}$$

Neither the wiggler nor the radiation field have components of $\mathbf{A}$ parallel to the beam and consequently,

$$H = \sqrt{(mc^2)^2 + c^2 p_{\|}^2 + c^2(\mathbf{p}_{\perp} + e\mathbf{A}_{\mathrm{w}} + e\mathbf{A}_{\mathrm{rf}})^2}. \tag{7.2.3}$$

As before, the conservation of the transverse canonical momentum $(p_{\perp} = 0)$ implies

$$H = \sqrt{(mc^2)^2 + c^2 p_{\|}^2 + c^2 e^2(\mathbf{A}_{\mathrm{w}} + \mathbf{A}_{\mathrm{rf}})^2}, \tag{7.2.4}$$

whereas the linearization of the Hamiltonian in the radiation field reads

$$\begin{aligned} H &= H_0 + H_1 \\ &= mc^2\gamma + \frac{e^2}{m\gamma}\mathbf{A}_{\mathrm{w}} \cdot \mathbf{A}_{\mathrm{rf}} \\ &= mc^2\gamma - \frac{e^2 A_0 B_{\mathrm{w}}}{mk_{\mathrm{w}}\gamma}\cos[\omega t - (k + k_{\mathrm{w}})z]. \end{aligned} \tag{7.2.5}$$

From this expression, we learn that the first order perturbation is proportional to the scalar product of the wiggler and radiation vector potentials. This part of the Hamiltonian determines the so-called pondermotive force (subscript p)

$$\begin{aligned} F_{\mathrm{p}} &= -\frac{\partial H_1}{\partial z} \\ &= \frac{e^2 A_0 B_{\mathrm{w}}}{m\gamma k_{\mathrm{w}}}(k + k_{\mathrm{w}})\sin[\omega t - (k + k_{\mathrm{w}})z]. \end{aligned} \tag{7.2.6}$$

For what follows, it is convenient to adopt a phasor notation

$$\begin{aligned} F_p \rightarrow \bar{F}_\mathrm{p} &= -j\frac{e^2 A_0 B_\mathrm{w}}{m\gamma k_\mathrm{w}}(K+K_\mathrm{w})\exp\{j[\omega t-(k+K_\mathrm{w})z]\} \\ &= \tilde{F}_\mathrm{p}\exp\{j[\omega t-(k+k_\mathrm{w})z]\}, \end{aligned} \tag{7.2.7}$$

where

$$\tilde{F}_\mathrm{p} \equiv -jeA_0\frac{eB_\mathrm{w}}{m\gamma}\frac{k+k_\mathrm{w}}{k_\mathrm{w}}. \tag{7.2.8}$$

Next we linearize the Liouville equation i.e., assume that the distribution function f has the form

$$f(z,t;p) = f_0(p) + f_1(z,t;p), \tag{7.2.9}$$

where f_0 is considered to be known and f_1 is linear in the pondermotive force hence

$$\left(j\omega + \mathrm{v}\frac{\partial}{\partial z}\right)f_1 = -\tilde{F}_\mathrm{p}\exp\{j[\omega t-(k+k_\mathrm{w})z]\}\frac{\mathrm{d}f_0}{\mathrm{d}p}. \tag{7.2.10}$$

A solution of this expression, assuming that the right hand side is known, can be formally written as

$$\begin{aligned} f_1(z,t) &= -\frac{\mathrm{d}f_0}{\mathrm{d}p}\frac{\tilde{F}_\mathrm{p}}{\mathrm{v}}\exp\left[j\omega\left(t-\frac{z}{\mathrm{v}}\right)\right]\frac{\exp(j\delta kz)-\exp\left(-j\delta k\frac{D}{2}\right)}{j\delta k}, \\ &= \frac{1}{p}\frac{\mathrm{d}f_0}{\mathrm{d}p}\frac{e^2A_0B_\mathrm{w}(k+k_\mathrm{w})}{k_w}\exp\left[j\omega\left(t-\frac{z}{\mathrm{v}}\right)\right] \\ &\quad\times\frac{\exp(j\delta kz)-\exp\left(-j\delta k\frac{D}{2}\right)}{\delta k}, \end{aligned} \tag{7.2.11}$$

where D is the length of the interaction region which starts at $z=-D/2$ and $\delta k \equiv \omega/\mathrm{v} - k - k_\mathrm{w}$. With this expression for the distribution function, we can define the macroscopic current density and in particular, its transverse components read

$$\begin{aligned} J_\perp &= -en_0\int \mathrm{d}p\mathrm{v}_\perp f_1 \\ &= \frac{1}{2}n_0\left(\frac{e^2B_\mathrm{w}}{k_w}\right)^2\frac{A_0(k+k_\mathrm{w})}{m}(\mathbf{1}_x - j\mathbf{1}_y) \\ &\quad\times\int\mathrm{d}p\tfrac{1}{\gamma p}\tfrac{\mathrm{d}f_0}{\mathrm{d}_p}\exp\left[j\omega\left(t-\tfrac{z}{\mathrm{v}}\right)\right]\frac{\exp(j\delta kz)-\exp\left(-j\delta k\frac{D}{2}\right)}{\delta k}. \end{aligned} \tag{7.2.12}$$

Here we used the explicit expression for the transverse velocity in (7.1.4) and ignored terms that vary rapidly in space; n_0 is the average density of the particles in the absence of the radiation. Since we calculated the current density generated by a known electric field, the next step is to calculate the power

$$P = S_{\mathrm{el}} \int_{-D/2}^{D/2} \mathrm{d}z \frac{1}{2} \mathrm{Re}\left(E_\perp \cdot J_\perp^*\right), \tag{7.2.13}$$

where S_{el} is the beam cross-section and

$$E_\perp = -j\omega A_{\mathrm{rf}} = -j\omega(\mathbf{1}_x - j\mathbf{1}_y)A_0 \exp(j\omega t - jkz). \tag{7.2.14}$$

Note that it has been tacitly assumed here that the effect of the beam on the radiation field is negligible. Substituting in (7.2.13) we obtain

$$P = \frac{1}{4} n_0 \omega (k + k_{\mathrm{w}}) S_{\mathrm{el}} D^2 \left(\frac{e^2 B_{\mathrm{w}} A_0}{k_{\mathrm{w}}}\right)^2 \int \mathrm{d}p \frac{\mathrm{v}}{p^2} \frac{\mathrm{d}f_0}{\mathrm{d}p} \mathrm{sinc}^2\left(\frac{1}{2}\delta k D\right). \tag{7.2.15}$$

At this point we can evaluate the last integral for two extreme regimes: (1) *cold* beam approximation and (2) *warm* beam approximation. In the former case it is assumed that the initial distribution function f_0 is much sharper than the sinc function hence by integration by parts we get

$$P = \frac{1}{2\gamma_0^5 \beta_0^4} \frac{eI\eta_0}{mc^2} \left[\frac{(\omega A_0 D)^2}{2\eta_0}\right] \left(\frac{eB_{\mathrm{w}}}{mck_{\mathrm{w}}}\right)^2 (k + k_{\mathrm{w}}) D \left[-\frac{1}{2} \frac{\mathrm{d}}{\mathrm{d}\xi} \mathrm{sinc}^2 \xi\right]_{\xi = \delta k D/2}, \tag{7.2.16}$$

or

$$P = \left[\left(\frac{\omega_{\mathrm{p}} D}{c}\right)^2 \frac{1}{(\gamma_0 \beta_0)^3}\right] \left[\frac{eB_{\mathrm{w}}}{mck_{\mathrm{w}}} \frac{1}{\gamma}\right]^2 \times \left[\frac{(\omega A_0)^2 S_{\mathrm{el}} D (k + k_{\mathrm{w}})}{4\eta_0}\right] \left[-\frac{1}{2} \frac{\mathrm{d}}{\mathrm{d}\xi} \mathrm{sinc}^2 \xi\right]_{\xi = \delta k D/2}, \tag{7.2.17}$$

where we used $f_0(p) = \delta(p - p_0)$. These expressions clearly indicate that the power is inversely proportional to the γ^5 out of which the γ^3 term is due to the longitudinal bunching, and γ^2 is due to the transverse oscillation in the wiggler. In addition, note that the power of the *coherent* radiation emitted is proportional to the derivative of the *spontaneous* emission line shape.

The second regime of interest is when the sinc function is much sharper than the distribution of particles and the power is

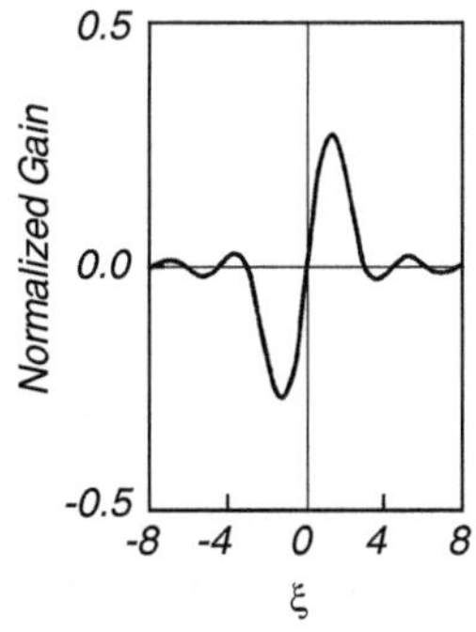

Fig. 7.1 Normalized gain. In the weak Compton regime when operating exactly at resonance the gain vanishes and it peaks $\zeta = 1.303$

$$
\begin{aligned}
P &\propto \int \mathrm{d}p \frac{\mathrm{v}}{p^2} \frac{\mathrm{d}f_0}{\mathrm{d}p} \mathrm{sinc}^2\left(\frac{1}{2}\delta kD\right) \propto \frac{\mathrm{v}}{p^2} \int \mathrm{d}p \frac{\mathrm{d}f_0}{\mathrm{d}p} \mathrm{sinc}^2\left(\frac{1}{2}\delta kD\right), \\
&\propto \frac{\mathrm{v}}{p^2} \left\{ \int \mathrm{d}p \frac{\mathrm{d}}{\mathrm{d}p} \left[f_0 \mathrm{sinc}^2\left(\frac{1}{2}\delta kD\right) \right] - \int \mathrm{d}p f_0 \frac{\mathrm{d}}{\mathrm{d}p} \mathrm{sinc}^2\left(\frac{1}{2}\delta kD\right) \right\}, \\
&\propto -\mathrm{sinc}^2(\pi D/L) \simeq 0,
\end{aligned}
\tag{7.2.18}
$$

which indicates that in the low-gain Compton regime, "warm" electrons do not generate coherent radiation.

The coherent radiation generated in the course of the electrons' motion in the wiggler as revealed by (7.2.16), (7.2.17) is illustrated in Fig. 7.1 where the normalized gain is $-\frac{1}{2}\frac{\mathrm{d}}{\mathrm{d}\xi}\mathrm{sinc}^2(\xi)$. We observe that when the velocity of the electrons is larger than the phase velocity of the pondermotive force i.e., $\mathrm{v} > \omega/(k + k_\mathrm{w})$ meaning negative ξ, the normalized gain is negative thus energy is transferred from the electrons to the wave. And when the electrons are slower, they are accelerated by the pondermotive force. Maximum gain does not occur at resonance but for $|\xi| = 1.303$ in which case the absolute value of the normalized gain is 0.27.

7.3 High-Gain Compton Regime

In the previous section, the collective effect of the particles was neglected in the sense that the effect of the beam on the radiation field was ignored and the gain was a result of an ensemble of dipoles oscillating coherently due to the common excitation of the external field. This interpretation is supported by the expression for the current density in (7.2.12) which indicates that the electrons are organized in bunches. At low currents the effect of these bunches on the radiation field is indeed small but as the current is elevated, their effect becomes more and more significant. In parallel, as the modulation increases, the quasi-electrostatic forces between particles also increases and space-charge effects have to be accounted for. These will be represented here by a scalar electric potential Φ and the longitudinal component of the magnetic vector potential A_z. Following the same approach as previously, the dynamics of the distribution function satisfies

$$\left[j\omega + \mathrm{v}\frac{\partial}{\partial z} + \left(F_{\mathrm{p}} + e\frac{\mathrm{d}\Phi}{\mathrm{d}z} + j\omega e A_z\right)\frac{\partial}{\partial p}\right] f = 0. \tag{7.3.1}$$

The electric scalar potential Φ is determined by the charge distribution via the non-homogeneous wave equation

$$\left[\frac{\mathrm{d}^2}{\mathrm{d}z^2} + \frac{\omega^2}{c^2}\right]\Phi = \frac{en_0}{\varepsilon_0}\left[\int \mathrm{d}pf - 1\right]. \tag{7.3.2}$$

This potential determines the longitudinal component of the magnetic vector potential since we have tacitly assumed the Lorentz gauge i.e.,

$$\frac{\mathrm{d}A_z}{\mathrm{d}z} + \frac{j\omega}{c^2}\Phi = 0. \tag{7.3.3}$$

Note that there is no magnetic field associated with this potential since it is dependent only on the z coordinate. In addition to these two potentials, the distribution of particles determines the transverse current density

$$J_{\perp}(z,\omega) = -en_0 \int \mathrm{d}p\mathrm{v}_{\perp} f, \tag{7.3.4}$$

which in turn governs the magnetic vector potential via the non-homogeneous wave equation as in (2.1.39). In this case we ignore transverse effects therefore we integrate the wave equation over the transverse dimensions. Assuming that the effective area of the electromagnetic field is S_{em} and that of the electron beam is S_{el}, we define the filling factor term $F_{\mathrm{f}} \equiv S_{\mathrm{el}}/S_{\mathrm{em}}$ by whose means the 1D wave equation for the magnetic vector potential reads

$$\left[\frac{\mathrm{d}^2}{\mathrm{d}z^2} + \frac{\omega^2}{c^2}\right]A_{\perp} = -\mu_0 F_{\mathrm{f}} J_{\perp}; \tag{7.3.5}$$

this filling factor is assumed to be known.

This is the set of equations, which describes the interaction in a free-electron laser in the high-gain Compton regime. Before we proceed to a solution of this set of equations it is instructive to examine the same set of equations when instead of the the *Lorentz* gauge we use the *Coulomb* gauge. In this case the equations read

$$
\begin{aligned}
&\left[j\omega + \mathrm{v}\frac{\partial}{\partial z} + \left(F_p + e\frac{\mathrm{d}\Phi}{\mathrm{d}z}\right)\frac{\partial}{\partial p}\right] f = 0,\\
&\frac{\mathrm{d}^2\Phi}{\mathrm{d}z^2} = \frac{en_0}{\varepsilon_0}\left[\int \mathrm{d}pf - 1\right],\\
&A_z = 0,\\
&J_\perp(z,\omega) = -en_0 \int \mathrm{d}p\mathrm{v}_\perp f,\\
&\left[\frac{\mathrm{d}^2}{\mathrm{d}z^2} + \frac{\omega^2}{c^2}\right] A_\perp = -\mu_0 F_\mathrm{f} J_\perp.
\end{aligned}
\tag{7.3.6}
$$

In principle the physical result should not be affected by the gauge choice. From the point of view of the particles' dynamics what is important is the acting field and the latter is independent of the gauge choice. This is in particular easy to show in this 1D case: when using the Coulomb gauge, for a given source term $\rho = -en$, assuming functional dependence of the form $\exp(j\omega t - jkz)$, the Poisson equation dictates

$$
\Phi = -\frac{en}{\varepsilon_0 k^2}, \tag{7.3.7}
$$

and since the longitudinal component of magnetic vector potential vanishes the electric field reads

$$
E_z = -\frac{\mathrm{d}\Phi}{\mathrm{d}z} = -jk\frac{en}{\varepsilon_0 k^2}. \tag{7.3.8}
$$

When choosing a Lorentz gauge, the non-homogeneous wave equation dictates

$$
\Phi = \frac{en/\varepsilon_0}{(\omega/c)^2 - k^2}, \tag{7.3.9}
$$

and since Φ is related to A_z by $A_z = (\omega/c^2 k)\Phi$, the electric field reads

$$
E_z = -j\omega A_z + jk\Phi = \frac{c^2}{j\omega}\left[\frac{\omega^2}{c^2} - k^2\right]\frac{\omega}{c^2 k}\Phi. \tag{7.3.10}
$$

Substituting (7.3.9) in (7.3.10) we obtain the same expression for the electric field as in (7.3.8) i.e.,

$$
E_z = \frac{c^2}{j\omega}\left[\frac{\omega^2}{c^2} - k^2\right]\frac{\omega}{c^2 k}\Phi = -jk\frac{en}{\varepsilon_0 k^2}. \tag{7.3.11}
$$

In both cases, the magnetic field (associated with the space-charge) is zero: in the case of the Coulomb gauge it vanishes since A_z is zero and in the Lorentz gauge case, because A_z depends only on the z coordinate. Following the same procedure, it can be shown that the force term in Liouville equations [(7.3.1) and (7.3.6)] is identical in both cases.

7.3.1 The Dispersion Relation

The set of equations established previously will be analyzed in this subsection in order to quantify the energy exchange process. For this purpose, it is convenient to adopt a phasor notation for all linearized quantities. According to (7.2.7) F_p is proportional to $\exp[j\omega t - j(k + k_\mathrm{w})z]$. However, since in this case the Hamiltonian is time dependent, γ is not conserved and therefore we redefine $\tilde{F}_\mathrm{p}$ such that it does not include the γ term i.e.,

$$F_\mathrm{p} = -j\tilde{F}_\mathrm{p}\frac{1}{\gamma}\exp[j\omega t - j(k + k_\mathrm{w})z], \quad \tilde{F}_\mathrm{p} = \frac{e^2 A_0 B_\mathrm{w}(k + k_\mathrm{w})}{mk_\mathrm{w}}. \tag{7.3.12}$$

Consequently, assuming that f_0 does not vary in time and in space, a similar dependence as F_p is anticipated for f_1. Furthermore, since according to (7.3.2) Φ is linear in f_1, a similar dependence is expected for Φ and A_z:

$$\begin{aligned} f_1 &= \tilde{f}_1\exp(j\omega t - jKz), \\ \Phi &= \tilde{\Phi}\exp(j\omega t - jKz), \\ \mathbf{A} &= A_0(\mathbf{1}_x - j\mathbf{1}_y)\exp(j\omega t - jKz), \end{aligned} \tag{7.3.13}$$

where $K \equiv k + k_\mathrm{w}$. Substituting in (7.3.2) we obtain

$$\tilde{\Phi} = \frac{en_0/\varepsilon_0}{(\omega/c)^2 - K^2}\int \mathrm{d}p\tilde{f}_1, \tag{7.3.14}$$

and in a similar way we substitute in (7.3.1) to get

$$\tilde{f}_1 = \frac{1}{\delta\omega}\left[\frac{1}{\gamma}\tilde{F}_\mathrm{p} + \frac{e}{K}\left(K^2 - \frac{\omega^2}{c^2}\right)\tilde{\Phi}\right]\frac{\mathrm{d}f_0}{\mathrm{d}p}, \tag{7.3.15}$$

where $\delta\omega \equiv \omega - \mathrm{v}(k + k_\mathrm{w})$ is the resonance term. Substituting the latter into (7.3.14) we have, for the potential,

$$\tilde{\Phi} = \frac{m\omega_\mathrm{p}^2}{\varepsilon(\omega, K)}\frac{\tilde{F}_\mathrm{p}}{(\omega/c)^2 - K^2}\int \mathrm{d}p\frac{1}{\gamma\delta\omega}\frac{\mathrm{d}f_0}{\mathrm{d}p}, \tag{7.3.16}$$

where $\varepsilon(\omega, K)$ is the dielectric coefficient of the beam defined by

$$\varepsilon(\omega, K) = 1 + \frac{m\omega_{\mathrm{p}}^2}{K} \int \mathrm{d}p \frac{1}{\delta\omega} \frac{\mathrm{d}f_0}{\mathrm{d}p}. \tag{7.3.17}$$

The explicit expression for the amplitude of the scalar electric potential can be substituted into (7.3.15) which results in an expression for $\tilde{f}_1$:

$$\tilde{f}_1 = \frac{1}{\delta}\omega\left[\frac{1}{\gamma} - \frac{m\omega_{\mathrm{p}}^2}{K\varepsilon(\omega, K)}\left(\int \mathrm{d}p' \frac{1}{\gamma'\delta\omega'} \frac{\mathrm{d}}{\mathrm{d}p'} f_0(p')\right)\right] \frac{\mathrm{d}f_0}{\mathrm{d}p} \tilde{F}_{\mathrm{p}}. \tag{7.3.18}$$

The particles' density defines the current density via (7.3.4) whose linear term (in the radiation field) is $J_\perp = -en_0 \int \mathrm{d}p \mathrm{v}_\perp f_1$ or explicitly,

$$\begin{aligned} \mathbf{J}_\perp &= \frac{1}{2}\omega_{\mathrm{p}}^2 \varepsilon_0 \frac{B_{\mathrm{w}}}{k_w} \tilde{F}_{\mathrm{p}} (\mathbf{1}_x - j\mathbf{1}_y) \exp(j\omega t - jkz) \\ &\times \int \mathrm{d}p \frac{1}{\delta\omega} \frac{\mathrm{d}f_0}{\mathrm{d}p} \left[\frac{1}{\gamma} - \frac{m\omega_{\mathrm{p}}^2}{K\varepsilon(\omega, K)} \int \mathrm{d}p' \frac{1}{\gamma'\delta\omega'} \frac{\mathrm{d}}{\mathrm{d}p'} f_0(p')\right], \end{aligned} \tag{7.3.19}$$

where off-resonance terms of the form $\omega - \mathrm{v}(k - k_{\mathrm{w}})$ were neglected and $\delta\omega' \equiv \omega - \mathrm{v}'(k + k_{\mathrm{w}})$. The current distribution from the above is the source term to the wave equation in (7.3.5) which, after being substituted, gives the dispersion relation

$$\begin{aligned} \left(\frac{\omega}{c}\right)^2 - k^2 &= -\frac{1}{2}\left(\Omega_{\mathrm{w}}\omega_{\mathrm{p}}\right)^2 F_{\mathrm{f}} km \\ &\times \int \mathrm{d}p \frac{1}{\gamma\delta\omega} \frac{\mathrm{d}f_0}{\mathrm{d}p} \left[\frac{1}{\gamma} - \frac{m\omega_{\mathrm{p}}^2}{K\varepsilon(\omega, K)} \int \mathrm{d}p' \frac{1}{\gamma'\delta\omega'} \frac{\mathrm{d}}{\mathrm{d}p'} f_0(p')\right]; \end{aligned} \tag{7.3.20}$$

Ω_{w} was defined in the context of (7.1.25). For a given initial distribution of particles, $f_0(p)$, this expression determines the relation between the ω and k in the system. In the remainder of this section we assume that the system operates as an amplifier which means that the frequency, ω, is set externally and the interaction determines the wave-number k.

The integrals in the dispersion relation indicate that there are two critical functions: (1) the resonance term $\delta\omega^{-1}$ and (2) the distribution function $f_0(p)$. At the simplest approximation one considers a distribution of particles which can be represented by the first two moments namely the average (longitudinal) momentum ($\langle p\rangle$) and its spread $\Delta p \equiv \sqrt{\langle p^2\rangle - \langle p\rangle^2}$. On the other hand, the "sharpness" of the resonance term is determined by the imaginary part of the wave-number – which is basically the gain and a priori, its value is not known. For solving the dispersion relation, it is instructive to consider two extreme regimes:

the first when the distribution of particles is much sharper than the resonance line. This will be referred to as the "cold beam operation" $[\mathrm{Im}(k)/|k| \gg \Delta \mathrm{v}/\langle \mathrm{v}\rangle]$. The other case corresponds to the opposite regime i.e., $\mathrm{Im}(k)/|k| \ll \Delta \mathrm{v}/\langle \mathrm{v}\rangle$ referred to as the "warm beam operation". We discuss the two in the following two subsections.

7.3.2 Cold Beam Operation

In the framework of the cold beam operation, we consider the extreme case namely, a Dirac delta function distribution,

$$f_0(p) = \delta(p - p_0). \tag{7.3.21}$$

With this distribution in mind, we can evaluate the three integrals in the dispersion relation. The integral

$$\int \mathrm{d}p \frac{1}{\gamma^2 \delta\omega} \frac{\mathrm{d}f_0}{\mathrm{d}p}, \tag{7.3.22}$$

can be simplified by integration by parts to read

$$\frac{1}{\gamma_0^2}\left[\int \mathrm{d}p \frac{\mathrm{d}}{\mathrm{d}p}\left(\frac{f_0}{\delta\omega}\right) - \int \mathrm{d}p f_0 \frac{\mathrm{d}}{\mathrm{d}p}\left(\frac{1}{\delta\omega}\right)\right]. \tag{7.3.23}$$

In this expression, it was assumed that γ^{-2} varies slower than the other two functions. The first term in (7.3.23) is zero and in the second, the distribution function varies slower than the resonance term, thus

$$\int \mathrm{d}p \frac{1}{\gamma^2 \delta\omega} \frac{\mathrm{d}f_0}{\mathrm{d}p} \simeq -\frac{K}{(\delta\omega)^2} \frac{1}{m\gamma_0^5}. \tag{7.3.24}$$

In a similar way,

$$\int \mathrm{d}p \frac{1}{\gamma \delta\omega} \frac{\mathrm{d}f_0}{\mathrm{d}p} \simeq -\frac{K}{(\delta\omega)^2} \frac{1}{m\gamma_0^4}, \quad \int \mathrm{d}p \frac{1}{\delta\omega} \frac{\mathrm{d}f_0}{\mathrm{d}p} \simeq -\frac{K}{(\delta\omega)^2} \frac{1}{m\gamma_0^3}. \tag{7.3.25}$$

With these results the dispersion relation reads

$$\left(\frac{\omega}{c}\right)^2 - k^2 = \frac{1}{2} \frac{\Omega_\mathrm{w}^2}{\gamma_0^2} \frac{\omega_\mathrm{p}^2 F_\mathrm{f}}{\gamma_0^3} \frac{K^2}{\varepsilon(\omega, K)} \frac{1}{(\delta\omega)^2}, \tag{7.3.26}$$

where

$$\varepsilon(\omega, K) = 1 - \frac{\omega_p^2}{\gamma_0^3(\delta\omega)^2} \tag{7.3.27}$$

and F_f is the filling factor defined in the context of (7.3.5). For a solution of this equation we examine two cases: firstly, when the space-charge effects are neglected and secondly, when their effect is taken into consideration.

No Space-Charge Effects. Firstly, we ignore the effect of the space-charge term in the dielectric coefficient of the beam. Following the same approach as in the case of the interaction in a slow-wave structure, we assume that the change, due to the interaction, of the vacuum solution is small and it is denoted by δk i.e., $\omega \gg c|\delta k|$. With this assumption, the dispersion relation is identical with the one in a traveling-wave amplifier and it reads

$$\delta k(\delta k - \Delta k)^2 = -K_0^3, \tag{7.3.28}$$

only that in this case

$$\begin{aligned} K_0^3 &= \frac{1}{4}\left(\frac{\Omega_w k_w}{\gamma_0}\right)^2 \left(\frac{\omega_p^2 F_f}{c^2 \beta_0^2 \gamma_0^3}\right)\left(1 + \frac{\omega}{ck_w}\right)^2 \left(\frac{\omega}{c}\right)^{-1}, \\ \Delta k &= \frac{\omega}{c}\frac{1}{\beta_{\|}} - \frac{\omega}{c} - k_w. \end{aligned} \tag{7.3.29}$$

Kroll (1978) was the first to point out the full equivalence between a free-electron laser and a traveling-wave amplifier. As in TWT, assuming that Δk and K_0 are independent, maximum gain occurs at resonance ($\Delta k = 0$) and it is given by

$$\mathrm{Im}(k) = \frac{\sqrt{3}}{2}K_0 = \frac{\sqrt{3}}{2}\left[\frac{1}{4}\left(\frac{\Omega_w k_w}{\gamma_0}\right)^2 \left(\frac{\omega_p^2 F_f}{c^2 \beta_0^2 \gamma_0^3}\right)\left(1 + \frac{\omega}{ck_w}\right)^2 \left(\frac{\omega}{c}\right)^{-1}\right]^{1/3}. \tag{7.3.30}$$

If we compare this result, as it stands, with the gain in a slow-wave structure we observe that the main difference is the fact that here K_0^3 is proportional to γ_0^{-5} and in the former it was proportional to γ_0^{-3}. However, for relativistic electrons, assuming that $\omega \gg ck_w$ and bearing in mind that at resonance

$$\frac{\omega}{c} \simeq k_w \frac{2\gamma_0^2}{1 + \Omega_w^2}, \tag{7.3.31}$$

we find that the coupling wave-number K_0 is

$$K_0^3 \simeq \frac{1}{2}\frac{\Omega_w^2}{1 + \Omega_w^2}\frac{F_f}{\beta_0^2 \gamma_0^3}\frac{\omega_p^2}{c^2}k_w, \tag{7.3.32}$$

which for a strong wiggler, $\Omega_{\mathrm{w}} \gg 1$, reads

$$K_0^3 \simeq \frac{1}{2} \frac{\omega_{\mathrm{p}}^2 F_{\mathrm{f}}}{c^2 \beta_0^2 \gamma_0^3} k_{\mathrm{w}} \simeq \frac{1}{2} \frac{eI\eta_0 F_{\mathrm{f}}}{mc^2} \frac{k_{\mathrm{w}}}{S_{\mathrm{el}}} \frac{1}{(\gamma_0 \beta_0)^3}. \tag{7.3.33}$$

This result indicates that for a given periodicity and strong wiggler the growth-rate *scales* with particle's energy (γ_0) as in a traveling-wave amplifier but still their numerical value can differ quite substantially. A difference between this expression and the one in (4.1.18) is that ω/c was replaced here by k_{w}. However, in slow-wave structures driven by relativistic electrons, the two are of the same order of magnitude. A more important difference regards the interaction impedance: in this case, it is simply

$$Z_{\mathrm{int}} = \eta_0 F_{\mathrm{f}}, \tag{7.3.34}$$

and since this might be substantially smaller than in a traveling-wave amplifier (based on metallic periodic structure) the gain per unit length in a FEL is typically smaller.

Before we consider the space-charge effect it is important to emphasize that the assumption $\Omega_{\mathrm{w}} \gg 1$ which leads to (7.3.33) should be considered only within the limited framework of the comparison with the traveling wave amplifier otherwise too large wiggler amplitude in a FEL has a detrimental effect on its performance which is clearly revealed when examining (7.1.23) since it reduces the operating frequency.

Space-Charge Effect. When the current density is high enough such that its effect on the dielectric coefficient $[\varepsilon(\omega, K)]$ of the beam is significant, we can simplify the dispersion for the forward propagating waves to read

$$\left(k - \frac{\omega}{c}\right)\left[\left(k + k_{\mathrm{w}} - \frac{\omega}{c}\frac{1}{\beta}\right)^2 - K_p^2\right] = -K_0^3, \tag{7.3.35}$$

where $K_p^2 \equiv \omega_{\mathrm{p}}^2 / \mathrm{v}_0^2 \gamma_0^3$. The space-charge waves in this case are characterized by

$$F(k) \equiv \left(k + k_{\mathrm{w}} - \frac{\omega}{c}\frac{1}{\beta}\right)^2 - K_p^2 = 0. \tag{7.3.36}$$

Thus expanding this expression in conjunction with the FEL resonance condition, we obtain

$$\begin{aligned} F(k) &\simeq F\left(k = \frac{\omega}{c}\right) + \left(k - \frac{\omega}{c}\right)\left[\frac{\mathrm{d}}{\mathrm{d}k} F(k)\right]_{k=\omega/\mathrm{v}_0 - k_{\mathrm{w}} + K_p} \\ &\simeq 2K_p\left(k - \frac{\omega}{c}\right). \end{aligned} \tag{7.3.37}$$

This simplifies the dispersion relation to

$$\left(k-\frac{\omega}{c}\right)^2 \simeq -\frac{1}{2}\frac{K_0^3}{K_p}, \tag{7.3.38}$$

and the spatial growth rate is

$$\mathrm{Im}(k) = \frac{\sqrt{2}}{2}\sqrt{\frac{K_0^3}{K_p}} = \frac{\sqrt{2}}{2}\left[\frac{1}{2}\frac{\Omega_\mathrm{w}^2}{1+\Omega_\mathrm{w}^2}\frac{F_\mathrm{f}}{\beta_0\gamma_0^{3/2}}\frac{\omega_\mathrm{p}}{c}k_\mathrm{w}\right]^{1/2}. \tag{7.3.39}$$

The main difference between this regime and the former is that here the gain scales as $I^{1/4}$ compared to $I^{1/3}$. In addition, here the gain scales with energy of the electrons like $\gamma^{-7/4}$ compared to the $\gamma^{-5/3}$ in the former case.

7.3.3 *Warm Beam Operation*

So far, we have investigated the dispersion relation in an FEL with a mono-energetic beam of electrons. In this subsection, we examine the operation of the FEL with a warm beam as defined in the context of (7.3.20). For this case, we have to evaluate the integral

$$\int \mathrm{d}p \frac{1}{\delta\omega}\frac{1}{\gamma^2}\frac{\mathrm{d}}{\mathrm{d}p}f_0(p), \tag{7.3.40}$$

only that the resonance term varies more rapidly than the distribution term. In the evaluation of the integral we assume that k is a complex quantity i.e.,

$$k = k_\mathrm{r} + jk_\mathrm{i}, \tag{7.3.41}$$

hence

$$\int \mathrm{d}p \frac{\gamma^{-2}\mathrm{d}f_0(p)/\mathrm{d}p}{\omega - \mathrm{v}(k_\mathrm{r}+k_\mathrm{w}) - j\mathrm{v}k_\mathrm{i}} = \int \mathrm{d}p \frac{\gamma^{-2}(\mathrm{d}f_0/\mathrm{d}p)}{[\omega-\mathrm{v}(k_\mathrm{r}+k_\mathrm{w})]^2+[\mathrm{v}k_\mathrm{i}]^2}[\omega-\mathrm{v}(k_\mathrm{r}+k_\mathrm{w})+j\mathrm{v}k_\mathrm{i}]. \tag{7.3.42}$$

The main contribution is from the region where the resonance term peaks i.e., $\mathrm{v} = \mathrm{v}_\mathrm{res} \equiv \omega/(k_\mathrm{r}+k_\mathrm{w})$. This allows us to extract the slow varying term out of the integral such that we are left with

$$\left[\gamma^{-2}\frac{\mathrm{d}}{\mathrm{d}p}f_0(p)\right]_{\mathrm{v}=\mathrm{v}_{\mathrm{res}}}\int \mathrm{d}p\frac{\omega-\mathrm{v}(k_{\mathrm{r}}+k_{\mathrm{w}})}{[\omega-\mathrm{v}(k_{\mathrm{r}}+k_{\mathrm{w}})]^2+[\mathrm{v}k_{\mathrm{i}}]^2}$$
$$+\left[\gamma^{-2}\frac{\mathrm{d}}{\mathrm{d}p}f_0(p)\right]_{\mathrm{v}=\mathrm{v}_{\mathrm{res}}}\int \mathrm{d}p\frac{j\mathrm{v}k_{\mathrm{i}}}{[\omega-\mathrm{v}(k_{\mathrm{r}}+k_{\mathrm{w}})]^2+[\mathrm{v}k_{\mathrm{i}}]^2}. \tag{7.3.43}$$

The contribution of the first term (near resonance) vanishes because of the asymmetry of the integrand relative to $\mathrm{v}=\mathrm{v}_{\mathrm{res}}$ and the second's can be evaluated analytically,

$$\int \mathrm{d}p\frac{1}{\delta\omega\gamma^2}\frac{\mathrm{d}f_0}{\mathrm{d}p}\simeq\left[\gamma\frac{\mathrm{d}f_0}{\mathrm{d}p}\right]_{v=v_{res}}\frac{j\pi m}{k_{\mathrm{r}}+k_{\mathrm{w}}}. \tag{7.3.44}$$

In a similar way

$$\begin{aligned}\int \mathrm{d}p\frac{1}{\delta\omega\gamma}\frac{\mathrm{d}f_0}{\mathrm{d}p}&\simeq\left[\gamma^2\frac{\mathrm{d}f_0}{\mathrm{d}p}\right]_{v=v_{res}}\frac{j\pi m}{k_{\mathrm{r}}+k_{\mathrm{w}}},\\ \int \mathrm{d}p\frac{1}{\delta\omega}\frac{\mathrm{d}f_0}{\mathrm{d}p}&\simeq\left[\gamma^3\frac{\mathrm{d}f_0}{\mathrm{d}p}\right]_{v=v_{res}}\frac{j\pi m}{k_{\mathrm{r}}+k_{\mathrm{w}}}.\end{aligned} \tag{7.3.45}$$

With these integrals the dispersion relation reads

$$\left(\frac{\omega}{c}\right)^2-k^2=-\frac{1}{2}\left(\Omega_{\mathrm{w}}\frac{\omega_{\mathrm{p}}}{c}\right)^2F_{\mathrm{f}}.$$
$$\times\left\{\frac{j\left[\pi(mc)^2\gamma(\mathrm{d}f_0/\mathrm{d}p)\right]_{\mathrm{v}=\mathrm{v}_{\mathrm{res}}}}{1+j(\omega_{\mathrm{p}}/c)^2(\omega/c+k_{\mathrm{w}})^{-2}\left[\pi(mc)^2\gamma^3(\mathrm{d}f_0/\mathrm{d}p)\right]_{\mathrm{v}=\mathrm{v}_{\mathrm{res}}}}\right\} \tag{7.3.46}$$

As in the previous subsection, the gain without space-charge effects is calculated neglecting the plasma frequency term in the denominator and it reads

$$\mathrm{Im}(k)\simeq\frac{1}{4}\left(\frac{\omega}{c}\right)^{-1}\left(\Omega_{\mathrm{w}}\frac{\omega_{\mathrm{p}}}{c}\right)^2F_{\mathrm{f}}\left[\pi(mc)^2\gamma\frac{\mathrm{d}f_0}{\mathrm{d}p}\right]_{\mathrm{v}=\mathrm{v}_{\mathrm{res}}}. \tag{7.3.47}$$

When the space-charge effect is significant, the growth rate is given by

$$\mathrm{Im}(k)\simeq\frac{1}{4}\left(\frac{\omega}{c}\right)^{-1}\left\{\left(\Omega_{\mathrm{w}}\frac{\omega_{\mathrm{p}}}{c}\right)^2F_{\mathrm{f}}\left[\pi(mc)^2\gamma\frac{\mathrm{d}f_0}{\mathrm{d}p}\right]_{\mathrm{v}=\mathrm{v}_{\mathrm{res}}}\right\}$$
$$\times\left[1+\left(\frac{\omega_{\mathrm{p}}^2}{c^2K^2}\left[\pi(mc)^2\gamma^3\frac{\mathrm{d}f_0}{\mathrm{d}p}\right]_{\mathrm{v}=\mathrm{v}_{\mathrm{res}}}\right)^2\right]^{-1}, \tag{7.3.48}$$

which also determines a quantitative criterion for the regime when the space-charge effect is negligible i.e.,

$$\left(\frac{\omega_{\mathrm{p}}^2}{c^2 K^2}\left[\pi (mc)^2 \gamma^3 \frac{\mathrm{d}f_0}{\mathrm{d}p}\right]_{\mathrm{v}=\mathrm{v}_{\mathrm{res}}}\right)^2 \ll 1. \tag{7.3.49}$$

Note that the spatial growth rate in these two cases is proportional to the current and if we consider Gaussian-like electrons' distribution i.e., $f_0(p) \simeq \exp\left[-(p-p_0)^2/\Delta p^2\right]$, then the gain vanishes when the resonance velocity corresponds to the peak value of the distribution function.

7.4 Macro-Particle Approach

Electrons that experience an electric field have a momentum which is either larger or smaller than the average momentum of the beam. Since the system is designed to operate as an amplifier, the number of electrons which have energy below the average of the beam is larger than those which are faster and the energy difference is transferred to the electromagnetic field. In addition, exactly as the gain is associated with the imaginary part of the wave-number, its real part changes the effective phase velocity of the wave and after a certain interaction length the electrons may be out of phase. Consequently, electrons, which at the beginning of the interaction region were decelerated are now accelerated and vice versa. At the point in space where the slow electrons start to be accelerated because of the phase slip, they drain energy from the electromagnetic field whose growth saturates and beyond it, the gain decreases. In order to avoid this situation it is required to adjust the relative phase between the wave and the electrons. In the FEL this can be done by adjusting the wiggler period or amplitude (or both). Because of the large energy spread, fluid or kinetic approaches are inadequate and we then use the macro-particle approach, which will be presented in this section. For free-electron lasers this approach was initially developed by Kroll et al. (1981) and in this section we follow the essentials of their approach.

7.4.1 *Basic Formulation*

Assuming that space-charge effects are negligible, the scalar electric potential and the longitudinal magnetic vector potential can be omitted from the expression for the Hamiltonian, thus

$$H = \sqrt{(p_\perp + eA_\perp)^2 c^2 + p_\parallel^2 c^2 + (mc^2)^2} = mc^2\gamma \tag{7.4.1}$$

The transverse magnetic vector potential has two components: the wiggler and radiation field i.e.,

$$\mathbf{A}_{\perp} = \mathbf{A}_{\mathrm{w}} + \mathbf{A}_{\mathrm{rf}}. \tag{7.4.2}$$

Since it will be necessary to adapt the wiggler parameters to the local conditions in order to keep the electron in resonance, we consider a wiggler with variable amplitude and wave-number namely,

$$\mathbf{A}_{\mathrm{w}} = -A_{\mathrm{w}}(z)\left[\mathbf{1}_{\mathrm{x}}\cos\left(\int_0^z \mathrm{d}z' k_{\mathrm{w}}(z')\right) + \mathbf{1}_{\mathrm{y}}\sin\left(\int_0^z \mathrm{d}z' k_{\mathrm{w}}(z')\right)\right]; \tag{7.4.3}$$

in a similar way, the magnetic vector potential which describes the radiation field has an amplitude $A_0(z)$ which varies in space and so does its wave-number:

$$\mathbf{A}_{\mathrm{rf}} = A_0(z)\left[\mathbf{1}_x\cos\left(\omega t - \int_0^z \mathrm{d}z' k(z')\right) + \mathbf{1}_y\sin\left(\omega t - \int_0^z \mathrm{d}z' k(z')\right)\right]. \tag{7.4.4}$$

The latter has two components, the wave-number in vacuum (ω/c) and the effect of the interaction $[\theta(z)]$. Therefore, $\int_0^z \mathrm{d}z' k(z') = (\omega/c)z + \theta(z)$; this is to say that θ is the phase accumulated by the wave due to the interaction. As in the previous sections the wiggler is assumed to be uniform in the transverse direction therefore the canonical momentum in these directions is conserved; for simplicity it will be assumed to be zero ($p_{\perp} = 0$) hence

$$\mathrm{v}_{\perp} = \frac{e}{m\gamma}A_{\perp} \simeq \frac{e}{m\gamma}A_{\mathrm{w}} \tag{7.4.5}$$

where in the second expression it is assumed that the contribution of the radiation field to the transverse motion is negligible.

After substituting the expressions for the magnetic vector potentials into the Hamiltonian we obtain

$$H = mc^2\sqrt{\mu^2 + \left(\frac{p_{\parallel}}{mc}\right)^2 - 2a_{\mathrm{w}}(z)a_{\mathrm{rf}}(z)\cos\psi} = mc^2\gamma, \tag{7.4.6}$$

where

$$\begin{aligned} &a_{\mathrm{w}}(z) = \frac{eA_{\mathrm{w}}(z)}{mc}, \quad a_{\mathrm{rf}}(z) = \frac{eA_{\mathrm{rf}}(z)}{mc}, \quad K(z) = k(z) + k_{\mathrm{w}}(z), \\ &\mu^2(z) = 1 + a_{\mathrm{w}}^2(z) + a_{\mathrm{rf}}^2(z), \quad \psi(z,t) = \omega t - \int_0^z \mathrm{d}z' K(z'). \end{aligned} \tag{7.4.7}$$

The last expression represents the phase between the wave and the particle, when the presence of the wiggler is accounted for. At the transition from the Hamiltonian in (7.4.1)–(7.4.6) no approximations were made other than $p_\perp = 0$. Note that μ^2 plays the role of a normalized effective mass of the electron, which is z-dependent but not time dependent. This Hamiltonian enables us to determine a relatively simple expression for the longitudinal velocity of the particle; this is given by

$$\begin{aligned} \mathrm{v}_\parallel &= \frac{\mathrm{d}z}{\mathrm{d}t} = \frac{\partial H}{\partial p_\parallel} = \frac{p_\parallel}{mc^2\sqrt{\mu^2 + \left(p_\parallel/mc\right)^2 - 2a_\mathrm{w}(z)a_\mathrm{rf}(z)\cos\psi}} \\ &= \frac{c}{\gamma}\sqrt{\gamma^2 - \mu^2 + 2a_\mathrm{w}(z)a_\mathrm{rf}(z)\cos\psi} \end{aligned} \tag{7.4.8}$$

where in the last expression we used (7.4.6) to express $p_\parallel$ in terms of γ.

Since we are interested in the operation of the system as an amplifier, it is assumed that the frequency is determined by the external source and only spatial variations are allowed. Consequently, we follow the particle in space and we consider the time it takes the ith particle to reach a point z starting from $z = 0$; this time interval is denoted by $\tau_i(z)$. Regarding the phase dynamics the situation seems at a first glance more complicated by the three dimensional motion of the electron (compared to 1D in the slow-wave structure). However, in practice, we need only the projection of the motion along the wave propagation and this fact simplifies the calculation substantially as will be shown next.

The phase between the wave and the particle is given by $\psi_i(z) = \omega\tau_i(z) - \int_0^z \mathrm{d}z' K(z')$ thus the dynamics of the phase ψ in space reads

$$\frac{\mathrm{d}\psi_i(z)}{\mathrm{d}z} = \omega\frac{\mathrm{d}\tau_i}{\mathrm{d}z} - K(z). \tag{7.4.9}$$

Now, the derivative of τ with respect to z is inversely proportional to the *longitudinal* component of the velocity as determined in (7.4.8) hence,

$$\frac{\mathrm{d}\psi_i(z)}{\mathrm{d}z} = \frac{\omega}{c}\frac{\gamma_i}{\sqrt{\gamma_i^2 - \mu^2 + 2a_\mathrm{w}a_\mathrm{rf}\cos\psi_i}} - K(z). \tag{7.4.10}$$

As the velocity of the particle varies in space, so does its energy which satisfies

$$mc^2\mathrm{v}_\parallel\frac{\mathrm{d}\gamma_i}{\mathrm{d}z} = -e\mathbf{v}\cdot\mathbf{E}, \tag{7.4.11}$$

and since (in phasor notation) $E_\perp = -j\omega\mathbf{A}_\mathrm{rf}$ we can use (7.4.8) to write

$$\frac{\mathrm{d}\gamma_i}{\mathrm{d}z} = -\frac{1}{2}\left[\frac{j\frac{\omega}{c}a_\mathrm{rf}a_\mathrm{w}\exp(j\psi_i)}{\sqrt{\gamma_i - \mu^2 + 2a_\mathrm{w}a_\mathrm{rf}\cos\psi_i}} + \mathrm{c.c.}\right]. \tag{7.4.12}$$

Next we determine the dynamics of the amplitude of the radiation field. The starting point is the non-homogeneous wave equation in (7.3.5). Its source term is the current density which in the framework of the present approach is

$$\mathbf{J}_{\perp}(\mathbf{r},t) = -e\sum_i \mathbf{v}_{\perp,i}\delta[x-x_i(t)]\delta[y-y_i(t)]\delta[z-z_i(t)]; \tag{7.4.13}$$

in particular, we can substitute the explicit expression for the magnetic vector potential and from the x-component of the wave equation we obtain

$$\left[\frac{\partial^2}{\partial z^2}-\frac{1}{c^2}\frac{\partial^2}{\partial t^2}\right]A_0(z)\cos\psi(z,t) = A_0(z)\cos\psi(z,t)\left[\frac{\omega^2}{c^2}-\left(\frac{\omega}{c}+d_z\theta\right)^2\right] + 2d_zA_0(z)\sin\psi(z,t)\left(\frac{\omega}{c}+d_z\theta\right). \tag{7.4.14}$$

In a similar way we substitute the explicit expression for $\mathbf{v}_{\perp}$ from (7.4.5) in the x component of the current density and write

$$J_x(z,t) = ec\sum_i \frac{a_{\rm w}}{\gamma_i}\cos\left[\int_0^z \mathrm{d}z' k_{\rm w}(z')\right]\frac{F_{\rm f}}{S_{\rm el}}\delta[z-z_i(t)], \tag{7.4.15}$$

where we have already averaged out over the beam cross-section and the filling factor was included [see (7.3.5)]. Note that second derivatives of A_0 and θ were neglected in (7.4.14).

The coefficients of the trigonometric functions are time independent and therefore, we use the orthogonality of the trigonometric functions to average the wave equation over one period (T) of the wave. First, we take advantage of the orthogonality of the $\cos(\omega t\cdots)$ function to obtain

$$-2\frac{\omega}{c}\left(\frac{d\theta}{dz}\right)A_0\frac{1}{2} = -\mu_0F_{\rm f}\frac{1}{S_{\rm el}}\frac{1}{T}\int \mathrm{d}t\cos\left[\omega t-\int \mathrm{d}z' k(z')\right] \times\left\{ec\sum_i \frac{a_{\rm w}}{\gamma_i}\cos\left[\int_0^z \mathrm{d}z' k_{\rm w}(z')\right]\delta[z-z_i(t)]\right\}, \tag{7.4.16}$$

and second, using the orthogonality of $\sin(\omega t\cdots)$ we have

$$2\frac{\omega}{c}\frac{dA_0}{dz}\frac{1}{2} = -\mu_0F_{\rm f}\frac{1}{S_{\rm el}}\frac{1}{T}\int \mathrm{d}t\sin\left[\omega t-\int \mathrm{d}z' k(z')\right] \times\left\{ec\sum_i \frac{a_{\rm w}}{\gamma_i}\cos\left[\int_0^z \mathrm{d}z' k_{\rm w}(z')\right]\delta[z-z_i(t)]\right\}. \tag{7.4.17}$$

Assuming that no electrons are reflected backwards, the time integral can be readily evaluated using the Dirac delta function and if only slowly varying (resonant) terms are kept, then (7.4.16) reads

$$-\frac{\omega}{c}\left(\frac{d\theta}{dz}\right)A_0 = -\frac{e^2\mu_0}{cmT}\frac{F_f}{S_{el}}A_w\frac{1}{2}\sum_i \frac{1}{\gamma_i\beta_{\|,i}}\cos\psi_i(z). \tag{7.4.18}$$

The summation in this case is over all particles in one period of the wave and assuming that there are N such particles, we can write $\sum_i \cdots = N\langle\cdots\rangle$. Since the average beam density is given by $n_0 = N/S_{el}cT$, we can finally write

$$A_0\frac{d\theta}{dz} = \frac{1}{2}\frac{\omega_p^2}{c^2}F_f A_w\left(\frac{\omega}{c}\right)^{-1}\left\langle\frac{\cos\psi_i(z)}{\gamma_i\beta_{\|,i}}\right\rangle. \tag{7.4.19}$$

Following exactly the same procedure, we have for (7.4.17)

$$\frac{dA_0}{dz} = -\frac{1}{2}\frac{\omega_p^2}{c^2}F_f A_w\left(\frac{\omega}{c}\right)^{-1}\left\langle\frac{\sin\psi_i(z)}{\gamma_i\beta_{\|,i}}\right\rangle. \tag{7.4.20}$$

If d is the total length of the interaction region, it is convenient to use the following set of normalized quantities: $\zeta = z/d$, $\Omega = (\omega/c)d$, $K_w = k_w d$, $\bar{a}_{rf} = a_{rf}\exp(j\theta)$ and $\alpha = \frac{1}{2}(\omega_p d/c)^2 F_f$. With these definitions, there are two equivalent ways to formulate the interaction: either in terms of complex variables

$$\begin{aligned}
\frac{d}{d\zeta}\bar{a}_{rf} &= -j\frac{\alpha}{\Omega}a_w\left\langle\frac{e^{-j\chi_i}}{\gamma_i\beta_{\|,i}}\right\rangle,\\
\frac{d}{d\zeta}\gamma_i &= -\frac{1}{2}\left[\frac{j\Omega\bar{a}_{rf}a_w\exp(j\chi_i)}{\gamma_i\beta_{\|,i}} + \text{c.c.}\right],\\
\frac{d}{d\zeta}\chi_i &= \Omega\frac{1}{\beta_{\|,i}} - \Omega - K_w,\\
\beta_{\|,i} &= \frac{1}{\gamma_i}\sqrt{\gamma_i^2 - \mu^2 + a_w[\bar{a}_{rf}\exp(j\chi_i) + \text{c.c.}]},
\end{aligned} \tag{7.4.21}$$

Comment 7.1. One can average over the equation of motion of the particles and substitute the amplitude equation to get

$$\frac{d}{d\zeta}\left[\langle\gamma\rangle + \frac{1}{2\alpha}(\Omega a)^2\right] = 0, \tag{7.4.22}$$

which is the global energy conservation.

7.4.2 Resonant Particle Solution

Now that we have determined the equations which govern the dynamics of the electrons and the electromagnetic field in the presence of a quasi-periodic wiggler, we should be able to solve them provided that the initial conditions are known as well as the wiggler's parameters. However, we shall now make one step further and ask what the wiggler should be, for a given initial distribution of particles and electromagnetic field, that maximizes the energy extraction from the electrons. A general solution of this problem is difficult and practically impossible with analytical techniques. However, if the distribution of electrons occupies only a small region of the phase-space then the problem can be treated analytically.

For this purpose consider an ideally bunched beam such that we assume that all particles in one bunch move together forming a single macro-particle whose shape is preserved along the entire interaction region. Based on the equations of motion, the condition for maximum energy extraction is to keep it in correct phase with the wave i.e., maintain it in resonance. Assuming that at the input the macro-particle is in phase with the wave, the resonance along the interaction region will be defined as $\mathrm{d}\psi_{\mathrm{r}}/\mathrm{d}\zeta = 0$ and it translates into

$$\gamma_{\mathrm{r}}^2 \simeq \frac{1}{2}\frac{\Omega}{(K_{\mathrm{w}} + \theta')}\mu^2, \tag{7.4.23}$$

where the subscript r indicates resonance conditions. This expression becomes exact if we choose the resonance phase to be $\psi_{\mathrm{r}} = \pm\pi/2$ since $d\theta/d\zeta = 0$. In this subsection we consider an amplifier configuration so we take $\psi_{\mathrm{r}} = -\pi/2$ in which case, $\beta_{\|,\mathrm{r}} = \sqrt{1-(\mu/\gamma_{\mathrm{r}})^2}$. Substituting in the equation for the amplitude we obtain

$$\frac{\mathrm{d}}{\mathrm{d}\zeta}a = \frac{\alpha}{\Omega}\frac{a_{\mathrm{w}}}{\sqrt{\gamma_{\mathrm{r}}^2 - 1 - a^2 - a_{\mathrm{w}}^2}}. \tag{7.4.24}$$

Bearing in mind that the total energy is conserved i.e., $\gamma_{\mathrm{r}}(\zeta) + [\Omega a(\zeta)]^2/2\alpha = \varepsilon \equiv \gamma_{\mathrm{r}}(0) + [\Omega a(0)]^2/2\alpha$, we substitute the expression for γ_{r} to get

$$\frac{\Omega}{\alpha}\frac{1}{a_{\mathrm{w}}}\int_{a(0)}^{a(\zeta)}\mathrm{d}x\sqrt{\left[\varepsilon - \frac{1}{2}\frac{\Omega^2}{\alpha}x^2\right]^2 - 1 - x^2 - a_{\mathrm{w}}^2} = \zeta, \tag{7.4.25}$$

This equation is solved numerically for a constant a_{w} (but variable k_{w}) assuming a 3 mm beam radius which carries 100 A current, the filling factor being $F_{\mathrm{f}} = 0.1$. The total interaction length is 5 m, at the entrance the wiggler period is $L = 2\mathrm{cm}$ and its amplitude is $B_{\mathrm{w}} = 0.2\mathrm{T}$. The initial energy of the electrons is $4.6MeV \times [\gamma(0) = 10]$ and the normalized amplitude of the radiation field is $a_{\mathrm{r}}(0) = 8.4 \times 10^{-5}$. The result is illustrated in Fig. 7.2: the upper left frame shows the way in which the amplitude should grow in space. The variation in space of γ (upper

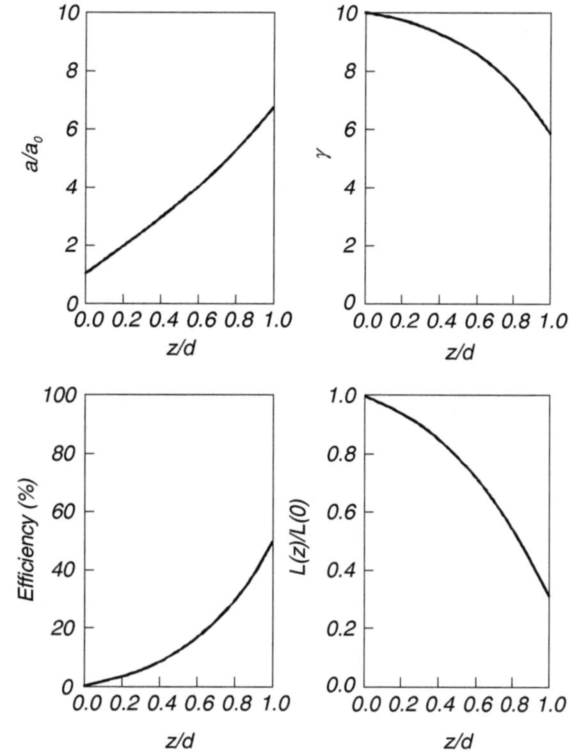

Fig. 7.2 Variation in space of the amplitude, energy, efficiency and period. All correspond to the resonant particle model

right frame) is calculated from the global energy conservation and the efficiency is illustrated in the lower left frame. The last frame illustrates the required variation in space of the period of the wiggler. It indicates that in order to achieve a 50% efficiency at 100 μm, the period of the wiggler has to be reduced from 2 cm to 0.6 cm and the intensity of the magnetic field enhanced to almost 0.7 T.

7.4.3 Buckets

In practice, any bunch has a finite spread in energy and phase. Let us denote the deviations from the resonant phase by $\delta\psi_i \equiv \psi_i - \psi_\mathrm{r}$ and from resonant energy by $\delta\gamma_i \equiv \gamma_i - \gamma_\mathrm{r}$. Based on the equations developed in the last subsection these two quantities satisfy

$$\frac{\mathrm{d}}{\mathrm{d}\zeta}\delta\psi_i = -\frac{(1+\beta_{\|,\mathrm{r}})K_\mathrm{w}}{\gamma_\mathrm{r}\beta_{\|,\mathrm{r}}}\delta\gamma_i, \tag{7.4.26}$$

and

$$\frac{\mathrm{d}}{\mathrm{d}\zeta}\delta\gamma_i = \Omega a_{\mathrm{w}} a_{\mathrm{r}} \frac{1}{\gamma_{\mathrm{r}}\beta_{\|,\mathrm{r}}} \cos\psi_{\mathrm{r}}\delta\psi_i, \tag{7.4.27}$$

provided that the deviations are small. For $\cos\psi_{\mathrm{r}} > 0$ the trajectories are stable and oscillate around the resonance point $(\psi_{\mathrm{r}}, \gamma_{\mathrm{r}})$ in the phase-space at a (spatial) "frequency"

$$\mathcal{D} = \frac{K_{\mathrm{w}}(1+\beta_{\|,\mathrm{r}})}{\mu\beta_{\|,\mathrm{r}}} \sqrt{a_{\mathrm{r}} a_{\mathrm{w}} \cos\psi_{\mathrm{r}}}. \tag{7.4.28}$$

These equations and the last result indicate that there is an entire range of trajectories around the resonance condition, which are stable. However, the analysis was limited to small deviation from resonance. We now reformulate the dynamics for the case when large deviations of the phase (ψ_i) are permitted. The phase equation has a similar form as (7.4.26)

$$\frac{\mathrm{d}}{\mathrm{d}\zeta}\psi_i = -\frac{(1+\beta_{\|,\mathrm{r}})K_{\mathrm{w}}}{\gamma_{\mathrm{r}}\beta_{\|,\mathrm{r}}}\delta\gamma_i, \tag{7.4.29}$$

but the equation for $\delta\gamma_i$ reads

$$\frac{\mathrm{d}}{\mathrm{d}\zeta}\delta\gamma_i = \frac{\Omega a_{\mathrm{w}} a_{\mathrm{r}}}{\gamma_{\mathrm{r}}\beta_{\|,\mathrm{r}}}[\sin\psi_i - \sin\psi_{\mathrm{r}}]. \tag{7.4.30}$$

It is convenient to redefine the phase as $\overline{\psi} \equiv -\psi$ and regard it as the canonical coordinate whereas $\delta\gamma_i$ is the canonical momentum. With these definitions, the Hamiltonian of the system reads

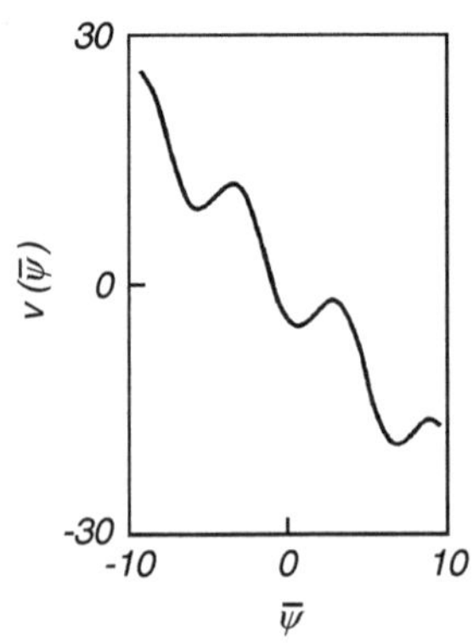

Fig. 7.3 Effective potential in whose minima electrons can be trapped

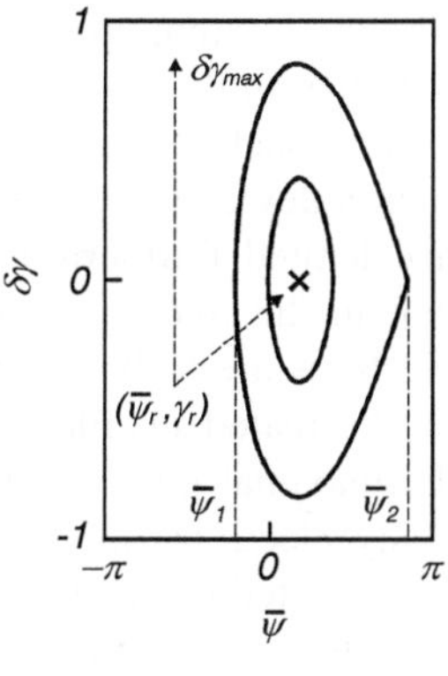

Fig. 7.4 The bucket limit (*outer curve*), a typical bucket shape (*inner curve*) and in the center the coordinates of the resonant particle are illustrated

$$H = \frac{1}{2}\frac{K_w(1+\beta_{\parallel,r})}{\gamma_r\beta_{\parallel,r}^2}\delta\gamma_i^2 - \Omega a_w a_r \frac{1}{\gamma_r\beta_{\parallel,r}}\left[\cos\bar{\psi}_i + \bar{\psi}_i \sin\bar{\psi}_r\right]$$
$$\equiv \frac{1}{2\mathcal{M}}\delta\gamma_i^2 - \frac{1}{2}\mathcal{K}\left[\cos\bar{\psi}_i + \bar{\psi}_i \sin\bar{\psi}_r\right] \qquad (7.4.31)$$

and it corresponds to a particle whose mass, $\mathcal{M}^{-1} \equiv K_w(1+\beta_{\parallel,r})/\gamma_r\beta_{\parallel,r}^2$, is z dependent which moves in a potential $V(\bar{\psi})$,

$$V(\bar{\psi}) = -\mathcal{K}\left[\cos\bar{\psi} + \bar{\psi}\sin\bar{\psi}_r\right], \qquad (7.4.32)$$

where $\mathcal{K} \equiv \Omega a_w a_r/\gamma_r\beta_{\parallel,r}$.

This potential is illustrated in Fig. 7.3 and it shows that particles can be trapped in its minima according to their initial conditions. The maximum stable trajectory of the particles is determined by the extrema of the potential in (7.4.32) and there are two sets of solutions: $\bar{\psi} = \bar{\psi}_r \pm 2\pi n$ or $\bar{\psi} = -\bar{\psi}_r + \pi \pm 2\pi n$. It is the latter which corresponds to the local maxima and thus represents the maximal value of a "bound state". Assuming that $\psi_r>0$ and that at the extremum the (canonical) momentum is zero, we find that the maximum value of H, for which the trajectories are still expected to be stable is given by

$$H_{\max} = \frac{1}{2}\mathcal{K}\left[\cos\bar{\psi}_r + (\bar{\psi}_r - \pi)\sin\bar{\psi}_r\right]. \qquad (7.4.33)$$

If ψ_r is negative, then π in this equation reverses its sign. Figure 7.4 illustrates the limits of the stable trajectories region (bucket) and a typical stable trajectory. Based on the maximal value of the Hamiltonian, one can also determine the maximal $\delta\gamma_{max}$ permissible for stable trajectory. It occurs at $\bar{\psi} = \bar{\psi}_r$ and it is given by

$$\delta\gamma_{\max} = 2\sqrt{\mathcal{MK}}\sqrt{\cos\bar{\psi}_r + \left(\bar{\psi}_r - \frac{\pi}{2}\right)\sin\bar{\psi}_r}. \qquad (7.4.34)$$

The two extreme phases of the bucket $(\bar{\psi}_1, \bar{\psi}_2)$ are determined in a similar way, only that in this case $\delta\gamma = 0$.

The bucket method infers that there is an inherent limit on the efficiency of such a device since only those electrons, which are trapped in the bucket, can be decelerated. Furthermore, in the context of a traveling-wave amplifier, it was shown that in the interaction process the area of the phase-space increases and if the electrons are to be "recycled" (ring configuration FEL), there are two conditions to be satisfied. (1) Elevate to maximum the bucket at the entrance in order to capture the maximal number of electrons and (2) minimize the energy spread, otherwise many electrons are lost in the next cycle. This kind of design was thoroughly investigated by Kroll et al. (1981). In what follows we investigate some additional aspects of the interaction and its manifestation in the phase-space.

7.4.4 Energy Spread

The set of equations as introduced in (7.4.21) can be reformulated in the case of a constant $a_{\rm w}$. From the equations of motion of the particles, we conclude that the quantity $\tilde{a} \equiv j\Omega\bar{a}_{\rm rf}a_{\rm w}$ is the effective (normalized) longitudinal electric field which acts on a single particle. Substituting this definition in the amplitude equation it is natural to redefine the normalized coupling coefficient α to read $\tilde{\alpha} \equiv \alpha a_{\rm w}^2$. With these definitions we have

$$
\begin{aligned}
\frac{\mathrm{d}}{\mathrm{d}\zeta}\tilde{a} &= \tilde{\alpha}\left\langle \frac{\mathrm{e}^{-j\chi_i}}{\gamma_i\beta_{\parallel,i}} \right\rangle, \\
\frac{\mathrm{d}}{\mathrm{d}\zeta}\gamma_i &= -\frac{1}{2}\left[\frac{\tilde{a}\exp(-j\chi_i)}{\gamma_i\beta_{\parallel,i}} + \mathrm{c.c.}\right], \\
\frac{\mathrm{d}}{\mathrm{d}\zeta}\chi_i &= \Omega\frac{1}{\beta_{\parallel,i}} - \Omega - K_{\rm w}, \\
\beta_{\parallel,i} &= \frac{1}{\gamma_i}\sqrt{\gamma_i^2 - \mu^2 + [-j\tilde{a}\exp(j\chi_i) + \mathrm{c.c.}]/\Omega}.
\end{aligned} \tag{7.4.35}
$$

From the second equation, we can develop the equation for the energy spread: $\Delta\gamma^2 \equiv \langle\gamma_i^2\rangle - \langle\gamma_i\rangle^2$. This is done by firstly averaging over all particles

$$
\frac{\mathrm{d}}{\mathrm{d}\zeta}\langle\gamma_i\rangle = -\frac{1}{2}\left[\tilde{a}\left\langle\frac{\exp(-j\chi_i)}{\gamma_i\beta_{\parallel,i}}\right\rangle + \mathrm{c.c.}\right]; \tag{7.4.36}
$$

secondly, we multiply by the local average $\langle\gamma_i\rangle$ to get

$$\frac{1}{2}\frac{\mathrm{d}}{\mathrm{d}\zeta}\langle\gamma_i\rangle^2 = -\frac{1}{2}\left[\tilde{a}\left\langle\frac{\exp(-j\chi_i)}{\gamma_i\beta_{\|,i}}\right\rangle\langle\gamma_i\rangle + \text{c.c.}\right]. \tag{7.4.37}$$

Next we repeat these steps but in the opposite order: we multiply the single particle equation of motion by γ_i and then average over the particles ensemble; the result is

$$\frac{1}{2}\frac{\mathrm{d}}{\mathrm{d}\zeta}\langle\gamma_i^2\rangle = -\frac{1}{2}\left[\tilde{a}\left\langle\frac{\exp(-j\chi_i)}{\beta_{\|,i}}\right\rangle + \text{c.c.}\right]. \tag{7.4.38}$$

Subtracting from the last expression (7.4.40) and using the definition of the energy spread, we obtain

$$\frac{\mathrm{d}}{\mathrm{d}\zeta}\Delta\gamma^2 = -\left\{\tilde{a}\left[\left\langle\frac{\exp(-j\chi_i)}{\gamma_i\beta_{\|,i}}\gamma_i\right\rangle - \left\langle\frac{\exp(-j\chi_i)}{\gamma_i\beta_{\|,i}}\right\rangle\langle\gamma_i\rangle\right] + \text{c.c.}\right\}. \tag{7.4.39}$$

We know from our analysis of traveling-wave amplifiers that the energy spread increases in the interaction process since part of the electrons are accelerated and others are decelerated. It is the same electromagnetic field, which causes the average deceleration (in an amplifier), and at the same time it accelerates a fraction of the particles causing the energy spread at the output. It was Madey (1979) who initially showed that in the low-gain Compton regime the energy spread at the output is directly related to the gain; at the input the energy spread is assumed to be negligible. Here we quote a result, which was revised, by Kroll et al. (1981) and it relates the gain $\langle\gamma_i(1)\rangle - \langle\gamma_i(0)\rangle$ to the energy spread at the output

$$\bar{\gamma}_1 - \bar{\gamma}_0 = \frac{1}{2}\frac{\mathrm{d}}{\mathrm{d}\bar{\gamma}_0}\Delta\gamma^2(1) \tag{7.4.40}$$

where $\bar{\gamma}_1 = \langle\gamma_i(1)\rangle$ and $\bar{\gamma}_0 = \langle\gamma_i(0)\rangle$. We will show now that the Madey theorem as formulated above for the low-gain Compton regime is related to the equation which describes the energy spread (7.4.39).

First, we note that in addition to $\Delta\gamma(\zeta)$, there are two other macroscopic quantities $\langle\gamma(\zeta)\rangle$ and $|\tilde{a}(\zeta)|$ which describe the system. Second, we bear in mind that these two are related via the energy conservation i.e., $\frac{\mathrm{d}}{\mathrm{d}\zeta}\left[\langle\gamma\rangle + |\tilde{a}|^2/2\tilde{\alpha}\right] = 0$, therefore we can use one of the two as an independent variable instead of ζ hence

$$\frac{\mathrm{d}}{\mathrm{d}\zeta}\Delta\gamma^2 = \left(\frac{\mathrm{d}\bar{\gamma}}{\mathrm{d}\zeta}\right)\frac{\mathrm{d}}{\mathrm{d}\bar{\gamma}}\Delta\gamma^2 \tag{7.4.41}$$

Comment 7.2. Although ζ does not occur explicitly in the right-hand side of the equation, it is implicitly there since we consider the values of $\Delta\gamma$ and $\bar{\gamma} \equiv \langle\gamma(\zeta)\rangle$ at the same location ζ.

Using the energy conservation, we obtain

$$\frac{\mathrm{d}}{\mathrm{d}\zeta}\Delta\gamma^2 = -\frac{1}{\bar{\alpha}}|\tilde{a}|\left(\frac{\mathrm{d}|\tilde{a}|}{\mathrm{d}\zeta}\right)\left(\frac{\mathrm{d}}{\mathrm{d}\bar{\gamma}}\Delta\gamma^2\right), \tag{7.4.42}$$

and for simplicity we define $\tilde{a}\mathrm{e}^{j\chi_i} \equiv |\tilde{a}|\mathrm{e}^{j\tilde{\chi}_i}$. With this definition and using the amplitude equation in (7.4.35) we obtain

$$\begin{aligned} \frac{\mathrm{d}}{\mathrm{d}\zeta}\Delta\gamma^2 &= -|\tilde{a}|\left\langle \cos(\tilde{\chi}_i)/(\gamma_i\beta_{\|,i})\right\rangle \frac{\mathrm{d}}{\mathrm{d}\bar{\gamma}}\Delta\gamma^2, \\ &= -2|\tilde{a}|\left[\left\langle \frac{\cos\tilde{\chi}_i}{\gamma_i\beta_{\|,i}}\gamma_i\right\rangle - \left\langle \frac{\cos\tilde{\chi}_i}{\gamma_i\beta_{\|,i}}\right\rangle\bar{\gamma}\right], \end{aligned} \tag{7.4.43}$$

where in the second expression we used (7.4.39). From the two right-hand side expressions we conclude that

$$\frac{1}{2}\frac{\mathrm{d}}{\mathrm{d}\bar{\gamma}}\Delta\gamma^2 = \bar{\gamma} - \frac{\langle\cos(\tilde{\chi}_i)/\beta_{\|,i}\rangle}{\langle\cos(\tilde{\chi}_i)/\gamma_i\beta_{\|,i}\rangle}, \tag{7.4.44}$$

and the resemblance with Madey's theorem becomes apparent. However, in contrast with the latter this relation is exact at any point in the interaction region. Furthermore, it is also valid in the high-gain Compton regime.

Under the simplifying conditions of low-gain Compton regime we may approximate the second expression in the right hand side with its value at the input

$$\frac{\langle\cos(\tilde{\chi}_i)/\beta_{\|,i}\rangle}{\langle\cos(\tilde{\chi}_i)/\gamma_i\beta_{\|,i}\rangle} \simeq \left[\frac{\langle\cos(\tilde{\chi}_i)/\beta_{\|,i}\rangle}{\langle\cos(\tilde{\chi}_i)/\gamma_i\beta_{\|,i}\rangle}\right]_{\zeta=0} = \bar{\gamma}(0). \tag{7.4.45}$$

thus

$$\frac{1}{2}\frac{\mathrm{d}}{\mathrm{d}\bar{\gamma}(1)}\Delta\gamma^2(1) \simeq \bar{\gamma}(1) - \bar{\gamma}(0). \tag{7.4.46}$$

Bearing in mind that in the low-gain Compton regime the energy transfer is small, we can replace $\bar{\gamma}(1)$ on the left-hand side with $\bar{\gamma}(0)$ to obtain

$$\frac{1}{2}\frac{\mathrm{d}}{\mathrm{d}\bar{\gamma}(0)}\Delta\gamma^2(1) \simeq \bar{\gamma}(1) - \bar{\gamma}(0), \tag{7.4.47}$$

which is exactly Madey's theorem as formulated in (7.4.40).

Although the Madey theorem relates the moments of the electrons' distribution function, it does not actually help us to calculate them and for this purpose we have to go back to the equations of motion. These were solved for a typical FEL set of parameters and the question we address now is how does the energy spread vary at the output of the device when the only other parameter that is changed is the energy spread at the input. The result of our simulation is presented in Fig. 7.5: the energy spread at the output $\Delta\gamma(1)$ as a function of the energy spread at the input, $\Delta\gamma(0)$ decreases for values of $\Delta\gamma(0)$ smaller than $\Delta\gamma(1)$. We observe that $\Delta\gamma(1)$ starts from a high value when the energy spread at the input is virtually zero. By increasing the latter we cause $\Delta\gamma(1)$ to decrease as does the gain. The latter vanishes when the energy spread at the output equals its value at the input. Any further increase of $\Delta\gamma(0)$ beyond this level does not change the gain and since the beam traverses the interaction region almost unaffected, $\Delta\gamma(1)$ increases linearly with $\Delta\gamma(0)$.

Another interesting aspect of the energy spread that we examine next is revealed when comparing the operation of an FEL and a TWT. In Sect. 7.3 it was shown that the dispersion relation of a free-electron laser is similar to that of a traveling wave tube but it was pointed out that the γ dependence of the gain is different in the two cases – a fact which may cause some differences in the operation of the two devices. In order to emphasize the similarities and the differences, we have summarized, in Table 7.1, the equations of the free-electron laser and the traveling-wave tube (TWT). In these equations each quantity which plays a similar role is denoted

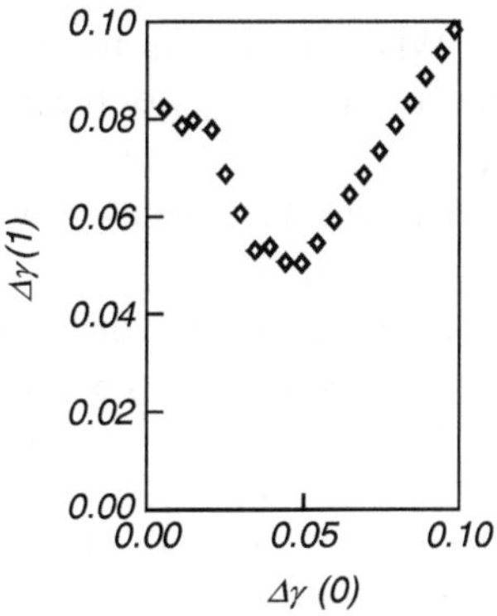

Fig. 7.5 Energy spread at the output as a function of the energy spread at the input

Table 7.1 Comparison of the TWT equation to these of an FEL

	TWT	FEL
Amplitude dynamics	$\frac{d}{d\zeta}a = \alpha\langle\exp(-j\chi_i)\rangle$	$\frac{d}{d\zeta}a = \alpha\left\langle\frac{\exp(-j\chi_i)}{\gamma_i\beta_{\parallel,i}}\right\rangle$
Equation of motion	$\frac{d}{d\zeta}\gamma_i = -\mathrm{Re}[a\exp(j\chi_i)]$	$\frac{d}{d\zeta}\gamma_i = -\frac{1}{\gamma_i\beta_{\parallel,i}}\mathrm{Re}[a\exp(j\chi_i)]$
Phase equation	$\frac{d}{d\zeta}\chi_i = \frac{\Omega}{\beta_i} - K$	$\frac{d}{d\zeta}\chi_i = \frac{\Omega}{\beta_i} - K$
Global energy conservation	$\frac{d}{d\zeta}\left[\langle\gamma\rangle + \frac{1}{2\alpha}\lvert a\rvert^2\right] = 0$	$\frac{d}{d\zeta}\left[\langle\gamma\rangle + \frac{1}{2\alpha}\lvert a\rvert^2\right] = 0$
Spatial growth rate	$q = \frac{\sqrt{3}}{2}\left[\frac{\alpha\Omega}{2}\left\langle(\gamma_i\beta_i)^{-3}\right\rangle\right]^{\frac{1}{3}}$	$q = \frac{\sqrt{3}}{2}\left[\frac{\alpha\Omega}{2}\left\langle(\gamma_i\beta_i)^{-5}\right\rangle\right]^{\frac{1}{3}}$

(intentionally) with the same notation: γ_i represents the energy of the ith particle and χ_i is its relative phase to the wave. In both cases a represents the normalized amplitude of the electromagnetic field; however in the TWT this is the longitudinal electric field ($a = eE_z d/mc^2$) whereas in the FEL it is the transverse field $[a = ja_w a_{rf} \omega d/c]$. The normalized coupling coefficient is denoted with α and it consists of different quantities: for the FEL α used here is $\alpha = \frac{1}{2}(\omega_p d/c)^2 F_f a_w^2$ and for the TWT $\alpha = (eIZ_{int}/mc^2)(d^2/\pi R^2)$. In the phase equation the longitudinal velocity is denoted by β_i and in the TWT case it is related to γ_i via $\beta_i = \sqrt{1 - 1/\gamma_i^2}$ whereas in the FEL case $\beta_i = \sqrt{1 - \mu^2/\gamma_i^2 + [-ja\exp(j\chi_i) + c.c.]/\Omega\gamma_i^2}$ and $\mu^2 = 1 + a_w^2 + |a|^2/a_w^2\Omega^2$. In addition, $\Omega = \omega d/c$ and $K = kd$ in the TWT case and $K = (k + k_w)d$ in the FEL.

The general form of the equations is similar for both devices. In fact, the form of the global energy conservation law is identical. The major difference is the momentum term which occurs in the phase terms and which is not there in the TWT case. It was indicated previously that this term originates in the transverse oscillation that the wiggler forces the electrons to undergo and associated with that is an "effective relativistic transverse mass" of the electron which is $m\gamma$ in contrast to $m\gamma^3$ associated with the longitudinal motion.

When comparing the TWT and FEL three parameters have to be the same: (1) the average energy of the electrons $\langle\gamma\rangle$, (2) the electromagnetic energy per particle, $|a|^2/2\alpha$ and (3) the total gain. Two cases have been examined. In the first, the total length and the spatial growth rate were assumed to be the same but at the input $a_{TWT} = a_{FEL}/\langle\gamma\beta\rangle$ and $\alpha_{TWT} = \alpha_{FEL}/\langle\gamma\beta\rangle^2$ in order to satisfy the conditions above. In the linear regime, the two devices operated practically the same and Fig. 7.5 also represents the energy spread of the TWT.

In the second case examined, it was assumed that a at the input is the same in both devices and consequently, from assumption (2), so is α. As a result, the spatial growth rate is smaller in the FEL by (roughly) a factor of $1/(\gamma\beta)^{2/3}$. Therefore, in order to satisfy the constraint in (3), we increase the length of the FEL by a factor of $(\gamma\beta)^2$ such that $\Omega_{FEL} = \Omega_{TWT}\langle\gamma\beta\rangle^2$ and $K_{FEL} = K_{TWT}\langle\gamma\beta\rangle^2$ the result is illustrated in Fig. 7.6 where we plotted the gain as a function of the energy spread at the input. The

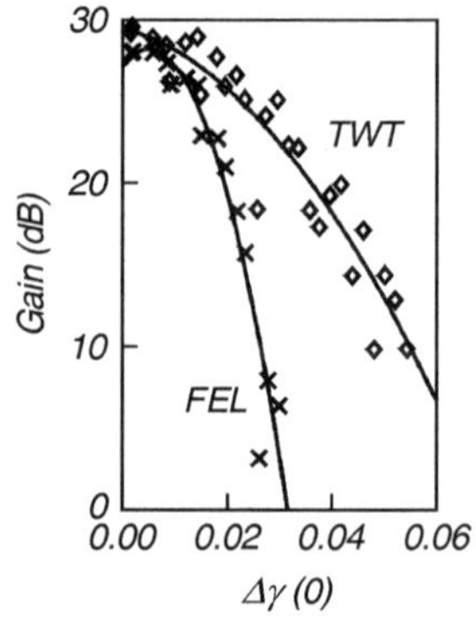

Fig. 7.6 Gain as a function of the energy spread at the input in a free-electron laser and in a traveling wave tube

simulation reveals a clear sensitivity of the free-electron laser to initial energy spread compared to the TWT in the conditions determined above. The situation is even worse at higher energies.

7.5 Other FEL Schemes

One of the major advantages of free-electron lasers is the fact that no external means are required to confine the radiation. In fact Scharlemann et al. (1985) have shown that under certain conditions the beam acts like an optical fiber and guides the radiation. Later Sprangle et al. (1987) formulated the three-dimensional problem introducing the source-dependent expansion technique. Beam guidance may become crucial in two cases: in the case of very high power radiation where the Ohm loss of walls makes the contact with the intense radiation field prohibitive. And in the case of very high frequency e.g., Ultra Violet or X-rays, where even if contact of the radiation with a metallic surface is permissible, from the perspective of the power levels, the reaction of the surface is not as regular as at low frequencies (visible and below). That is to say, that the surface quality is poor since micro-perturbations are of the same size as the radiation wavelength. For the reasons mentioned above, the FEL has the potential to generate coherent and tunable radiation at short wavelengths such as UV and X-rays. There are, however, two major obstacles in its way: the beam quality, which is a major limitation and the wiggler. As for the first, it was shown in this chapter (Sects. 7.2 and 7.3) that the gain depends strongly on the temperature of the beam. This problem becomes acute at high frequencies. In addition, transverse beam effects (emittance), which were not discussed here, start to play an important role. The constraints imposed on the wiggler are also stringent. Obviously, the shorter the period the better. However, a short period implies a low intensity magnetic field that in turn implies a long interaction length. When a large number of magnets are involved, two problems occur: alignment and statistical errors in the intensity of each pole. While the first can be minimized, the latter is unavoidable. In this section, we briefly review alternative configurations that aim to overcome some of these difficulties.

7.5.1 Gas Loaded FEL

In order to release some of the constraints on the beam and wiggler Pantell et al. (1986) and Feinstein et al. (1986) have shown that there are substantial advantages to slowing down the phase velocity of the wave by loading the FEL with gas since at the frequency of interest the refraction coefficient, n, is larger than unity. The resonance condition in this case is

$$\frac{\omega}{\mathrm{v}_{\parallel}} = n(\omega)\frac{\omega}{c} + k_{\mathrm{w}}, \tag{7.5.1}$$

and the resonance frequency reads

$$\frac{\omega}{c} = k_{\mathrm{w}}\left[\frac{1}{\sqrt{1-\mu^2/\gamma^2}} - n\right]. \tag{7.5.2}$$

In order to emphasize the effect we shall examine two cases. Firstly we consider a vacuum system ($n = 1$) with $\mu = 1 + (eB_0/mck_{\mathrm{w}})^2 = 1.2$ and $\gamma = 100$. If the period of the wiggler is 5 cm then the radiation wavelength is 3.6 μm. If the refraction coefficient of the gas is $n - 1 = 7 \times 10^{-5}$ then the radiation wavelength is 1,000 Å. Therefore, the presence of the medium caused a frequency shift from infra red to UV.

7.5.2 *Longitudinal Wiggler FEL*

Another free-electron laser configuration which was considered by McMullin and Bekefi (1981, 1982) consists of a longitudinal rather than a transverse wiggler. In this case the guiding magnetic field is rippled and it is approximately given by

$$\mathbf{B} = B_0\mathbf{1}_z + B_1\mathrm{I}_0(k_{\mathrm{w}}r)\sin(k_{\mathrm{w}}z)\mathbf{1}_z - B_1\mathrm{I}_1(k_{\mathrm{w}}r)\cos(k_{\mathrm{w}}z)\mathbf{1}_r. \tag{7.5.3}$$

Both the wigglers as well as the guiding field control the transverse motion of the electrons. Consequently, the resonance condition reads

$$\omega = \frac{\Omega_c}{\gamma} + (k + k_{\mathrm{w}})\mathrm{v}_{\parallel}, \tag{7.5.4}$$

where $\Omega_c = eB_0/m$ is the non-relativistic cyclotron frequency. Assuming that the interaction only slightly affects the TEM mode i.e., $k \simeq \omega/c$, the resonant frequency is

$$\omega = \gamma^2 \frac{1+\beta_{\parallel}}{1+(\gamma\beta_{\perp})^2}\left(\Omega_c\frac{1}{\gamma} + k_{\mathrm{w}}c\beta_{\parallel}\right), \tag{7.5.5}$$

and it can be readily seen that without the guiding field ($\Omega_c = 0$) the resonance corresponds to that of a transverse wiggler FEL. Therefore, the guiding field causes an effective increase in the wave number of the wiggler by the factor $\Omega_c/c\gamma$ – which can be significant.

7.5.3 Rippled-Field Magnetron

The basic configuration of free-electron lasers discussed so far was co-linear in the sense that the dominant component of the electrons' velocity was in the longitudinal direction. Bekefi (1982) has suggested constructing a smooth bore magnetron where the interaction is facilitated by a wiggler rather than a slow-wave structure as is generally the case. The system consists of two cylindrical electrodes, an insulating magnetic field along the axis and a wiggler that is azimuthally periodic but its magnetic field is in the radial direction. A positive voltage V is applied on the anode. Electrons emitted from the cathode form a Brillouin flow around the cylinder provided that the intensity of the insulating magnetic field exceeds the critical value

$$B_z > B_{cr} = \frac{mc}{e\delta R}\sqrt{\gamma^2 - 1}, \tag{7.5.6}$$

where $\gamma = 1 + eV/mc^2$ and δR is the anode-cathode gap. This equilibrium is altered by a wiggler, which can be approximated by

$$B_r = B_0 \cos(N\phi), \tag{7.5.7}$$

where N is the number of magnetic poles and ϕ is the azimuthal coordinate. Conceptually the interaction is similar to the co-linear case however, the cylindrical configuration complicates the detailed analysis. Destler et al. (1985) tested the concept experimentally and good agreement with theoretical predictions was found.

7.5.4 Wiggler and Guiding Magnetic Field

In many cases, the electron beam is immersed in a guiding magnetic field even before it enters the wiggler field. It is therefore reasonable to calculate the trajectories of the electrons in a configuration which combines the two magnetic fields. Friedland (1980) calculated these trajectories for an idealized magnetic field and Freund et al. (1983) has improved the model for a more realistic configuration. The guiding field causes an increase in the transverse velocity of the electrons but the effect on the gain is strongly dependent on the parameters of the wiggler according to detailed trajectory of the electrons – two types were initially emphasized. Conde and Bekefi (1991) discovered that by inverting the direction of the guiding field a substantial improvement of the efficiency could be achieved.

7.5.5 Electromagnetic Wiggler

In principle the magneto-static wiggler can be replaced by an intense electromagnetic wave which propagates anti-parallel to the beam. This was in fact the original concept considered by Kapitza and Dirac back in 1933, however they investigated the interaction with a standing wave (wave-number was perpendicular to the beam).

In the process of interacting with the wave, one gains a factor of 2 in the frequency of the emitted radiation. Since the wiggler field varies in time (ω_{w}) the resonance condition reads

$$\omega - \omega_{\mathrm{w}} = (k + k_{\mathrm{w}})\mathrm{v}_{\|}. \tag{7.5.8}$$

Furthermore, both wiggler and emitted fields behave as free waves ($k \simeq \omega/c$ and $\omega_{\mathrm{w}} \simeq ck_{\mathrm{w}}$) therefore the resonance frequency reads

$$\omega = ck_{\mathrm{w}} \frac{1 + \sqrt{1 - \mu^2/\gamma^2}}{1 - \sqrt{1 - \mu^2/\gamma^2}} \simeq ck_{\mathrm{w}} \frac{4\gamma^2}{\mu^2}. \tag{7.5.9}$$

This concept was demonstrated by Carmel et al. (1983) when a high power microwave pulse was generated with a backward-wave oscillator and used in a second stage as an electromagnetic wiggler for a free-electron laser. We discuss this scheme in more detail in the next chapter.

7.5.6 Electrostatic Wiggler

One of the difficulties with magneto-static wigglers is that their period, for substantial field intensity, is of the order of cm's, therefore highly relativistic particles are required in order to achieve optical (or shorter) wavelengths. Even if the electrons with this energy are available, disregarding the cost of their acceleration, we still confront another problem which is: the scaling of the gain with γ. In a regular FEL, the electron undergoes a transverse motion under the influence of the transverse wiggler and since the coupling coefficient is quadratic in the wiggler field, this motion contributes a γ^{-2} term to this parameter. The pondermotive force modulates the beam in the longitudinal direction and this motion gives rise to an additional factor of $(\gamma\beta)^{-3}$. As a result, the coupling coefficient in the high-gain Compton regime is proportional to $(\gamma\beta)^{-5}$. For comparison in TWT the coupling coefficient is proportional to $(\gamma\beta)^{-3}$.

It is relatively easy to make a short period electrostatic wiggler with a period of a few microns and even shorter using photolithography techniques. However, the problem is that the gain is proportional to $\gamma^{-9}\beta^{-5}$. Originally, Gover (1980) did the calculation for the *low-gain* Compton regime. Here we present the analysis of the *high-gain* Compton regime.

In order to prove our previous statement let us consider an electrostatic potential of the following form

$$\phi(z, r \simeq 0) = \phi_0 \cos(k_w z), \tag{7.5.10}$$

and since all the discussion so far in this chapter considered only magnetic wigglers we shall present the analysis of this scheme in more detail.

The electron's motion has three components: the major one is the "dc", β, the second is due to this electrostatic potential (β_w) and the third is proportional to the radiation field, $\delta\beta$. If we ignore momentarily the radiation field then the motion of the electrons is longitudinal and it is given by

$$\beta_w = \frac{e\phi_0}{mc^2} \frac{1}{\beta\gamma^3} \cos(k_w z). \tag{7.5.11}$$

As in the regular FEL we expect the "resonance" motion to be determined by the product of the wiggler induced motion and the magnetic vector potential of the radiation field. The Hamiltonian of the system can be approximated by

$$H = H_0 + H_1 = mc^2\gamma + ec\beta_w A_{rf}, \tag{7.5.12}$$

and the pondermotive force $F_p = -\partial H_1/\partial z$. Assuming that the rf field is given by $A_{rf} = A\cos(\omega t - kz)$, neglecting off resonance terms and using a phasor notation we have

$$H_1 \simeq h \exp(j\omega t - jKz) \tag{7.5.13}$$

where $h = (ecA)(e\phi_0/2mc^2\beta\gamma^3)$ and $K = k + k_w$. The phasor of the oscillatory motion is therefore

$$\delta\beta = \frac{Kc}{\omega - K\mathrm{v}} \frac{h}{mc^2\gamma^3}. \tag{7.5.14}$$

Next we use the continuity equation to determine the particles' density and, as above, keeping only terms which may contribute to the resonant process, we have

$$\delta n = n_0 \frac{Kc}{\omega - K\mathrm{v}} \delta\beta. \tag{7.5.15}$$

The longitudinal current density is given by $J_z = -ec(n_w\delta\beta + \beta_w\delta n)$ and since the motion induced by the wiggler does not contribute to the net current we have $n_w = -n_0\beta_w/\beta$ which allows us to use the following approximation for the current density

$$J_z = -ecn_0 \frac{Kc}{\omega - K\mathrm{v}} \beta_\mathrm{w} \delta\beta,$$
$$\simeq -(ec)^2 n_0 \frac{(Kc)^2}{(\omega - K\mathrm{v})^2} \left(\frac{e\phi_0}{2mc^2\beta\gamma^3} \right)^2 \frac{1}{mc^2\gamma^3} A\exp(j\omega t - jkz). \tag{7.5.16}$$

This current density drives the magnetic vector potential which satisfies

$$\left[\frac{\mathrm{d}^2}{\mathrm{d}z^2} + k_0^2 \right] A_\mathrm{rf} = -\mu_0 F_\mathrm{f} J_z, \tag{7.5.17}$$

where F_f is the filling factor and k_0 is the wavenumber of the interacting wave (harmonic) in the absence of the beam. Substituting the expression for the current density, assuming resonance and that $k = k_0 + \delta k$ we obtain

$$\delta k^3 = -K_0^3 = -\frac{I\eta_0 F_\mathrm{f}}{8mc^2 S_{el}} \frac{(k_0 + k_\mathrm{w})^2}{k_0} \left(\frac{e\phi_0}{mc^2} \right)^2 \frac{1}{\beta^5\gamma^9}. \tag{7.5.18}$$

In this expression I is the total current carried by the beam and S_el denote its cross-section. It clearly reveals what we indicated previously that the coupling coefficient is proportional to $\gamma^{-9}\beta^{-5}$. If we assume the same frequency and compare the coupling coefficient in this case with that of a regular FEL [see (7.3.29)] we get

$$\frac{K_\mathrm{elec}^3}{K_\mathrm{mag}^3} \propto \frac{(\gamma^5\beta^4)_\mathrm{mag}}{(\gamma^9\beta^4)_\mathrm{elec}} \left(\frac{e\phi_0 k_\mathrm{w}}{eB_\mathrm{w}c} \right)^2. \tag{7.5.19}$$

If the two systems generate 5,000 Å radiation in the magnetic wiggler case for a period of 10 cm the electrons must have $\gamma_\mathrm{mag} \sim 3 \times 10^2$. In order to generate the same radiation with an electrostatic wiggler of 5 μm periodicity, the electrons must have $\gamma_\mathrm{elec} = 2.4 (\beta = 0.9)$. If on a metallic surface the amplitude of the first harmonic is $\phi_0 \sim 50\,\mathrm{V}$ and the intensity of the magnetic field is $B_\mathrm{w} = 0.1\,\mathrm{T}$, then the ratio of the two coupling coefficients is 10. However, in (7.5.19) the two filling factors were assumed to be the same and this is not generally the case. Only those electrons which are within 5 μm from the surface do interact; therefore, if the beam radius is $R_\mathrm{b} \sim 2\,\mathrm{mm}$, then the filling factor is $\sim 6 \times 10^{-5}$. Within 2–3 μm from the surface there is an exponential decay in the amplitude of the field by a factor of 10, thus a factor of 100 in the coupling coefficient. We may therefore expect the latter to be smaller by almost a factor of $K_\mathrm{elec}^3 / K_\mathrm{mag}^3 \sim 10^{-6}$. In principle one can increase the amplitude of the electrostatic wiggler to 50 kV in which case the two growth rates are comparable. Unfortunately, in this case we run into a breakdown problem since 50 kV applied on a structure of 5 μm period generate (dc) gradients of the order of 10 GV/m or higher. Once breakdown occurs, the wiggler is short circuited and the gain vanishes. Consequently, the design of such a system is a trade-off between high voltage requirements dictated by the high gain requirement and low voltage regime imposed by the need to avoid breakdown.

7.5.7 Channeling Radiation

All the wigglers mentioned so far were on the macroscopic scale i.e., order of cm's and down to the micron level. However, the lattice of a crystalline material forms a natural periodic electrostatic wiggler. If a beam of relativistic electrons is injected parallel into one of the symmetry planes of a lattice then it sees two periodicities. One in the longitudinal direction that is negligible since the intensity of the potential is too weak in order to bunch the beam, which is to say that the longitudinal momentum of the particle is many orders of magnitude larger than the quanta of lattice momentum. The other periodic system is in the transverse direction (Berry 1971). If the electrons have a small momentum in this direction, they are "reflected" by the lattice plane and they undergo an oscillatory motion (Kumakhov 1976) – therefore they may emit radiation. The radiation is a direct result of the transverse momentum relative to the symmetry plane of the crystal. Terhune and Pantell (1977) and later Pantell and Alguard (1979) discussed the effect from the quantum mechanical perspective: the transverse potential of the lattice as seen by the moving electron consists of a set of "bound states". When the electrons are injected parallel to the plane of symmetry only the lowest state is populated. When the beam is tilted to this plane, higher states are populated. As in normal bound states, electrons can jump from a high state to a lower state emitting a photon in the process. It is interesting to note that from the point of view of the electron wave function propagation, it is completely analogous to the propagation of an electromagnetic wave in an optical fiber (Schächter 1988). Spontaneous emission of this process has been observed (Swent et al. 1979) but the condition imposed on the emittance of the beam is very stringent and to the best of our knowledge no stimulated radiation was observed so far. Friedman et al. (1988) compiled an extended review of the quantum picture of interaction of free electrons with radiation.

7.6 X-Ray FEL

Although the FEL, in principle, may be employed in a wide range of frequencies, probably its major advantage is the possibility of generating controllable coherent and obviously non-coherent radiation at wavelengths, which are virtually inaccessible by other sources. It plays a crucial role in the fourth generation of advanced light-sources developed in recent years and in various stages of construction. Light-sources are facilities based on powerful accelerators that inject multi-GeV electrons in an FEL generating short-wavelength (nano-meter and below) in femptosecond pulses of many GigaWatts of power. Lack of sources in this spectral region, even at much lower power levels entails that the coherent radiation is achieved by self-amplified spontaneous emission (SASE) in a high-gain FEL. As of today, there are two operating machines: (1) The DESY machine (FLASH), employs a 1 GeV accelerator which injects 2 kA of current to generate 50 fs of 6.9 nm (fundamental) radiation at a power of up to 5 GW – corresponding to up to 10^{13} photons per bunch. The average (peak) brilliance is 10^{19} photons/sec/mm^2/mrad^2/0.1% (10^{30} photons/sec/mm^2/mrad^2/0.1%) and it is facilitated by the presence of 15 m long wiggler of 2.73 cm periodicity and a peak magnetic field of 0.47 T. Users have first used radiation (32 nm) from this machine in the first quarter of 2006. (2) The second operational machine is the Linear-accelerator Coherent Light Source (LCLS) which employs the Stanford Linear Collider (SLC) injecting a 13.6 GeV, 3.4 kA, 70 fs (FWHM) electron beam in a 3 cm periodicity, peak magnetic field 1.25 T and 110 m long wiggler. The first light (1.5A) from this machine was produced in April 2009 – Emma et al. (2010). Several other machines are in various stages of design or construction. Among those it warrants to mention the Energy Recovery Linac (ERL) at Cornell University, Jefferson Laboratory and Argonne National Laboratory in the United States as well as European project, X-FEL at DESY.

It is virtually impossible in the framework of one section to cover in a reasonable mathematical detail all the important concepts involved in the design of such a complex facility therefore, whatever will be chosen to be presented in such a framework, is a subset of a larger list of topics. Since, earlier in this chapter we have presented the essentials of the 1D FEL theory, we choose to focus in this section on aspects which we conceive to be important to the fourth generation of light-sources. A reader interested in more professional details should consult the review of the subject compiled by Huang and Kim (2007). For a more detailed tutorial the reader is referred to Attwood (2000) and Schmüser et al. (2008) that are recommended for a deeper elaboration of the subject.

7.6.1 Seeding Techniques

In the VUV, X-ray and beyond, there are no sources to feed the free electron laser therefore, the coherent radiation has to grow from noise. Two main approaches have been contemplated in the design of the next generation light sources are the self-amplified spontaneous emission (SASE) and the high-gain harmonic generation (HGHG) scheme. One of the main advantages of the HGHG over the SASE FEL is that, by using up-frequency conversion of the initial seed signal, HGHG allows us to produce not only transversely, but also temporally coherent pulses. In contrast, the SASE radiation starts from initial shot noise in the beam, with the resulting radiation having an excellent spatial coherence, but a rather poor temporal one. Standard HGHG allows only a limited frequency multiplication factor in a single stage leading to *multi-stage* approach for X-ray production seeded at an ultraviolet wavelength accompanied by a significant complication in the overall design. Moreover, generation of the 10th harmonic (or higher) requires a large energy modulation of the beam and deteriorates the beam properties as a "lasing" medium.

Self-seeded FEL consists of two undulators separated by a mono-chromator and a magnetic chicane. The interaction in the first undulator is started by shot noise and is interrupted well before saturation. While the SASE radiation is sent through a monchromator, the electron beam passes through a magnetic chicane which destroys the micro-bunching introduced by the SASE and compensates the delay introduced by the monochromator. Further, the monochromatic radiation and the demodulated electron beam are injected into the second undulator where the interaction between the two reaches saturation. Since the self-seeded FEL does not rely on an external radiation source to seed the FEL process and it can be scaled, in principle, to any arbitrary wavelength.

In the *High-Gain Harmonic Generation* (HGHG) the FEL is seeded with a sub-harmonic of the output wavelength. It consists of two undulators separated by a magnetic chicane. The first undulator, the "modulator", is seeded by an external coherent source. In the first undulator an energy modulation in the electron beam occurs while the dispersive section transforms the energy modulation into a density modulation on higher harmonics of the seed wavelength. The second undulator ("radiator") is tuned to one of these harmonics and the bunches generated by the dispersive section drive it. It is interesting to note that the approach resembles the severed TWT (Kuang et al. 1993) and to some extent a klystron. A HGHG FEL with a seed consisting of high harmonics generated by a laser in a gas has been reported recently by Lambert et al. (2008). Marinelli et al. (2010) presented a comparative study between self-seeded FEL and HGHG.

Stupakov (2009) has suggested a novel concept related to the *echo* effect in circular accelerators whereby it was demonstrated that modulating the beam energy with the frequency ω_1 and, after some delay with frequency ω_2 leads, after a build-up time, to an echo signal oscillating at a frequency $m\omega_1 + n\omega_2$ where m and n are integers.

In the remainder of this sub-section, we employ the 1D approach developed previously in order to clarify SASE. Obviously this is a very crude approximation since at the very foundations of this 1D-model is the assumption that the system operates at a single frequency and when dealing with growth from noise this clearly is an oversimplification. Nevertheless, we push this model beyond its limits in order to convey the essentials of the process and not for the accuracy of its predictions. Our stating point is the ultra-relativistic version of the 1D model as summarized in Table 7.1 or explicitly,

$$\begin{aligned}\frac{da}{d\zeta} &= \alpha\left\langle\frac{\exp(-j\chi_i)}{\gamma_i}\right\rangle_i,\\ \frac{d\gamma_i}{d\zeta} &= -\frac{1}{2\gamma_i}[a\exp(j\chi_i) + c.c.],\\ \frac{d\chi_i}{d\zeta} &= \frac{\Omega}{2\gamma_i^2}.\end{aligned} \tag{7.6.1}$$

We consider first the second derivative of the field-amplitude

$$\frac{d^2a}{d\zeta^2} = \alpha\left\langle -j\frac{\Omega}{2\gamma_i^2}\frac{\exp(-j\chi_i)}{\gamma_i}\right\rangle_i - \alpha\left\langle\frac{d\gamma_i}{d\zeta}\frac{\exp(-j\chi_i)}{\gamma_i}\right\rangle_i, \tag{7.6.2}$$

and substituting the equation for the single particle energy dynamics. Ignoring fast phase-variations we get

$$\frac{d^2a}{d\zeta^2} = \frac{\alpha}{2}\left\{-j\Omega\left\langle\frac{\exp(-j\chi_i)}{\gamma_i^3}\right\rangle_i + a\left\langle\frac{1}{\gamma_i^3}\right\rangle_i\right\}, \tag{7.6.3}$$

We now repeat the procedure for the third derivative but terms quadratic in the amplitude are ignored therefore we obtain

$$\frac{d^3a}{d\zeta^3} + j\Omega\frac{3\alpha}{4}\left\langle\frac{1}{\gamma_i^5}\right\rangle a = \frac{\alpha^2}{2}\left\langle\frac{1}{\gamma_i^3}\right\rangle\left\langle\frac{\exp(-j\chi_i)}{\gamma_i}\right\rangle - \frac{\alpha}{4}\Omega^2\left\langle\frac{\exp(-j\chi_i)}{\gamma_i^5}\right\rangle \tag{7.6.4}$$

This equation is equivalent to the result developed in Sect. 7.3 [(7.3.28)] but for a synchronous beam. Contrary to the homogeneous model employed earlier, here we clearly observe the source term in the right hand side which is dependent on the beam current.

In the left hand side, we clearly identify the kind of expression that describes the exponential spatial-growth associated with traveling-wave interaction (see Table 7.1). Explicitly, the spatial-growth may be derived from the solution of the characteristic third-order polynomial of the last equation (7.3.28). In the right hand side, we realize that if, on the scale of X-ray wavelength the electrons emerging from the accelerator have a *uniform* phase-distribution and this distribution is not correlated to the electrons energy distribution, then the source term is zero.

In order to understand the self-amplified spontaneous emission we need to take into consideration the fact that there are *fluctuations* due to the fact that there are a *finite number of electrons* in one period of the radiation field. Further assuming that the phase and energy distributions are not correlated we obtain

$$\frac{d^3a}{d\zeta^3} + j\Omega\frac{3\alpha}{4}\left\langle\frac{1}{\gamma_i^5}\right\rangle a = \left[\frac{\alpha^2}{2}\left\langle\frac{1}{\gamma_i^3}\right\rangle\left\langle\frac{1}{\gamma_i}\right\rangle - \frac{\alpha}{4}\Omega^2\left\langle\frac{1}{\gamma_i^5}\right\rangle\right]\langle\exp(-j\chi_i)\rangle \tag{7.6.5}$$

Assuming that there are N_λ electrons in one period of the radiation field then the phase term may be approximated by

$$\langle\exp(-j\chi_i)\rangle \simeq \frac{1}{\sqrt{N_\lambda}} \tag{7.6.6}$$

and also ignoring the energy spread we get

$$\frac{d^3a}{d\zeta^3} + j\Omega\frac{3\alpha}{4\bar{\gamma}^5}a = \left[1 - \frac{\Omega^2}{2\alpha\bar{\gamma}}\right]\frac{\alpha^2}{2\bar{\gamma}^4\sqrt{N_\lambda}} \simeq -\frac{\alpha\Omega^2}{4\bar{\gamma}^5\sqrt{N_\lambda}} \tag{7.6.7}$$

Here we have tacitly assumed that the energy exchanged is negligible and the average energy of incoming electrons is denoted by $\bar{\gamma}$. Let us renormalize the coordinate

$\bar{\zeta} = \zeta\left(3\alpha\Omega/4\bar{\gamma}^5\right)^{1/3}$as well as redefining the "noise-source" as $a_{\rm n} = \Omega/3\sqrt{N_\lambda}$ thus (7.6.7) reads

$$\frac{d^3 a}{d\bar{\zeta}^3} + ja \simeq -a_{\rm n} \tag{7.6.8}$$

subject to zero boundary conditions at $\bar{\zeta} = 0$. Following the same approach as in Sect. 4.1 the solution has the form

$$a(\bar{\zeta}) = ja_{\rm n} + \sum_{v=1}^{3} A_v \exp(-ju_v\bar{\zeta}) \tag{7.6.9}$$

wherein u_v are the three solutions of the third order polynomial $u^3 + 1 = 0$namely

$$u_1 = -1, \;\; u_2 = \exp(-j\pi/3), \;\; u_2 = \exp(j\pi/3) \tag{7.6.10}$$

from the boundary conditions we have

$$\begin{bmatrix} 1 & 1 & 1 \\ u_1 & u_2 & u_3 \\ u_1^2 & u_2^2 & u_3^2 \end{bmatrix} \begin{pmatrix} A_1 \\ A_2 \\ A_3 \end{pmatrix} = \begin{pmatrix} -ja_{\rm n} \\ 0 \\ 0 \end{pmatrix} \Rightarrow A_1 = A_2 = A_3 = -ja_{\rm n}/3 \tag{7.6.11}$$

which finally implies

$$a(\bar{\zeta}) = ja_{\rm n}\left[1 - \frac{1}{3}\sum_{v=1}^{3} \exp(-ju_v\bar{\zeta})\right]. \tag{7.6.12}$$

Evidently, far away from the input the normalized power is

$$\begin{aligned} \frac{1}{2\alpha}\left|a(\bar{\zeta} \gg 1)\right|^2 &\simeq \frac{1}{2\alpha}|a_{\rm n}|^2 \exp\left(\sqrt{3}\bar{\zeta}\right) \\ &\simeq \frac{1}{2\alpha}\left(\frac{\Omega^2}{9N_\lambda}\right) \exp\left[\sqrt{3}\left(\frac{3\alpha\Omega}{4\bar{\gamma}^5}\right)^{1/3} \frac{z}{d}\right] \end{aligned} \tag{7.6.13}$$

demonstrating the exponential growth from shot noise. The exponential gain eventually stops as the beam loses enough energy to upset the resonant condition. Both the radiation intensity and the electron beam micro-bunching reach a maximum saturation level as discussed both in this Chapter as well as in Chaps. 4 and 6.

Before we conclude this sub-section one aspect warrants consideration. In Chap. 4 we investigated the two extremes of any radiation source (amplifier and oscillator). Contrary to the microwave and optical regimes where reflections may be detrimental to the operation of an amplifier, in the X-ray range lack of natural reflectors (mirrors) is a significant impediment to the development of X-ray sources. Huang and Ruth (2006) have suggested to employ narrow-band Bragg crystals to form an X-ray feedback loop and thus implement a real oscillator configuration.

7.6.2 *Self Focusing*

In an X-ray FEL the magnetic field between the poles may be represented by $B_y = B_{\rm w}\cos(k_{\rm w}z)$. Not too far off axis a reasonable representation of a planar undulator with wide and flat pole faces is

$$\begin{aligned} B_y &= B_{\rm w}\sin(k_{\rm w}z)\cosh(k_{\rm w}y), \\ B_z &= B_{\rm w}\cos(k_{\rm w}z)\sinh(k_{\rm w}y). \end{aligned} \tag{7.6.14}$$

A three dimensional solution of the equations of motion reveals that in the horizontal plane (x) the force vanishes whereas in the vertical plane (y) the force is focusing

$$\frac{d^2y}{dz^2} \simeq -\frac{a_{\rm w}^2 k_{\rm w}^2}{2\bar{\gamma}^2}y. \tag{7.6.15}$$

For focusing also horizontal plane, one can shape the pole faces to be parabolic. Note that this is virtually identical to the focusing periodic magnetic field discussed in Sect. 3.4.2.

7.6.3 *Quantum Recoil*

Although in the Introduction of this chapter we indicated that quantum analysis of the FEL interaction was performed throughout the years by several authors, quantum effects have never had a central role in any of the operating FEL's. This picture may change in future light sources. As an electron emits an X-ray photon, its energy is reduced due to the quantum recoil. If the fractional energy change is on the order of or larger than the FEL gain *bandwidth*, the quantum recoil may significantly degrade the FEL gain. For typical short-wavelength, high-gain FELs using *magneto-static* field and high-energy electron beams, the typical FEL gain bandwidth is on the order of 10^{-3}, while the fraction energy change after a photon emission is about three orders of magnitude smaller therefore, the quantum recoil is negligible. However, it may become significant in the future when an extremely bright and low-energy electron beam interacts with an *electromagnetic* field (laser), as the fractional energy change due to an X-ray photon emission may be comparable to or exceed the FEL bandwidth [Schächter (1987)].

7.6.4 Harmonic Generation

In zero order, the motion of an electron in a periodic magnetic field is itself periodic however, as the required efficiency is elevated, the motion is no longer purely sinusoidal and as a result, odd harmonics are generated. Accounting for energy spread, emittance, radiation diffraction and guiding, a 3D analysis of nonlinear harmonic generation Huang and Kim (2000) have shown that the properties of the first few harmonics are governed by those of the fundamental after the latter is well in the exponential regime.

A 1D SASE simulation (Saldin et al. 2006) reveals that the maximum third-harmonic power at saturation (for a cold beam) is about 2% of the fundamental level. The relative spectral bandwidth is independent of the harmonic number. For all practical purposes, the third nonlinear harmonic radiation is the most significant harmonic component and can naturally extend the wavelength reach of the X-ray FEL by a factor of three.

7.6.5 Undulator Errors

Demonstrating that X-ray radiation may be generated assuming an ideal model of an undulator, is a necessary condition but it is by no means sufficient. Evidently, a periodic magnetic field with a well defined period and amplitude, is an idealization of a practical system in which each cell has minute defects associated with the manufacturing process or alignment. Since the emerging spectrum is expected to be sensitive to periodicity variations, we examine here a simple model that focuses on miniscule variations in the amplitude and consider their effect on the gain.

In order to have a rough estimate on the role of undulator's error on the output power let us return to (7.6.13) and rewrite it at $z = d$

$$\frac{1}{2\alpha}|a(\zeta = 1)|^2 \simeq \frac{1}{2\alpha}\left(\frac{\Omega^2}{9N_\lambda}\right)\exp\left[\frac{1}{d}\int_0^d dz'\sqrt{3}\left(\frac{3\alpha\Omega}{4\bar{\gamma}^5}\right)^{1/3}\right]. \quad (7.6.16)$$

As already indicated, in this expression, it was explicitly assumed that the magneto-static field is ideally periodic in other words, all the undulator's cells are identical. A relatively simple assessment is possible if we consider a situation whereby there is no error in the period of the field but its amplitude may vary randomly – the average error being zero but the standard deviation is non-zero. Obviously the question is what is the effect on the gain.

Bearing in mind that the coupling parameter is proportional to the square of the undulator's amplitude, ($\alpha \propto B_0^2$) we replace $B_0 \to B_0 + \delta B(z)$ and this error δB is assumed to be much smaller than $|\delta B| \ll B_0$ hence

$$\frac{1}{d}\int_0^d dz B_0^{2/3} \to \frac{1}{d}\int_0^d dz[B_0 + \delta B]^{2/3} = B_0^{2/3}\frac{1}{d}\int_0^d dz\left[1 + \frac{\delta B}{B_0}\right]^{2/3}. \tag{7.6.17}$$

A Taylor expansion up to the second order of the last term entails,

$$B_0^{2/3}\frac{1}{d}\int_0^d dz\left[1 + \frac{\delta B}{B_0}\right]^{2/3} \simeq B_0^{2/3}\frac{1}{d}\int_0^d dz\left[1 + \frac{2}{3}\frac{\delta B}{B_0} - \frac{1}{9}\left(\frac{\delta B}{B_0}\right)^2\right] \tag{7.6.18}$$

and since the average error is zero, we get that the power is reduced by a factor of

$$\frac{1}{2\alpha}|a(\zeta = 1)|^2 \simeq \frac{1}{2\alpha}\left(\frac{\Omega^2}{9N_\lambda}\right)\exp\left[\sqrt{3}\left(\frac{3\alpha\Omega}{4\bar{\gamma}^5}\right)^{1/3}\left(1 - \frac{\langle\delta B^2\rangle}{9B_0^2}\right)\right]. \tag{7.6.19}$$

Consequently, the pole field error tolerance seems fairly relaxed (second order) provided the gain parameter is not too large. To envision this fact let us define the gain parameter $g = \sqrt{3}\left(3\alpha\Omega/4\bar{\gamma}^5\right)^{1/3}$ leading to the following expression for the relative change in power

$$1 - \exp\left[-g\frac{\langle\delta B^2\rangle}{9B_0^2}\right]. \tag{7.6.20}$$

Obviously, if the gain is large $g\langle\delta B^2\rangle/9B_0^2 > 2$ the effect of poles errors becomes significant. In fact, these errors are being continuously amplified.

Comment 7.3. Note that this approach may be extended to include the effect of minute variations in the undulator's periodicity since actually the coupling coefficient is proportional to $\alpha \propto a_\mathrm{w}^2 = (eB_0/mck_\mathrm{w})^2$. Explicitly, in the gain term one should replace $\langle\delta B^2\rangle/B_0^2 \to \langle\delta a_\mathrm{w}^2\rangle/a_{\mathrm{w},0}^2$. The difficulty with this approximation is that in the gain expression (7.6.16), it was tacitly assumed that the system operates at resonance that, in turn, assumes a uniform wiggler.

7.6.6 Roughness and Resistive Wall Effects

Two parasitic effects we have discussed in this text are important in high brightness light sources: the most important one is the wake generated by the resistive wall and the second in importance is the wake due to finite roughness of the beam chamber. In both cases, there are two effects: long-range effects whereby wakes generated by one bunch affects trailing bunches and short-range wakes generate weak energy variations along the same bunch. In the latter case, the main impact of the wake is to change the *central energy* of various portions of the bunch. Consequently, it shifts its resonant wavelength along the interaction region. Such an energy shift is detrimental to the achievable power at saturation. This depends on the tapering along the undulator that, as was earlier shown, depends on the local resonant condition assuming that all portions of the bunch have the same energy. However, obviously, if the wake affects differently various portions of the bunch, the taper is no longer optimal and saturation occurs after a shorter interaction length implying lower radiation at the output. One possible remedy is reduction of the charge in each bunch weakening the wake yet facilitating higher output power by reducing the deviation from resonance as imposed by the tapering.

Exercises

7.1. Show that the equivalent to (7.4.44) in the case of a traveling-wave tube reads

$$\frac{1}{2}\frac{\mathrm{d}}{\mathrm{d}\bar{\gamma}}\Delta\gamma^2 = \bar{\gamma} - \frac{\langle\gamma_i \cos(\tilde{\chi}_i)\rangle}{\langle\cos(\tilde{\chi}_i)\rangle},$$

but the Madey theorem has the same form as in (7.4.47).

7.2. Use the formulation of the interaction in a tapered wiggler to calculate the power generated in the low-gain Compton regime in a uniform wiggler with stochastic errors in its parameters. Assume that the errors follow a Gaussian distribution.

7.3. Calculate the spontaneous emission emitted by a particle moving in a periodic electrostatic potential as in Sect. 7.5.6. Compare the spectrum with that of the Smith-Purcell effect. Calculate the effective decelerating force which acts on the particle and compare it with the result in (7.1.30) and with the decelerating force in the Smith-Purcell case.

7.4. Calculate the stable trajectories of a particle in a combination of wiggler and uniform magnetic field. Draw $\beta_{\|}$ as a function of B_{w}.

7.5. Based on Sects. 7.2 and 7.3 make a summarizing table which will include the gain and the condition (say on the current and beam temperature) that the system will operate in a particular regime. Discuss the transition from one regime to another.

Chapter 8
Basic Acceleration Concepts

One of the important systems where beam-wave interaction in periodic structures plays a crucial role is the particle accelerator. The latter provides us with a unique tool to test, on earth, the different models that describe the constituents of matter. Accelerators have undergone a great progress in the last sixty years and it seems that they still have a long way to go in order to meet the requirements necessary to test the present theoretical models (Richter 1985). Over the years, in addition to High-Energy Physics, several other disciplines learned to harness the unique characteristics of accelerators and today they are widely used in chemistry, biology and medicine. Regardless the application, it is virtually impossible to present a thorough presentation of the topic in a single chapter and the reader is encouraged to consult textbooks dedicated to this topic such as Wiedemann (1999a, b) and Lee (2004) as well as collection of selected of lectures by Chao et al. (2002) or collection of selected articles Pellegrini and Sessler (1995).

This chapter has two parts. In the first part (Sect. 8.1), we discuss the basics of linear accelerator (linac) concepts with particular emphasis on the beam-wave interaction (Loew and Tolman 1983). Our discussion will be limited to electron linear accelerators of the type operational today at SLAC National Accelerator Laboratory. The second part (Sects. 8.2–8.6) is a collection of brief overviews of different alternative schemes of acceleration which are in their early stages of research. In these sections the discussion is in general limited to the basic concepts and the figure of merit which characterizes their application i.e. the achievable gradient. For discussions that are more detailed the reader is referred to articles by the experts of each scheme. In addition, the reader is encouraged to consult the Proceedings of the Advanced Acceleration Workshop organized by the United States Department of Energy and published by the American Institute of Physics (AIP) every 2 years since 1982.

L. Schächter, *Beam-Wave Interaction in Periodic and Quasi-Periodic Structures*, Particle Acceleration and Detection, DOI 10.1007/978-3-642-19848-9_8,

8.1 Basic Linear Acceleration Concepts

The field of linear accelerators is very broad and comprises many subfields which have been covered in books and presented in detail in summer schools [e.g. Lapostolle and Septier (1970)], and an entire set of proceedings of summer schools or meetings which will be referred to in what follows. When such a subject is presented in a single chapter, compromises have to be made as of what to include and what to leave out. Since this text focuses on periodic and quasi-periodic structures we concentrate on the acceleration structure itself. Therefore, we ignore topics associated with RF generation, pulse compression, injection and extraction of the RF. Furthermore, we do not discuss beam generation or cooling.

8.1.1 Constant Gradient and Constant Impedance Structures

The basic configuration of an acceleration section is conceptually identical to the coupled-cavity structure as discussed in the context of closed periodic structure in Chap. 5 except that in this case, the effect of the beam on the radiation field is small. Moreover, the ohm loss has to be considered since it causes amplitude and phase variations that affect the acceleration process. This loss will be the focus of the present subsection. In Chap. 2 we defined several impedances which have been used throughout the text. At this point, we introduce an additional concept namely, the *shunt impedance*, and several other related quantities that are important to establish the dynamics of electrons in an acceleration structure. The shunt impedance (Z_{sh}) is important in accelerators since large amounts of electromagnetic power flows in the system and any change in this power affects the electron dynamics. This quantity is a measure of the ohm power loss in a unit length in terms of the electric field $[E(z)]$ that acts on the electrons. It is given by

$$Z_{sh} \equiv |E(z)|^2 \left(-\frac{dP}{dz} \right)^{-1}, \tag{8.1.1}$$

where $P(z)$ is the power that flows along the structure. According to this definition, the shunt impedance has units of impedance per unit length and it is related to the quality factor of a waveguide defined by

$$Q \equiv \frac{\omega W}{-dP/dz}, \tag{8.1.2}$$

through

$$\frac{Z_{sh}}{Q} = \frac{|E(z)|^2}{\omega W}; \tag{8.1.3}$$

here W is the average electromagnetic energy per unit length. This quantity is also related to the spatial decay associated with the Ohm loss $\alpha_{\rm ohm} = (-{\rm d}P/{\rm d}z)/2P$, which in turn is related to the skin resistance $R_{\rm s} \equiv (\sigma\delta)^{-1}$ where σ is the conductivity of the metal and δ is the skin-depth defined by $\delta = \sqrt{2/\omega\sigma\mu_0}$. For a cylindrical resonator of radius R and length d the shunt impedance is

$$Z_{\rm sh} = \eta_0\left(\frac{\eta_0}{R_s}\right)\frac{d}{\pi R(R+d){\rm J}_1^2(p_1)}, \tag{8.1.4}$$

and

$$Q = \frac{\eta_0}{R_{\rm s}}\frac{dp_1}{2(R+d)}. \tag{8.1.5}$$

Based on (8.1.2) and the fact that the energy velocity is $v_{\rm en} \equiv P/W$, it can be shown that $\alpha_{\rm ohm} = \omega/2v_{\rm en}Q$.

With the exception of the injection section, the electrons move in a typical acceleration section at almost c, therefore in what regards the phase velocity, there is no need to taper the structure. However, in order to have a feeling on the effect of the ohm loss we can readily understand that part of the energy is absorbed by the walls and consequently, the field experienced by the electrons decreases in space. Therefore, if the motion is calculated for a lossless system, the variation in amplitude or phase due to lossy material may cause the electron to slip out of phase. Let us now calculate the energy transferred to the electron as a function of the shunt impedance and the electromagnetic power injected at the input. For this purpose, we first assume that the electron is "riding on the crest of the wave" and it gains in a length D a kinetic energy

$$\delta E_{\rm kin} = e\int_0^D {\rm d}zE(z); \tag{8.1.6}$$

the effect of the phase will be considered in Sect. 8.1.3. The energy gain will be calculated for two different acceleration structures: (1) constant impedance, in which case the shunt impedance is constant and the gradient varies in space. In addition, (2) constant gradient structure, in which case primarily the group velocity varies and consequently, the interaction impedance changes.

Constant Impedance. The shunt impedance in this case is constant along the structure and this entails constant geometry and consequently uniform $\alpha_{\rm ohm}$. Bearing in mind that the power along the structure is given by $P(z) = P(0)\exp(-2\alpha_{\rm ohm}z)$ and

$$E(z) = \sqrt{2\alpha_{\rm ohm}Z_{\rm sh}P(z)}, \tag{8.1.7}$$

we can readily conclude that when the beam loading effect is ignored the change in the kinetic energy is given by

$$\delta E_{\mathrm{kin}} = e[P(0)Z_{\mathrm{sh}}D]^{1/2}(2\xi)^{1/2}\frac{1-\exp(-\xi)}{\xi}, \tag{8.1.8}$$

where $\xi \equiv \alpha_{\mathrm{ohm}}D$.

Constant Gradient. Here α_{ohm} is tapered in such a way that the gradient remains constant along the structure. In order to avoid unnecessary complications we assume that we can ignore spatial variations in Q as well as in the shunt impedance and that the geometric variations affect primarily the group (energy) velocity and consequently, the interaction impedance. From the definition of the shunt impedance $E^2(z) = -Z_{\mathrm{sh}}(\mathrm{d}P/\mathrm{d}z)$, we conclude that $\mathrm{d}P/\mathrm{d}z$ has to be constant, which implies

$$P(z) = A + Bz. \tag{8.1.9}$$

Since we know the power at the input $[P(0)]$ and at the output $P(0)\exp(-\zeta)$ where

$$\zeta \equiv \int_0^D \mathrm{d}z\alpha_{\mathrm{ohm}}(z), \tag{8.1.10}$$

we conclude that the variation of the power in space is given by

$$P(z) = P(0)\left\{1 - \frac{z}{D}[1-\exp(-\zeta)]\right\}. \tag{8.1.11}$$

Using its formal definition, we find that

$$\alpha_{\mathrm{ohm}} = \frac{1}{2D}\frac{1-\exp(-2\zeta)}{1-[1-\exp(-2\zeta)]z/D} \tag{8.1.12}$$

which subject to our assumptions, dictates the group velocity based on $\alpha_{\mathrm{ohm}} = \omega/2\mathrm{v}_{\mathrm{en}}Q$. Since the gradient is constant, the kinetic energy gained by one electron is $\delta E_{\mathrm{kin}} = eE(0)D$ which in terms of the input power (8.1.8), (8.1.11) reads

$$\delta E_{\mathrm{kin}} = e[P(0)Z_{\mathrm{sh}}D]^{1/2}[1-\exp(-2\zeta)]^{1/2}. \tag{8.1.13}$$

In either one of the two cases, maximum energy gain occurs for maximum shunt impedance. This conclusion leads us to the choice of the phase advance per cell. Recall that in Chap. 5, when solving the dispersion relation of a disk-loaded structure, we assumed a certain phase advance per cell that together with the resonance condition determined the periodicity of the structure. No reason has been given for this particular choice and it will be explained in what follows.

Now that we have concluded that for maximum energy gain, one has to maximize the shunt impedance, we can ask what number of disks in one period of the wave that satisfies this condition. It is intuitive that the larger the number of disks, the greater the total Ohm loss and consequently for a given $E(z)$ the shunt impedance decreases. The same phenomenon occurs at the other extreme, since for a small number of disks, the electric field that acts on the electrons, is expected to be small (for a given Ohm loss). Simulations indicate that maximum shunt impedance occurs for three disks in one period of the wave and for this reason, traveling-wave accelerating structures are designed with a phase advance per cell of $2\pi/3$.

Beam Loading. Up to this point, we ignored the energy transferred to the beam. In order to consider the effect of the beam we may argue that in the absence of the beam and for constant impedance the electric field decays exponentially with a coefficient α_{ohm} therefore it satisfies

$$\frac{\mathrm{d}E}{\mathrm{d}z} = -\alpha_{\text{ohm}} E. \tag{8.1.14}$$

In the presence of the beam, an additional change in the field amplitude occurs $E \to E + Z_{\text{sh}} I$ whose variation in space is given by

$$\frac{\mathrm{d}E}{\mathrm{d}z} = -\alpha_{\text{ohm}} (E + Z_{\text{sh}} I). \tag{8.1.15}$$

Here I represents the current carried by the beam in a narrow bunch. The solution of this equation is $E = A\exp(-\alpha_{\text{ohm}} z) - Z_{\text{sh}} I$ and since at the input the loading effect is expected to be zero we have, according to (8.1.7), $E(z = 0) = \sqrt{2\alpha_{\text{ohm}} Z_{\text{sh}} P(0)}$ hence

$$E(z) = \sqrt{2\alpha_{\text{ohm}} Z_{\text{sh}} P(0)}\exp(-\alpha_{\text{ohm}} z) - Z_{\text{sh}} I[1 - \exp(-\alpha_{\text{ohm}} z)]. \tag{8.1.16}$$

Consequently, the kinetic energy gain of a single particle is given by

$$\delta E_{\text{kin}} = e[D Z_{\text{sh}} P(0)]^{1/2} (2\xi)^{1/2} \frac{1 - \mathrm{e}^{-\xi}}{\xi} - e Z_{\text{sh}} I D \left(1 - \frac{1 - \mathrm{e}^{-\xi}}{\xi}\right); \tag{8.1.17}$$

clearly the second term represents the beam loading effect.

For a constant gradient structure, when the beam loading is ignored it implies that

$$\frac{\mathrm{d}E}{\mathrm{d}z} = 0, \tag{8.1.18}$$

thus when the beam effect is included, in analogy with (8.1.15), we have

$$\frac{\mathrm{d}E}{\mathrm{d}z} = -\alpha_{\text{ohm}} Z_{\text{sh}} I. \tag{8.1.19}$$

Following the same procedure as previously, we have

$$\delta E_{\text{kin}} = e[P(0)Z_{\text{sh}}D]^{1/2}\left(1-e^{-2\zeta}\right)^{1/2} - \frac{1}{2}eIZ_{\text{sh}}D\left[1-\frac{2\zeta\exp(-2\zeta)}{1-\exp(-2\zeta)}\right]; \quad (8.1.20)$$

as in (8.1.17) the second term represents the beam loading effect. Note that subject to the present assumptions, we expressed the gain in kinetic energy of the bunch in terms of a few "global" parameters ($Z_{\text{sh}}, \alpha_{\text{ohm}}$ etc.) determined in turn by the geometric and electrical parameters which may vary from one module to another.

8.1.2 Auxiliary Coupling

The design of an acceleration structure is a continuous process of trade-offs. In the previous subsection, we indicated that the phase advance per cell in a traveling-wave structure should be $2\pi/3$ in order to maximize the shunt impedance. There are, however, additional considerations that come into play. Strictly speaking from the electrons' point of view, the best choice would be a π-mode i.e., standing-wave configuration, since the gradient for a given input power reaches its maximum. But a π-mode is unacceptable from the electromagnetic wave perspective since the group velocity at this point is zero, the filling time is long and if the structure is sufficiently long, the mode is unstable due to proximity of the various longitudinal modes near the crest of the dispersion relation. If we examine the same problem from the field aspect then the best choice would be a $\pi/2$-mode since in this case the group velocity is the largest. Unfortunately, in this case half of the cavities do not contribute to acceleration.

Knapp et al. (1965) suggested a way to break this vicious circle. His basic idea is to satisfy both the electrons and the electromagnetic field: the former sees a π-mode and the electromagnetic wave sees a $\pi/2$-mode. This is possible because the beam occupies only a small fraction of the transverse dimension of the structure whereas the electromagnetic wave fills the entire volume. Implementation of this concept is possible by making each cell of two cavities: one cavity is the regular pillbox cavity of a disk-loaded waveguide while the second is recessed and its aperture is on the external wall coupling two adjacent pillbox cavities. In this way, one can design the structure such that the electron sees a series of pillbox cavities operating at the π-mode whereas the electromagnetic wave actually operates in a $\pi/2$-mode. Schematically this configuration is presented in Fig. 8.1.

8.1.3 Phase Dynamics

In Sect. 8.1.1 we assumed that the particle "rides" on the crest of the wave and we examined the energy transfer assuming it stays on the crest along the entire interaction region. In general, this is not the case since as the particle is accelerated,

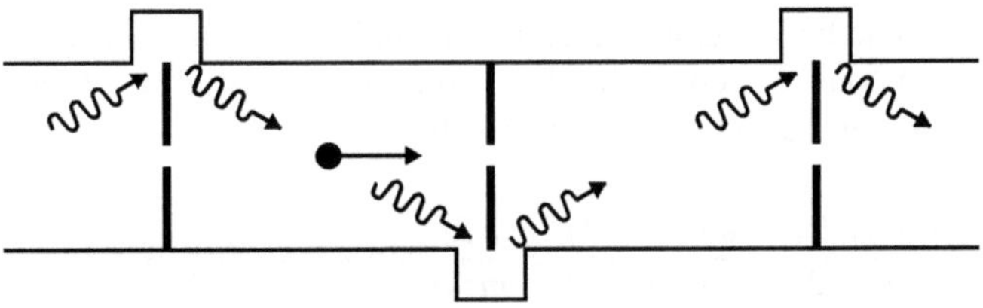

Fig. 8.1 Accelerating structure with side couplers. The electromagnetic wave undergoes a $\pi/2$ advance per cell but the accelerated electron experiences a π-phase advance per cell

and as a result, its relative phase varies. In this subsection we consider the phase dynamics assuming that the amplitude of the electric field is constant and so is the phase-velocity (c). In addition, we disregard the transverse motion of the electrons. Subject to these assumptions, the dynamics of the particle is given by

$$\frac{\mathrm{d}}{\mathrm{d}t}\gamma\beta = -\frac{eE_0}{mc}\cos\left[\omega t - \frac{\omega}{c}z(t)\right]. \tag{8.1.21}$$

It is convenient to define the phase of the particle relative to the wave as

$$\chi(t) = \chi(0) + \omega t - \frac{\omega}{c}z(t), \tag{8.1.22}$$

and since β is always smaller than 1, the normalized velocity is expressed as

$$\beta = \cos\psi. \tag{8.1.23}$$

With these definitions, we can write two equations that describe the dynamics of the particles as:

$$\begin{aligned} \frac{-1}{\sin^2\psi}\frac{\mathrm{d}}{\mathrm{d}t}\psi &= -\frac{eE_0}{mc}\cos\chi, \\ \frac{\mathrm{d}}{\mathrm{d}t}\chi &= \omega(1-\cos\psi), \end{aligned} \tag{8.1.24}$$

in addition, since we limited the motion to the longitudinal direction we can replace

$$\frac{\mathrm{d}\chi}{\mathrm{d}t} = \left(\frac{\partial\chi}{\partial\psi}\right)\frac{\mathrm{d}\psi}{\mathrm{d}t}. \tag{8.1.25}$$

Substituting, the two expressions in (8.1.24) we obtain,

$$\mathrm{d}\psi\,\frac{1-\cos\psi}{\sin^2\psi} = \frac{eE_0}{mc\omega}\cos\chi\mathrm{d}\chi. \tag{8.1.26}$$

This expression can be integrated analytically and then re-arranged. Denoting with indexes "in" and "out" the corresponding values of the variables at the input of the interaction region and at its output; the result reads

$$\frac{\sin(\psi_{\text{out}}/2)}{\cos(\psi_{\text{out}}/2)} - \frac{\sin(\psi_{\text{in}}/2)}{\cos(\psi_{\text{in}}/2)} = \frac{eE_0}{mc\omega}[\sin\chi_{\text{out}} - \sin\chi_{\text{in}}]. \tag{8.1.27}$$

Finally, this relation can be rewritten in terms of the familiar β and the phase χ as follows

$$\sin\chi_{\text{out}} = \sin\chi_{\text{in}} + \frac{m\omega c}{eE_0}\left(\sqrt{\frac{1-\beta_{\text{out}}}{1+\beta_{\text{out}}}} - \sqrt{\frac{1-\beta_{\text{in}}}{1+\beta_{\text{in}}}}\right). \tag{8.1.28}$$

Expression (8.1.28) determines, for given initial conditions, the relation between the phase of the particle and its energy. Under certain circumstances, the particles are trapped as may be readily seen since once trapped they are accelerated and we may assume that they reach high γ. Explicitly if, $\beta_{\text{out}} \sim 1$ we have

$$\sin\chi_{\text{out}} = \sin\chi_{\text{in}} - \frac{m\omega c}{eE_0}\sqrt{\frac{1-\beta_{\text{in}}}{1+\beta_{\text{in}}}}, \tag{8.1.29}$$

which indicates that the value of the phase at the output is determined only by the initial values and it is independent of the particle's energy at the output. Since the trigonometric functions together are of order 1, we conclude that the condition for particles to become trapped is

$$E_0 > E_{\text{cr}} \equiv \frac{mc^2}{e}\frac{\omega}{c}\sqrt{\frac{1-\beta_{\text{in}}}{1+\beta_{\text{in}}}}. \tag{8.1.30}$$

For an initial energy of 300 keV the field intensity, assuming operation at 10 GHz, is 38 MV/m. Increasing the initial energy to 400 keV, lowers the required field to 33 MV/m. Figure 8.2 illustrates the trapping process for $E_0 = 60$ MV/m, $f = 11.424$ GHz, $\chi_{\text{in}} = 95°$ and the initial energy of the particles is 400 keV. We observe that beyond 5 MeV, in practice, the phase does not change.

Another direct result of (8.1.28) is the fact that if the initial particle is also highly relativistic, implying that we can write

$$\begin{aligned}\sin\chi_{\text{out}} &= \sin\chi_{\text{in}} + \frac{m\omega c}{eE_0}\left(\frac{1}{\gamma_{\text{out}}(1+\beta_{\text{out}})} - \frac{1}{\gamma_{\text{in}}(1+\beta_{\text{in}})}\right)\\ &\simeq \sin\chi_{\text{in}} + \frac{m\omega c}{2eE_0}\left(\frac{1}{\gamma_{\text{out}}} - \frac{1}{\gamma_{\text{in}}}\right),\end{aligned} \tag{8.1.31}$$

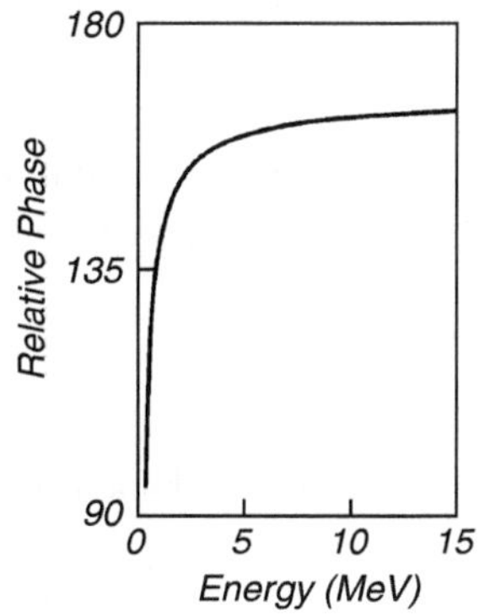

Fig. 8.2 Relative phase as a function of the energy. For an initial energy of 0.4 MeV the phase virtually remains constant when the electron's energy exceeds the 5 MeV level

and the change in γ is small relative to the initial value i.e., $\Delta\gamma = \gamma_{out} - \gamma_{in} \ll \gamma_{in}$ then the phase remains reasonably stable. In order to demonstrate this statement note that according to the single particle energy conservation, the energy change is

$$\Delta\gamma \simeq -\frac{eE_0 D}{mc^2}\cos\chi_{in}, \tag{8.1.32}$$

therefore, the phase-shift varies according to

$$\sin\chi_{out} - \sin\chi_{in} \simeq \frac{1}{\gamma_{in}^2}\left(\frac{\omega}{c}D\right)\cos\chi_{in}, \tag{8.1.33}$$

where D is the interaction length. For a 2 m long structure, $f = 11.4$ GHz (SLAC) and $\gamma_{in} \sim 600$ the term in the right-hand side is on the order of 0.001.

Finally, (8.1.28) indicates that a bunch of finite (phase) spread is actually compressed in the acceleration process. For example consider a 20° bunch i.e., $130° > \chi_{in} > 110°$ while the other parameters are $\gamma_{in} = 2.0$, $f = 11.424$ GHz and $E_0 = 60$ MV/m. The asymptotic values ($\gamma \gg 1$) of the phase are $166° > \chi_{out} > 156°$ which is one-half of the initial phase distribution. This calculation disregards the space-charge effect.

8.1.4 Transverse Effects: Panofsky-Wenzel Theorem

Being relativistic ($\gamma_{in} \gg 1$), the bunch does not spread in phase and the same reason works in our favor with regard to the transverse motion. So far, we have considered only the longitudinal motion of the electrons assuming that the beam width is very small on the scale of the transverse variations of the electromagnetic field. However, its width is finite and the transverse components of the electromagnetic field may affect the bunch. Panofsky and Wenzel (1956) were the first to point out that the transverse kick on a relativistic bunch which traverses cavity is zero in the case of a symmetric TE mode and non-zero for a symmetric TM mode. In order to examine the effect we first adopt an intuitive approach (Palmer 1986) followed by a generalized formulation of what is known as the Panofsky-Wenzel theorem.

Consider a particle moving with a velocity v_0 and a TM wave propagating at c. On axis the longitudinal electromagnetic wave is

$$E_z(r,z;\omega) = E_0 \exp\left(-j\frac{\omega}{c}z\right), \tag{8.1.34}$$

and outside the bunch (but in its close vicinity) Maxwell equations imply

$$\frac{1}{r}\frac{\partial}{\partial r} rE_r(r,z;\omega) + \frac{\partial}{\partial z}E_z(r,z;\omega) = 0. \tag{8.1.35}$$

Solving for the radial component we obtain

$$E_r(r,z;\omega) = j\left(\frac{1}{2}\frac{\omega}{c}r\right)E_0 \exp\left(-j\frac{\omega}{c}z\right), \tag{8.1.36}$$

and since $H_\phi = E_r/\eta_0$, the radial component of the force that acts on the envelope of the bunch is

$$F_r = -e\{E_r[r,z(t);t] - v_0\mu_0 H_\phi[r,z(t);t]\} = e\left(\frac{1}{2}\frac{\omega}{c}r\right)\frac{1}{(1+\beta)\gamma^2}E_0 \sin\chi, \tag{8.1.37}$$

where $\chi = \chi_0 + \omega[t - z(t)/c]$. Expression (8.1.37) indicates that off-axis, the radial force is by a factor of γ^2 smaller than the longitudinal force. Let us, for the sake of simplicity, ignore variation in γ, which is to say that we examine the transverse motion in a relatively short section where the change in the energy of the electrons is small. For a relativistic particle, the radial motion is governed by

$$\left[\frac{\mathrm{d}^2}{\mathrm{d}t^2} - \frac{1}{\gamma^3}\frac{eE_0\omega\sin\chi}{4mc}\right]r = 0; \tag{8.1.38}$$

the azimuthal motion was neglected here. From this last expression, we conclude that: (1) the radial motion scales as γ^{-3} therefore, for relativistic particles in a symmetric TM mode propagating at c, the transverse motion is expected to be stable. (2) If in the longitudinal direction the electromagnetic field is accelerating the particle i.e.,

$$F_z = -eE_0\cos\chi, \tag{8.1.39}$$

is positive (say $\chi = 135°$), then the transverse motion is unstable.

Now that we have a general feeling as to the transverse processes that occur when a bunch traverses an acceleration structure, we shall introduce a systematic way to deal with the problem. In their original work, Panofsky and Wenzel (1956) defined the transverse momentum experienced by a relativistic bunch as it traverses an interaction of length D by

$$p_{\perp} = \frac{-e}{\mathrm{v}_0} \int_0^D \mathrm{d}z(\mathbf{E} + \mathbf{v_0} \times \mathbf{B})_{\perp, t=z/\mathrm{v}_0}. \tag{8.1.40}$$

This definition takes into account the effect of the field generated by the particle on itself via the structure (the self-field is obviously excluded). We have indicated that as a particle traverses a slow-wave structure or a cavity, it leaves behind a broad spectrum of electromagnetic waves – this was also referred to as electromagnetic wakefield. Since this field may affect bunches trailing far behind the generating one, it is convenient to define the so-called transverse wake potential as

$$W_{\perp}(s,r) = -e \int_{-\infty}^{\infty} \mathrm{d}z[\mathbf{E} + \mathbf{v_0} \times \mathbf{B}]_{\perp}(r, z, t = (z+s)/\mathrm{v}_0), \tag{8.1.41}$$

which can be conceived as a generalization of (8.1.40). In an equivalent way one can define the longitudinal wake potential as

$$W_{\mathrm{L}}(s,r) = -e \int_{-\infty}^{\infty} \mathrm{d}z E_z[r, z, t = (z+s)/\mathrm{v}_0]. \tag{8.1.42}$$

The Fourier transform of these two potentials

$$\begin{aligned} Z_{\perp}(\omega, r) &= \frac{j}{\mathrm{v}_0 e^2} \int_{-\infty}^{\infty} \mathrm{d}s W_{\perp}(s,r) \exp\left(-j\frac{\omega}{\mathrm{v}_0}s\right), \\ Z_{\mathrm{L}}(\omega, r) &= \frac{1}{\mathrm{v}_0 e^2} \int_{-\infty}^{\infty} \mathrm{d}s W_{\mathrm{L}}(s,r) \exp\left(-j\frac{\omega}{\mathrm{v}_0}s\right), \end{aligned} \tag{8.1.43}$$

determines the longitudinal and transverse impedances; several cases of interest were considered by Heifets and Kheifets (1990).

Since the two wake potentials previously introduced are determined by various components of the electromagnetic field, which are inter-dependent via Maxwell's equation, we may expect the two wake potentials to be also inter-dependent. The relation between the two can be shown based on Faraday law, assuming a steady state regime and having the symmetric TM mode in mind,

$$\frac{\partial}{\partial r}E_z - \frac{\partial}{\partial z}E_r = j\omega\mu_0 H_{\phi}. \tag{8.1.44}$$

We firstly calculate

$$\frac{\partial}{\partial r}W_{\mathrm{L}}(s,r) = -e\int_{-\infty}^{\infty} \mathrm{d}z\frac{\partial}{\partial r}E_z[r,z,t=(z+s)/\mathrm{v}_0], \tag{8.1.45}$$

and secondly,

$$\begin{aligned}\frac{\partial}{\partial s}W_{\perp}(s,r) &= -e\int_{-\infty}^{\infty} \mathrm{d}z\frac{\partial}{\partial s}\left[E_r - \mathrm{v}_0\mu_0 H_\phi\right](r,z,t=(z+s)/\mathrm{v}_0),\\ &= -e\int_{-\infty}^{\infty} \mathrm{d}z\left[j\frac{\omega}{c}\frac{1}{\beta}E_r(r,z,t) - j\omega\mu_0 H_\phi(r,z,t)\right]_{t=(z+s)/\mathrm{v}_0}.\end{aligned} \tag{8.1.46}$$

Bearing in mind that

$$\begin{aligned}\frac{\partial}{\partial s}E_r[r,z,t=(z+s)/\mathrm{v}_0] &= \left[\frac{\partial E_r}{\partial z}\right]_{t=(z+s)/\mathrm{v}_0} + \frac{1}{c\beta}\left[\frac{\partial E_r}{\partial t}\right]_{t=(z+s)/\mathrm{v}_0}\\ &= \left[\frac{\partial E_r}{\partial z}\right]_{t=(z+s)/\mathrm{v}_0} + \left[\frac{1}{c\beta}j\omega E_r\right]_{t=(z+s)/\mathrm{v}_0},\end{aligned} \tag{8.1.47}$$

and assuming that for $z \to \pm\infty$ the transverse electric field vanishes, i.e., $E_r[r,z,t=(z+s)/\mathrm{v}_0] \to 0$, we have

$$\frac{\partial}{\partial s}W_{\perp}(s,r) = -e\int_{-\infty}^{\infty} \mathrm{d}z\left[\frac{\partial E_r}{\partial z} - j\omega\mu_0 H_\phi\right]_{t=(z+s)/\mathrm{v}_0}. \tag{8.1.48}$$

Adding (8.1.45), (8.1.48) we obtain

$$\frac{\partial}{\partial r}W_{\mathrm{L}}(s,r)+\frac{\partial}{\partial s}W_{\perp}(s,r) = -e\int_{-\infty}^{\infty} \mathrm{d}z\left[\frac{\partial E_z}{\partial r} - \frac{\partial E_r}{\partial z} - j\omega\mu_0 H_\phi\right]_{t=(z+s)/\mathrm{v}_0} \tag{8.1.49}$$

which by virtue of Faraday law (8.1.44) implies

$$\frac{\partial}{\partial s}W_{\perp}(s,r) = -\frac{\partial}{\partial r}W_{\mathrm{L}}(s,r). \tag{8.1.50}$$

This relation is the formal notation of the Panofsky-Wenzel theorem phrased above. Equivalently, this theorem can be formulated in terms of the longitudinal and transverse impedances introduced in (8.1.43) as:

$$\frac{\partial}{\partial r}Z_{\mathrm{L}} = -\frac{\omega}{c}\frac{1}{\beta}Z_{\perp}. \tag{8.1.51}$$

To reiterate, an acceleration structure is a *defocusing element* and for maintaining stable motion over extended length, it is strictly necessary to incorporate a focusing element in an acceleration module. For a thorough discussion on electron optics, magnetic lenses and periodic focusing systems (e.g. FODO lattice) the reader is referred to Chaps. 4–6 in Wiedemann (1999a, b).

8.1.5 Beam Break-up

Even with focusing elements in place, the transverse motion may become unstable. Already in the 1960s, it was observed at SLAC that the transmitted electron-pulse appears to shorten if the total current exceeds a certain threshold. This effect was attributed to a radial instability called beam break-up (BBU) which is due to the coherent interaction of the electrons with a *hybrid mode*, that is to say a mode which possesses properties of both TM and TE modes. In particular, we can conceive a cavity in which the TM_{110} mode is excited. Longitudinal variations are ignored in this case ($\partial/\partial z \sim 0$) but we allow radial and azimuthal variations such that the non-zero component of the magnetic vector potential reads

$$A_z(r,\phi) = A\mathrm{J}_1\left(s_1 \frac{r}{R}\right) \cos\phi, \tag{8.1.52}$$

where s_1 is the first zero of the Bessel function of the first kind and first order i.e., $\mathrm{J}_1(s_1) \equiv 0$. Consequently, the non-zero components of the electromagnetic field are

$$\begin{aligned} E_z(r,z) &= -j\omega A\mathrm{J}_1\left(s_1 \frac{r}{R}\right) \cos\phi, \\ B_r(r,z) &= -\frac{1}{r} A\mathrm{J}_1\left(s_1 \frac{r}{R}\right) \sin\phi, \\ B_\phi(r,z) &= -A\frac{s_1}{R}\mathrm{J}_1'\left(s_1 \frac{r}{R}\right) \cos\phi, \end{aligned} \tag{8.1.53}$$

where $\mathrm{J}_1'(u) = \mathrm{J}_0(u) - \mathrm{J}_1(u)/u$ and the eigen-frequency of this cavity is $\omega = s_1 c/R$. Near the axis the non-zero components are

$$\begin{aligned} E_z &\sim E\frac{x}{R}, \\ B_y &\sim \frac{1}{j\omega R}E. \end{aligned} \tag{8.1.54}$$

An electron, which traverses this cavity, experiences a deflection even if it is perfectly aligned because of the $\mathbf{v} \times \mathbf{B}$ term. Specifically, the change in the transverse momentum due to the excitation of this mode is

$$\Delta p_x = -e\int_0^D \mathrm{d}z \frac{1}{\mathrm{v}_0}(-\mathrm{v}_0 B_y) \sim \frac{e}{\omega}\frac{D}{R}\mathrm{Re}(-jE). \tag{8.1.55}$$

Consistency with the prior assumptions forced us to assume a uniform field in the z direction – which is not the case in general but it is a reasonable approximation on the scale of a single cavity. Furthermore, from this expression we learn that the transverse deflection is proportional to the longitudinal electric field and therefore, if we now consider a set of such coupled cavities, then the mode may grow in space and after a certain interaction region it dumps the beam to the wall. The corresponding mode is called hybrid electric magnetic mode and in our particular case, it is HEM_{11}.

BBU can be divided into two different types: beam break-up that occurs in the scale of a single acceleration structure because of feedback (either due to backward-wave interaction or due to reflections) and then the condition for BBU occurs as the threshold condition is reached – as in an oscillator. This is called *regenerative* beam break-up. In the other case, the cumulative beam break-up, the "information" is carried by the beam and it occurs on the scale of many acceleration sections (which are electromagnetically isolated). Panofsky and Bander (1968) developed based on Panofsky-Wenzel theorem a model which fits the basic features of long range BBU occurring on the scale of many acceleration sections. Helm and Loew (1970) have given a good tutorial of the various BBU mechanisms. Lau (1989) has proposed a framework from which the various BBU regimes can be derived.

While in Sect. 4.5 we have discussed in some detail the BBU in the context of coherent hybrid mode coupled to the beam and the way to suppress it, in the previous paragraphs we have briefly discussed the deflection of a single bunch by a hybrid mode in a cavity. In the remainder of this subsection we describe qualitatively the BBU that may occur along a single bunch due to the *wake-field* itself generates. Chao (1993) has developed a simple model to describe the motion of the tail of a bunch in terms of the wake generated by the head of the bunch. This is a two-particles model whereby both particles follow an *off-axis* betatron motion however, while the motion of the head-particle is determined solely by the external field, the tail-particle is affected also by the wake-field the former generates. Because the external field experienced by both particles is identical, the contribution of the wake-field to the motion of the tail-particle is in resonance with external force. Consequently, the transverse motion is unstable. If the acceleration is ignored, the growth rate of the instability is inversely proportional to γ and therefore, if the acceleration is accounted for, the growth rate is somewhat reduced. Assuming a linear acceleration

$$\gamma(z) = \gamma_{\text{in}} \left(1 + \frac{z}{L} \frac{\gamma_{\text{f}} - \gamma_{\text{in}}}{\gamma_{\text{in}}} \right) \tag{8.1.56}$$

thus

$$\frac{1}{\gamma} \to \frac{1}{L} \int_0^L dz \frac{1}{\gamma(z)} = \frac{1}{\gamma_{\text{f}} - \gamma_{\text{in}}} \ln\left(\frac{\gamma_{\text{f}}}{\gamma_{\text{in}}} \right). \tag{8.1.57}$$

In order to suppress this resonant instability, Balakin, Novokhatsky and Smirnov (BNS) have suggested (Balakin et al. 1983) to decouple between the two resonances. In other words, the external betatron field experienced by the tail-particle is slightly different than that experienced by the head-particle. In this way, under certain idealized conditions (discussed in detail by Chao (1993)) the instability can be eliminated. One example of such a shift is to introduce a time-dependent betatron field.

8.2 Advanced Accelerator Concepts: Brief Overview

The need for more powerful accelerators has triggered an extensive search for various other schemes to accelerate electrons. These schemes rely on either entirely new concepts as is the case in the plasma-based accelerators or on new technologies which is the case in optical accelerator. Although energy is the major parameter of interest in accelerators, it is not the only critical one. Other parameters such as emittance, repetition rate and number of particles per bunch are also of great importance. Therefore, the test of each one of the methods that will be discussed in the following sections is not only by the gradient that they are generating but also in the potential of being incorporated in a large system taking into consideration all the other parameters mentioned above. At this stage of research, the zero order parameter of comparison remains the accelerating gradient and this will be the basis for our discussion.

Before we start describing the various schemes, it is important to point the reader to the first characterization compiled by J.D. Lawson (1979) for acceleration of electrons with a laser field. Throughout the years this study has evolved to the so-called *Lawson-Woodward Theorem.* It specifies the conditions when electrons *cannot gain energy* from a laser field: (1) the laser field is in vacuum with no walls or boundaries present. (2) The electron is highly relativistic ($v \sim c$)along the acceleration path. (3) No static electric or magnetic fields are present. (4) The region of interaction is infinite, and (5) ponderomotive effects (nonlinear forces, e.g. $\mathbf{v} \times \mathbf{B}$ force) are neglected.

The various schemes can be categorized according to several criteria: (1) the energy-source and (2) the facilitating structure/medium. In the former category, we can list microwave, laser, a different electron beam or an active-medium whereas the second includes slow-wave structures, plasmas and wigglers. In the past 30 years, many of the possible combinations of the two categories have been explored in one way or another and the reader can consult the series of proceedings of the Advanced Accelerator Concepts workshops published since 1982; the credit for funding this research goes virtually exclusively to the US Department of Energy.

Two examples are in place at this point: First let us consider the plasma-based acceleration whereby space-charge waves are excited and they eventually accelerate the particles. Typically either they are generated by a short electron-bunch or a short but intense laser pulse or two-long medium-power laser pulses that the difference between the two frequencies equals the background plasma frequency – the latter is called plasma beat-wave acceleration.

When electrons are injected with a velocity close to that of the space-charge wave (beat wave), they can be trapped and therefore accelerated. Since no external walls are involved there is no problem of breakdown; however, it does not mean that the gradients are "infinite" since once the density modulation becomes significant (larger than 10%) non-linear effects dominate and the acceleration is altered. In order to have a rough estimate as of the typical gradients achievable we quote here the numerical example presented by Joshi et al. (1993): the longitudinal electric field that develops in plasma as a result of a density modulation $\delta n/n_0$ where n_0 the density of the unperturbed background plasma is, $E \sim 100\sqrt{n_0}\delta n/n_0$ (V/m). If we take a 10% modulation on a background density of $10^{21} < n_0(m^{-3}) < 10^{23}$ then the achievable gradients are between 0.3 and 3 GV/m.

A difficulty that might be raised is the scattering of the accelerated electrons with the background plasma. Apparently, the dominant scattering mechanism for over MeV electrons, is scattering by the plasma nuclei. Estimates made at UCLA (Katsouleas and Dawson 1989) indicate that even at a density of 10^{26} m^{-3} the mean free path of a relativistic electron (few MeV) is 2 km and it increases with the energy of the accelerated electrons. Historically, Clayton et al. (1993) reported the first clear indication of actual acceleration explicitly, externally injected 2.1-MeV electrons were accelerated by a laser beat-wave driven relativistic plasma wave. Electrons with energies up to the detection limit of 9.1 MeV were detected when such a plasma wave was resonantly excited using a two-frequency laser in about 1 cm long interaction region. Throughout the years, several groups around the world have investigated the concept in Japan, Kitagawa et al. (1992) in France, Amiranoff et al. (1992) and others. Recently, Tochitsky et al. (2004) demonstrated a laser plasma beat-wave acceleration of almost 40 MeV in a 3 cm long plasma channel.

A second example is the two-beam accelerator. Conceptually, an accelerator can be conceived, as a transformer in the sense that high current – low voltage beams form the primary and low-current high-voltage constitutes the secondary. In all large machines operating today, multiple microwave sources form the primary and a single accelerated beam is the secondary. However, in principle, it is possible to have a single beam that generates microwave radiation in a series of output structures. The radiation is guided into an acceleration structure that accelerates a different charged beam. In this case, the primary consists of a single high-current low-voltage whereas the secondary is a low-current high-voltage. We discuss this concept in more detail in the next section. Subsequently we discuss additional advanced acceleration concepts.

8.3 Two Beam Accelerator

Conceptually a linear accelerator consists of many modules of acceleration structures each one fed by one or more klystrons. Each klystron in turn is driven by an electron beam generated separately therefore thousands of beams form the primary and they accelerate a single beam which, as indicated in the previous example, is the secondary of a large "transformer". Almost 30 years ago, Sessler (1982) has suggested replacing all these discrete beams by a single driving beam carrying all the required energy – this primary is still driven by discrete sources. It came in parallel to the substantial progress in understanding the operation of the free electron laser. In particular, the fact that electrons could be trapped [see Sect. 7.4 as originally shown by Kroll et al. (1981)] and their energy extracted without substantial energy spread, suggested that after extraction in an FEL, the electrons could be re-accelerated. In addition, the operation of a klystron beyond X-band becomes problematic because of the small structure required whereas the free electron laser can generate high power levels at high frequencies without inherent structure limitations but with substantial constraint upon the beam quality.

Initially, this original approach of *power generation*, called two-beam acceleration (TBA) – see Fig. 8.3 – was contemplated to start with a medium-energy (3 MeV) high-current (1 kA) beam, extract power in each segment and compensate the driving beam for the lost energy in a re-acceleration unit. Thus, each section consists of three units: the acceleration unit, extraction unit and re-acceleration unit (Hübner 1993).

At CERN, the original approach was somewhat different (Schnell 1991): the initial energy of the electrons is three orders of magnitude higher, in the GeV range, and at least in the preliminary experimental stages (2 $\times$ 250 GeV) no re-acceleration was planned – as illustrated in Fig. 8.4. In the conceptually future system (2 TeV) conceived at CERN, a few superconducting re-acceleration cavities are included in the design. Traveling-wave structures were planned to extract electromagnetic power on the order of 40 MW from a pre-bunched beam at 30 GHz. In order to have the correct perspective of the performance of each section we should note that the 40 MW of power generated at 30 GHz produce almost the same gradient as 400 MW at 11.4 GHz as is the case in the Choppertron (Haimson 1992). In spite of the clear advantage of operation at high frequency with regard to the acceleration gradient, the wake-fields are correspondingly high.

If, in order to reduce wake effects, the frequency is reduced from 30 to 24 GHz the longitudinal wake effects are reduced to about 60% and the transverse wakes to 50% but this comes at the expense of the accelerating gradient which is lowered from 80 to 50 MV/m. The scaling law behind this result can be readily retrieved

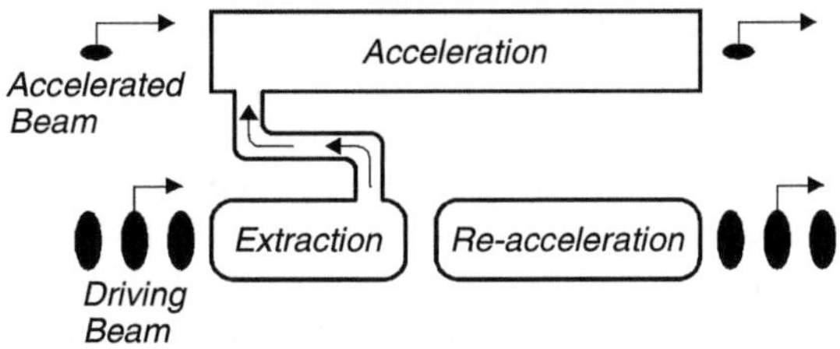

Fig. 8.3 Conceptual set up of the two-beam accelerator as proposed by Sessler (1982)

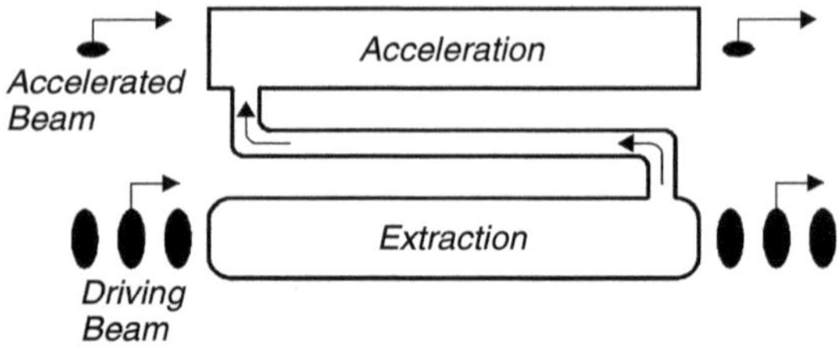

Fig. 8.4 Conceptual set up of the two-beam accelerator originally designed at CERN

bearing in mind that in a uniform waveguide the relation between the power carried by a single TM_{01} mode and the electric field on axis is

$$\frac{P}{|E|^2} \propto \beta_{gr} R^2 \left(\frac{\omega}{c} \frac{R}{p_1} \right)^2 . \tag{8.3.1}$$

If we require a single mode operation, $\omega R/c$ is limited by the cutoff of the second mode i.e., $\omega < p_2 c/R$. Consequently, for a given group velocity $\beta_{gr} c$ and a given gradient, the power (in the accelerating structure) is expected to be inversely proportional to the frequency i.e.,

$$P \times \omega^2 \sim \text{const.} \tag{8.3.2}$$

Although the beam energy in both schemes is larger than in the klystrons of the "conventional" scheme, the amount of current required to provide the power goes up too. Consequently, the amount of charge that propagates is larger and this may deteriorate the bunching via space-charge forces. In spite the apparent advantages of high-frequency operation the drawbacks associated with such an operation, have forced the designers in 2007 to return to X-band operation (12 GHz) and reduce the requirement from an accelerating of 150 MV/m at 30 GHz to 100 MV/m. In 2008, the CERN design was built at KEK and tested at SLAC. In the remainder of this section, we highlight some of the design parameters of the Compact Linear Collider (CLIC). A detailed account has been recently reported by Tomas (2010).

Two alternative linear collider projects are developed, the International Linear Collider (ILC), based on superconducting technology and CLIC that relies on the two-beam acceleration and harnesses room-temperature RF technology. Both designs aim to the TeV range and as a first phase the goal is to demonstrate 0.5 TeV operation and if successful, in Phase II, upgrades will aim to 1 TeV or higher – see a comparison of the various parameters corresponding to the Phase I (0.5 TeV) in Table 8.1.

Let us trace the conceptual design (3 TeV), as illustrated in Fig. 8.5, from top to bottom: two 2.38 GeV 1 km long conventional accelerators driven by 326 klystrons each; the 1GHz klystron generates a 139 μs-long pulse of 33 MW. In the next stage the train of bunches is compressed such that the peak current intensity is elevated from 4 to 100 A and the spacing between the bunches is reduced from 60 cm to 2.5 cm. Each train is 239 ns long and there are 24 trains with 5.8 μs spacing – see

Table 8.1 CLIC and ILC parameters

	CLIC	ILC
Center of Mass Energy [TeV]	0.5	0.5
Luminosity [10^{34}]	1.4	2.0
Main linac rf frequency [GHz]	12	1.3
Gradient [MV/m]	80	31.5
Linac repetition rate [Hz]	50	5
Number of particles per bunch [10^9]	6.8	15
Number of bunches per train	354	2,670
Beam power [MW]	4.9	10.8
Total site AC power [MW]	130	230
Norm. hor. emittance [mm × mrad]	2.4	10
Norm. vert. emittance [mm × mrad]	0.025	0.04

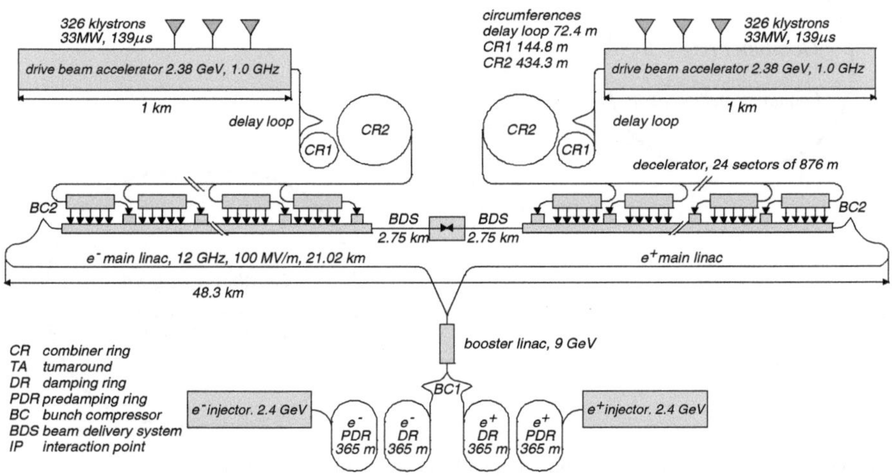

Fig. 8.5 Schematic of the recent design of Compact Linear Collider (CLIC) at CERN based on two-beam acceleration paradigm

Fig. 8.6. The trains of bunches are injected into power extraction and transfer structures distributed along the 2 × 21 km long linear accelerator generating a gradient of the order of 100 MV/m at 12 GHz; the initial energy of the electrons/positrons is 9 GeV.

In the context of the topics discussed earlier in this monograph, there are a few "critical items" that need to be demonstrated experimentally in order for this design to materialize. (a) Accelerating structures that sustain gradients of the order of 100 MV/m for a fraction of a microsecond. (b) Power extraction and transfer structures (PETS) that generate the corresponding power level for the required duration and transfer this power to the acceleration structure; integration of the two. (c) Generation of the 100 A drive beam with 12 GHz bunch frequency, meeting the phase, energy, and intensity stability tolerances. As of today, each one of these

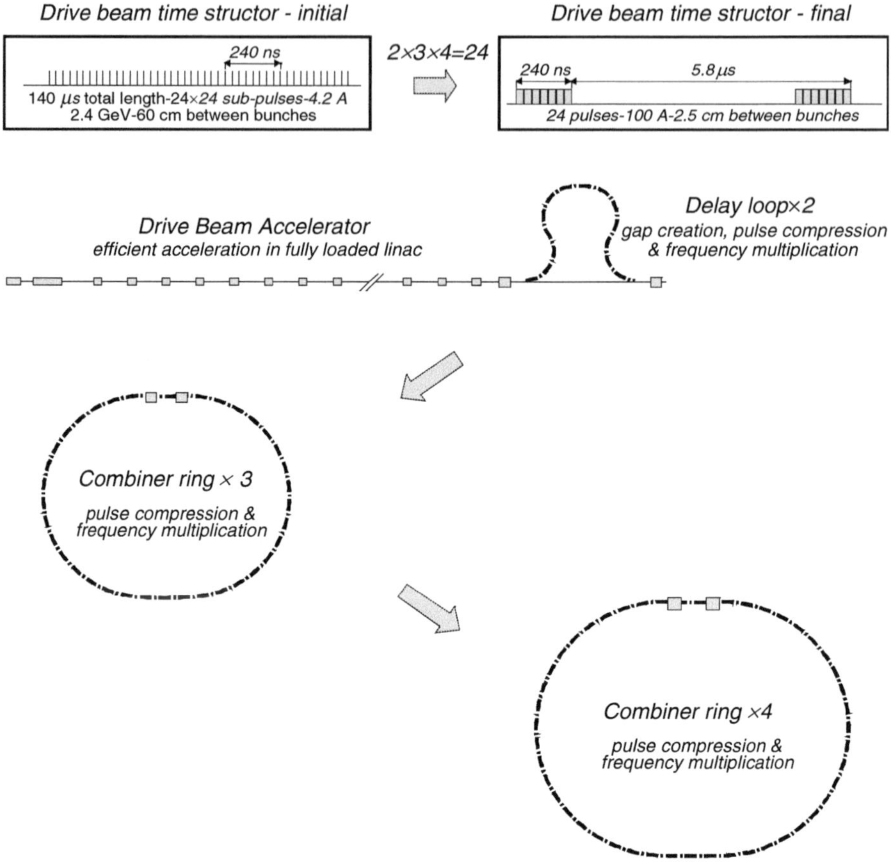

Fig. 8.6 The scheme of pulse compression and frequency multiplication (×24)

topics is in various demonstration stages and only time will tell whether the conceptual or technical obstacles will be conquered – Tomas (2010).

8.4 Plasma-Based Acceleration

Wakefields of intensities relevant for acceleration may be generated in plasma by either an intense laser pulse or a short bunch of electrons. Regardless the method, the pulse duration needs to be half a plasma wavelength long as shown by Tajima and Dawson (1979) – or shorter. For plasma densities in the range 10^{23}–10^{25} m^{-3}, this wavelength corresponds to laser pulse duration of the order of 15–150 fs – this scheme is also known as laser wake-field accelerator (LWFA). Six years later Chen et al. (1985) published the second paradigm whereby a particle beam drives the plasma wakes. Plasma electrons are pushed off axis either by the transverse force of the light pulse or by the repelling electrostatic force. As the pulse passes, in both cases, the plasma electrons snap back toward the back of the drive pulse because of the restoring force exerted by the immobile plasma ions, overshoot, and generate in the process the wake-field oscillation. In both cases, the wake-field propagates virtually at the speed of light and the bunch to be accelerated may be placed 1.5 plasma wavelengths behind the driving bunch.

8.4.1 Laser Wake Field Acceleration

Observations of a wake produced by a single short laser pulse were first published in 1996 by the Ecole Polytechnique (Marqués et al. 1996) and University of Texas at Austin (Siders et al. 1996) groups. In both of these experiments, the laser was focused to a spot size much smaller than the wavelength of the plasma oscillation and consequently, the oscillation was dominated primarily by the radial motion of the electrons and measurements of these oscillations provided the first indications associated with the wake-field.

In the Ecole Polytechnique experiments, the cylindrical plasma wake-field was excited by a 130 fs Ti:sapphire laser – Amiranoff et al. (1998). By injecting a 3 MeV electron beam into the wake, a maximum energy gain of 1.6 MeV was measured; the interaction length was of the order of 1 mm therefore the energy gain corresponds to a maximum longitudinal field of 1.5 GeV/m. This experiment was followed a series of experiments that demonstrated that plasma electrons could be trapped by the wake and accelerated to a broad range of energies that peaks at 200 MeV – see Malka et al. (2002). In practice, in all these experiments the laser pulse length is longer than half of the plasma wavelength. Nevertheless, at the high plasma densities and laser intensities employed, the laser pulse rapidly self-squeezes due to the non-linear processes involved. These initially longer laser pulses propagated through the plasma, they become shorter and excite the wake-field in the plasma.

Such a broad spectrum of electrons is not practical for acceleration application where $\Delta\gamma/\gamma$ smaller than one thousandth are required. Therefore, researchers started to look into the possibility of generating "mono-energetic" bunches. In order to understand the broadening mechanism we should bear in mind that the ions-bubble generated by the laser, moves close to the speed of light but still accelerated electrons may outrun it in a distance known as the *de-phasing* distance. While this limits the maximum energy gain, it may generate an electron beam with a relatively narrower energy spread.

For this purpose, consider the fraction of the electrons blown out by the drive pulse and are first trapped by the spike of the accelerating field. It so happens that a significant number of electrons become trapped so that the wake is beam-loaded and as a result, the accelerating field drops in amplitude. At this stage, no more electrons can be trapped and they are accelerated as a collective. However, the electrons initially have a spread of energies yet those in the front de-phase and begin to lose energy whereas the electrons in the back continue to gain energy. This phase-space rotation generates a relatively low energy-spread bunch.

It is imperative for this kind of process to terminate the acceleration close to the de-phasing distance by having a plasma-vacuum boundary. Otherwise, this phase-space squeezing process terminates. Such relatively low energy-spread beams have been observed experimentally in a plasma channel experiment at Lawrence Berkeley National Laboratory – Geddes et al. (2005). Shortly afterwards, the same group used a 3.3 cm long capillary discharge to produce a hydrogen plasma channel with a density of $4 \times 10^{24}[\mathrm{m}^{-3}]$. Injecting a 40 TW laser pulse through this channel, an average energy of 1 GeV was measured with a relative spread $\Delta\gamma/\gamma < 0.1$ – see Leemans et al. (2006). For a review of the various theoretical and experimental developments the reader is referred to Esarey et al. (2009) who compiled a detailed summary of Laser Wake-field Acceleration.

8.4.2 Beam-Driven Plasma Wake-Field Acceleration (PWFA)

The University of Wisconsin group – Rosenzweig et al. (1988), reported first demonstration of the excitation of a wakefield by a relativistic beam at Argonne National Laboratory. Its essence was to measure the change in energy of a low-intensity witness beam (15 MeV) with a variable delay for mapping the wake-field induced by a high-intensity the drive beam (21 MeV) in plasma. A peak acceleration gradient in excess of MeV/m and more importantly, the experiment clearly showed the wakefield persisting for several plasma wavelengths. Beyond this pioneering work, there was a major effort that demonstrated gradients of tens of GeV/m. A detailed review was compiled by Joshi (2007) here we skip to discuss one of the most exciting results achieved in recent years in the field of Advanced Acceleration Concepts.

In the mid nineties it was suggested to employ the 30–50 GeV beam of the Stanford Linear Accelerator Center to demonstrate 1 GeV energy gain in a one meter long plasma column of Lithium $[10^{18} - 10^{23}\mathrm{m}^{-3}]$ for both electrons and positrons. Since the amplitude of the wake is inversely proportional to the square of the bunch length one of the major contributions to the successful results to be reported next, is the ability to compress the 5 ps long bunch by two orders of magnitude (50 fs). In two different experiments at SLAC, the energy of a part of the electrons in the bunch was doubled. First, the energy of trailing electrons of an initial 28.5 GeV electron bunch was doubled at the expense of the front electrons; the plasma column is 60 cm long. Extending the plasma column to a total of 85 cm and using a 42 GeV bunch, the maximum energy that electrons have reached was 85 GeV – Blumenfeld et al. (2007).

E157 and E162 were carried out with 28.5 GeV electron and positron beams and even for the same number of particles and identical plasmas, the wake-field excitation is different for electron and positron bunches. Contrary to the electrons beam driver whereby the plasma electrons are completely expelled by the head of the beam, in the positrons case, the background electrons are pulled in from a plasma region that has a much larger radius than the beam radius. This leads to a factor 2 smaller wakes for identical conditions.

8.5 Other Wake-Field Acceleration Schemes

Wake-field generated by a bunch of electron in plasma is not the only configuration whereby an accelerating field is generated by a bunch. As indicated in Chaps. 3 and 5 wakes may occur in dielectric loaded waveguide or in periodic waveguide. Common to all is the fact that all the modes propagate with the same speed as the driving bunch and acceleration is achieved by "synchronizing" the radiation pulse and the accelerated bunch such that the latter trails behind and it sees an accelerating gradient.

8.5.1 Dielectric Wake-Field Accelerator

In order to understand the principles of the Dielectric Wake-Field Accelerator (DWFA) it is convenient to go back to Sect. 2.4.2 where we examined the Cerenkov radiation emitted by a point-charge as it traverses a dielectric loaded waveguide along its axis. We found that provided the velocity of the particle is higher than the phase velocity of the plane wave in the medium, then the magnetic vector potential is given by

$$A_z(r,z,t) = \frac{-q}{4\pi\varepsilon_0 R^2}\frac{\beta^2}{n^2\beta^2-1} \times \sum_{s=1}^{\infty}\frac{4J_0(p_s r/R)}{J_1^2(p_s)|\Omega_s|}\sin\left[|\Omega_s|\left(t-\frac{z}{v_0}\right)\right]h\left(t-\frac{z}{v_0}\right), \tag{8.5.1}$$

see (2.4.46) and the corresponding definitions. It is interesting to note that although the waveguide is dispersive, all the electromagnetic waves that belong to the wake travel at the particle's velocity v_0 – though they may trail far behind. It also explains why a "broad" spectrum signal can still provide net acceleration.

With this expression we can calculate the longitudinal electric field acting on a test particle lagging behind – see basic configuration in Fig. 8.7. On axis, E_z is

$$E_z(r=0,z,t) = \frac{-q}{4\pi\varepsilon_0 R^2}\sum_{s=1}^{\infty}\frac{4}{J_1^2(p_s)}\cos\left[|\Omega_s|\left(t-\frac{z}{v_0}\right)\right]h\left(t-\frac{z}{v_0}\right), \tag{8.5.2}$$

and since in practice the dielectric coefficient is frequency dependent the summation is only on these modes for which the Cerenkov radiation is satisfied i.e.,

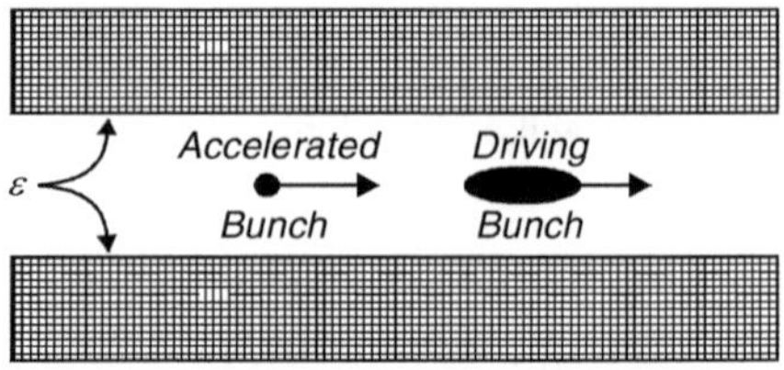

Fig. 8.7 Conceptual set up of the dielectric wake-field accelerator

$\beta^2 > 1/\varepsilon(\omega = \Omega_s)$. For simplicity sake, let us assume that only the first two modes contribute, thus the normalized force that acts on a negative point charge e is

$$F(\tau) \equiv F_z \left[\frac{eq}{4\pi\varepsilon_0 \varepsilon R^2}\right]^{-1} = \frac{4h(\tau)}{J_1^2(p_1)} \cos\left(p_1 \frac{c\tau}{\bar{n}R}\right) + \frac{4h(\tau)}{J_1^2(p_2)} \cos\left(p_2 \frac{c\tau}{\bar{n}R}\right), \tag{8.5.3}$$

and it is plotted in Fig. 8.8.; $\tau \equiv t - z/\mathrm{v}_0,\ \bar{n} = \sqrt{\varepsilon - \beta^{-2}}$.

As anticipated, in the close vicinity of the particle, the force is decelerating since "naturally" the negative charge repels another negative charge. However, if the test particle is located adequately behind the leading bunch, the trailing one will be accelerated. This field distribution also helps us to envision the bunch compression, which is a byproduct of this process.

For this purpose consider a uniform distribution of particles which spread between the zero acceleration point and the crest of the wave. Electrons that experience zero acceleration preserve their relative location in the bunch while all the others are accelerated. Even if the accelerated electrons bypass the first group, they immediately reach a deceleration region that pulls them back. Obviously, space-charge effects, disregarded in this discussion, limit this process. Conceptually, one can regard the system as a transformer with a low voltage and high current (say 10 MV, 1 kA) primary and the secondary is a high-voltage pulse of low current (say 1 GV, 1 A). DWFA was tested experimentally at Argonne National Laboratory by Gai et al. (1988) and analyzed theoretically by Rosing and Gai (1990).

Park and Hirshfield (2000) published an extension of the theoretical models beyond what was presented in Chaps. 3 and 5. In addition, a variety of reports describing experimental work has been published. Here we intend to briefly focus on two: the ANL group (Power et al. 2000) has designed a structure to have its TM_{0n} modes nearly equally spaced so that the modes generated by a single short electron bunch constructively interfere in the neighborhood of integral multiples of the fundamental wavelength producing large acceleration gradients. Since the space-charge force limits the amount of charge concentrated in one bunch, it is possible to split the charge into a train of micro-bunches. In the ANL experiment, trains of 4–5 nC electron bunches, separated by 760 ps, were injected into a 60 cm long dielectric-lined cylindrical waveguide. Use of a train of drive bunches spaced by one wavelength reinforced the accelerating wake-field – the gradient was in excess of 1 MV/m.

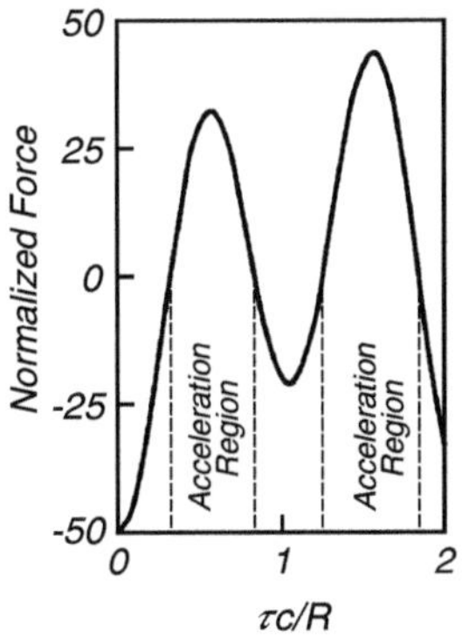

Fig. 8.8 Normalized force on a test particle which lags behind the driving bunch

The Omega-P group at BNL has recently performed a second demonstration of the superposition of wakefields. The wake was excited by 50 MeV bunches which travel 50 cm along the axis of a cylindrical waveguide loaded with alumina (Shchelkunov et al. 2006). The bunches were prepared by splitting a single laser pulse prior to focusing onto the cathode of an RF gun into two pulses and inserting an optical delay in the path of one of them. Wakefields from two short (5–6 ps) 0.15–0.35 nC bunches were superimposed and the energy loss of each bunch was measured as the separation between the bunches is varied to encompass approximately one wakefield period (21cm). A spectrum of 40 TM_{0n} eigenmodes is excited by the bunch. A substantial retarding wakefield (2.6 MV/m for the first bunch) develops because of the short bunch length and the narrow vacuum channel diameter (3 mm) through which they move. The energy loss of the second bunch exhibits a narrow peak when the bunch spacing is varied by only 4 mm (13.5 ps).

Gai and Jing (2006) compiled a comprehensive review of dielectric-loaded acceleration (DLA) structures including theoretical analysis of accelerating modes and wake-fields as well as experimental aspects. From the theoretical perspective, two computational methods used to compute the wake-fields of the DLA structure are described in detail. In the experimental context, multi-pactoring and dielectric joint breakdown are two major concerns for DLA structures which are subject to high EM fields and physical models and possible amelioration methods for both phenomena are considered. A multi-layered DLA structure based on the Bragg reflection waveguide principle was analyzed by Mizrahi and Schächter (2004a, b). It was demonstrated that the concept can reduce the RF power attenuation and improve the shunt impedance significantly compared to a structure consisting of a simple dielectric tube.

8.5.2 Wake-Field Acceleration in a Periodic Structure

The concept in this case is very similar to the case of the dielectric structure namely, a driving pulse generates a wake in the periodic structure that in turn accelerates the

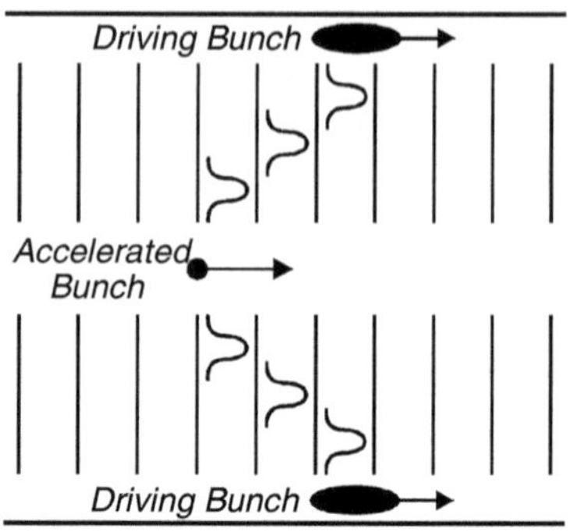

Fig. 8.9 Conceptual set up of the periodic structure wake-field accelerator

trailing bunch. Mathematical complexity of the calculations involved is substantially higher because the boundary conditions, however the outcome is similar to the dielectric case. Voss and Weiland (1982) (at DESY, Germany) suggested an annular configuration that is to say that the driving beam forms a ring which excites a wake-field. The latter propagates toward the axis and in the process its amplitude increases. As it reaches the axis, it accelerates a trailing bunch. Figure 8.9 illustrates this concept.

A different implementation of the same concept involves electro-optic switches: acceleration of a bunch requires a gradient at the (momentary) location of the particle. The way this gradient is accomplished has no importance to the longitudinal motion however, the transverse wake may vary from one scheme to another. Progress in optical switching of semiconductor devices (Lee 1984), facilitates the generation of fast voltage pulses that in turn can be used for acceleration. The essence of this concept is to optically switch a *radial transmission line* connected to a relatively high voltage source and benefit from the transformer effect as the voltage pulse propagates inwards to accelerate the electrons on axis. Bamber (1983) at the University of Rochester experimentally demonstrated the concept. Wilson (1988) and Cooper (1988) compiled detailed reviews of periodic wake-field in periodic structures.

8.6 Inverse of Radiation Effects

In all the effects where coherent radiation is generated by bunches of electrons, these are located such that the electromagnetic field, which acts at their location, is decelerating theses bunches. In principle, by virtue of the reciprocity theorem (see Sect. 2.1.8), we can place a similar bunch distribution to be in *anti-phase* with an illuminating wave, in which case the latter accelerate the electrons. The illuminating wave needs to have the similar characteristics as the far-field generated by the bunches.

8.6.1 Inverse FEL

In principle, the same mechanism that facilitates generation of coherent radiation in a periodic transverse wiggler as discussed in Chap. 7, allows to accelerate a bunch of electrons. In fact, Palmer (1972) suggested this concept several years before the renewed interest in the free electron laser as a radiation source (Elias et al. 1976). An intense laser pulse interacts with a beam of electrons in the presence of a transverse and periodic magnetic field. As a result, electrons may be accelerated – see Courant et al. (1985). The advantages and disadvantages of the free electron lasers as a radiation source, discussed in Chap. 7, apply also to its operation as an accelerator. In addition, Courant et al. (1985) pointed out that the decelerating gradient (E_{dec}) due to the emission of spontaneous radiation has to be smaller than the accelerating gradient associated with the laser field (E_{acc}). According to (7.1.30) and (7.2.6) this can be formulated as $E_{acc} > E_{dec}$ thus

$$\omega A_0 \frac{1}{\gamma} \frac{eB_w}{mck_w} > \frac{1}{2}\beta\gamma^2 \frac{e}{4\pi\varepsilon_0/k_w^2}\left(\frac{eB_w}{mck_w}\right)^2. \tag{8.6.1}$$

This expression determines a critical laser intensity I_{cr} which has to be exceeded in order to obtain net acceleration i.e.,

$$I \equiv \frac{1}{2\eta_0}|\omega A_0|^2 > I_{cr} \equiv \left[\frac{1}{2\eta_0}\left(\frac{e}{4\pi\varepsilon_0 k_w^{-2}}\right)^2\right]\left(\frac{eB_w}{2mck_w}\right)^2\gamma^6. \tag{8.6.2}$$

If we take $B_w = 1\,\mathrm{T}$ and $2\pi/k_w = 10\,\mathrm{cm}$ then $I_{cr} \sim 10^{-16}\gamma^6 \mathrm{W/cm^2}$ and for a typical γ of the order of 10^4 the required laser intensity is $10^8 \mathrm{W/cm^2}$ as shown next for several other energies:

$$\begin{aligned}
\gamma = 10^4 &\rightarrow I_{cr} = 10^8 \mathrm{W/cm^2},\\
\gamma = 10^5 &\rightarrow I_{cr} = 10^{14} \mathrm{W/cm^2}\\
\gamma = 10^6 &\rightarrow I_{cr} = 10^{20} \mathrm{W/cm^2},\\
\gamma = 10^7 &\rightarrow I_{cr} = 10^{26} \mathrm{W/cm^2}.
\end{aligned}$$

This list indicates that with laser intensities of $10^{20}\mathrm{W/cm^2}$ one may accelerate electrons up to a few TeV but at least with the present technology, this seems to be the limit.

During the past decade, the I-FEL was instrumental in demonstrating two fundamental questions: Let us assume that in one module a laser pulse accelerates electrons, is it reasonable to expect that electrons entering a second module keep their shape after a few meters long drift region? Staging of two laser-driven I-FEL's has been demonstrated in a proof-of-principle experiment performed by Kimura et al. (2001) at ATF. Two distinct and serial laser accelerators acted on an electron beam in a coherently cumulative manner. Output from a CO_2 laser was split into two beams to drive two inverse free electron lasers separated by 2.3 m. The first IFEL served to bunch the electrons into 3 fs micro-bunches, which were "rephased" with the laser wave in the second IFEL.

A second fundamental question that needs to be addressed concerns the efficiency of trapping at optical frequencies. Development of practical and efficient laser linear accelerators requires accelerating a large ensemble of electrons together while keeping their energy spread small. Kimura et al. (2004) was the first to demonstrate high-trapping efficiency (80%) and narrow energy spread (0.36%) via laser acceleration. All these accomplishments (staging, high-trapping efficiency and narrow energy-spread) were achieved in the framework of the Staged Electron Laser Acceleration (STELLA) program conducted by Wayne Kimura at ATF – Brookhaven National Laboratory.

8.6.2 Inverse Cerenkov

Edighoffer et al. (1981) at Stanford demonstrated experimentally the feasibility of the inverse Cerenkov effect for acceleration of electrons. The idea here is to illuminate an electron moving in a dielectric medium (gas) at the Cerenkov angle with a laser beam at the adequate frequency. Later Fontana and Pantell (1983) proposed an improved setup for the same purpose by ensuring an extended and symmetric interaction region with axicon lens – see Fig. 8.10. The lens generates a symmetric longitudinal electric field on axis when illuminated by a radially-polarized laser beam and the gas slows down the electromagnetic wave which in turn, intersects the electron trajectory at an angle θ; the longitudinal wave-number is $(\omega/c)n\cos\theta$ where n is the refraction coefficient. Consequently, the resonance occurs when the phase velocity equals the velocity of the particle i.e.,

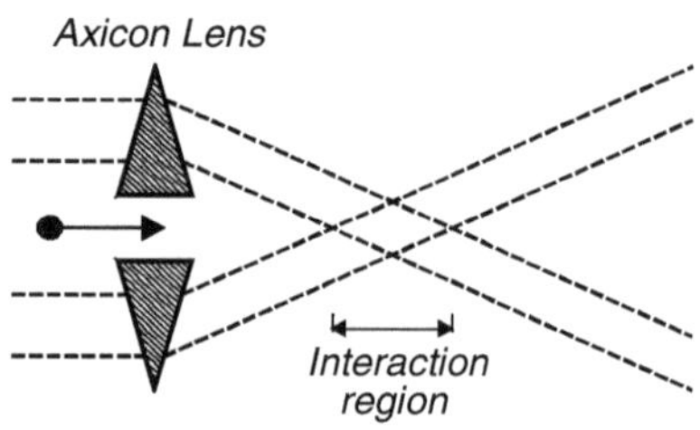

Fig. 8.10 Conceptual set-up of the inverse Cerenkov accelerator

$$v_{ph} \equiv \frac{c}{n\cos\theta} = v_0. \tag{8.6.3}$$

The concept was tested experimentally (Kimura et al. 1995) when electrons were illuminated by a laser focused by an axicon lens have gained an energy which corresponds to a gradient of 31 MV/m in the interaction region. Design of such a system is a trade-off between a (relatively) large refractive coefficient that requires high pressure and a long mean free path, which implies low pressure. A possible solution, suggested in this context by Steinhauer and Kimura (1990) [previously suggested in the context of FEL by Feinstein et al. (1986)], is to operate close to the resonant frequencies of the gas. This facilitates the required refraction coefficient but at low pressure.

8.6.3 Open Structure Accelerator

In Chap. 5 we discussed the electromagnetic characteristics of an open periodic structure and two main conclusions were emphasized: (1) the number of eigen-modes is finite and their number is controlled by the geometry of the structure. Each such mode consists of an infinite spectrum of harmonics whose phase velocity in the pass band is smaller than c; these harmonics correspond to evanescent waves and they do not carry power in the transverse direction. Another conclusion we reached was that (2) a particle moving in the proximity of an open structure emits radiation (Smith-Purcell effect). In principle, one can use this effect to accelerate electrons by illuminating the grating at the adequate angle and wavelength. However, contrary to the inverse Cerenkov effect, where the use of the radiation field in the interaction region can be fairly efficient, in the grating case, the incident wave is scattered in a spectrum of harmonics, part of which are radiative therefore a substantial fraction of the energy is lost.

Alternatively, we may use the eigen-modes of an open structure to accelerate electrons (Palmer 1982; Kroll 1985). However, one can immediately realize that a wave with phase velocity c is not supported by the kind of symmetric structure we discussed in Chap. 5 since if $k_z = \omega/c$ is parallel to the direction in which the electron moves then at least one harmonic of the mode does not decay exponentially. A simple solution of this problem is to enforce field variation in the third direction. Pickup (1985) has analyzed a grating which is periodic in the z direction, the y direction is perpendicular to the surface and in the x direction two metallic plates were placed at a distance D one from the other. Therefore, if the grating is designed such that in the z direction $k_z = \omega/c$ and in the x direction $k_x = \pi/D$ (lowest mode) then according to the homogeneous wave equation, $k_x^2 + k_y^2 + k_z^2 - (\omega/c)^2 = 0$, we have

$$\frac{\pi^2}{D^2} + k_y^2 + \frac{\omega^2}{c^2} - \frac{\omega^2}{c^2} = 0 \quad \rightarrow \quad k_y^2 = -\frac{\pi^2}{D^2}. \tag{8.6.4}$$

Consequently, the wave decays exponentially perpendicular to the grating.

Although the problem of the radiation confinement was solved, there still is a problem of stability of the beam. To illustrate this problem let us consider the following electric field

$$\begin{aligned} E_x(\mathbf{r},t) &= 0, \\ E_y(\mathbf{r},t) &= E\frac{\omega}{cq}\sin[\omega(t-z/c)]\sin(qx)\exp(-ky), \\ E_z(\mathbf{r},t) &= E\cos[\omega(t-z/c)]\sin(qx)\exp(-ky), \end{aligned} \tag{8.6.5}$$

where $q = \pi/D$. We can substitute in Maxwell's equation and obtain the magnetic field:

$$\begin{aligned} H_x(\mathbf{r},t) &= \frac{q^2-(\omega/c)^2}{\omega\mu_0 q}E\sin[\omega(t-z/c)]\sin(qx)\exp(-ky) \\ H_y(\mathbf{r},t) &= \frac{q}{\omega\mu_0}E\sin[\omega(t-z/c)]\cos(qx)\exp(-ky) \\ H_z(\mathbf{r},t) &= \frac{1}{\eta_0}E\cos[\omega(t-z/c)]\cos(qx)\exp(-ky). \end{aligned} \tag{8.6.6}$$

With these field components, we can calculate the transverse force, which acts on the particle. For this purpose, we assume that the particle's motion is around $x \simeq D/2 + \delta x$ and $y = 0 + \delta y$ consequently,

$$\begin{aligned} &\left[\frac{\mathrm{d}^2}{\mathrm{d}t^2} + \frac{qeE\sin\phi}{m\gamma}\right]\delta x = 0, \\ &\left[\frac{\mathrm{d}^2}{\mathrm{d}t^2} - \frac{qeE\sin\phi}{m\gamma}\frac{qc}{\omega}\right]\delta y = -\frac{eE}{m}\frac{qc}{\omega}\sin\phi, \\ &\frac{\mathrm{d}}{\mathrm{d}t}\gamma = -\frac{eE}{mc}\cos\phi; \end{aligned} \tag{8.6.7}$$

here it was assumed that on the scale of the transverse motion variations in γ can be ignored, $\gamma \gg 1$ and $\phi = \omega[t - z(t)/c]$. The first two expressions indicate that even if, in the x direction, the motion is stable in the y direction the beam diverges and vice versa. Pickup (1985) suggested a solution whose essence is of rotating the orientation of the grating relative to the axis as illustrated in Fig. 8.11. Alternatively, the phase ϕ can be switched periodically as suggested by Kim and Kroll (1982).

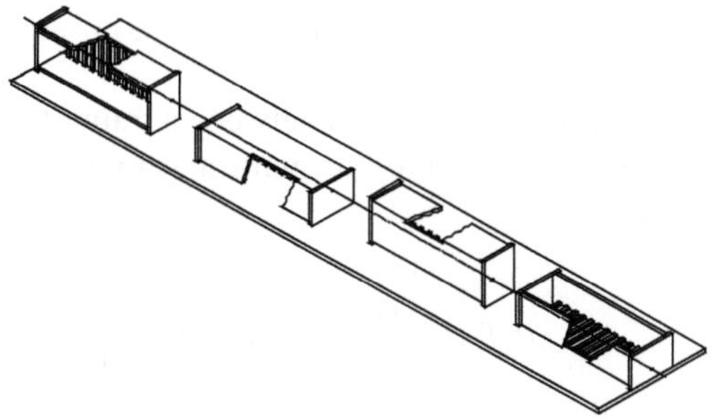

Fig. 8.11 Conceptual set up of the grating accelerator (inverse Smith-Purcell effect)

8.6.4 PASER: Particle Acceleration by Stimulated Emission of Radiation

In Sect. 2.4.5 we calculated the decelerating force that acts on a single electron as it moves in a vacuum channel surrounded by lossy material. At the end of that section we indicated that if the conductivity of the material is negative which is to say that the medium is active, then the moving electron is accelerated (Schächter 1995).

Motivated by this result, let us now take a closer look at the interaction of moving charged particles with an active medium. For this purpose, consider an ensemble of atoms, each one modeled by a *two-states* quantum system and a moving electron whose primary field consists of a broadband spectrum of evanescent waves – including the resonance frequency of this two-level system. These waves may be conceived as a spectrum of virtual photons continuously emitted and absorbed by the electron. When a virtual photon corresponding to the resonance frequency impinges upon an excited atom, its effect is identical to that of a real photon. Therefore, it stimulates the atom and as a result, two identical phase correlated photons are emitted.

Being phase correlated, the moving electron can absorb the stimulated photon, causing the latter's acceleration. The inverse process is also possible: if the virtual photon encounters an atom in the ground state and excites it, then the moving electron loses energy namely, the electron decelerates. We may expect a nonzero net acceleration only if the number of atoms in the excited state is larger than that in the lower state; i.e., the population is inverted. From the description above, the acceleration force is a result of a stimulated radiation and accordingly the acronym PASER stands for particle acceleration by stimulated emission of radiation.

Efficient interaction occurs only in the close vicinity of the resonance. Therefore, from the perspective of a single moving electron, it is quite clear that since in the laboratory frame of reference its spectrum is broad, the effect of the medium on the electron's energy is modest. In order to overcome this difficulty, it was suggested to inject a train of microbunches rather than a single macrobunch – its periodicity being identical to the resonance frequency of the medium. In this way, the projection of the train's spectrum on the resonance frequency of the medium becomes dominant.

The power exchanged between a train of M bunches of $N_{\rm el}$ electrons, separated by a distance λ_0 moving with a velocity v_0 and a dielectric active medium is given by

$$\begin{aligned} P(\beta) &= \frac{Q^2 v_0}{4\pi\varepsilon_0\lambda_0^2}\frac{2}{\pi}\int_0^\infty d\Omega \mathrm{Re}\left\{ j\Omega\left[1-\frac{1}{\beta^2\varepsilon_{\rm r}(\Omega)}\right]F_\perp(u)\right\} \\ &\quad\times \mathrm{sinc}^2\left(\frac{1}{2}\frac{\Omega}{\beta}\frac{\Delta}{\lambda_0}\right)\frac{\mathrm{sinc}^2\left(\frac{1}{2}\frac{\Omega}{\beta}M\right)}{\mathrm{sinc}^2\left(\frac{1}{2}\frac{\Omega}{\beta}\right)} \\ F_\perp(u) &\equiv \frac{2}{u^2}\left[1-2\mathrm{I}_1(u)\mathrm{K}_1(u)\right] \\ u &\equiv \frac{\Omega}{\beta}\frac{R_{\rm b}}{\lambda_0}\sqrt{1-\beta^2\varepsilon_{\rm r}(\Omega)}, \end{aligned} \tag{8.6.8}$$

wherein in non-normalized units the dielectric function is

$$\varepsilon_r(\omega > 0) \equiv 1 + \frac{\omega_p^2}{\omega_0^2 - \omega^2 + 2j\omega/T_2}, \tag{8.6.9}$$

subject to the condition $\varepsilon_r(\omega<0)=\varepsilon_r^*(-\omega)$. It is tacitly assumed that the medium has a single resonance frequency ω_0 chosen to correspond to the macro-bunch modulation i.e., $\omega_0 = 2\pi c/\lambda_0$; $\omega_p^2 \equiv e^2 n/m\varepsilon_0$ is the "plasma" frequency with m being the rest mass of the electron, and n representing the population density atoms that store a photon. For an excited medium, when the relevant population density is inverted, the plasma frequency is negative ($\omega_p^2<0$); T_2 being the relaxation time.

The density of the energy stored in the medium at the resonance frequency, assuming its population is inverted, $n_{\rm ph}$ denoting the photons' density stored in the medium, is $w_{\rm act} = -n_{\rm ph}\hbar\omega_0$ and $\hbar = 1.05457\times10^{-34}[\mathrm{J\cdot s}]$ is the Planck constant. Further simplification of (8.6.8) is possible subject several assumptions: (1) the bandwidth associated with the frequency response of the medium is much narrower than the spectrum associated with the finite-length of the macro-bunch ($M\lambda_0/cT_2 \ll 1$). In other words, the frequency response of the medium is sharper than the longitudinal form factor. (2) The bandwidth of the resonance is very narrow compared to the resonance itself i.e., $\lambda_0/cT_2 \ll 2\pi$. Based on these assumptions, we can take the longitudinal form factor out of the integral and evaluate the latter only at the poles of the dielectric function $\varepsilon_r(\Omega)=0$. (3) The bunches are initially relativistic ($\gamma \gg 1$); (4) the optical gain is small on the scale of one wavelength $\left(\omega_p\lambda_0/c \ll 1\right)$; (5) although the number of bunches is large, still $\gamma^2 \gg M$; (6) beam's radius is much smaller than the resonant wavelength $2\pi R_b/\lambda_0 \gg 1$. Based on these assumptions the relative increase in the kinetic energy of the train along an interaction length d is

$$\begin{aligned} \overline{\Delta E_k} &\equiv \frac{-Pd/v_0}{N_{el}mc^2(\gamma-1)} \\ &\simeq \frac{4N_{el}d\left(\pi r_e^2\right)}{(\gamma-1)}\frac{w_{act}}{\hbar\omega_0}\text{sinc}^2\left(\pi\frac{\Delta}{\lambda_0}\right)F_\perp\left(2\pi\frac{R_b}{\lambda_0}\right), \\ &\simeq \frac{4N_{el}d\left(\pi r_e^2\right)}{(\gamma-1)}\frac{w_{act}}{\hbar\omega_0}\text{sinc}^2\left(\pi\frac{\Delta}{\lambda_0}\right)2\left(2\pi\frac{R_b}{\lambda_0}\right)^{-2} \end{aligned} \tag{8.6.10}$$

N_{el} is the total number of electrons in the train ($Q = N_{el}e$), $r_e \equiv e^2/4\pi\varepsilon_0 mc^2$ is the classical radius of the electron.

A proof-of-principle experiment has been carried out (Banna et al. 2006a) at the accelerator test facility (ATF) of the Brookhaven National Laboratory. It harnessed the photocathode-driven microwave linear accelerator and a high-peak power CO_2 laser. A wiggler in which a laser beam interacts with a macro-bunch of electrons traveling through a periodic magnet array, is used to impart a sinusoidal energy modulation on the electron beam (e-beam). After a long drift region the velocity modulation translates into density-modulation. The resulting train of micro-bunches enters a CO_2 PASER cell follows shortly after by a spectrometer. A schematic layout of the experiment is presented in Fig. 8.12.

More specifically, a quasi-mono-energetic (45 MeV) electrons macro-bunch of 5 ps duration and consisting of at least 7×10^8 particles, was injected into a wiggler where it was bunched into about 150 micro-bunches by its interaction with a high-power CO_2 laser pulse (200 ps, 0.5 GW) operating at a wavelength of 10.6 μm. A 2.5 m long drift region separates the wiggler from the PASER cell. Along this drift region, the velocity modulation emerging from the wiggler becomes a density modulation at the entrance to the cell. The former is controlled by the intensity of the CO_2 laser pulse, and in this particular set-up, a $\sim 1.5\%$ peak-to-peak energy modulation at the wiggler was found to generate an optimal density modulation in the PASER cell. Either stronger or weaker modulation at the wiggler, lead to less

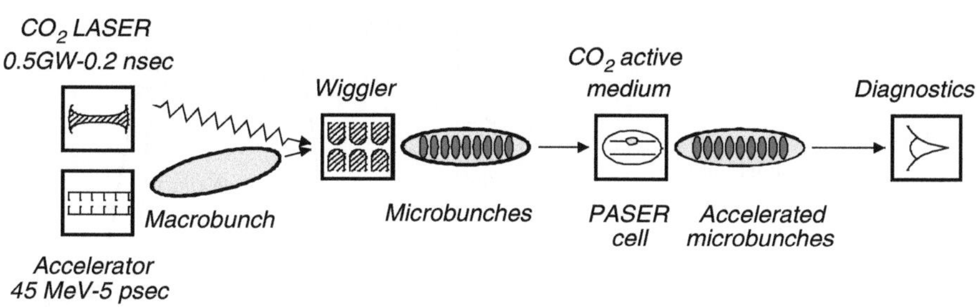

Fig. 8.12 Schematic layout for the PASER experiment. The distance separating the wiggler and the PASER cell is about 2.5 m

than optimal modulation, at the location of the interaction with the active medium and thus, smaller acceleration.

The train of micro-bunches enters next the PASER cell that contains a mixture of CO_2 [CO_2:N_2:He(2:2:3)], held at a pressure of 0.25 atm, activated by a discharge driven by a 130 nF low-inductance capacitor, initially charged to 30 kV. The discharge is facilitated by two 40 cm $\times$ 12 cm aluminum electrodes, which are 2.5 cm apart. Two diamond windows, of 1 mm diameter and 2 μm thickness each, are attached to both ends of the cell in order to maintain the pressure in the cell and at the same time to allow the train to propagate through the cell.

For the typical values mentioned above electrical measurements (voltage and current) of the discharge, indicate that the total energy density stored in the mixture is at the most of the order of 0.1 J/cm^3. Only a small fraction of this energy density is associated with the resonance of the CO_2 molecule at 10.2 μm. Therefore, assuming a potential efficiency (as an amplifier) of 1%, we estimate the energy-density available, at 10.6 μm, to be of the order of 1 mJ/cm^3. Based on this estimate, in the volume covered by a beam of a radius of 150 mm, the available energy is of the order of 70 mJ. However, the field associated with a relativistic bunch covers an area which effectively is γ^2 larger than the geometric beam cross section. In practice, in the vertical dimension the expansion is limited by the electrode spacing, and hence, the available energy is about 200 mJ. This value should be compared to 5 mJ kinetic energy of the train.

In order to demonstrate the PASER effect, during the experiment, pairs of shots with discharge on and off were recorded for different peak-to-peak energy modulations (1–3%). A jitter of up to 50 keV in the energy spread of the e-beam was observed at the spectrometer. As the PASER effect manifested itself as an increase of the energy spread of the macro-bunch as measured at the spectrometer, in the presence of the discharge in the cell, any increase beyond the 50 keV in the energy spread is considered as an acceleration via the PASER scheme.

The energy spectra as monitored by the spectrometer's camera are presented Fig. 8.13. The spectrum is illustrated for the two relevant cases: discharge-off and on. Evidently, the energy spectrum when the discharge is on, is broadened

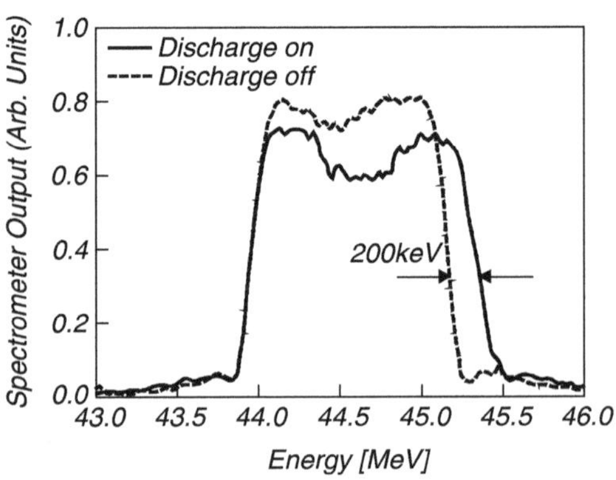

Fig. 8.13 The energy spectra of a single pair of shots with and without discharge, and with $\sim$1.5% peak-to-peak energy modulation. Comparison of the energy spectra discharge off/on, $\sim$200 keV energy increase is observed

comparing to the "off"-case by about 0.45% i.e., 200 keV. A detailed theoretical and experimental account was presented by Banna et al. (2006b) which also shows that the simple analytic result in (8.6.10) is in good agreement with the experimental results.

8.7 Optical Accelerators

The progress in laser-technology in the past twenty year has triggered the question to what extent can lasers replace the microwave sources as the drivers for particles accelerators. While they still have a long way to go before reaching the wall-plug to radiation efficiency of microwave sources (>70%), they possess a few inherent advantages which can not be ignored. On the top of the list is the compactness. Reducing the wavelength from tens of centimeters to microns and at the same time increasing the accelerating gradient from a few tens of MV/m to 1GV/m leads to reduction of almost two orders of magnitude in the length of a linear accelerator either for high-energy physics or medical applications.

8.7.1 Optical Linear Collider

The first proposal for an optical linear accelerator has been suggested by Robert Byer in the mid nineties (Huang et al. 1996) and in the past fifteen years some major experimental progress have been made in collaboration with SLAC [e.g. Plettner et al. (2005)]. While the initial configuration relied on a series of prisms that allowed to focus a laser beam along the trajectory of relativistic electrons, about a decade ago, a complex 2D quasi-periodic structure has been analyzed (Lin 2001) at SLAC following the regular approach of designing a slow-wave structure. This is to say that at the laser wavelength, the mode propagates at the speed of light and it has a significant longitudinal component. In this case the structure was (quasi) periodic in the transverse direction and uniform in the longitudinal direction. Some improvement was later found (Cowan 2003) when the designed structure was periodic in the longitudinal direction. Zhang et al. (2005) have also proposed to employ periodic structures to couple power into these structures from the transverse direction. A significant limitation of these structures is the fact that the periodic "obstacles", which facilitate the proper mode, are smaller that the wavelength and thus impose some stringent constraints on the power that could be injected. Contrary to RF acceleration structures, the problem is not breakdown but rather non-linear effects in dielectric material. The latter in turn is the only option in the optical regime since dielectric loss is by far lower comparing to ohm loss. For reducing the difficulties associated with these miniscule obstacles in 2D periodic structures, a simple 1D Bragg structure was investigated theoretically in great detail by Mizrahi and Schächter (2004b). Some of the fundamentals have been discussed in some detail in Sect. 5.4. At this point, we wish to emphasize a few aspects that are common to all optical acceleration structures mentioned above.

The second significant difference between an optical structure comparing to an RF one is the behavior of the wake-field and in particular, the long range wakes. Being made of dielectric obstacles or layers the structure acts like a filter. It confines radiation of frequencies close to the central frequency but it allows most of other frequencies to leak out. This is dramatically different comparing to metallic structures where all the spectrum generated by a single bunch is confined to the structure's volume – see Schächter (2003).

A third aspect that is conceptually different in optical accelerators is the structure of the bunch. As specified in the description of CLIC in this chapter, at RF frequencies, the number of electrons that are being accelerated are of the order of 10^{11} and for simplicity sake, let us assume that the volume they occupy is of the order of 1 mm^3 – thus the typical density is of the order of $10^{20}\ \text{m}^{-3}$. In the optical regime (say 1μm) the vacuum tunnel has a vertical dimension that is smaller than one wavelength (1/5) and obviously, the bunch needs to be much shorter than one wavelength (say 1/20). In the horizontal plane, we may conceive a sheet-beam configuration and we assume it to be 100 wavelengths wide. Assuming a similar density as specified, in one period of the pulse there are only 100 electrons!

Regardless the acceleration scheme the number of electrons per second that need to be accelerated, for either high energy physics applications or medicine is of the order of 10^{14}[1/s]. Assuming a conservative rep-rate of the driving laser – say 10 MHz then the number of electrons in one micro-bunch needs to be of the order of 10^7. There are very promising experimental results (Hommelhoff et al. 2006) indicating that by the time such a machine will get to the drawing board, densities of $10^{21}\ \text{m}^{-3}$ will become available at reasonable low energy, implying that each micro-bunch may contain order of 10^3 electrons and there should be 10^4 micro-bunches in one train. While at a first glance this seems a complexity, it is actually an advantage since this train of micro-bunches generates a coherent wake that can be re-circulated – see Schächter (2004) – leading to a dramatic enhancement in the *efficiency* of the acceleration process. However, there still remains one deficiency associated with the different wake-field experienced by various micro-bunches along the train. To compensate for this effect, it was suggested to taper the shape of the external laser pulse whereas the wake generated by one train may be amplified and harnessed for the amplification of the trailing train – see Fig. 8.14. Another option is to reverse (front to back) the wake-field before being fed back into the acceleration structure.

8.7.2 All Optical Light Source

While the discussion so far was primarily motivated by high energy physics, not less appealing is the optical accelerator compactness when applied to light sources in general and specially to medical accelerators. In the latter case, the electrons'

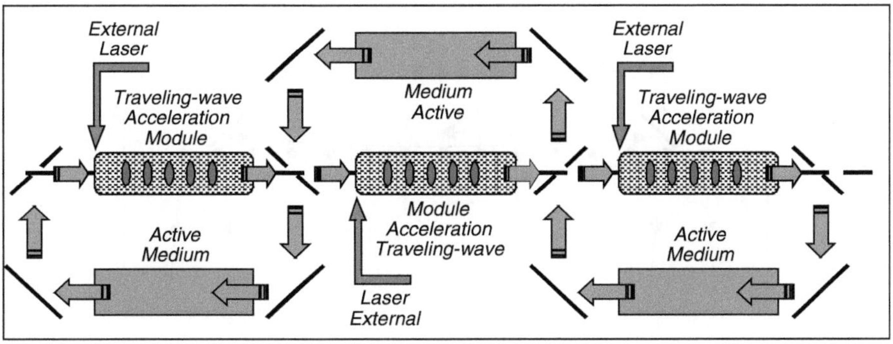

Fig. 8.14 Energy recovery in an optical linear accelerator

energy is limited to about 10 MeV, which implies that if the gradient is of the order of 1 GV/m, the acceleration structure could be 1 cm long; this should be compared to the current 50 cm long structures. However, the accelerator is only half the benefit. The other half, at least if quasi-monochromatic radiation is required, is the possibility to generate such radiation in a relatively efficient and compact way. Quasi-monochromatic X-ray radiation has been generated by focusing a laser beam against a relativistic beam of electrons. Inverse Compton effect facilitates X-ray radiation – see Sect. 7.5.5. Now that we have a vague idea of how FEL's and accelerators work, we may extend the discussion we started in that section.

X-Ray generation has been undergoing a steady revolution for the past five decades. The brightness of these sources, measured in photons·s^{-1} mm^{-2} $mrad^{-2}$/ 0.1%-bandwidth, has been rising exponentially, starting from X-ray tubes (10^7) on to bending magnets (10^{10}–10^{14}), wigglers (10^{13}–10^{16}) and undulators (10^{15}–10^{22}) – see Attwood (2000). The distinction between wigglers and undulators is set by the deflection parameter $a_{w,u} = eB/mck_{w,u}$ that is a measure as of how strongly the e-beam is deflected because of the interaction with the magnetic field. In wigglers the magnetic field is stronger than in undulators and the magnets are more widely spaced (λ_U is larger). Accordingly, the e-beam deflection is typically much larger than in an undulator and consequently, wiggler based sources are less bright. Optimal deflection parameter for maximal brightness is $a_u = \sqrt{2}$, and this is approximately where modern undulator based sources operate. For wigglers the deflection parameter is typically measured in tens. In terms of this deflection parameter κ, and accounting for the double Doppler shift, the wavelength of the emitted radiation is given by $\lambda_{X-ray} \simeq (\lambda_U/2\gamma^2)(1+\kappa^2/2)$. In order to Doppler shift from a few centimeters to X-Ray, the e-beam has to be accelerated to GeV (γ in the thousands). This requires relatively large and expensive accelerator. In fact, the size of undulators is not exactly negligible.

In an effort to overcome the stringent (GeV) acceleration requirements of undulators and wigglers, X-ray sources based on Compton scattering have emerged. These sources harness the electromagnetic field of a laser as a

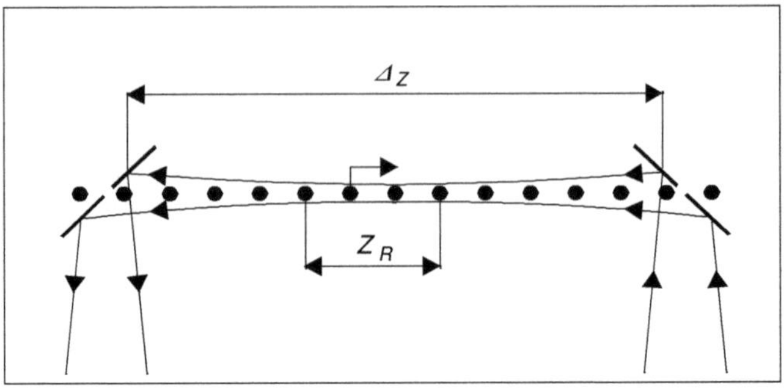

Fig. 8.15 Conventional free space Compton scattering setup. An interaction between a laser and an e-beam, propagating towards each other generates X-Ray emission in the +z direction. The laser is typically focused to a ten-micron spot, which corresponds to Rayleigh length z_R of the order of millimeter. Δ_z is the overall interaction length

replacement for the static magnetic field of an undulator. A typical configuration is a 180° incidence between the e-beam and the laser pulse, as described in Fig. 8.15

For an electromagnetic wiggler $\kappa = eE\lambda_L/2\pi m_0 c^2$ and the radiation's wavelength on axis is $\lambda_{\text{X-ray}} = (\lambda_L/4\gamma^2)(1+\kappa^2/2)$; E being the amplitude of the electric field and λ_L the laser wavelength. Comparing the two expressions, reveals an extra factor of 2 in the relativistic double Doppler shift term $4\gamma^2$, which is due to the fact that the field is no longer static and the laser pulse propagates towards the e-beam. While this factor is in our favor, the main benefit stems from the fact that the laser wavelength λ_L is 4 orders of magnitude smaller than the undulator period λ_U and consequently, the e-beam acceleration energy requirement is reduced by two orders of magnitude from GeV to tens MeV. This is the main advantage of Compton scattering sources over undulators. The deflection parameter of Compton sources is of the order of $\kappa \sim 1$ to the very most, provided that high power laser pulses (TW) are used (Babzien et al. 2006; Schwoerer et al. 2006).

In recent years, several groups have reported successful X-Ray generation via Compton scattering from a laser pulse. In 2000, at the Brookhaven National Laboratory Accelerator Test Facility (BNL ATF) collaborators reported (Pogorelsky et al. 2000) generation of 6.5 keV photons by scattering a 10.6 μm CO_2 laser from a 60 MeV e-beam. A collaboration at the Lawrence Livermore National Laboratory, at the PLEIADES facility, demonstrated generation of 78 keV X-Ray photons (Gibson et al. 2004) using a 57 MeV e-beam and an 820 nm Ti:sapphire laser. An all-optical setup was reported more recently (Schwoerer et al. 2006), employing an 800 nm Ti:sapphire laser split into two pulses: one used for the acceleration of electrons (5 MeV) and the second, counter-propagating pulse, used for Compton scattering. Photons emerged in the energy range between 0.4 to 2 keV.

In typical Compton scattering experiments the laser pulse, propagating in free-space, and the e-beam are focused to a spot size of tens of microns in diameter – aiming to enhance the local power density which, in turn, facilitates higher X-ray brightness. Obviously, focusing the laser comes at the expense of the interaction

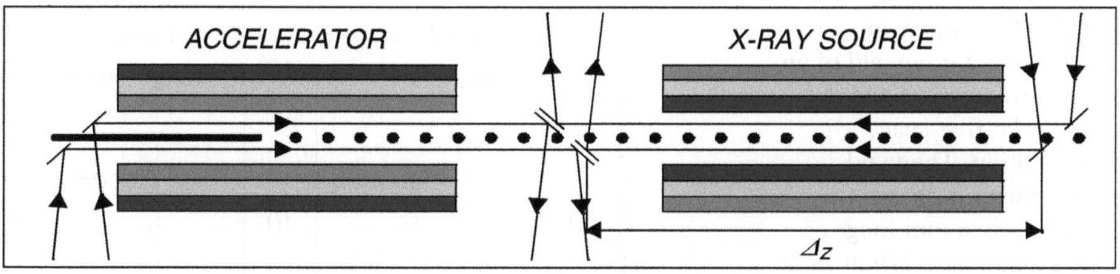

Fig. 8.16 A two-stage all optical setup, the first stage being optical acceleration by TM_{01} mode guided in an optical Bragg accelerator. The second stage is inverse Compton scattering inside a Bragg structure with a different mode design (TEM)

length between the e-beam and the laser – which is approximately two Rayleigh lengths. Moreover, a shorter interaction length implies a broader X-ray spectrum – low monochromaticity. Compromising one at the expense of the other is of limited appeal, and a setup that facilitates both high-power density and long interaction is desirable.

The easiest conceptual method to decouple focusing from interaction length is by introducing a wave-guiding structure. For this purpose, rather than using a Bragg structure that supports a TM_{01} mode for acceleration purposes, we employ a different design. It facilitates propagation of a TEM mode (see Sect. 5.4.4) which has a uniform transverse profile in the interaction region and the latter, in principle, can extend at will. Obviously, in practice, emittance and defocusing of the e-beam will limit the interaction length. Schematics of the all-optical structure is illustrated in Fig. 8.16: the first stage is an optical Bragg acceleration structure that harnesses the laser field to accelerate the electrons as the two propagate parallel to each other. In the second stage, the electrons propagate anti-parallel to the laser pulse and X-ray radiation is generated parallel to the beam – for detailed analysis see Karagodsky et al. (2010).

Relying on a Bragg structure for a Compton scattering process significantly improves the overall operation of an X-ray source by decoupling focusing and interaction length. In order to quantitatively assess the improvement, we proceed to a comparison between the two configurations: (1) Compton scattering based on free-space Gaussian laser-beam (2) Compton scattering in a Bragg structure. For adequate comparison, we assume that both systems have the same e-beam characteristics and the laser injected into both systems is identical in terms of power and polarization (linear).

For simplicity, we make the comparison in a two-dimensional regime. This means that in both configurations the laser profile is focused along the x-axis only and is uniform along the y-axis. This also implies that the e-beam has a sheet-beam shape. For a typical set of parameters [see Karagodsky et al. (2010)] and an interaction length of $\Delta_z = 10^3\lambda_L$ the minimal enhancement in the emitted energy is 38 whereas for $\Delta_z = 10^4\lambda_L$ the minimal enhancement is 119 – see Fig. 8.17.

The main reason for such enhancement is decoupling focusing and interaction length by introducing a wave-guiding structure. This destructive trade-off is

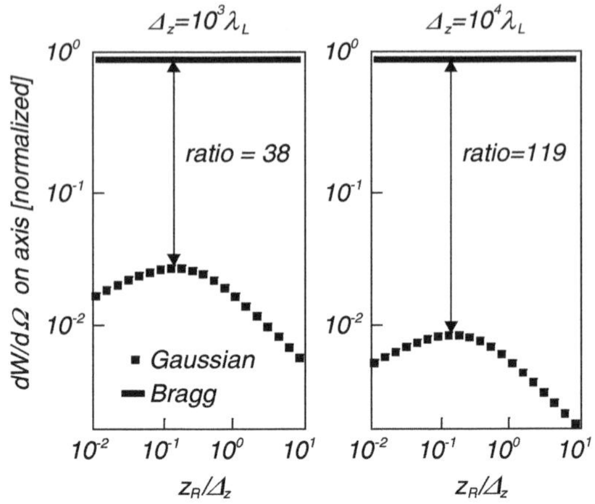

Fig. 8.17 Comparison between the X-ray yield of an optical Bragg setup and the yield of the free-space configuration. The yield enhancement is shown to exceed 38 in a mm long configuration and 119 in a cm long configuration (assuming $\lambda_L = 1\ \mu m$)

graphically described by the free-space curve: if z_R is too small, the interaction is too short, and if z_R is too large, the focusing is too weak – in both cases the yield falls off. Recently, Plettner and Byer (2008) suggested a similar concept but the laser illumination is perpendicular to the e-beam.

Exercises

8.1. Calculate the two wake potentials, (8.1.42)–(8.1.43), as an electron traverses two parallel plates separated by a distance D. Consider only the region between the plates.

8.2. Calculate the two wake potentials, (8.1.42)–(8.1.43), as an electron traverses a pill-box cavity of radius R and length D; the electron moves along the axis. Consider only the internal region. Compare your result with that in Exercise 8.1.

8.3. Consider a uniform cylindrical waveguide of radius R which is infinitely long; ignore walls loss. Between $z = 0$ and $z = D$ the waveguide is filled with a dielectric material ε which is frequency independent; otherwise the waveguide is empty. Calculate the two wake potentials, (8.1.42)–(8.1.43), as an electron traverses this system along its axis. Compare your result with that in Exercises 8.1 and 8.2.

References

Abramowitz, M., Stegun, I.A.: Handbook of Mathematical Functions, 5th edn. Dover Publications, New York (1968)

Alyokhin, B.V., Dubinov, A.E., Selemir, V.D., Shamro, O.A., Shibalko, O.A., Stepanov, N.V., Vatrunin, V.E.: Theoretical and experimental studies of virtual cathode microwave devices. IEEE Trans. **PS-22**, 945 (1994)

Amiranoff, F., Laberge, M., Marques, J.R., Moulin, F., Fabre, E., Cros, B., et al.: Observation of modulational instability in Nd-laser beat-wave experiments. Phys. Rev. Lett. **68**, 3710 (1992)

Amiranoff, F., Baton, S., Bernard, D., Cros, B., Descamps, D., Dorchies, F., Jacquet, F., Malka, V., et al.: Observation of laser wakefield acceleration of electrons. Phys. Rev. Lett. **81**, 995 (1998)

Andrews, H.L., Boulware, C.H., Brau, C.A., Jarvis, J.D.: Super-radiant emission of Smith-Purcell radiation. Phys. Rev. Spec. Top. Accel. Beams **8**, 110702 (2005)

Ashcroft, N.W., Mermin, N.D.: Solid State Physics. Saunders College, Philadelphia (1976)

Attwood, D.: Soft-X-Rays and Extreme Ultra-Violate Radiation. Cambridge University Press, Cambridge, UK (2000)

Babzien, M., Ben-Zvi, I., Kusche, K., Pavlishin, I.V., Pogorelsky, I.V., Siddons, D.P., Yakimenko, V., Cline, D., et al.: Observation of the second harmonic in thomson scattering from relativistic electrons. Phys. Rev. Lett. **96**, 054802 (2006)

Baird, J.M.: Gyrotron theory. In: Granatstein, V.L., Alexeff, I. (eds.) High Power Microwave Sources, p. 103. Artech House, Boston (1987)

Balakin, V., Novokhatsky, A., Smirnov, V.: VLEPP: Transverse Beam Dynamics. Proceedings of the 12th International Conference on High-Energy Accelerators, Fermilab, p. 119 (1983)

Bamber, C., Donaldson, W.R., Lincke, E., Melissinos, A.C.: A pulsed power electron accelerator using laser driven photoconductive switches. In: Wurtele, J.S. (ed.) Advanced Accelerator Concepts, Port Jefferson, NY 1992. AIP Conference Proceedings 279, p. 802 (1983)

Bane, K.L.F., Chao, A., Weiland, T.: A simple model for the energy loss of a bunched beam traversing a cavity. IEEE Trans. Nucl. Sci. **NS-28**, 2605 (1981)

Bane, K.L.F., Stupakov, G.V.: Impedance of a rectangular beam tube with small corrugations. Phys. Rev. Spec. Top. Accel. Beams **6**, 024401 (2003)

Banna, S., Schächter, L., Nation, J.A., Wang, P.: Coupling of symmetric and asymmetric modes in a high-power, high-efficiency traveling wave amplifier. Phys. Rev. E **64**(4), 4445 (2000a). See also: Schächter L., Nation J.A.: Beam-quality and guiding magnetic field requirements for a high-power traveling wave amplifier operating at 35 GHz. Phys. Rev. E **57**, 7176–7183 (1998)

Banna, S., Nation, J.A., Schächter, L., Wang, P.: The interaction of symmetric and asymmetric modes in a high-power traveling wave amplifier. IEEE Plasma Sci. Special Issue **28**, 798–811 (2000b)

Banna, S., Schieber, D., Schächter, L.: Electromagnetic wake-field due to surface roughness in an optical structure. J. Appl. Phys. **Vol. 95**(8), 4415 (2004)

L. Schächter, *Beam-Wave Interaction in Periodic and Quasi-Periodic Structures*, Particle Acceleration and Detection, DOI 10.1007/978-3-642-19848-9,

Banna, S., Berezovsky, V., Schächter, L.: Experimental observation of direct particle acceleration by stimulated emission of radiation. Phys. Rev. Lett. **97**, 134801 (2006a)

Banna, S., Berezovsky, V., Schächter, L.: Particle acceleration by stimulated emission of radiation: Theory and experiment. Phys. Rev. E **74**, 046501 (2006b)

Beck, A.H.W.: Space Charge Waves and Slow Electromagnetic Waves. Pergamon Press, Inc., New York (1958)

Bekefi, G.: Rippled-field magnetron. Appl. Phys. Lett. **40**(7), 578 (1982)

Berry, M.V.: Diffraction in crystals at high energies. J. Phys. C Solid State Phys. **4**, 697 (1971)

Blumenfeld, I., Clayton, C.E., Decker, F.J., Hogan, M.J., Huang, C., Ischebeck, R., et al.: Energy doubling of 42 GeV electrons in a meter-scale plasma wake-field accelerator. Nature **445**, 741 (2007)

Born, M., Wolf, E.: Principles of Optics, 6th edn. Pergamon Press, Oxford (1984)

Brillouin, L.: Wave guides for slow waves. J. Appl. Phys. **19**, 1023 (1948)

Brillouin, L.: Periodic Structures: Electric Filters and Crystal Lattices, 2nd edn. Dover, New York (1953)

Bugaev, S.P., Cherpenin, V.A., Kanavets, V.I., Klimov, A.I., Kopenkin, A.D., Koshelev, V.I., Popov, V.A., Slepov, A.I.: Relativistic multiwave Cerenkov generator. IEEE Trans. **PS-18**, 525 (1990)

Byrd, J.M., Leemans, W.P., Loftsdottir, A., Marcelis, B., Martin, M.C., McKinney, W.R., Sannibale, F., Scarvie, T., Steier, C.: Observation of broadband self-amplified spontaneous coherent terahertz synchrotron radiation in a storage ring. Phys. Rev. Lett. **89**(22), 224801 (2002)

Carmel, Y., Nation, J.A.: Instability of an unneutralized relativistic electron beam. Phys. Rev. Lett. **31**, 286 (1973)

Carmel, Y., Granatstein, V.L., Gover, A.: Demonstration of two-stage backward-wave-oscillator free electron laser. Phys. Rev. Lett. **51**, 566 (1983)

Carmel, Y., Minami, K., Kehs, R.A., Destler, W.W., Granatstein, V.L., Abe, D., Lou, W.L.: Demonstration of efficiency enhancement in a high power backward wave oscillator by plasma injection. Phys. Rev. Lett. **62**, 2389 (1989)

Caryotakis, G.: High power microwave tubes: in the laboratory and on-line. IEEE Trans. **PS-22**, 683 (1994)

Chang, D.B., McDaniel, J.C.: Compact short-wavelength free-electron laser. Phys. Rev. Lett. **63**, 1066 (1989)

Chen, P., Dawson, J.M., Huff, R.W., Katsouleas, T.: Acceleration of electrons by the interaction of a bunched electron beam with a plasma. Phys. Rev. Lett. **54**, 693 (1985)

Chao, A.W.: Physics of Collective Beam Instabilities in High Energy Accelerators, pp. 1–126. Wiley, New York (1993)

Chao, A.W., Moser, H.O., Zhao, Z.: Accelerator Physics, Technology and Applications, 2nd edn. World Scientific, Singapore (2002)

Chao, A.W., Tigner, M.: Handbook of Accelerator Physics and Engineering. World Scientific, Singapore (1998)

Chatterton, P.A.: A theoretical study of field emission initiated vacuum breakdown. Proc. Phys. Soc. **88**, 231 (1966)

Chodorow, M., Susskind, C.: Fundamentals of Microwave Electronics. McGraw-Hill Book Company, New York (1964)

Chu, J.L., Jackson, J.D.: Field theory of traveling wave tubes. Proc. IRE **36**, 853 (1948)

Clayton, C.E., Marsh, K.A., Dyson, A., Everett, M., Lal, A., Leemans, W.P., Williams, R., Joshi, C.: Ultrahigh-gradient acceleration of injected electrons by laser-excited relativistic electron plasma waves. Phys. Rev. Lett. **70**, 37 (1993)

Conde, M.E., Bekefi, G.: Experimental study of a 33.3 GHz free electron laser amplifier with a reversal axial guide magnetic field. Phys. Rev. Lett. **67**, 3082 (1991)

Cooper, R.K.: Wake fields: limitations and possibilities. In: Hyder, A.K., Rose, M.F., Guenther, A.H. (eds.) High Brightness Accelerators. NATO ASI Series, vol. 178, p. 157. Plenum Press, New York (1988)

Courant, E.D., Pellegrini, C., Zakowicz, W.: High energy inverse free electron laser accelerator. Phys. Rev A **32**, 2813 (1985)

Cowan, B., Javanmard, M., Siemann, R.: Photonic crystal laser accelerator structures Proceedings of the Particle Accelerator Conference, PAC 2003, **3**, 1855–1857 (2003)

Curtin, M., Bennett, G., Burke, R., Bhowmik, A., Metty, P., Benson, S., Madey, J.M.J.: First demonstration of a free-electron laser driven by electrons from a laser-irradiated photocathode. Nucl. Instrum. Methods Phys. Res. A **296**, 127 (1990)

Davis, T.J., Nation, J.A., Schächter, L.: Results from an X-band coaxial extended length cavity. IEEE Trans. Plasma Sci. **22**, 504 (1994)

DeSanto, J.A.: Scattering from a periodic corrugated structure: thin comb with soft boundary conditions. J. Math. Phys. **12**, 1913 (1971)

DeSanto, J.A.: Scattering from a periodic corrugated structure: thin comb with hard boundary conditions. J. Math. Phys. **13**, 337 (1972)

Destler, W.W., Aghimir, F.M., Boyd, D.A., Bekefi, G., Shefer, R.E., Yin, Y.Z.: Experimental study of millimeter wave radiation from a rotating electron beam in a rippled magnetic field. Phys. Fluids **28**(6), 1962 (1985)

Dome, G.: Wake potentials of a relativistic point charge crossing a beam-pipe gap: an analytical approximation. IEEE Trans. Nucl. Sci. **Vol. NS-32**, 2531 (1985)

Dome, G., Palumbo, L., Vaccaro, V.G., Verolino, L.: A method for computing the longitudinal coupling impedance of circular apertures in a periodic array of infinite planes. Part. Accel. **36**(1–3), 161–76 (1991)

Doucas, G., Mulvey, J.H., Omori, M., Walsh, J.E., Kimmitt, M.F.: First observation of Smith-Purcell radiation from relativistic electrons. Phys. Rev. Lett. **69**, 1761 (1992)

Edighoffer, J.A., Kimura, W.D., Pantell, R.H., Piestrup, M.A., Wang, D.Y.: Observation of inverse Cerenkov interaction between free electrons and laser light. Phys. Rev. A **23**, 1848 (1981)

Elachi, C.: Waves in active and passive periodic structures: a review. Proc. IEEE **64**, 1666 (1976)

Elias, L.R., Fairbank, W.M., Madey, J.M.J., Schwettman, H.A., Smith, T.I.: Observation of stimulated emission of radiation by relativistic electrons in a spatially periodic transverse magnetic field. Phys. Rev. Lett. **36**, 717 (1976)

Emma, P., et al.: First lasing and opration of an angstrom-wavelength free-electron laser. Nat. Photonics (2010). doi:10.1038/NPHOTON.2010.176

Esarey, E., Schroeder, C.B., Leemans, W.P.: Physics of laser-driven plasma-based electron accelerators. Rev. Mod. Phys. **81**, 1229 (2009)

Feinstein, J., Pantell, R.H., Fauchet, A.M.: Prospects for visible and VUV free electron lasers using dielectric resonance. IEEE Trans. Quantum Electr. **QE-22**, 587 (1986)

Flaygin, V.A., Gaponov, A.V., Petelin, M., Yulpatov, V.K.: The Gyrotron. IEEE Trans. **MTT-25**, 514 (1977)

Fontana, J.R., Pantell, R.H.: A high - energy, laser accelerator for electrons using the inverse Cherenkov effect. J. Appl. Phys. **54**, 4285 (1983)

Fowler, R.H., Nordheim, L.: Electron emission in intense electric fields. Proc. R. Soc. Lond. **119**(781), 173–181 (1928)

Freund, H.P., Johnston, S., Sprangle, P.: Three-dimensional theory of free electron lasers with an axial guide field. IEEE J. Quant. Electron. **QE-19**, 322 (1983)

Freund, H.P., Antonsen, T.: Principles of Free Electron Lasers. Chapman and Hall, London (1992)

Friedland, L.: Electron beam dynamics in combined guide and pump magnetic fields for free electron laser applications. Phys. Fluids **23**, 2376 (1980). See also Friedland L., Hirshfield J.L.: Free electron laser with a strong axial magnetic field. Phys. Rev. Lett. **44**, 1456 (1980)

Friedman, A., Gover, A., Kurizki, G., Ruschin, S., Yariv, A.: Spontaneous and stimulated emission from quasi-free electrons. Rev. Mod. Phys. **60**(2), 471 (1988)

Friedman, M., Serlin, V.: Modulation of intense relativistic electron beams by an external microwave source. Phys. Rev. Lett. **55**, 2860 (1985)

Gai, W., Schoessow, P., Cole, B., Konecney, R., Norem, J., Rosenzweig, J., Simpson, J.: Experimental demonstration of wake-field effects in dielectric structures. Phys. Rev. Lett. **61**, 2756 (1988)

Gai, W., Jing, C.: Dielectric-loaded accelerating structures. In: Bozzi, M., Perregrini, L. (eds.) Periodic Structures, p. 345. Research Signpost, Trivandrum, India (2006)

Gaponov, A.V.: (Translation of) "Interaction of nonlinear electron beams with electromagnetic waves in transmission lines". Soviet Radiophys. **2**, 837 (1959)

Geddes, C.G.R., Cs, T., van Tilborg, J., Esarey, E., Schroeder, C.B., Cary, J., Leemans, W.P.: Guiding of relativistic laser pulses by preformed plasma channels. Phys. Rev. Lett. **95**, 145002 (2005)

Gibson, J., Anderson, S.G., Barty, C.P.J., Betts, S.M., Booth, R., Brown, W.J., Crane, J.K., Cross, R.R., Fittinghoff, D.N., Hartemann, F.V., et al.: PLEIADES: a picosecond Compton scattering x-ray source for advanced backlighting and time-resolved material studies. Phys. Plasmas **11**, 2857 (2004)

Gilmour Jr., A.S.: Microwave Tubes. Artech House, Norwood (1986)

Glyavin, MYu, Luchinin, A.G., Golubiatnikov, GYu: Generation of 1.5-kW, 1-THz coherent radiation from a gyrotron with a pulsed magnetic field. Phys. Rev. Lett. **100**, 015101 (2008)

Goldstein, H.: Classical Mechanics. Addison-Wesley Publishing Company Inc., Reading Massachusetts (1950)

Gorgy, W., Smith, W.V., Trambarulo, R.F.: Microwave Spectroscopy, pp. 95–96. Dover Publications Inc., New York (1966). 191–192

Gover, A.: An analysis of stimulated longitudinal electrostatic bremsstrahlung in a free electron laser structure. Appl. Phys. **23**, 295 (1980)

Granatstein, V.L.: Gyrotron experimental studies. In: Granatstein, V.L., Alexeff, I. (eds.) High Power Microwave Sources, p. 103. Artech House, Boston (1987)

Haimson, J.: Suppression of beam-induced pulse-shortening modes in high power RF generator TW output structures. In: Brandt, H.E. (ed.) Intense Microwave and Particle Beams III. SPIE Proceedings vol. 1629, p. 209. Los Angeles, CA (1992)

Hasegawa, A.: Free electron laser. Bell Syst. Tech. J. **57**, 3069 (1978)

Haus, H.A.: Signal and noise propagation along electron beams. In: Smullin, L.D., Haus, H.A. (eds.) Noise in Electron Devices, p. 77. John Wiley & Sons, New York (1959)

Haus, H.A., Melcher, J.R.: Electromagnetic Fields and Energy. Prentice Hall, Englewood Cliffs, New Jersey (1989)

Heifets, S., Kheifets, S.: Coupling impedance in modern accelerators. SLAC-PUB-5297 (1990). See also: Heifets, S.: Broadband impedances of accelerating structures: perturbation theory. SLAC-PUB-5792 (1992)

Heifets, S.A., Kheifets, S.A.: Rev. Mod. Phys. **63**, 631–673 (1991)

Helm, R.H., Loew, G.A.: Beam breakup. In: Lapostolle, P.M., Septier, A.L. (eds.) Linear accelerators, p. 173. North-Holland Pub. Company, Amsterdam (1970)

Hirshfield, J., Granatstein, V.L.: The electron cyclotron maser – an hystorical survey. IEEE Trans. **MTT-25**, 522 (1977)

Hommelhoff, P., Sortais, Y., Aghajani-Talesh, A., Kasevich, M.A.: Field emission tip as a nanometer source of free electron femtosecond pulses. Phys. Rev. Lett. **96**, 077401 (2006)

Huang, Y.C., Zheng, D., Tulloch, W.M., Byer, R.L.: Proposed structure for a crossed-laser beam, GeV per meter gradient, vacuum electron linear accelerator. Appl. Phys. Lett. **68**(6), 753 (1996)

Huang, Z., Kim, K.J.: Three-dimensional analysis of harmonic generation in high-gain free-electron lasers. Phys. Rev. E **62**, 7295 (2000)

Huang, Z., Ruth, R.: Fully coherent X-ray pulses from a regenerative-amplifier free electron laser. Phys. Rev. Lett. **96**, 144801 (2006)

Huang, Z., Kim, K.J.: Review of x-ray free-electron laser theory. Phys. Rev. **10**, 034801 (2007)

Hübner, K.: Two-beam linear colliders. In: Rossbach, J. (ed.) HEACC'92, XV International Conference on High Energy Accelerators, Hamburg, Germany. Int. J. Mod. Phys. A, vol. 2B, p. 791. World Scientific, River Edge, NJ (1993)

Hull, A.W.: The effect of a uniform magnetic field on the motion of electrons between coaxial cylinders. Phys. Rev. **18**(1), 31–57 (1921a)

Hull, A.W.: The magnetron. J. Am. Inst. Electr. Eng. **40**(9), 715–723 (1921b)
Hutter, R.G.E.: Beam and wave electronics in microwave tubes. D. Van Nostrand Company Inc., Princeton, NJ (1960)
Ivers, J.D., Advani, R., Kerslick, G.S., Nation, J.A., Schächter, L.: Electron beam using ferroelectric cathodes. J. Appl. Phys. **73**, 2667 (1993)
Joshi, C., Clayton, C.E., Marsh, K.A., Dyson, A., Everett, M., Lal, A., Leemans, W.P., Williams, R., Katsouleas, T., Mori, W.B.: Acceleration of injected electrons by the plasma beat wave accelerator. In: Wurtele, J.S. (ed.) Advanced Accelerator Concepts. AIP Conference Proceedings 279, Port Jefferson, NY 1992, p. 379 (1993)
Joshi, C.: The development of laser- and beam-driven plasma accelerators as an experimental field. Phys. Plasmas **14**, 055501 (2007)
Jackson, J.D.: Classical Electrodynamics, p. 470. John Wiley and Sons, New York (1962)
Kapitza, P.L., Dirac, P.A.M.: The reflection of electrons from standing light waves. Proc. Camb. Phil. Soc. **29**, 297 (1933)
Karagodsky, V., Schieber, D., Schächter, L.: Enhancing X-ray generation by electron-beam-laser interaction in an optical bragg structure. Phys. Rev. Lett. **104**(2), 024801 (2010) (4)
Kartikeyan, M.V., Borie, E., Thumm, M.K.A.: Gyrotrons: High Power Microwave an Millimeter Wave Technology. Springer-Verlag, Berlin (2004)
Katsouleas, T., Dawson, J.M.: Plasma acceleration of particle beams. In: Month, M., Dienes, M. (eds.) Physics of Particle Accelerators. AIP Conference proceedings, vol. 184, p. 1799. AIP, New York (1989)
Kim, K.-J., Kroll, N.M.: Some effects of the transverse stability requirement on the design of a grating linac. In: Channell, P.J. (ed.) Laser Acceleration of Particles, Los Alamos 1982. AIP Conference Proceedings, vol. 91, p. 190. AIP, New York (1982)
Kim, K.J.: RF and space-charge effects in laser driven rf electron guns. Nucl. Instrum. Methods Phys. Res. **A275**, 201–218 (1989)
Kwang-Je, K., Kumar, V.: Electron beam requirements for a three-dimensional Smith-Purcell backward-wave oscillator for intense terahertz radiation. Phys. Rev. Spec. Top. Accel. Beams **10**, 080702 (2007)
Kimura, W.D., Kim, J.H., Romea, R.D., Steinhauer, L.C., Pogoreisky, I.V., Kusche, K.P., Fernow, R.C., Wano, X., Liu, Y.: Laser acceleration of relativistic electrons using the inverse Cerenkov effect. Phys. Rev. Lett. **74**, 546 (1995)
Kimura, W.D., van Steenbergen, A., Babzien, M., Ben-Zvi, I., Campbell, L.P., Dilley, C.E., Cline, D.B., Gallardo, J.C., Gottschalk, S.C., He, P., Kusche, K.P., Liu, Y., Pantell, R.H., Pogorelsky, I.V., Quimby, D.C., Skaritka, J., Steinhauer, L.C., Yakimenko, V.: First staging of two laser accelerators. Phys. Rev. Lett. **86**, 4041–4043 (2001)
Kimura, W.D., Babzien, M., Ben-Zvi, I., Campbell, L.P., Cline, D.B., Dilley, C.E., Gallardo, J.C., Gottschalk, S.C., Kusche, K.P., Pantell, R.H., Pogoresky, I.V., Quimby, D.C., Skaritka, J., Steinhauer, L.C., Yakimenko, V., Zhou, F.: Demonstration of high-trapping efficiency and narrow energy spread in a laser-driven accelerator. Phys. Rev. Lett. **92**, 054801 (2004)
Kitagawa, Y., Matsumoto, T., Manamihata, S.K., Matsuo, K., Mima, K., et al.: Beat-wave excitation of plasma wave and observation of accelerated electrons. Phys. Rev. Lett. **68**, 48 (1992)
Kittel, C.: Introduction to Solid State Physics, 2nd edn. John Wiley & Sons, New York (1956)
Knapp, B.C., Knapp, E.A., Lucas, G.L., Potter, J.M.: Accelerating Structures for high current proton linacs. IEEE Trans. Nucl. Sci. **NS-12**, 159 (1965)
Korbly, S.E., Kesar, A.S., Sirigiri, J.R., Temkin, R.J.: Observation of frequency-locked coherent terahertz Smith-Purcell radiation. Phys. Rev. Lett. **94**, 054803 (2005)
Kroll, N.M.: The free electron laser as a traveling wave amplifier. In: Jacobs, S.F., Sargent, M., Scully, M.O. (eds.) Novel Sources of Coherent Radiation. Physics of Quantum Electronics, vol. 5, p. 115. Addison Wesley Pub. Corp., Massachusettes (1978)
Kroll, N.M., Morton, P.L., Rosenbluth, M.N.: Free-electron lasers with variable wigglers. IEEE Quant. Electron. **QE-17**, 1436 (1981)

Kroll, N.M.: General features of accelerating modes in open structures. In: Joshi, C., Katsouleas, T. (eds.) Laser Acceleration of Particles, Malibu 1985. AIP Conference Proceedings, vol. 130, p. 253. AIP, New York (1985)

Kuang, E., Davis, T.J., Kerslick, G.S., Nation, J.A., Schächter, L.: Transit time isolation of a high power microwave TWT. Phys. Rev. Lett. **71**, 2666 (1993)

Kumakhov, M.A.: On the theory of electromagnetic radiation of charged particles in a crystal. Phys. Lett. A **57**, 17 (1976)

Kurennoy, S.S.: Beam coupling impedance of obstacles protruding into a beam pipe. Phys. Rev. **E 55**, 3529 (1997)

Lambert, G., Hara, T., Garzella, T., Tanikawa, D., Labat, M., Carre, H., Kitamura, B., Shintake, T., Bougeard, M., Inoue, S., Tanaka, Y., Salieres, P., Merdji, H., Chubar, O., Gobert, O., Tahara, K., Couprie, M.E.: Injection of harmonics generated in gas in a free-electron laser providing intense and coherent extreme-ultraviolet light. Nat. Phys. **4**, 296 (2008)

Landau, L.D., Lifshitz, E.M.: Mechanics. Pergamon Press, Oxford (1960)

Langmuir, I.: The effect of space charge and residual gases on thermionic currents in high vacuum. Phys. Rev. **2**, 450–486 (1913)

Lapostolle, P.M., Septier, A.L. (eds.): Linear Accelerators. North-Holland Pub. Company, Amsterdam (1970)

Lapostolle, P.M.: Possible emittance increase through filamentation due to space-charge effects. IEEE Trans. Nucl. Sci. **NS-18**, 1101 (1971)

Lawson, J.D.: Lasers and accelerators. IEEE Trans. Nucl. Sci. **26**(3), 4217 (1979)

Lawson, J.D.: The Physics of Charged Particles Beams, 2nd edn. Clarendon Press, Oxford (1988)

Lau, Y.Y.: Classification of beam breakup instabilities in linear accelerators. Phys. Rev. Lett. **63**, 1141 (1989)

Lau, Y.Y., Friedman, M., Krall, J., Serlin, V.: Relativistic klystron amplifiers driven by modulated intense beams. IEEE Trans. **PS-18**, 553 (1990)

Lee, C.H. (ed.): Picosecond Optoelectronic Devices. Academic Press, New York (1984). In particular see Chap. 7 by Mourou G., Knox, W.H. and Williamson, S.

Lee, S.Y.: Accelerator physics, 2nd edn. World Scientific, Singapore (2004)

Leemans, W.P., Nagler, B., Gonsalves, A.J., Cs, T., Nakamura, K., Geddes, C., Esarey, E., Schroeder, C.B., Hooker, S.M.: GeV electron beams from a centimeter-scale accelerator. Nat. Phys. **2**, 696 (2006)

Lewin, L.: Theory of Waveguides. John Wiley & Sons, New York (1975)

Lewis, T.J.: High field electron emission from irregular cathode surfaces. J. Appl. Phys. **26**, 1405 (1955)

Lin, X.E.: Photonic band gap fiber accelerator. Phys. Rev. Spec. Top. Accel. Beams **4**, 051301 (2001)

Loew, G.A., Tolman, R.: Lectures on the Elementary Principles of Linear Accelerators. In: Month, M. (ed.) Physics of High Energy Particle Acceleration, SLAC Summer School 1982. AIP Conference Proceeding, vol. 105, p. 1. AIP, New York (1983)

Madey, J.M.J.: Stimulated emission of bremsstrahlung in a periodic magnetic field. J. Appl. Phys. **42**, 1906 (1971)

Madey, J.M.J.: Relationship between mean radiated energy, mean square radiated energy and spontaneous power spectrum in a power series expansion of the equations of motion in a free electron laser. Il Nuovo Cimento **50B**(1), 64 (1979)

Marinelli, A., Pellegrini, C., Giannessi, L., Reiche, S.: Comparative study of nonideal beam effects in high gain harmonic generation and self-seeded free electron lasers. Phys. Rev. Spec. Top. Accel. Beams **13**, 070701 (2010)

Malka, V., Fritzler, E., Lebvre, M.-M., Aleonard, M.M., Burgy, F., Chambaret, J.P., Chemin, J.F., Krushelnick, K., Malka, G., Mangles, S.P.D., Najmudin, D., Pittman, M., Rousseau, J.P., Scheurer, J., Walton, B., Dangor, A.E.: Electron acceleration by a wake field forced by an intense ultrashort laser pulse. Science **22**(298), 1596–1600 (2002)

Marqués, J.R., Geindre, J.P., Amiranoff, F., Audebert, P., Gauthier, J.C., Antonetti, A., Grillon, G.: Temporal and spatial measurements of the electron density perturbation produced in the wake of an ultrashort laser pulse. Phys. Rev. Lett. **76**, 3566 (1996)

Marshall, T.C.: Free Electron Lasers. Macmillan Pub. Comp, New York (1985)

McDonald, K.T.: Design of the Laser-Driven RF Electron Gun for the BNL Accelerator Test Facility, DOE/ER/3072-43 (Princeton University) (1988)

McMullin, W.A., Bekefi, G.: Coherent radiation from a relativistic electron beam in a longitudinal periodic magnetic field. Appl. Phys. Lett. **39**(10), 845 (1981)

McMullin, W.A., Bekefi, G.: Stimulated emission from relativistic electrons passing through a spatially periodic longitudinal magnetic field. Phys. Rev. **A 25**(4), 1826 (1982)

Michalke, A., Piel, H., Sinclair, C.K., Michelato, P.: First operation of high-quantum efficiency Photo-cathodes inside superconductive cavities, EPAC '92, Berlin, Germany, March 24–28, p. 1014 (1992)

Miller, H.C.: Change in field intensification factor of an electrode projection (whisker) at short gap lengths. J. Appl. Phys. **38**, 4501 (1967)

Miller, H.C.: Influence of gap length on the field increase factor of an electrode projection. J. Appl. Phys. **55**, 158 (1984)

Miller, R.B.: An Introduction to the Physics of Intense Charged Particle Beams. Plenum Press, New York (1982)

Mittra, R., Lee, S.W.: Analytical Techniques in the Theory of Guided Waves. Macmillan, New York (1971)

Mizrahi, A., Schächter, L.: Bragg reflection waveguides with a matching layer, 12, Optics Express 3156 (2004a).

Mizrahi, A., Schächter, L.: Optical Bragg accelerator. Phys. Rev. E **70**, 016505 (2004)

Mizrahi, A., Schächter, L.: Electromagnetic forces on the dielectric layers of the planar optical Bragg acceleration structure. Phys. Rev. E **74**, 036504 (2006)

Mostacci, A., Ruggiero, F., Angelici, M., Migliorati, M., Palumbo, L., Ugoli, S.: Wakefields due to surface waves in a beam pipe with a periodic rough surface. Phys. Rev. Spec. Top. Accel. Beams **5**, 044401 (2002)

Motz, H.: Applications of the radiation from fast electron beams. J. Appl. Phys. **22**, 527 (1951)

Nation, J.A.: On the coupling of a high-current relativistic beam to a slow wave structure. Appl. Phys. Lett. **17**, 491 (1970)

Naqvi, S.A., Nation, J.A., Kerslick, G.S., Schächter, L.: Resonance shift in relativistic traveling wave amplifiers. Phys. Rev. **E 53**(4), 4229 (1996)

Ng, K.Y.: Resonant impedance in a toroidal beam pipe. Part. Accel. **25**, 153 (1990)

Nodvick, J., Saxon, D.: Suppression of coherent radiation by electrons in a synchrotron. Phys. Rev. **96**, 180 (1954)

Nusinovich, G.S.: Introduction to the Physics of Gyrotrons. Johns Hopkins University Press, Baltimore (2004)

Orzechowsky, T.J., Anderson, B.R., Fawley, W.M., Prosnitz, D., Scharlemann, E.T., Yarema, S.M., Hopkins, D., Paul, A.C., Sessler, A.M., Wurtele, J.: Microwave radiation from a high gain FEL amplifier. Phys. Rev. Lett. **54**, 889 (1985)

O'Shea, P.G., Bender, S.C., Byrd, D.A., Early, J.W., Fortgang, C.M., Goldstein, J.C., Newman, B.E., Sheffield, R.L., Warren, R.W., Zaugg, T.J.: Ultraviolet free-electron laser driven by a high brightness 45-MeV electron beam. Phys. Rev. Lett. **71**, 3661 (1993)

Palmer, R.B.: Interaction of relativistic particles and free electromagnetic waves in the presence of a static helical magnet. J. Appl. Phys. **43**, 3014 (1972)

Palmer, R.B.: Near field accelerators. In: Channel, P.J. (ed.) Laser Acceleration of Particles, Los Alamos 1982. AIP Conference Proceedings, vol. 91, p. 179. AIP, New York (1982). See also Palmer, R.B., Baggett, N., Claus, J., Fernow R., Stumer, I., Figueroa, H., Kroll, N.M., Funk, W., Lee-Whiting, G., Pickup, M., Goldstone, P., Lee, K., Corkum, P., Himel, T.: Report of Near Field Group. In: Joshi C., Katsouleas T. (eds) Laser Acceleration of Particles, Malibu 1985. AIP Conference Proceedings 130, p. 234. AIP, New York (1985)

Palmer, R.B.: An introduction to acceleration mechanisms. In: Month, M., Turner, S. (eds.) Frontiers of Particle Beams, p. 607. Springer-Verlag, Berlin (1986)
Panofsky, W.K.H., Bander, M.: Asymptotic theory of beam break-up in linear accelerators. Rev. Sci. Instrum. **39**, 206 (1968)
Panofsky, W.K.H., Wenzel, W.A.: Some considerations concerning the transverse deflection of charged particles in radio frequency fields. Rev. Sci. Instrum. **27**, 967 (1956)
Pantell, R.H., Soncini, G., Puthoff, H.E.: Stimulated photon-electron scattering. J. Quant. Electron. **QE-4**, 905 (1968)
Pantell, R.H., Alguard, M.J.: Radiation characteristics of planar channeled positrons. J. Appl. Phys. **50**, 798 (1979)
Pantell, R.H.: Interactions between electromagnetic fields and electrons. In: Carrigan, R.A., Huson, F.R., Month, M. (eds) Physics of High Energy Particle Accelerators, Fermilab Summer School July 13–24 1981. AIP Conference Proceedings No. 87, p. 864 (1981)
Pantell, R.H., Feinstein, J., Fisher, A.L., Deloney, T.L., Reid, M.B., Grossman, W.M.: Benefits and costs of the gas-loaded free electron laser. Nucl. Instrum. Methods Phys. Res. A **250**, 312 (1986)
Park, S.Y., Hirshfield, J.L.: Theory of wakefields in a dielectric-lined waveguide. Phys. Rev. E **62**, 1266–1283 (2000)
Pauli, W.: Theory of Relativity. Pergamon Press, London (1958)
Pellegrini, C., Sessler, A.M.: The Development of Colliders. AIP Press, New York (1995)
Petit, R.: Electromagnetic Theory of Gratings. Springer-Verlag, Berlin (1980)
Phillips, R.M.: The Ubitron, a high power traveling wave tube based on a periodic beam interaction in unloaded waveguide. Trans. IRE Electron Dev. **7**, 231 (1960)
Pickup, M.: A grating linac at microwave frequencies. In: Joshi, C., Katsouleas, T. (eds.) Laser Acceleration of Particles, Malibu 1985. AIP Conference Proceedings, vol. 130, p. 281. AIP, New York (1985)
Pierce, J.R.: Theory of the beam-type traveling wave tube. Proc. IRE, 111 (1947)
Pierce, J.R.: Traveling-Wave Tubes. D. van Nostrand Company Inc., Princeton, NJ (1950)
Plettner, T., Byer, R.L., Colby, E., Cowan, B., Sears, C., Spencer, J.E., Siemann, R.H.: Visible-laser acceleration of relativistic electrons in a semi-infinite vacuum. Phys. Rev. Lett. **95**, 134801 (2005)
Plettner, T., Byer, R.L.: Proposed dielectric-based microstructure laser-driven undulator. Phys. Rev. Spec. Top. Accel. Beams **11**, 030704 (2008)
Pogorelsky, I., Ben-Zvi, T., Hirose, S., Kashiwagi, V., Yakimenko, K., Kusche, P., Siddons, J., Skaritka, T., Kumita, A.T., et al.: Demonstration of 8×10^{18}photons/second peaked at 1.8 Å in a relativistic Thomson scattering experiment. Phys. Rev. Spec. Top. Accel. Beams **3**, 090702 (2000)
Power, J., Conde, M.E., Gai, W., Konecny, R., Schoessow, P., Kanareykin, A.D.: Measurements of the longitudinal wakefields in a multimode, dielectric wakefield accelerator driven by a train of electron bunches. Phys. Rev. Spec. Top. Accel. Beams **3**, 101302 (2000)
Ramo, S., Whinnery, J.R., Van Duzer, T.: Fields and Waves in Communication Electronics. John Wiley & Sons, New York (1965)
Richter, B.: Requirements for very high energy accelerators. In: Joshi, C., Katsouleas, T. (eds.) Laser Acceleration of Particles, Malibu California. AIP Conference Proceedings, vol. 130, p. 8. AIP, New York (1985)
Riege, H.: Electron emission from ferroelectrics – a review. Nucl. Instrum. Methods Phys. Res. **340**, 80 (1993)
Roberson, C.W., Sprangle, P.: A review of free-electron lasers. Phys. Fluids **B1**, 3 (1989)
Rosenzweig, J.B., Cline, D.B., Cole, B., Figueroa, H., Gai, W., Konecny, R., Norem, J., Schoessow, P., Simpson, J.: Experimental observation of plasma wake-field acceleration. Phys. Rev. Lett. **61**, 98 (1988)
Rosing, M., Gai, W.: Longitudinal and transverse wake field effects in dielectric structures. Phys. Rev. D **42**, 1829 (1990)

Russell, S.J.: Overview of high-brightness, high-average-current photo-injectors for FELs. Nucl. Instrum. Methods Phys. Res. **A 507**, 304 (2003)

Sakamoto, K., Kasugai, A., Takahashi, K., Minami, R., Kobayashi, N., Kajiwara, K.: Achievement of robust high-efficiency 1 MW oscillation in the hard-self-excitation region by a 170 GHz continuous-wave gyrotron. Nat. Phys. **3**, 411 (2007)

Saldin, E., Schneidmiller, E., Yurkov, M.: Properties of the third harmonic of the radiation from self-amplified spontaneous emission free electron laser. Phys. Rev. Spec. Top. Accel. Beams **9**, 030702 (2006)

Salisbury, W.W.: Generation of light from free electrons. J. Opt. Soc. Am. **60**, 1279 (1970)

Sarakinos, K., Alami, J., Konstantinidis, S.: High power pulsed magnetron sputtering: a review on scientific and engineering state of the art. Surf. Coat. Technol. **204**(11), 1661–1684 (2010)

Schächter, L.: Relativistic quantum mechanical analysis of a free electron laser. J. Appl. Phys. **61**(8), 2718–2728 (1987)

Schächter, L.: Remarks on channeling radiation. J. Appl. Phys. **63**, 712 (1988)

Schächter, L., Nation, J.A., Shiffler, D.A.: Theoretical studies of high power cerenkov amplifiers. J. Appl. Phys. **70**, 114 (1991)

Schächter, L., Nation, J.A.: Slow wave amplifiers and oscillators: a unified study. Phys. Rev. A. **45**, 8820 (1992)

Schächter, L.: PASER: Particle acceleration by stimulated emission of radiation. Phys. Lett. A **205**, 355 (1995)

Schächter, L.: Wake-field in dielectric acceleration structures. Phys. Rev. **E 68**, 036502 (2003)

Schächter, L.: Energy recovery in an optical linear collider. Phys. Rev. **E 70**, 016504 (2004)

Scharlemann, E.T., Sessler, A.M., Wurtele, J.S.: Optical guiding in free electron laser. Phys. Rev. Lett. **54**(17), 1925 (1985)

Schieber, D.: Electromagnetic Induction Phenomena. Springer-Verlag, Berlin (1986)

Schiff, L.: Production of particle energies beyond 200 MeV. Rev. Sci. Instrum. **17**, 6 (1946)

Schmüser, P., Dohlus, M., Rossbach, J.: Ultraviolet and Soft Xray Free Electron Lasers: Introduction to Physical Principles, Experimental Results and Technical Challanges. Springer-Verlag, Heidleberg, Germany (2008)

Schneider, J.: Stimulated emission of radiation by relativistic electrons in a magnetic field. Phys. Rev. Lett. **2**, 504 (1959)

Schnell, W. (1991): The CERN Study of a Linear Collider in the TeV Range. CERN Div. Rep. SL/91-49

Schwinger, J.: On the classical radiation of accelerated electrons. Phys. Rev. **75**, 1912 (1949)

Schwoerer, H., Liesfeld, B., Schlenvoigt, H.P., Amthor, K.U., Sauerbrey, R.: Thomson-backscattered X rays from laser-accelerated electrons. Phys. Rev. Lett. **96**, 014802 (2006)

Sessler A.M.: The FEL as a power source for a high gradient accelerating structure. In: Channel, P.J. (ed.) AIP Conference Proceedings 91, p. 154 (1982)

Serlin, V., Friedman, M.: Development and optimization of the relativistic klystron amplifier. IEEE Tran. **PS-22**, 692 (1994)

Shchelkunov, S.V., Marshall, T.C., Hirshfield, J.L., Babzien, M.A., LaPointe, M.A.: Experimental observation of constructive superposition of wakefields generated by electronbunches in a dielectric-lined waveguide. Phys. Rev. Spec. Top. Accel. Beams **9**, 011301 (2006)

Shiffler, D.A., Nation, J.A., Wharton, C.B.: High-power traveling wave amplifier. Appl. Phys. Lett. **54**, 674 (1989)

Shiffler, D.A., Ivers, J.D., Kerslick, G.S., Nation, J.A., Schächter, L.: A high power two stage traveling wave amplifier. J. Appl. Phys. **70**, 106 (1991)

Shin, Y.-M., So, J.-K., Jang, K.-H., Won, J.-H., Srivastava, A., Park, G.S.: Superradiant terahertz Smith-Purcell radiation from surface plasmon excited by counter streaming electron beams. Appl. Phys. Lett. **90**, 031502 (2007)

Shintake, T.: The choke mode cavity. Jpn. J. Appl. Phys. **31**, 1567 (1992)

Siders, C.W., Le Blanc, S.P., Fisher, D., Tajima, T., Downer, M.C., et al.: Laser wakefield excitation and measurement by femtosecond longitudinal interferometry. Phys. Rev. Lett. **76**, 3570 (1996)

Siggins, T., Sinclair, C., Bohn, C., Bullard, D., Douglas, D., Grippo, A., Gubeli, J., Krafft, G.A., Yunn, B.: Performance of a DC GaAs photocathode gun for the Jefferson lab FEL. Nucl. Instrum. Methods Phys. Res. **A 475**, 549–553 (2001)

Slater, J.C.: Microwave Electronics. D. Van Nostrand Company Inc., New York (1950)

Smith, S.J., Purcell, E.M.: Visible light from localized charges moving across a grating. Phys. Rev. **92**, 1069 (1953)

Sotnikov, G.V., Marshall, T.C., Hirshfield, J.L.: Coaxial two-channel high-gradient dielectric wakefield accelerator. Phys. Rev. Spec. Top. Accel. Beams **12**, 061302 (2009)

Sprangle, P., Esarey, E., Ting, A.: Nonlinear interaction of intense laser pulses in plasmas. Phys. Rev. A **41**, 4463 (1990)

Sprangle, P., Ting, A., Tang, C.M.: Radiation focusing and guiding with application to the free electron laser. Phys. Rev. Lett. **59**, 202 (1987)

Sprangle, P., Esarey, E., Krall, J., Joyce, G., Ting, A.: Electron acceleration and optical guiding in the laser wake field accelerator. In: Wurtele, J.S. (ed.) Advanced Accelerator Concepts, Port Jefferson, New York 1992. AIP conference proceedings 279, p. 490 (1993)

Steinhauer, L.C., Kimura, W.D.: High-γ inverse cerenkov acceleration in resonant media. J. Appl. Phys. **68**, 4929 (1990)

Stratton, J.A.: Electromagnetic Theory. McGraw-Hill, New York (1941)

Strickland, D., Moureau, G.: Compression of amplified chirped optical pulses. Opt. Commun. **56**, 219 (1985)

Stupakov, G.V.: Impedance of Small Obstacles and Rough Surfaces. Phys. Rev. Spec. Top. Accel. Beams **1**, 064401 (1998)

Stupakov, G.: Using the beam-echo effect for generation of short-wavelength radiation. Phys. Rev. Lett. **102**, 074801 (2009)

Sullivan, D.J., Walsh, J.E., Coutsias, E.A.: Virtual cathode oscillator theory. In: Granatstein, V.L., Alexeff, I. (eds.) High-Power Microwave Sources, p. 441. Artech House, Boston (1987)

Swent, R.L., Pantell, R.H., Alguard, M.J., Berman, B.L., Bloom, S.D., Datz, S.: Observation of channeling radiation from relativistic electrons. Phys. Rev. Lett. **43**, 1723 (1979)

Tajima, T., Dawson, J.M.: Laser electron accelerator. Phys. Rev. Lett. **43**, 267 (1979)

Terhune, R.W., Pantell, R.H.: X-ray and γ-ray emission from channeled relativistic electrons and positrons. Appl. Phys. Lett. **30**, 265 (1977)

Tochitsky, S.Ya, Narang, R., Filip, C.V., Musumeci, P., Clayton, C.E., Yoder, R.B., Marsh, K.A., Rosenzweig, J.B., Pellegrini, C., Joshi, C.: Enhanced acceleration of injected electrons in a laser-beat-wave-induced plasma channel. Phys. Rev. Lett. **92**, 095004–1 (2004)

Tomas, R.: Overview of the compact linear collider. Phys. Rev. Spec. Top. Accel. Beams **13**, 014801 (2010)

Toraldo di Francia, G.: On the theory of some cerenkovian effects. Il Nuovo Cimento **16**, 61 (1960)

Twiss, R.O.: Radiation transfer and the possibility of negative absorption in radio astronomy. Aust. J. Phys. **11**, 564 (1958)

Van Bladel, J.: Relativity and Engineering. Springer Verlag, Berlin (1984)

Van den Berg, P.M.: Smith-Purcell radiation from a line charge moving parallel to a reflection grating. J. Opt. Soc. Am. **63**, 689 (1973). See also, Van den Berg, P.M.: Smith-Purcell radiation from a point charge moving parallel to a reflection grating. J. Opt. Soc. Am. **62**, 1588 (1973)

Varian, R.H., Varian, S.F.: A high frequency oscillator and amplifier. J. Appl. Phys. **10**, 321 (1939)

Venturini, M., Warnock, R., Ruth, R., Ellison, J.A.: Coherent synchrotron radiation and bunch stability in a compact storage ring. Phys. Rev. Spec. Top. Accel. Beams **8**, 014202 (2005)

Voss, G.A., Weiland, T.: The wake field acceleration mechanism, DESY 82-074 (1982). See also, Voss, G.A., Weiland, T.: Particle acceleration by wake fields, DESY 82-10 (1982)

Walsh, J.E.: Cerenkov masers: experiment. In: Granatstein, V.L., Alexeff, I. (eds.) High Power Microwave Sources, p. 421. Artech House, Boston (1987)

Warnock, R., Morton, P.: Fields excited by a beam in a smooth toroidal chamber: part 1. Longitudinal coupling impedance. Part. Accel. **25**, 113 (1990)

Westenskow, G.A., Madey, J.M.A.: Microwave electron gun. Laser Part. Beams **2**, 223 (1984)
Wheeler, J.A., Feynman, R.P.: Interaction with the absorber as the mechanism of radiation. Rev. Mod. Phys. **17**, 157 (1945). See also Wheeler, J.A., Feynman, R.P.: Classical electrodynamics in terms of direct interparticle action. Rev. Mod. Phys. **21**, 425 (1949)
Wiedemann, H.: Particle Accelerator Physics I: Basic Principles and Linear Beam Dynamics, 2nd edn. Springer, Berlin (1999a)
Wiedemann, H.: Particle Accelerator Physics II: Nonlinear and Higher-order Beam Dynamics, 2nd edn. Springer, Berlin (1999b)
Yeh, P., Yariv, A.: Bragg reflection waveguides. Opt. Commun. **19**, 427–430 (1976)
Zhang, Z., Tantawi, S.G., Ruth, R.D.: Distributed grating-assisted coupler for optical all-dielectric electron accelerator. Phys. Rev. Spec. Top. Accel. Beams **8**, 071302 (2005)
Zotter, B.W., Kheifets, S.A.: Impedances and Wakes in High-Energy Particle Accelerators. World Scientific, Singapore (1998)

Index

MIX
Papier aus verantwortungsvollen Quellen
Paper from responsible sources
FSC® C105338
FSC
www.fsc.org

If you have any concerns about our products,
you can contact us on
ProductSafety@springernature.com

In case Publisher is established outside the EU,
the EU authorized representative is:
Springer Nature Customer Service Center GmbH
Europaplatz 3, 69115 Heidelberg, Germany

Printed by Libri Plureos GmbH
in Hamburg, Germany